Enzyme Dynamics and Regulation

P. Boon Chock, Charles Y. Huang,
C. L. Tsou, and Jerry H. Wang
Editors

Enzyme Dynamics and Regulation

With 166 Illustrations

Springer-Verlag
New York Berlin Heidelberg
London Paris Tokyo

P. Boon Chock and *Charles Y. Huang*, Laboratory of Biochemistry, National Heart, Lung, and Blood Institute, National Institutes of Health, Bethesda, Maryland 20892, USA
C. L. Tsou, Institute of Biophysics, Academia Sinica, Beijing, China
Jerry H. Wang, Department of Medical Biochemistry, The University of Calgary, Calgary, Alberta T2N 4NI, Canada

Library of Congress Cataloging-in-Publication Data
Enzyme dynamics and regulation.
Papers contributed to the International Symposium on the Dynamics of Soluble and Immobilized Enzyme Systems, held May 26–30, 1986 in Beijing, China, sponsored by the International Union of Biochemistry Interest Group on Kinetics and Mechanisms of Enzymes and Metabolic Networks and the Academia Sinica.
Includes bibliographies and index.
1. Enzyme kinetics—Congresses. 2. Immobilized enzymes—Congresses. I. Chock, P. Boon.
II. International Symposium on the Dynamics of Soluble and Immobilized Enzyme Systems (1986: Peking, China)
III. International Union of Biochemistry. Interest Group on Kinetics and Mechanisms of Enzymes and Metabolic Networks.
IV. Chung-kuo k'o hsüeh yüan.
QP601.3.E59 1987 574.19'25 87-16522

Typeset by Asco Trade Typesetting Ltd., North Point, Hong Kong.

9 8 7 6 5 4 3 2 1

ISBN-13:978-1-4612-8330-0 e-ISBN-13:978-1-4612-3744-0
DOI: 10.1007/978-1-4612-3744-0

Preface

Knowledge of enzymes is basic to practically every branch of biochemical investigation. The scope and direction of enzymology continue to expand and shift while its foundation gains an ever-stronger foothold in physico-chemical principles. Recent developments in concepts and techniques have brought enzyme research into a changing yet exciting stage. For instance, enzymes serve as indispensable tools in the phenomenal rise of molecular biology. In turn, the resultant biotechnology thrusts enzymology to new heights and territories. How does one utilize the long-established and newly acquired information to proceed? To provide a current overview of this field, the International Union of Biochemistry Interest Group on Kinetics and Mechanisms of Enzymes and Metabolic Networks and the Academia Sinica cosponsored the International Symposium on the Dynamics of Soluble and Immobilized Enzyme Systems. As indicated by its name, the symposium also sought to focus due attention on the kinetic, or time-dependent, element of enzyme-catalyzed or regulated processes and on enzyme systems entrapped within membranes or solid matrices. This volume collects papers contributed to the symposium, which was held May 26–30, 1986, in Beijing, China, and was attended by more than 100 leading scientists from 12 countries. The diversity of these papers is reflected in the seven categories listed in the Table of Contents. The success of the symposium was made possible by members of the Organizing Committee, composed of P. Boon Chock, Carl Frieden, Robert Y. Hsu, Charles Y. Huang, Jacques Ricard, C. L. Tsou, and Jerry H. Wang; the American Society of Biological Chemists; and colleagues and friends too numerous to single out. The success was especially aided by much-appreciated financial support from the following organizations: Chandra Djojonegoro, P.T. International Chemical Industrial Co., Ltd., Jakarta, Indonesia; CIBA-GEIGY Corporation, Summit, New Jersey; Hoechst AG, Frankfurt, West Germany; Merck Sharp & Dohme, Rahway, New Jersey; Miles Laboratories, Inc., Elkhart, Indiana; Miwon USA, Inc., Hoboken, New Jersey; Monsanto Company, St. Louis, Missouri; Smith Kline Beckman

Corporation, Philadelphia, Pennsylvania; and Springer-Verlag New York Inc., New York, New York.

P. Boon Chock
Charles Y. Huang
C. L. Tsou
Jerry H. Wang

Contents

Contributors and Participants

Donna J. Arndt-Jovin, Department of Molecular Biology, Max-Planck-Institut für Biophysikalische Chemie, D-3400 Göttingen, West Germany (190)

**R. Dean Astumian*, National Institutes of Health, Bethesda, Maryland 20892, USA (247)

Daphne Atlas, Department of Biological Chemistry, Institute of Life Sciences, Hebrew University of Jerusalem, Jerusalem, Israel

**David S. Auld*, Center for Biochemical and Biophysical Sciences and Medicine, Harvard University Medical School, Boston, Massachusetts 02115, USA (62)

**Wong Gi Bang*, National Institutes of Health, Bethesda, Maryland 20892, USA (136)

**P. A. Benkovic*, Department of Chemistry, The Pennsylvania State University, University Park, Pennsylvania 16802, USA (1)

**S. J. Benkovic*, Department of Chemistry, The Pennsylvania State University, 152 Davey Laboratory, University Park, Pennsylvania 16802, USA (1)

**A. Bertazzon*, Department of Biological Chemistry, Johns Hopkins University School of Medicine, Baltimore, Maryland 21205, USA (206)

**William A. Bridger*, Department of Biochemistry, University of Alberta, Medical Science Building, Edmonton, Alberta, Canada T6G 2H7 (6)

**José Bubis*, Department of Chemistry, University of California, San Diegeo, La Jolla, California 92093, USA (327)

**Lilia Calabrese*, Department of Biochemistry, La Sapienza University and CNR Center for Molecular Biology, Rome, Italy (84)

---

*Indicates contributor; number in parenthesis indicates beginning page of author's contribution.

Hui-Li Chen, Department of Biochemistry, Shanghai Medical University, Shanghai, China

Lan-Ying Chen, Cardiovascular Institute, Chinese Academy of Medical Sciences, Beijing, China

Shi-Gen Chen, Laboratory of Enzyme Engineering, Fudan University, Shanghai, China

Yu-Hua Cheng, Department of Molecular Biology, Jilin University, Changchun, Jilin, China

Cheng-Wu Chi, Shanghai Institute of Biochemistry, Academia Sinica, Shanghai, China

**P. Boon Chock*, National Institutes of Health, Bethesda, Maryland 20892, USA (128, 247)

**J. Colanduoni*, Department of Chemistry and Molecular and Cell Biology, The Pennsylvania State University, University Park, Pennsylvania 16802, USA (77)

**Roberta F. Colman*, Department of Chemistry, University of Delaware, Newark, Delaware 19716, USA (351)

**Pierre Douzou*, Institut de Biologie Physico-Chimique, 75005 Paris, France (312)

Jin-Zhu Du, Department of Biology, Beijing University, Beijing, China

**C. D. Eads*, Department of Chemistry and Molecular and Cell Biology, The Pennsylvania State University, University Park, Pennsylvania 16802, USA (77)

**Fritz Eckstein*, Max-Planck-Institut für Experimentelle Medizin, D-3400 Göttingen, West Germany (200)

**Rolf Eliasson*, Department of Biochemistry, Medical Nobel Institute, Karolinska Institute, 10401 Stockholm, Sweden (181)

**Blair Q. Ferguson*, Department of Chemistry, Georgetown University, Washington, DC 20057, USA (71)

**Alan R. Fersht*, Department of Chemistry, Imperial College of Science and Technology, South Kensington, London SW7 2AY, United Kingdom (405)

**Marc Fontecave*, Department of Biochemistry, Medical Nobel Institute, Karolinska Institute, 10401 Stockholm, Sweden (181)

**Carl Frieden*, Department of Biological Chemistry, Washington University School of Medicine, St. Louis, Missouri 63110, USA (268)

**Herbert J. Fromm*, Biochemistry and Biophysics Department, Iowa State University, Ames, Iowa 50011, USA (48)

B. N. Goldstein, USSR Academy of Sciences, Institute of Biological Physics, 142292 Pushchino, Moscow Region, USSR (322)

Lian Guan, Department of Biochemistry, Henan Medical University, Zhengzhou, Henan, China

**Gordon G. Hammes*, Department of Chemistry, Cornell University, Baker Laboratory, Ithaca, New York 14853, USA (226)

**W. F. Harrington*, Department of Biology, Johns Hopkins University School of Medicine, Baltimore, Maryland 21205, USA (206)

**Guy Hervé*, Institut d'Enzymologie, CNRS, 91190 Gif Sur Yvette, France (155)

**K. Hikichi*, Department of Polymer Science, Hokkaido University, Kita-ku, Saporo, Hokkaido 060, Japan (362)

**T. Hiraoki*, Department of Polymer Science, Hokkaido University, Kita-ku, Saporo, Hokkaido 060, Japan (362)

**Ching T. Hou*, Department of Microbial Biochemistry and Genetics, Squibb Institute for Medical Research, Princeton, New Jersey 08543, USA (261)

**Robert Y. Hsu*, Department of Biochemistry, State University of New York, Upstate Medical Center, Syracuse, New York 13210, USA (17)

**Guo-fu Hu*, Shanghai Institute of Biochemistry, Academia Sinica, Shanghai, China (55)

**Charles Y. Huang*, National Institutes of Health, Bethesda, Maryland 20892, USA (40)

Fen Huang, Institute of Biophysics, Academia Sinica, Beijing, Shanghai, China

**S. T. Huang*, Shanghai Institute of Biochemistry, Academia Sinica, Shanghai, China (367)

**M. Ikura*, Department of Polymer Science, Hokkaido University, Kita-ku, Saporo, Hokkaido 060, Japan (362)

**Haris Jamil*, Department of Biochemistry, University of Alberta, Edmonton, Alberta T6G 2H7, Canada (121)

**K. A. Johnson*, Department of Chemistry and Biochemistry, The Pennsylvania State University, University Park, Pennsylvania 16802, USA (1)

**Thomas M. Jovin*, Department of Molecular Biology, Max-Planck-Institut für Biophysikalische Chemie, D-3400 Göttingen, West Germany (190)

**Stewart R. Jurgensen*, National Institutes of Health, Bethesda, Maryland 20892, USA (128)

*E. T. Kaiser, Laboratory of Bioorganic Chemistry and Biochemistry, The Rockefeller University, New York, New York 10021-6399 USA (411)

*Robert P. Keenan, Department of Biochemistry, School of Medicine, Case Western Reserve University, Cleveland, Ohio 44106, USA (28)

*Debra A. Kendall, Laboratory of Bioorganic Chemistry and Biochemistry, The Rockefeller University, New York, New York 10021, USA (411)

*Ja Hyun Koo, National Institutes of Health, Bethesda, Maryland 20892, USA (136)

Ting-Yun Kuang, Institute of Botany, Academia Sinica, Beijing, Shanghai, China

*R. D. Kuchta, Department of Chemistry and Biochemistry, The Pennsylvania State University, University Park, Pennsylvania 16802, USA (1)

*B. I. Kurganov, All-Union Vitamin Research Institute, Moscow 117246, USSR (175)

*Marina Lanciotti, National Institutes of Health, Bethesda, Maryland 20892, USA (40)

*Alexander Levitzki, Department of Biological Chemistry, Institute of Life Sciences, Hebrew University of Jerusalem, Jerusalem, Israel (113)

Li-Ren Li, Shanghai Institute of Plant Physiology, Academia Sinica, Shanghai, China

Zhen-Hua Li, Department of Biology, Sichuan University, Chengdu, Sichuan, China

Nian-Ci Liang, Department of Biochemistry, Zhanjiang Medical School, Zhanjiang, Guangdong, China

Yong-Ning Lian, Institute of Biophysics, Academia Sinica, Beijing, Shanghai, China

Chi-Shui Lin, Shanghai Institute of Biochemistry, Academia Sinica, Shanghai, China

*S. X. Lin, Shanghai Institute of Biochemistry, Academia Sinica, Shanghai, China (367)

Zou-Guo Lin, Department of Biology, Wuhan University, Wuhan, Hubei, China

Li-Sheng Liu, Department of Biology, Nankai University, Tianjin, China

Rong Liu, Institute of Biophysics, Academia Sinica, Beijing, Shanghai, China

Shu-Sen Liu, Institute of Zoology, Academia Sinica, Beijing, Shanghai, China

*Wei Liu, Institute of Biophysics, Academia Sinica, Beijing, Shanghai, China (289)

*R. LoBrutto, Department of Chemistry and Molecular and Cell Biology, The Pennsylvania State University, University Park, Pennsylvania 16802, USA (77)

J. X. Lu, Academia Sinica, Beijing, Shanghai, China

Ying-Hua Lu, Department of Biology, Lanzhou University, Lanzhou, Gansu, China

Yong-Zhi Lu, Department of Biology, Fuzhou Normal University, Fuzhou, Fujian, China

*Neil B. Madsen, Department of Biochemistry, University of Alberta, Edmonton, Alberta T6G 2H7, Canada (121)

*G. Magni, Istituto di Chimica Biologica, Facolta di Medicina e Chirurgia, Ancona, Italy (377)

*Wilfried Merlevede, Afdeling Biochemie, Campus Gasthuisberg, B-3000 Leuven, Belgium (146)

*F. Miao, Shanghai Institute of Biochemistry, Academia Sinica, Shanghai, China (367)

*Kyung Hee Min, National Institutes of Health, Bethesda, Maryland 20892, USA (136)

*O. Minowa, Faculty of Science and Institute of Immunological Science, Hokkaido University, Kita-ku, Saporo, Hokkaido 060, Japan (362)

*V. Mizrahi, Department of Chemistry and Biochemistry, The Pennsylvania State University, University Park, Pennsylvania 16802, USA (1)

*P. Natalini, Dipartimento Biologia Cellulaire, Universita di Camerino, 62032 Camerino, Italy (377)

*Kenneth E. Neet, Department of Biochemistry, School of Medicine, Case Western Reserve University, Cleveland, Ohio 44106, USA (28)

*Georges Noat, CNRS Centre de Biochemie et de Biologie Moleculaire, 13402 Marseille Cedex 9, France (235)

*Dexter B. Northrop, School of Pharmacy, University of Wisconsin, Madison, Wisconsin 53706 USA (275)

*Marion H. O'Leary, Department of Chemistry and Biochemistry, University of Wisconsin, Madison, Wisconsin 53706, USA (301)

Yao-Hua Ou, Department of Biology, Qinghua University, Beijing, Shanghai, China

Ren-Rui Pan, Department of Biology, University of Science and Technology, Hefei, Anhui, China

**Sang Chul Park*, National Institutes of Health, Bethesda, Maryland 20892, USA (136)

**Michael H. Penner*, Department of Biological Chemistry, Washington University School of Medicine, St. Louis, Missouri 63110, USA (268)

**Richard N. Perham*, Department of Biochemistry, University of Cambridge, Tennis Court Road, Cambridge CB2 1QW, United Kingdom (92)

**Daniel L. Purich*, Department of Biochemistry and Molecular Biology, University of Florida, Gainesville, Florida 32605, USA (217)

Zheng-Wu Qi, Institute of Biochemistry, Academia Sinica, Beijing, Shanghai, China

**N. Raffaelli*, Dipartimento Biologia Cellulaire, Universita di Camerino, 62032 Camerino, Italy (377)

**Peter Reichard*, Department of Biochemistry, Medical Nobel Institute, Karolinska Institute, 10401 Stockholm, Sweden (181)

**Sue Goo Rhee*, National Institutes of Health, Bethesda, Maryland 20892, USA (128, 136)

**Jacques Ricard*, CNRS Centre de Biochimie et de Biologie Moleculaire, 13402 Marseille Cedex 9, France (235)

**Adelio Rigo*, Institute of Biochemistry, University of Padua, Padua, Italy (84)

**Giuseppe Rotilio*, Department of Biology, II Universita degli Studi di Roma, 00173 Roma, Italy (84)

**S. Ruggieri*, Dipartimento Biologia Cellulaire, Universita di Camerino, 62032 Camerino, Italy (377)

**I. Santarelli*, Dipartimento Biologia Cellulaire, Universita di Camerino, 62032 Camerino, Italy (377)

**Lakshmi D. Saraswat*, Department of Chemistry, University of California–San Diego, La Jolla, California 92092, USA (327)

**Robin M. Scaife*, Department of Chemistry, University of California, Santa Barbara, California 93106, USA (217)

**H. K. Schachman*, Department of Molecular Biology, University of California, Berkeley, California 94720, USA (394)

**Julie E. Scheffler*, Department of Biochemistry and Biophysics, Iowa State University, Ames, Iowa 50011, USA (48)

**Rajendra K. Sharma*, Department of Medical Biochemistry, The University of Calgary, Calgary, Alberta T2N 4N1, Canada (162)

J. P. Shi, Shanghai Institute of Biochemistry, Academia Sinica, Beijing, Shanghai, China (367)

**Zhang Shu-zheng*, Institute of Microbiology, Academia Sinica, Beijing, Shanghai, China (55)

**Janusz Sowadski*, Department of Chemistry, University of California–San Diego, La Jolla, California 92093, USA (327)

**Earl R. Stadtman*, National Institutes of Health, Bethesda, Maryland 20892, USA (128)

**Lubert Stryer*, Department of Cell Biology, Stanford University School of Medicine, Stanford, California 94305, USA (102)

Lian-Kui Sun, Department of Biology, Northwestern University, Xian, Shaanxi, China

Man-Ji Sun, Institute of Medical Sciences, Military Academy of Medical Sciences, Beijing, Shanghai, China

Palmer W. Taylor, Division of Pharmacology, Department of Medicine, University of California–San Diego, La Jolla, California 92093, USA

**Susan S. Taylor*, Department of Chemistry, University of California–San Diego, La Jolla, California 92093, USA (327)

Wei-Xi Tian, Graduate School of the University of Science and Technology, Beijing, Shanghai, China

**Peter S. Tippett*, Department of Biochemistry, Case Western Reserve University, Cleveland, Ohio 44106, USA (28)

**H. Toda*, Faculty of Science and Institute of Immunological Science, Hokkaido University, Kita-ku, Saporo, Hokkaido 060, Japan (362)

**Jean A. Toner*, Department of Chemistry, University of California–San Diego, La Jolla, California 92093, USA (327)

Tan-Jun Tong, Department of Biochemistry, Beijing Medical University, Beijing, Shanghai, China

**Tian Yow Tsong*, Department of Biological Chemistry, Johns Hopkins University School of Medicine, Baltimore, Maryland 21205, USA (206, 247)

**C. L. Tsou*, Institute of Biophysics, Academia Sinica, Beijing, Shanghai, China (289, 342)

**Bert L. Vallee*, Center for Biochemical and Biophysical Sciences and Medicine, Harvard University Medical School, Boston, Massachusetts 02115, USA (62, 224)

**Carline Vanden Abeele*, Afdeling Biochemie, Campus Gasthuisberg, B-3000 Leuven, Belgium (146)

*_Jackie R. Vandenheede_, Afdeling Biochemie, Campus Gasthuisberg, B-3000 Leuven, Belgium (128, 146)

*_J. J. Villafranca_, Department of Chemistry, The Pennsylvania State University, University Park, Pennsylvania 16802, USA (77)

*_A. Vita_, Dipartimento Biologia Cellulaire, Universita di Camerino, 62032 Camerino, Italy (377)

*_Chiayeng Wang_, Department of Medical Biochemistry, The University of Calgary, Calgary, Alberta, T2N 1N4, Canada (162)

*_Jerry H. Wang_, Department of Medical Biochemistry, The University of Calgary, Calgary, Alberta, Canada T2N 1N4 (162)

Yang-Sheng Wang, Institute of Microbiology, Academia Sinica, Beijing, Shanghai, China

*_Y. L. Wang_, Shanghai Institute of Biochemistry, Academia Sinica, Beijing, Shanghai, China (367)

*_F. C. Wedler_, Department of Chemistry and Molecular and Cell Biology, The Pennsylvania State University, University Park, Pennsylvania 16802, USA (77)

*_Theodore G. Wensel_, Department of Cell Biology, Stanford University School of Medicine, Stanford, California 94305, USA (102)

*_Susan R. Wente_, Department of Biochemistry and Molecular Biology, University of California, Berkeley, California 94720, USA (394)

*_Hans V. Westerhoff_, Section on Theoretical Molecular Biology, Laboratory of Molecular Biology, NIDDK, National Institutes of Health, Bethesda, Maryland 20892, USA (247)

*_Susan P. Williams_, Department of Biochemistry, The University of Alberta, Edmonton, Alberta T6G 2H7, Canada (6)

*_William T. Wolodko_, Department of Biochemistry, The University of Alberta, Edmonton, Alberta T6G 2H7, Canada (6)

Guo-Li Wu, Department of Biology, Beijing Normal University, Beijing, Shanghai, China

Zhao-Feng Wu, Department of Biochemistry, Western China Medical University, Chengdu, Sichuan, China

*_Qi-chang Xia_, Shanghai Institute of Biochemistry, Academia Sinica, Beijing, Shanghai, China (55)

*_Gen-jun Xu_, Harvard Medical School, Center for Biochemical and Biophysical Sciences and Medicine, Boston, Massachusetts 02115, USA (55)

*_K. Yagi_, Department of Chemistry, Faculty of Science, Hokkaido University, Kita-ku, Sapporo, Hokkaido 060, Japan (362)

*_Gu Ya-jun_, Institute of Microbiology, Academia Sinica, Beijing, Shanghai, China (385)

*_David C. H. Yang_, Department of Chemistry, Georgetown University, Washington, DC 20057, USA (71)

*_Ying R. Yang_, Department of Biochemistry and Molecular Biology, University of California, Berkeley, California 94720, USA (394)

*_M. Yazawza_, Department of Chemistry, Hokkaido University, Kita-ku, Saporo, Hokkaido 060, Japan (362)

*_M. C. Young_, Department of Chemistry and Biochemistry, The Pennsylvania State University, University Park, Pennsylvania 16802, USA (1)

Bin-Zhi Yu, Department of Biochemistry, China Medical University, Shenyang, Liaoning, China

*_Zeng Yu-cheng_, Institute of Microbiology, Academia Sinica, Beijing, Shanghai, China (385)

Ding Zeng, Department of Biology, Xiamen University, Xiamen, Fujian, China

*_Aile Zhang_, National Institutes of Health, Bethesda, Maryland 20892, USA (40)

Long-Xiang Zhang, Department of Biology, Beijing University, Beijing, Shanghai, China

*_Zhang Shu-zheng_, Institute of Microbiology, Academia Sinica, Beijing, Shanghai, China (385)

*_Fu-kun Zhao_, Shanghai Institute of Biochemistry, Academia Sinica, Beijing, Shanghai, China (55)

Guo-Lin Zhou, Department of Biology, Wuhan University, Wuhan, Hubei, China

Ting-Chong Zhou, Institute of Basic Medical Sciences, Beijing, Shanghai, China

De-Zu Zhu, Department of Biochemistry, Nanjing University, Nanjing, China

1
Mechanistic Studies on DNA Polymerase I

V. MIZRAHI, P. A. BENKOVIC, R. D. KUCHTA, M. C. YOUNG, K. A. JOHNSON, and S. J. BENKOVIC

The multifunctional DNA polymerase I (Pol I) of *Escherichia coli* has served as the most widely studied model for describing, at the molecular level, certain enzymatic processes involved in the replication of DNA (1). In addition to its polymerase activity, the enzyme also catalyzes DNA degradation by distinct 5′ → 3′ and 3′ → 5′ exonuclease activities, as well as by net pyrophosphorolysis. Extensive kinetic (2, 3) and stereochemical (4, 5) studies of the various activities have elucidated the important underlying features of the phosphodiester bond-forming and bond-breaking reactions. In addition, the availability of a 3-Å resolution x-ray structure of the large proteolytic (Klenow) fragment of Pol I (6) has generated considerable interest in the area of structure–function assignment (7).

In this chapter we present our mechanistic studies on the idling-turnover and the polymerization reactions catalyzed by the Klenow fragment (KF), which bear on the general problem of describing a unified mechanism for the interrelated activities of this enzyme.

Experimental Procedures

The description of the experimental protocols employed to obtain the results reported herein can be found elsewhere (2, 3, 8).

Results and Discussion

Mechanism of the Idling-Turnover Reaction

The conversion of a fraction of the available deoxynucleoside 5′-triphosphate (dNTP) pool into a corresponding monophosphate pool has provided evidence in support of alternating polymerase and 3′ → 5′ exonuclease expression during the course of DNA synthesis (9, 10). In view of the convincing evidence implicating the 3′ → 5′ exonuclease activity in ensuring fidelity (11–14), the extent of the dNTP → dNMP conversion may thus reflect the degree of proofreading accompanying replication by DNA polymerases possessing such an exonuclease activity (15). In the absence of the following complementary dNTP, which is required for normal polymerication, the turnover process is exaggerated as the enzyme is constrained to "idle" at the primer terminus until depletion of the available dNTP pool is complete (16).

The overall mechanism of the idling-turnover reaction is as follows:

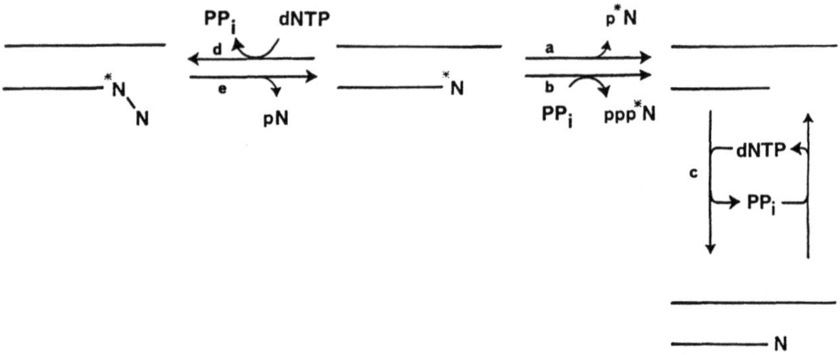

The reaction cycle involving steps a, b, and c is one of excision/incorporation, in which the 3′-terminal deoxynucleotide residue of the primer DNA strand is partitioned into its 5′-monophosphate (pathway a) and 5′-triphosphate (pathway b) derivatives, respectively (8). Our mechanistic studies suggest that the 5′-monophosphate product is formed in the first step by simple 3′ → 5′ exonucleolytic cleavage, converting the DNA to a form capable of reacting specifically with dNTP in a rapid polymerization step (pathway c), with concomitant release of PP_i. The dNTP in the exogenous pool becomes, after primer excision, complementary to the template. The use of isotope-trapping techniques was employed to demonstrate that the exonuclease → polymerase activity switch that is involved in going from pathway a → c occurs without an intervening dissociation of the enzyme from its DNA 'substrate. The released PP_i is free to react with the DNA in a pyrophosphorolysis step (pathway b) to yield a triphosphate product. The rate of this parallel excision pathway is comparable with that of the exonuclease pathway at extremely low levels of accumulated PP_i ($<0.5\ \mu M$). Furthermore, comparative studies on the pyrophosphorolysis kinetics of several DNA substrates have indicated an unexpected dependence of the reaction rate on the DNA sequence within the duplex region six to twelve bases upstream of the primer–template junction.

In addition to the excision/incorporation reaction cycle, the existence of an alternative DNA sequence-dependent misincorporation/excision cycle (pathways d and e) has been directly demonstrated by means of gel electrophoretic product analysis. The mismatched 3′-terminus generated via pathway d is a poor substrate for further polymerization. The mismatch is not excisable by

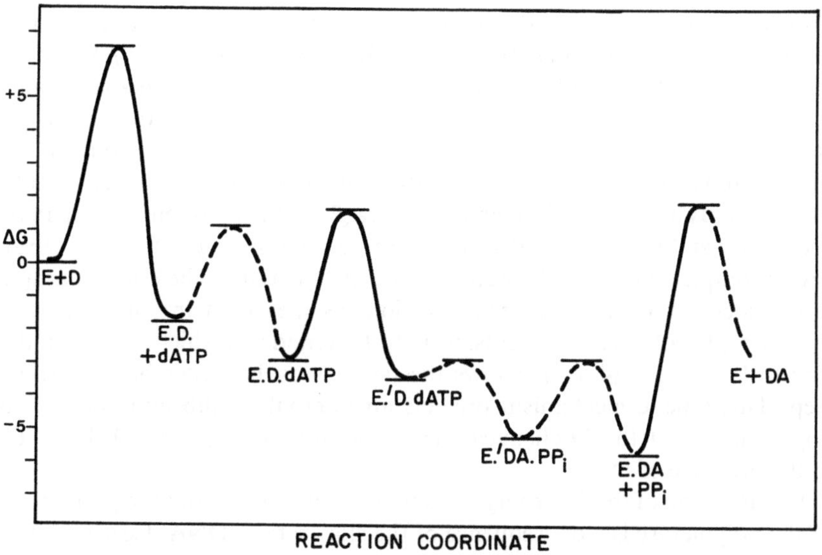

FIG. 1.1. Energy profile for the reaction D + dATP $\rightleftharpoons$ Da + PP_i catalyzed by the Klenow fragment. Rates of substrate binding were calculated assuming $k_{on}^{D} = 4 \times 10^6\ M^{-1}\ s^{-1}$; $k_{on}^{dATP} = k_{on}^{PP_i} = 10^7\ M^{-1}\ s^{-1}$ and [D] = 100 nM, [dATP] = [PP_i] = 20 μM. ΔG is in kcal mol^{-1}.

pyrophosphorolysis and accumulates despite the 3' → 5' exonuclease activity (pathway e). The extent of reaction via pathway d is determined by the 3'-terminal DNA sequence, and the frequency of the particular misinsertion.

Kinetic Mechanism of the Polymerase Activity

Steady-state and presteady-state kinetic studies have established the important features underlying processive synthesis of alternating copolymer and homopolymer template primers catalyzed by Pol I and the Klenow fragment (2, 3, 17). We have employed isotope trapping and rapid kinetic techniques to examine the individual steps involved in Klenow fragment-catalyzed single nucleotide incorporation and pyrophosphorolysis. For this purpose, a synthetic DNA substrate of defined sequence was employed in order to gain insight into the DNA sequence dependence of the kinetic mechanism.

$$\begin{matrix} 3' - \text{AGCGTCGGCAGGTTCCCAAA} \\ \text{TCGCAGCCGTCCA} \end{matrix} + \text{dATP}$$

$$\rightleftharpoons \begin{matrix} 3' - \text{AGCGTCGGCAGGTTCCCAAA} \\ \text{TCGCAGCCGTCCAA} \end{matrix} + PP_i$$

The minimal mechanism consistent with the results of our study is illustrated by the energy profile of Fig. 1.1. The binding parameters of the template–

primer substrate and product to Klenow fragment (KD = 5 nM and 15 nM, respectively), and of dATP (KD = 2.5 μM) and PP_i (KD = 40 μM) to the respective E · DNA complexes, were evaluated by means of isotope-partitioning techniques. The rapid-quench polymerization time course exhibited a pre-steady-state burst (10 s^{-1}) followed by a steady-state incorporation at a rate that is limited by the rate of dissociation of the E · DNA product complex (0.06 s^{-1}). By analogy with previous studies (2, 3), the rate-limiting step in the pre-steady-state is attributed to a conformational change of the E · DNA · dATP complex immediately preceding the chemical step. The kinetic parameters associated with the reverse reaction, as measured in single-turnover pyrophosphorolysis experiments, were fit to a model in which a rate-limiting first-order process (4 s^{-1}) follows the rapid phosphodiester bond-breaking step. This kinetic mechanism predicts an internal equilibrium constant of approximately 20 for the chemical step and approximately 800 for [PP_i]/[NTP] in the presence of DNA.

Examination of the free energy reaction profile reveals that the processivity of the polymerase is a consequence of the slow rate of DNA dissociation, so that in the presence of all four nucleotides the reaction cycles between the E · DNA and E · DNA^{+1} complexes. For the present substrate, simultaneous addition of dATP, dGTP, and dTTP results in the rapid synthesis of the 19 mer requiring that translocation probably proceeds at rates in excess of 1 s^{-1}. It is of considerable interest to determine if this reaction sequence and associated free energy changes apply to other polymerases used primarily in replication.

References

1. Kornberg, A. (1980) *DNA Replication.* W. H. Freeman, San Francisco.
2. Bryant, F. F., Johnson, K. A., and Benkovic, S. J. (1983) *Biochemistry* 22:3537–3546.
3. Mizrahi, V., Henrie, R. N., Marlier, J., Johnson, K. A., and Benkovic, S. J. (1985) *Biochemistry* 24:4010–4018.
4. Burgers, P. M. J., and Eckstein, F. (1979) *J. Biol. Chem.* 254:6889–6893.
5. Gupta, A. F., and Benkovic, S. J. (1985) *Biochemistry* 23:5874–5881.
6. Ollis, D. L., Brick, P., Hamlin, R., Xuong, N. G., and Steitz, T. A. (1955) *Nature* 313:762–766.
7. Joyce, C. M., Ollis, D. L., Rush, J., and Steitz, T. A. (1985) *UCLA Symposium Molecular and Cellular Biology*, New Series (Protein Structure, Folding & Design) (in press).
8. Mizrahi, V., Benkovic, P. A., and Benkovic, S. J. (1986) *Proc. Natl. Acad. Sci. USA* 83:231–235.
9. Loeb, L. A., Dube, D. K., Beckman, R. A., Koplitz, M., and Gopinathan, K. P. (1981) *J. Biol. Chem.* 256:3978–3987.
10. Gupta, A., DeBrosse, C., and Benkovic, S. (1982) *J. Biol. Chem.* 257:7689–7692.
11. Brutlag, D., and Kornberg, A. (1972) *J. Biol. Chem.* 247:241–248.

12. Bessman, M. J., Muzyczka, N., Goodman, M. F., and Schnaar, R. L. (1974) *J. Mol. Biol.* 88:409–421.
13. Galas, D. J., and Branscomb, E. W. (1978) *J. Mol. Biol.* 124:653–687.
14. Clayton, L. K., Goodman, M. F., Branscomb, E. W., and Galas, D. J. (1979) *J. Biol. Chem.* 254:1902–1912.
15. Fersht, A. R., Knill-Jones, J. W., and Tsui, W-C. (1982) *J. Mol. Biol.* 156:37–51.
16. Gupta, A. P., Benkovic, P. A., and Benkovic, S. J. (1984) *Nucleic Acids Res.* 12:5892–5911.
17. McClure, W. R., and Jovin, T. M. (1975) *J. Biol. Chem.* 250:4073.

2
Assembly and Catalytic Functions of the Subunits of Succinyl Coenzyme A Synthetase

WILLIAM A. BRIDGER, WILLIAM T. WOLODKO, and SUSAN P. WILLIAMS

Succinyl-CoA synthetase catalyzes the substrate level phosphorylation step of the tricarboxylic acid cycle via the intermediary participation of a phosphohistidine-containing enzyme form. This enzyme, as isolated from *Escherichia coli*, has an $\alpha_2\beta_2$ subunit structure, and it has been proposed that catalysis may involve cooperative interactions between the two halves of the molecule that are mediated by alternating action of the respective active sites. Evidence in favor of this concept includes that derived from hybrid enzyme formation, from ^{31}P-NMR (nuclear magnetic resonance) studies, and from the kinetic behavior of the phosphorylated intermediate. In keeping with this model for catalysis, *E. coli* succinyl-CoA synthetase shows half-site reactivity with respect to its phosphorylation. This property has been closely examined by stoichiometric measurements of the replacement of ^{35}S-PSO$_2$ by ^{32}P-PO$_3$ during treatment of thiophosphorylated enzyme with various combinations of ADP and labeled ATP. The sum of –PSO$_2$ plus –PO$_3$ groups never reaches significantly beyond one per enzyme tetramer, confirming the half-site reactivity of the active site histidine residue. Furthermore, preliminary positional isotope exchange experiments with the monothiophosphorylated tetramer suggest that the neighboring active site may not be capable of even transient, reversible phosphorylation.

In contrast to the tetrameric *E. coli* enzyme, succinyl-CoA synthetase from pig heart has been proposed to be an $\alpha\beta$ dimer based on its gel filtration properties. Particularly because the mammalian enzyme, if truly a dimer, could provide a natural control to aid in the evaluation of the contribution of alternating sites' cooperativity to catalysis, we have carried out a detailed study of the sedimentation and related hydrodynamic properties of both enzyme species. These experiments include determination of the rate of sedimentation of a boundary of enzyme through a chromogenic substrate solution, enabling estimation of the sedimentation velocities of active enzyme species at the equivalent of assay concentrations. Taken together, our sedimentation data confirm that the bacterial enzyme is a nondissociating $\alpha_2\beta_2$ tetramer, whereas the pig heart species is a nonassociating $\alpha\beta$ dimer.

We have extended our studies on the assembly of *E. coli* succinyl-CoA synthetase from its isolated unfolded subunits. This process requires that the phosphate binding site of the α subunit must be occupied by either $-PO_3$ or P_i. The rate of assembly of active enzyme shows first-order dependence on the concentrations of each of the subunits, suggesting that it is the rate of their mutual interaction, rather than the rate of refolding of one of them, that controls the rate of formation of native enzyme. The refolding subunits show remarkable ability to recognize their partners, thus obviating formation of unproductive complexes with other proteins.

Succinyl-CoA synthetase (SCS) catalyzes the substrate-level phosphorylation step of the tricarboxylic acid cycle:

$$\text{Succinyl-CoA} + \text{NDP} + P_i \rightleftarrows \text{succinate} + \text{NTP} + \text{CoA}$$

This enzyme, as isolated from two sources in particular, i.e., *E. coli* and pig heart mitochondria, has been the subject of most investigations of its structure and function. These enzymes may be regarded as prototypes of two distinct classes of SCS that differ in their nucleotide specificity and their subunit structures. For example, the enzyme from *E. coli* prefers adenine nucleotides as substrates, whereas heart SCS is specific for guanine nucleotides (for review, see ref. 1). *E. coli* SCS has been shown to be an $\alpha_2\beta_2$ tetramer, with an aggregate molecular weight of 142,000; the M_r values of the α and β subunits are 29,600 and 41,400, respectively (1–3).

A subject of continued interest in this laboratory has been the functional rationale for the subunit structure of this enzyme. Many lines of evidence indicate that the active site lies at the α–β interface, with binding sites for nucleotide in α and for CoA and succinate in β (1, 4, 5). The rationale for the dimer of $\alpha\beta$ dimers that constitutes the structure of the *E. coli* enzyme is less clear and more controversial, however. From consideration of the kinetics of exchange of ^{18}O exchange between succinate and phosphate (6), the properties of hybrid enzyme preparations (7), ^{31}P-NMR studies (8, 9), and the marked stimulatory effects of ATP on the discharge of thiophosphate from the enzyme in the presence of succinate and CoA (10), it has been proposed that the two $\alpha\beta$ halves of SCS function in alternating fashion, with exergonic phosphorylation of one site promoting more difficult partial reactions at the neighboring active center. Here we describe some experiments that led to further refinement of this model of catalytic cooperativity in the *E. coli* enzyme. We also report studies on the mechanism of assembly of this enzyme species from its isolated subunit and describe sedimentation experiments that confirm the fundamental differences in the quaternary structures of the two classes of SCS.

Experimental Procedures

Enzymes

Succinyl-CoA synthetase was isolated from both *E. coli* (Crooks strain grown on a phosphate-buffered succinate-based medium) and pig heart muscle. The *E. coli* enzyme was purified from a thawed cell suspension essentially as described by Leitzmann et al. (11). Significant modifications to this preparative procedure included the elimination of the heat treatment step and chromatography on hydroxylapatite, and the inclusion of a final purification step involving affinity chromatography on Blue Sepharose CL-6B [Pharmacia (Canada) Ltd.]. The pig enzyme was purified from fresh heart muscle by modifications to the method developed by Cha et al. (12). Preparations were carried out in a single process starting with 15 to 20 hearts. Further important deviations from their procedure were as follows: extraction with water instead of buffer, elimination of all calcium phosphate gel chromatography, and final purification by affinity chromatography on Affi-Gel Blue (BioRad Laboratories). Protein concentrations were determined spectrophotometrically at 280 nm using values of $E_{1\,\mathrm{cm}}^{1\%}$ equal to 5.0 for *E. coli* SCS (2) and 3.5 for pig heart SCS (13). Enzymatic activity was assayed by the direct spectrophotometric method (14) based on the increase in A_{232} accompanying succinyl-CoA formation. ATP was used as substrate with *E. coli* SCS, and GTP was used for pig heart SCS.

Positional Isotope Exchange Experiments

Adenosine triphosphate containing ^{18}O in all four of the oxygens bonded to P_γ was synthesized as follows. First, [^{18}O]*o*-phosphate was prepared from [^{18}O]H_2O (97 atom %) and PCl_5 as described by Risley and Van Etten (15). It was then allowed to react with ADP-morpholidate (16) to produce ATP containing ^{18}O in all four of the oxygens bonded to P_γ (82% [$^{18}O_4$]ATP, 17% [$^{18}O_3$, $^{18}O_1$]ATP). This labeled nucleotide was then used to test the ability of monothiophosphorylated SCS [prepared as described by Wolodko et al. (10)] to catalyze positional isotope exchange (see Fig. 2.3). Analysis was carried out by ^{31}P-NMR spectroscopy using a Nicolet NT 300WB spectrometer at 121.46 mHz at 21°C, 70° pulses, and 3.7 s acquisition times; a spectral width of ±1100 Hz were used. Samples for NMR contained 1 m*M* [^{18}O]ATP, 35 m*M* EDTA, 50 m*M* Tris-HCl, 50% D_2O pH 9.1 in a final volume of 4 ml. Spectra comprised 3000 scans and were recorded with additional line broadening of 0.1 Hz.

Reassembly of SCS from Isolated Subunits

Purified α and β subunits were prepared from *E. coli* SCS by methods developed by Pearson and Bridger (17). Their reassembly to form catalytically active enzyme was carried out essentially according to procedures previously described (18).

Active Enzyme Centrifugation

Details of the methods used are described elsewhere (19). The activity of the enzyme was monitored during centrifugation at 60,000 rpm by measuring the production of the free thiol of CoA from succinyl-CoA by its reaction with 5,5′-dithiobis(2-nitrobenzoate) (DTNB). Generation of CoA-SH is thus followed spectrophotometrically at 412 nm. A 10 $\mu\ell$ volume of the given enzyme was layered onto 300 $\mu\ell$ of assay medium, which contained 0.5 m*M* DTNB, 0.5 m*M* ADP (for *E. coli* SCS) or 0.5 m*M* GDP (for pig heart SCS), 10 m*M* $MgCl_2$, 0.48 m*M* succinyl-CoA in 50 m*M* potassium phosphate, 0.2 m*M* EDTA, pH 7.4.

Results and Discussion

Test for Transient *bis*-Phosphorylation

In the direction of synthesis of succinyl-CoA, the three steps of the SCS reaction are believed to be as follows:

$$E + ATP \rightleftarrows E\text{-}PO_3 + ADP \quad [1]$$

$$E\text{-}PO_3 + \text{succinate} \rightleftarrows E \cdot \text{succinyl-}PO_3 \quad [2]$$

$$E \cdot \text{succinyl-}PO_3 + CoA \rightleftarrows E + \text{succinyl-CoA} + P_i \quad [3]$$

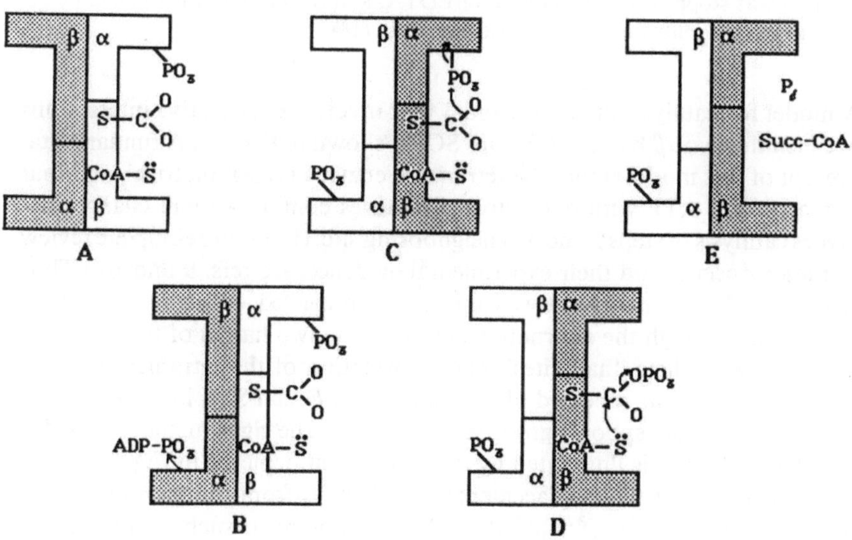

FIG. 2.1. Model for alternating sites of catalytic cooperativity in the action of *E. coli* SCS. The shaded and unshaded portions of the molecule represent $\alpha\beta$ dimers with different conformations. *A*: $E\text{–}PO_3 \cdot \text{succinate} \cdot CoA$ complex, not in the optimum configuration for catalysis of transfer of the phosphoryl group from the enzyme to succinate. *B*: ATP interacts with the neighboring active site, after which the two halves of the molecules undergo reciprocal change in configuration (*C*), producing an active site on the right that is favorable for phosphoryl transfer, the step of the reaction that is believed to be rate-limiting. *D*: Displacement of phosphate by CoA, and the products are released from the enzyme (*E*).

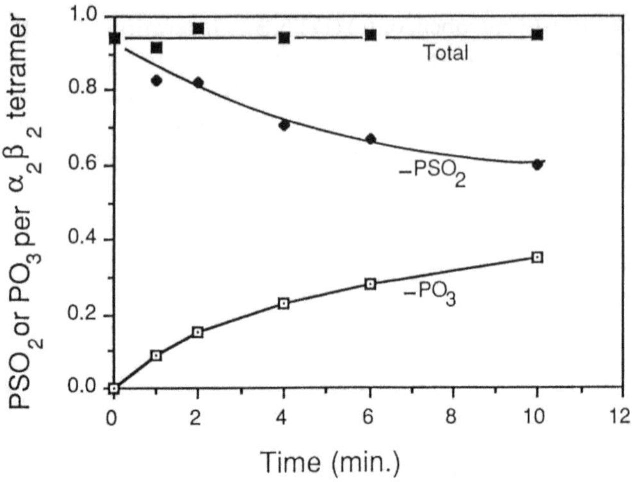

FIG. 2.2. Replacement of $-PSO_2$ by $-PO_3$. Monothiophosphorylated SCS (50 μg) was incubated with 20 μM [γ-^{32}P]ATP, 20 μM ADP, 1 mM $MgCl_2$, 50 mM Tris-HCl, 50 mM KCl, pH 7.4 at 25°, and samples were removed at the times indicated. The reaction was stopped by the addition of EDTA, and the protein was isolated by gel filtration in the centrifuge as described elsewhere (10).

A model for catalysis of steps 2 and 3 that involves cooperative interactions between the two $\alpha\beta$ halves of *E. colis* SCS is shown in Fig. 2.1. A fundamental precept of this model is that the tetrameric enzyme is asymmetrical, and that interaction of ATP with one active site causes conformational change that favors catalysis of steps 2 and 3 at neighboring site. (For more complete review of these concepts and their experimental evidence, see refs. 9 and 10.) Thus steps 1 and 2 plus 3 may be viewed as being disconnected, occurring at opposite active sites through the alternating action of the two halves of the tetramer. This model predicts "half-sites" phosphorylation of the tetramer and thus can account for our repeated observations that *E. coli* SCS incorporates no more than one phosphoryl group per molecule. The rigorous nature of this half-sites behavior is illustrated by the experiment depicted in Fig. 2.2. Here we have measured the displacement of [^{35}S]PSO_2 from the monothiophosphorylated enzyme by [^{32}P]ATP. If ADP is absent in such a mixture, no replacement is detectable (data not shown), but in the presence of 20 μM ADP we see the expected slow removal of $-PSO_2$ and its replacement by $-PO_3$. Significantly, under no conditions does the sum of these two residues exceed one group per $\alpha_2\beta_2$ enzyme molecule.

This kind of strict half-sites behavior could of course be interpreted in terms of a single active site with a single potential phosphohistidine residue in a tetramer in which one of the two α subunits was permanently inaccessible or otherwise nonreactive. This situation is not the case, however, as previous tracer and hybrid enzyme experiments (7) have clearly established that both

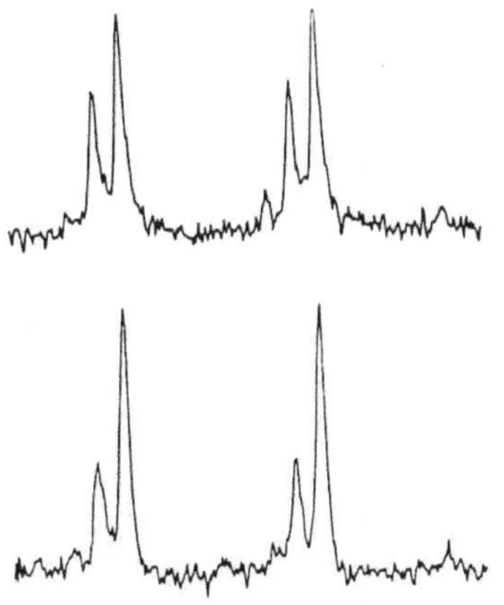

FIG. 2.3. ^{31}P-NMR spectra of the γ-phosphoryl group of [^{18}O]ATP after incubation with *E. coli* SCS. Fully thiophosphorylated SCS (one $-PSO_2$ per $\alpha_2\beta_2$) was incubated in the presence of *A*, [^{18}O]ATP, 0 min; *B*, [^{18}O]ATP, 2 min; *C*, [^{18}O]ATP, 6 min; and *D*, [^{18}O]ATP, plus 0.1 m*M* ADP, 6 min. The enzyme was separated from the nucleotides by ultrafiltration prior to ^{31}P-NMR. See text for further details.

sites participate in catalytic turnover, in keeping with the model depicted in Fig. 2.1. When considering this model, then, we wished to determine whether cooperative effects exerted at one site (i.e., the right site as drawn) were driven by noncovalent interactions with the neighboring (left) site with ATP or by covalent phosphorylation of that site. Therefore we have exploited the method of positional isotope exchange, as proposed by Midelfort and Rose (20), to answer this question. The results of a preliminary set of experiments are shown in Fig. 2.3. Here we have measured the scrambling of ^{18}O from the $\beta\gamma$ bridge position of ATP to the β nonbridge position as a test for transient phosphoryl transfer. As Midelfort and Rose noted, reversible phosphoryl

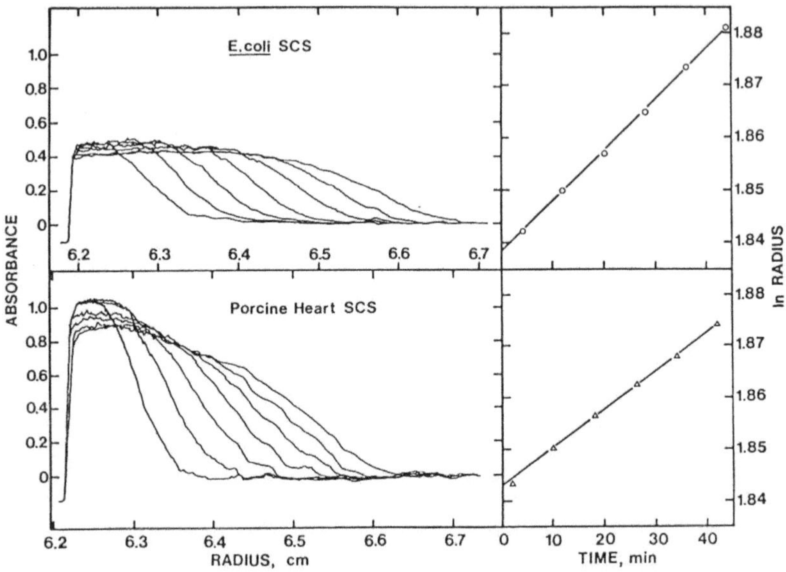

FIG. 2.4. Active enzyme centrifugation of SCS from *E. coli* and from pig. The left panels show tracings of superimposed successive scanner records taken at 8-min intervals. Sedimentation is from left to right. The right panels show plots of the logarithm of the radial position of the maxima of difference curves calculated from successive scanner records versus the mean time of corresponding scans.

transfer results in the scrambling of isotope from the bridge to nonbridge position, provided only that the transiently formed ADP is capable of undergoing O_α–P_β bond rotation. In Fig. 2.3 (spectra A to C) it may be seen that when fully thiophosphorylated enzyme is incubated with $[^{18}O_4]$ATP in the absence of ADP, no rapid bridge–nonbridge scrambling takes place that would be indicative of transient *bis*-phosphorylation. Instead, slow scrambling takes place that can be directly correlated to the slow exchange of $-PO_3$ for $-PSO_2$ at one site. Thus when ADP is included in the incubation mixture (Fig. 2.3D) increased scrambling takes place, consistent with the ability of ADP to promote the slow dethiophosphorylation of SCS akin to that seen in Fig. 2.2. As expected, when we began with enzyme having some of the thiophosphoryl groups quantitatively replaced by $[^{32}P]PO_3$, again no scrambling took place. These results allow us to refine further the model for subunit interactions in catalysis by *E. coli* SCS. Specifically, we have learned that covalent phosphorylation of one site does not even transiently occur when the neighboring site is occupied by either $-PSO_2$ or $-PO_3$. It will be of great interest to see the results when the substrates succinate and CoA are included in the scrambling experiment (S. P. Williams and W. A. Bridger, *unpublished data*), as it is conceivable that the enzyme's structural constraints may require coordinated phosphoryl transfer events at the two sites: Transfer of $-PO_3$ to ADP at one

TABLE 2.1. Comparison of the molecular size of SCS from *E. coli* and pig heart muscle

Parameter	*E. coli* SCS	Pig heart SCS
MW by sed. eqm. (~9 mg/ml)	139,000[a]	78,000[a]
M_r by gel filtration	136,000[b]	77,000[c]
MW calculated from a.a. sequence	142,068[d]	—
$s_{20,w}$	6.3 S (8 mg/ml)	4.3 S (9.3 mg/ml)
$s^{\circ}_{20,w}$	7.04 S	4.55 S
Ave. $s_{20,w}$ by A.E.C[e]	6.6 S	4.6 S

[a] Wolodko et al. (19).
[b] O'Connor (23).
[c] Bild et al. (6).
[d] Buck et al. (3).
[e] A.E.C.: active enzyme centrifugation

site may be accompanied by virtually simultaneous transfer to succinate at the neighboring site.

Active Enzyme Centrifugation

Characterization of the sizes of SCSs from *E. coli* and pig heart, performed more than a decade ago, indicated that the bacterial enzyme was an $\alpha_2\beta_2$ tetramer, whereas the mammalian enzyme behaved as a simple $\alpha\beta$ dimer (1). Despite this apparent fundamental difference in subunit structure, the two species of SCS have similar catalytic properties (10, 21). Thus, this led us to wonder if the differences in subunit structures extend to low concentrations compatible with activity measurement. We have therefore undertaken a complete study of the sedimentation and other hydrodynamic properties of these two species of SCS (19) that includes use of the technique of active enzyme centrifugation. This method involves centrifugation of a boundary of enzyme, at assay concentration, through a column of chromogenic substrate mixture; the measured sedimentation rate is therefore that of the active species of enzyme (see refs. 19 and 22 for further details). The results of such experiments are illustrated in Fig. 2.4, showing the expected difference in sedimentation rate of the two activities (6.6 S for *E. coli* SCS and 4.6 S for the pig heart enzyme). Taken together with the rest of the molecular size data that are summarized in Table 2.1, we conclude that *E. coli* SCS is a nondissociating $\alpha_2\beta_2$ tetramer, whereas pig heart SCS is a nonassociating $\alpha\beta$ dimer.

Reconstitution of SCS from Isolated Subunits

The α and β subunits of *E. coli* SCS, isolated by gel filtration in the presence of 6 *M* urea/5% acetic acid, may be reassembled to form active enzyme by refolding equimolar mixtures under appropriate conditions (16, 17). Either ATP or P_i must be present during the reconstitution process, suggesting that occupation of the phosphoryl binding region of the active site of α is required

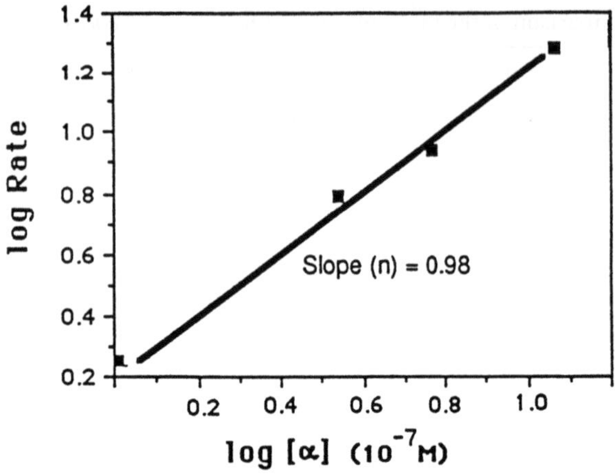

FIG. 2.5. Determination of the kinetic order of reassembly of *E. coli* SCS from its subunits. The isolated α and β subunits were mixed and allowed to refold to form active enzyme using methods described earlier (18). The final concentration of β was held constant at 0.98 μM, and that of α was varied between 0.10 and 1.0 μM as indicated. The initial rate of appearance of enzyme activity was measured by sampling at time intervals, and the log of that rate is plotted against the log of [α], giving a slope of 0.98. The corresponding experiment in which [α] was held constant at 1.0 μM and [β] was varied gave a straight line with slope = 0.97.

to achieve the correct configuration for productive refolding. As part of our study of the process of assembly of the enzyme, we have determined the kinetic order of reconstitution with respect to each of the subunits. In the experiment depicted in Fig. 2.5, for example, we have measured the initial rate of appearance of SCS activity from mixtures in which the concentration of α is varied and that of β is held constant. The slope of this log/log plot, equal to unity, indicates that the kinetics of the reconstitution process is first order with respect to α; an equivalent experiment (data not shown) indicates a first-order kinetic dependence on β. Thus we conclude that the rate of reassembly, second order overall, is likely to be controlled by the rate of contact between the two subunits to form the αβ dimers. Furthermore, we have shown that the refolding subunits can even discriminate between intact and damaged counterparts, and that the reassembled protein has the same size and specific activity as the native enzyme (Fig. 2.6) We believe that this in vitro reconstitution process mimics the assembly of the enzyme *in vivo*, as it is known that the two subunits are discrete gene products (3) and that the rate and extent of reconstitution are not affected by the presence of a crude cellular extract (18). The latter point, together with the results shown on Fig. 2.6, show that the α and β subunits, during their refolding from the denatured state, possess a high degree of mutual recognition.

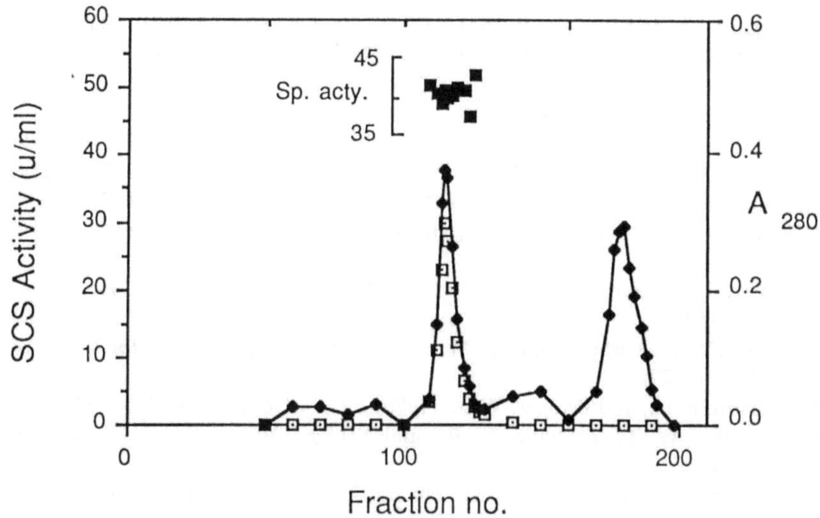

FIG. 2.6. Gel filtration of a preparation of *E. coli* SCS that has been reassembled from a mixture of isolated subunits. Equimolar amounts of α and β subunits, in acid urea solution, were allowed to refold to form active enzyme by dilution of the urea and neutralization as described elsewhere (18). The refolded mixture was then passed through a 1.5 × 108 cm column of Sephacryl S300 equilibrated with 50 m*M* Tris-HCl, 50 m*M* KCl, pH 7.2. Samples were analyzed as indicated. The native enzyme has been shown in a separate experiment to peak at fraction 114, corresponding to the position shown for the peak of reassembled enzyme and well separated from the peak of pig heart SCS (MW 78,000). The peak of absorbance near fraction 180 is largely nucleotide judging from its high $A_{260/280}$ ratio.

Acknowledgments. We thank Ed Brownie for providing the large amounts of purified succinyl-CoA synthetases required for these studies. S.P.W. is the recipient of a Fellowship and Research Allowance from the Alberta Heritage Foundation for Medical Research. This work was supported by a grant (MT-2805) from the Medical Research Council of Canada.

References

1. Bridger, W. A. (1974) *The Enzymes*, vol. 10, 3rd ed., edited by P. D. Boyer, pp. 581–606. Academic Press, New York.
2. Krebs, A., and Bridger, W. A. (1974) *Can. J. Biochem.* 52:594–598.
3. Buck, D., Spencer, M. E., and Guest, J. R. (1985) *Biochemistry* 24:6245–6252.
4. Pearson, P. H., and Bridger, W. A. (1975) *J. Biol. Chem.* 250:8524–8529.
5. Collier, G. E., and Nishimura, J. S. (1978) *J. Biol. Chem.* 253:4938–4943.
6. Bild, G. S., Janson, C. A., and Boyer, P. D. (1980) *J. Biol. Chem.* 255:8109–8115.
7. Wolodko, W. T., O'Connor, M. D., and Bridger, W. A. (1980) *Proc. Natl. Acad. Sci. USA* 78:2140–2144.

8. Vogel, H. J., and Bridger, W. A. (1982) *J. Biol. Chem.* 257:4834–4842.
9. Vogel, H. J., and Bridger, W. A. (1983) *Biochem. Soc. Trans.* 11:315–323.
10. Wolodko, W. T., Brownie, E. R., O'Connor, M. D., and Bridger, W. A. (1983) *J. Biol. Chem.* 258:14116–14119.
11. Leitzmann, C., Wu, J-Y., and Boyer, P. D. (1970) *Biochemistry* 9:2338–2346.
12. Cha, S., Cha, C-J. M., and Parks, R. E. (1967) *J. Biol. Chem.* 242:2577–2581.
13. Murakami, Y., and Nishimura (1974) *Biochim. Biophys. Acta* 336:252–263.
14. Bridger, W. A., Ramaley, R. F., and Boyer, P. D. (1969) *Methods Enzymol.* 13:70–75.
15. Risley, J. M., and Van Etten, R. L. (1978) *J. Lab. Compounds Radiopharm.* 15:533–537.
16. Wehrli, W. E., Verheyden, D. L. M., and Moffat, J. G. (1965) *J. Am. Chem. Soc.* 87: 2265–2277.
17. Pearson, P. H., and Bridger, W. A. (1975) *J. Biol. Chem.* 250:4451–4455.
18. Wolodko, W. T., Brownie, E. R., and Bridger, W. A. (1980) *J. Bacteriol.* 143:231–237.
19. Wolodko, W. T., Kay, C. M., and Bridger, W. A. (1986) *Biochemistry* 5420–5425.
20. Midelfort, C. F., and Rose, I. A. (1976) *J. Biol. Chem.* 251:5881–5887.
21. Nishimura, J. S., and Mitchell, T. (1985) *J. Biol. Chem.* 260:2077–2079.
22. Teller, D. (1973) *Methods Enzymol.* 27:346–441.
23. O'Connor, M. D. (1982) PhD thesis, University of Alberta.

3
Fatty Acid Synthetase of Chicken Liver: A Novel Active-Site Structure for Condensation Comprised of SH Groups from a Cysteine Residue and an Oscillating Phosphopantetheine Swinging Arm on Adjacent Subunits

ROBERT Y. HSU

The synthesis of palmitic acid by avian fatty acid synthetase occurs through successive elongation of the acetyl primer by two-carbon acyl residues from malonyl-CoA, and reduction of the resulting ketoacyl intermediates by NADPH to corresponding saturated acyl derivatives. After completion of seven elongation cycles, enzyme-bound palmitate is released as the free acid. The enzyme complex is a dimeric molecule comprised of identical subunits. Each subunit contains, in addition to active sites for component reactions, a phosphopantetheine SH group which serves as a carrier for the covalently attached acyl intermediates (1).

In the current study (2), evidence for the complementary of a phosphopantetheine SH group on one subunit and a cysteine SH group on the adjacent subunit is provided by results of hybridization experiments. Detailed studies on the reaction of the phosphopantetheine SH group with DTNB[1] (1) suggests that the synthetase molecule exists in two limiting conformation states differing in reactivity, and in the distance between this group and the complementary cysteine SH group. A model for these conformation states is presented.

The Complementarity of Phosphopantetheine and Cysteine SH Groups on Adjacent Subunits

The requirement of subunit interactions in synthetase function is accounted for by a "head-to-tail" model proposed by Stoops and Wakil (3), which depicts a condensation site consisted of juxtaposed Sp and Sc SH groups from

[1] The abbreviations used in this chapter are as follows: ClAcCoA = chloroacetyl-CoA; DTNB = 5,5′-dithiobis-(2-nitrobenzoic acid); FAS = fatty acid synthetase; IAM = iodoacetamide; Sp and Sc = the phosphopantetheine and active cysteine SH groups, respectively. The asterisk indicates a modified group.

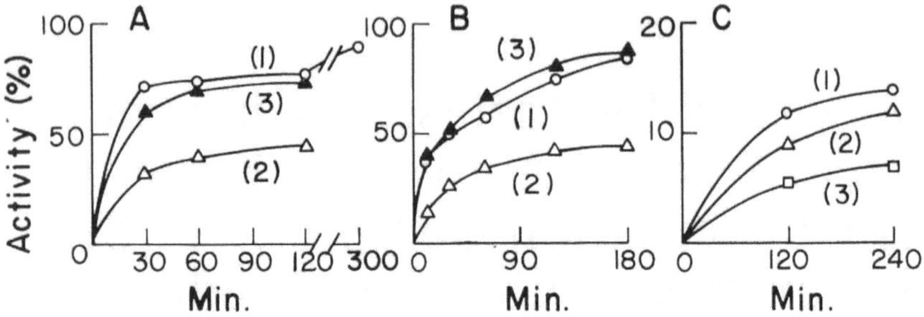

FIG. 3.1. Recovery of activity upon reconstitution of fatty acid synthetase subunits. The reconstitution experiments were performed in 0.2 *M* Na phosphate buffer, pH 7.0, at room temperature with FAS subunits in the following combinations. *A*: Native-FAS and IAM-FAS—(1) 1.38 μM native-FAS; (2) 0.14 μM native-FAS; and (3) 0.14 μM native-FAS and 1.46 μM IAM-FAS. *B*: Native-FAS and ClAcCoA-FAS—(1) 1.46 μM native-FAS; (2) 0.15 μM native-FAS; and (3) 0.15 μM native-FAS and 1.52 μM ClAcCoA-FAS. *C*: IAM-FAS and ClAcCoA-FAS—(1) 0.62 μM IAM-FAS and 0.66 μM ClAcCoA-FAS; (2) 0.24 μM IAM-FAS and 1.06 μM ClAcCoA-FAS; and (3) 1.0 μM IAM-FAS and 0.26 μM ClAcCoA-FAS. The activities in *A* and *B* are expressed as the percent of native-FAS in the incubation. The activities in *C* are the percent of specific activity relative to the purified enzyme.

separate subunits; the reaction then takes place at the subunit interface between the malonyl residue bound to the former and the acetyl residue bound to the latter. Following condensation, the Sp arm carrying enzyme bound ketoacyl intermediate is swung toward sites on the same subunit to affect reduction of this intermediate.

Hybridization experiments were carried out on chicken liver fatty acid synthetase variants obtained by chemical modification(2). For this purpose, IAM was used for the modification of Sc(3), and the affinity label ClAcCoA was used for the modification of Sp(4). The specificity of these reagents was confirmed in our laboratory by the use of ^{14}C-labeled compounds; the modified, inactive enzymes were hydrolyzed in acid, and analyzed by paper chromatography. The chromatograms showed that in each case, a major radioactive spot was seen, which for the IAM-modified enzyme co-chromatographed with authentic carboxymethyl cysteine and for the ClAcCoA-modified enzyme co-chromatographed with authentic carboxymethyl cysteamine. In hybridization experiments synthetase samples were inactivated by each reagent and dissociated in low ionic strength buffer at slightly alkaline pH. The enzyme subunits were then incubated in the following combinations at high ionic strength and neutral pH for reconstitution. (a) IAM-FAS subunit (Sc*-Sp) and native-FAS subunits (Sc-Sp); (b) ClAcCoA-FAS subunits (Sc-Sp*) and native-FAS subunits (Sc-Sp); and (c) IAM-FAS and ClAcCoA-FAS subunits. In each case the time course of activity regain was followed.

The results of typical experiments are shown in Figs. 3.1. Subunits of native-FAS reconstituted and reactivated rapidly (curve 1, Fig. 3.1A, B), yielding

89% and 85% of native FAS activity in 300 min (curve 1, Fig. 3.1A) and 180 min (curve 1, Fig. 3.1B), respectively. The rate of reactivation decreased significantly with a tenfold decrease in enzyme concentration (curve 2, Fig. 3.1A, B), as would be expected for a bimolecular reaction involving association of half-molecular-weight subunits. When total protein concentration was maintained at a relatively constant level by the addition of IAM-FAS subunits (curve 3, Fig. 3.1A) or ClAcCoA-FAS subunits (curve 3, Fig. 3.1b), the decreases in rates were not apparent. This result provides kinetic evidence for the formation, in addition to the inactive (Sc*-Sp)(Sc*-Sp) or (Sc-Sp*)(Sc-Sp*) and the fully active (Sc-Sp)(Sc-Sp) homodimers, of heterologous (Sc*-SP)(Sc-Sp) or (Sc-Sp*)(Sc-Sp) dimers containing a single viable active site and possessing 50% activity; furthermore, it may be deduced that the functional sites in the enzyme dimer behave independently without discernible cooperativity. The subunit interactions that account for this behavior may be conformational; i.e., the modified subunit stabilizes the catalytically competent state of the unmodified subunit. Alternatively, the condensation site is comprised of an Sc SH group from one subunit and an Sp SH group from the other as depicted by the "head-to-tail" model proposed on the basis of cross-linking studies (3). In the hybrid dimer, one of these groups is provided by the modified subunit. Independent evidence for the reversible dissociation of IAM-FAS under our conditions was obtained by sedimentation velocity analysis in a Beckman model E ultracentrifuge. The S_{20} values of the enzyme following modification, dissociation, and reconstitution were 12.5, 8.5, and 12.15, respectively.

Figure 3.1C shows results obtained by hybridization of IAM-FAS and ClAcCoA-FAS. Reconstitution of the subunits of IAM-FAS and ClAcCoA-FAS in approximately equal amounts (curve 1, Fig. 3.1C) induced regeneration of synthetase activity by 13.8% in 240 min, indicating the formation of an active (Sc*-Sp)(Sc-Sp*) hybrid dimer in addition to the inactive homodimers. Reconstitution of the modified enzymes in unequal amounts of 1.0:4.4 (curve 2, Fig. 3.1C) and 3.9:1.0 (curve 3, Fig. 3.1C) yielded lower recoverable activities of 12.0 and 7.0%, respectively, which may be accounted for by decreased probability of hybrid formation. Hybridization of IAM-FAS and ClAcCoA-FAS yields (Sc*-Sp)(Sc*-Sp), (Sc*-Sp)(Sc-Sp*), and (Sc-Sp*)(Sc-Sp*) dimers in a 1.0:2.0:1.0 ratio, assuming that subunit recombination is unaffected by chemical modification. Because the hybrid dimer is 50% active (see below), an overall recoverable activity of 25% is expected. Although the observed value of 13.8% falls short of quantitative recovery, it is reasoned that this loss was due to limited cross-reactions of IAM with Sp and, lesslikely, ClAcCoA with Sc, as well as possibly decreased subunit recombination.

Since the IAM-FAS subunit contains only an intact Sp SH group and the ClAcCoA-FAS subunit an intact Sc SH group, the demonstration of an enzymatically active hybrid comprising these subunits provides further evidence for the existence of two independent functional sites in the enzyme dimer and, moreover, direct evidence for the complementarity of these groups in the synthesis of palmitate. Such a requirement involving intersubunit transfer of acyl residues strongly supports the "head-to-tail" arrangement (3). It also explains the well documented inability of the monomeric enzyme to carry out the catalyzed reaction (5).

Presence of Two Discrete Conformational States of the Enzyme Differing in the Distance Between Complementary Phosphopantetheine and Cysteine SH Groups

Inactivation by DTNB

In the DTNB inactivation studies, the reaction of Sp with DTNB was used as a conformational probe for the enzyme (1). A large number of SH groups on the synthetase molecule were reactive with DTNB. However, only a single group reacted rapidly at a rate up to 2200 times that of simple thiol compounds, resulting in the loss of synthetase activity. This group was identified as the Sp rather than a cysteine by the following observations: (a) The inactivation was competitively protected by acetyl-CoA or malonyl-CoA, each binding covalently to the Sp SH group. (b) The reaction of this group was prevented by affinity labeling with ClAcCoA. (c) Reversible modification by DTNB protected the enzyme from inactivation by ClAcCoA but not by the Sc Reagent 1,3-dibromopropanone. (d) Inactivation by DTNB was followed by cross-linking of subunits. Intersubunit cross-linking was inhibited by either acyl substrate added before but not after inactivation as would be expected if the Sp binding site was blocked by reaction with DTNB (see below).

The inactivation of FAS by DTNB followed typical second-order kinetic behavior. The rate of inactivation was highly dependent on salt concentration, as shown by the plot in Fig. 3.2. The inactivation was maximal ($k_2 = 132$ mM^{-1} s^{-1}, filled circle) at the lowest salt concentration (3 mM EDTA, pH 7) tested; the addition of Na phosphate in increasing amounts to the reaction medium containing 6 mM EDTA caused an asymptotic decrease of the rate constant from 106 mM^{-1} s^{-1} to a value of 10.6 mM^{-1} s^{-1} at 186 mM. Furthermore, this effect was seen with other salts such as KCl and Tris-HCl. At 199 mM KCl, a rate constant of 8.3 mM^{-1} s^{-1} was obtained. In comparison, the rate constant for the reaction of CoA with DTNB was found to be 0.06 mM^{-1} s^{-1} at 6 mM EDTA and 0.3 mM^{-1} s^{-1} upon addition of 100 mM Na phosphate.

The strong dependence of the reactivity of Sp SH group on ionic strength can be attributed to a salt-induced conformational change affecting the local environment of this group. Although FAS is known to dissociate in low ionic strength buffer, dissociation requires hours for completion and is too slow to account for the change in reactivity. A plausible minimum model would

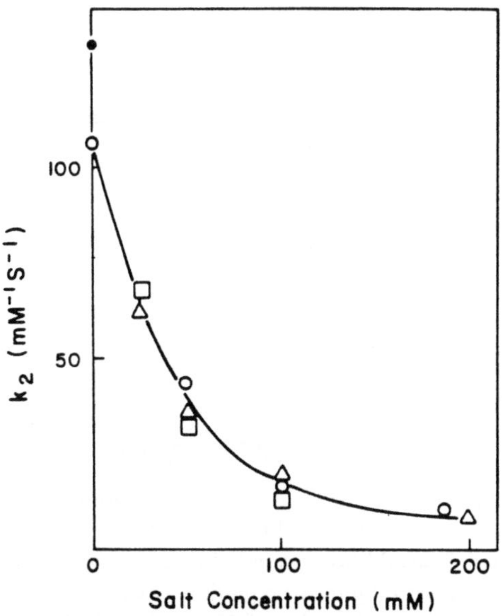

FIG. 3.2. Dependence of second-order rate constant of inactivation on salt concentration. The incubations contained FAS 0.13 μM, DTNB 0.49–5.70 μM in 6 mM EDTA, 1% (v/v) glycerol, pH 7.0, 25°C, and salt as indicated. The salts are as follows: (○) Na phosphate, (△) KCl, and (□) Tris-Cl. (●) is rate constant obtained at 3 mM EDTA, 1% (v/v) glycerol without added salt.

involve the existence of two conformational states of the enzyme: At very low salt concentration, the enzyme exists solely as conformer I in which the Sp SH group is highly reactive; the addition of salt induces a transition to conformer II, and the reactivity of this group is markedly reduced. This model is pictured as

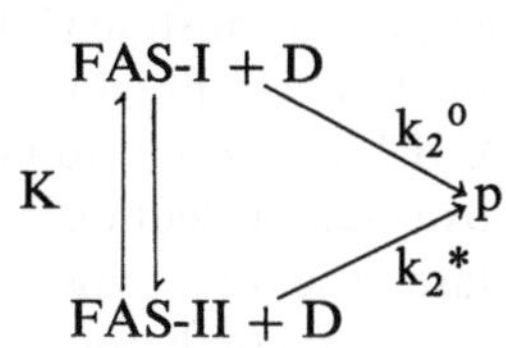

where D = DTNB

k_2^0, k_2^* = second-order rate constants of conformers I and II, respectively

P = inactive TNB enzyme

K = equilibrium constant of transition and equals FAS-I/FAS-II

Furthermore, this transition is rapid relative to the reaction with DTNB.

To facilitate interpretation of data, further experiments were carried out in

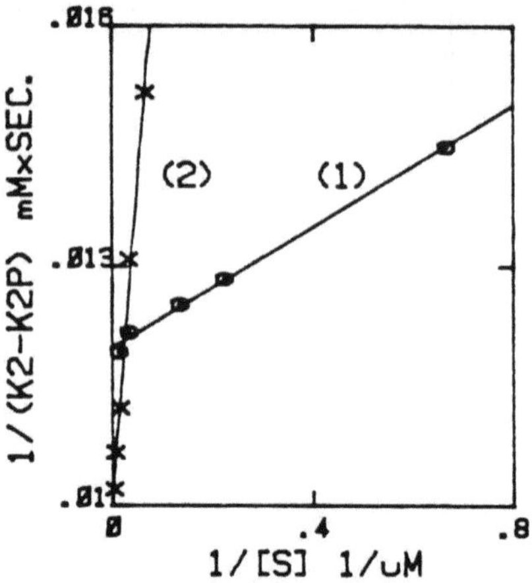

FIG. 3.3. Kinetics of nucleotide protection. The incubations contained FAS 0.13–0.20 μM, DTNB 0.49–5.70 μM in 6 mM EDTA, 1% (v/v) glycerol, pH 7.0, 25°C, and other additions as indicated. The additions were NADPH (curve 1) and $NADP^+$ (curve 2). The data were plotted according to the equation $1/(k_2 - k_2^{P}) = \{1/[(k_2 - k_2^{P'})\overline{K}]\} \cdot \{1/[S]\} + 1/(k_2 - k_2^{P'})$ using a k_2 value of 106 mM^{-1} s^{-1}. k_2 = rate constant in the absence of substrate [S]. k_2^{P}, $k_2^{P'}$, and $\overline{K}$ = experimentally determined second-order rate constant at a given substrate concentration, the limiting rate constant at infinite substrate concentration, and the association equilibrium constant of substrate, respectively.

6 mM EDTA, 1% (v/v) glycerol at pH 7.0 (low salt), and in the same medium in the presence of 100 mM Na phosphate (high salt), except where indicated.

We have shown that both $NADP^+$ and NADPH afforded noncompetitive protection of DTNB inactivation. The experimental plots at low salt are given in Fig. 3.3. Extrapolation of the linear plots yielded limiting second-order rate constants of 6.8 mM^{-1} s^{-1} at saturating $NADP^+$ and 22.5 mM^{-1} s^{-1} at saturating NADPH. The dissociation constant ($1/\overline{K}$) of 0.32 μM for NADPH was identical to the value of 0.29 μM previously reported by us (6). The corresponding value for $NADP^+$ was found to be 7.81 μM, indicating relatively weak binding for the oxidized nucleotide.

In the fatty acid synthetase reaction, pyridine nucleotides bind to the reductase domain distinct from the condensation site. The noncompetitive protection observed for nucleotides is consistent with this fact; moreover, such behavior suggests that a conformational change was involved. A common salt and nucleotide effect on conformational change (from conformer I to conformer II) was supported by the observation that the former effect was significantly smaller in the presence of either nucleotide. At low salt in the absence of nucleotide, the rate constant was 106 mM^{-1} s^{-1}. This value was reduced 6.5-fold to 16.4 mM^{-1} s^{-1} by high salt. The rate constants of 9.3 mM^{-1} s^{-1} at 300 μM $NADP^+$ and 23.0 mM^{-1} s^{-1} at 75 μM NADPH in low

TABLE 3.1. Rough estimates of equilibrium constants of conformer transition.

Enzyme	k_2 ($mM^{-1}\ s^{-1}$)	k
Low salt		
E	106.0	~4.00
E-NADPH	22.5	~0.14
E-$NADP^+$	6.8	0
High salt		
E	16.4	~0.08

Values of K were calculated according to the equation $K = (k_2 - k_2^*)/(k_2^0 - k_2)$ using a k_2^0 value of 132 $mM^{-1}\ s^{-1}$ and a k_2^* value of 6.8 $mM^{-1}\ s^{-1}$ (E-$NADP^+$ at low salt).

TABLE 3.2. Effects of incubation time on cross-linking.[a]

Time course	Incubation with DTNB (DTNB:FAS = 9:1)[b] (s)	Cross-linking (%)
a	−DTNB	0
b	15	6
c	60	22
d	180	31
e	600	32

[a] Incubations were performed at low salt.
[b] Inactivation was completed in <8 s.

salt, however, were reduced by high salt only 1.2- and 2.6-fold, respectively, to 8.0 and 8.7 $mM^{-1}\ s^{-1}$. If we assume that the nucleotide affects the equilibrium of conformer transition, rough estimates of the equilibrium constants may be obtaining as shown in Table 3.1.

Cross-Linking Studies

In the cross-linking experiments, chicken liver fatty acid synthetase was inactivated by DTNB and then incubated for an additional period to permit cross-linking. The enzyme sample was then subjected to electrophoresis in sodium dodecyl sulfate, and the degree of dimerization was calculated from densitometric reading of the stained protein monomer and dimer bands. Intersubunit cross-linking is a two-step reaction as depicted below.

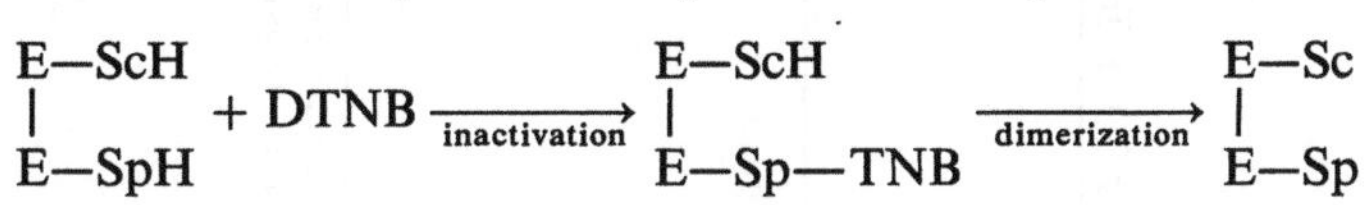

The effects of incubation conditions on dimerization are shown in Tables 3.2 and 3.3. The time course experiment (a–e) indicates that dimerization was slower than inactivation, which occurred in <8 sec. Maximum dimerization was 32% in e and 37% in f. In other experiments a value of 52% was obtained. The ⩽50% value for dimerization may be explained by the low phospho-

TABLE 3.3. Effects of ligand addition on cross-linking.[a]

	Primary incubation with DTNB (DTNB:FAS = 2:1)		Secondary incubation		
	Effector (μM)	Time (s)	Effector (μM)	Time (s)	Cross-linking (%)
f (control)	None	80[b]	None	0	37
g	None	15[b]	Na phosphate 0.1 M	80	21
h	None	25[b]	NADPH 180 μM	80	20
i	None	15[b]	$NADP^+$ 360 μM	80	20
j	None	15[b]	Malonyl-CoA 140 μM	75	36
k	None	15[b]	Acetyl-CoA 130 μM	75	35
l	Malonyl-CoA 140 μM	75	None	0	14
m	Acetyl-CoA 130 μM	75	None	0	18

[a] Incubations were performed at low salt concentration.
[b] Inactivation was completed in <15 to 25 s during the primary incubation.

pantetheine content (0.91–1.15 per enzyme dimer, data not shown) of the enzyme. A comparison of samples obtained in the absence (f) and presence of effectors shows that dimerization was unaffected by either malonyl-CoA (j) or acetyl-CoA (k) due to blockage of the Sp binding site during DTNB inactivation, but it was strongly inhibited by salt (g), NADPH (h), or $NADP^+$ (i). In samples where the acyl substrates were added before DTNB (1 and m), dimerization was reduced owing to the decreased rate of inactivation by DTNB.

Dimerization requires a structural arrangement involving close proximity of Sc and Sp SH groups on adjacent subunits. Because salt, NADPH, and $NADP^+$ also inhibit the preceding DTNB inactivation step and the inhibition has been attributed to a conformational transition from FAS-I to FAS-II, it may be suggested that FAS-I has the above structural arrangement favorable for dimerization and that in FAS-II the SH groups are more spatially separated. Because DTNB reacts more rapidly with FAS-I, it may be further reasoned that this enhancement is caused by proximal localization of the two SH groups, with the DTNB molecule bound initially to an affinity site on the Sc (rather than the Sp) bearing subunit. This interpretation is supported by results of an independent experiment which showed that dissociation of the enzyme was accompanied by the loss of the rapidly reacting Sp SH group.

The nature of subunit interactions responsible for the proximal arrangement of Sc and Sp SH groups in FAS-I is not entirely clear. Because such an arrangement is inhibited by salt, it may be suggested that these interactions are predominantly ionic, possibly involving the negatively charged Sp-bearing acyl carrier domain (7) of one subunit, with positive groups on the Sc-bearing domain of the other subunit. Possible geometry for the functional site of fatty acid synthetase conformers based on complementarity of essential SH groups on separate subunits in the condensation reaction (3) is depicted in scheme 1,

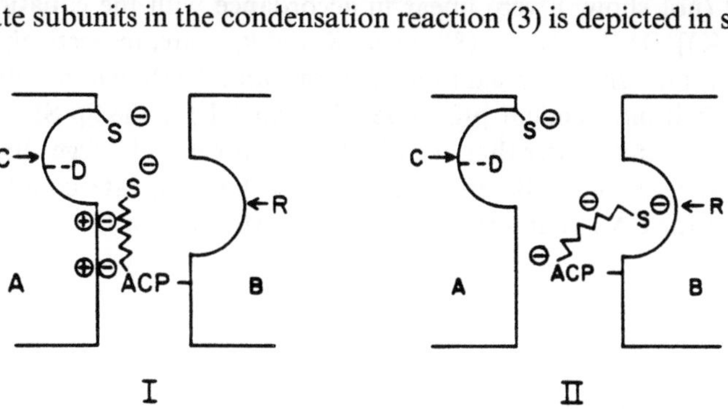

where A and B are the Sc- and Sp-bearing subunits and C and R are the condensation and reductase domains, respectively. In FAS-I, DTNB (D) bound noncovalently to A is poised for reaction with the adjacent Sp SH group on B. Such a structural arrangement, however, is absent in FAS-II. It is tempting to suggest that these conformational states may represent those

TABLE 3.4. pK values of the SH groups of fatty acid synthetase and coA.

SH compound	K	pK
CoA		
Low salt	2.6×10^{-8}	7.6
High salt	2.0×10^{-8}	7.7
FAS		
Low salt	1.2×10^{-6}	5.9
High salt	0.73×10^{-6}	6.1

actually occurring in catalysis. During condensation the enzyme assumes the FAS-I state, and the reaction between acetyl and malonyl residues bound to Sc and Sp, respectively, is facilitated; following condensation, the prothetic group bearing the ketoacyl product then moves to the reductase site on B for subsequent reactions, as shown for FAS-II. The enzyme thus oscillates seven times between these conformational states for the synthesis of palmitate. The induction of FAS-II by NADPH, the cofactor for reduction, is consistent with the above suggestion. The primary event triggering the transition from FAS-II to FAS-I may be malonyl-CoA binding. Such a possibility is being tested by additional experiments.

pK of Sp SH Group

The pH rate profile for the inactivation of chicken liver fatty acid synthetase was obtained by determining the rate constants in the pH range of 6.0 to 7.5, where the enzyme is highly stable. The $1/k_2$ versus $[H^+]$ plots at low salt and high salt (not shown) were linear in accordance with the equation $1/k_2 = [1/(k_{max} \cdot K)] \cdot [H^+] + 1/k_{max}$ (8), where K and k_{max} are, respectively, the ionization constant and the maximum rate constant. As shown in Table 3.4, the relatively salt-independent pK value of 5.9 to 6.1 for the Sp SH group was 1.5 to 1.8 pH units lower than that for CoA control. This large change in pK explains the 23-fold difference between the limiting rate constant of 6.8 $mM^{-1}\ s^{-1}$ for FAS-II and the value of 0.3 $mM^{-1}\ s^{-1}$ for CoA.

Summary

Enzymatically inactive variants of chicken liver fatty acid synthetase were prepared by modification of the enzyme with chloroacetyl-CoA, which reacts with the phosphopantetheine SH group or with iodoacetamide, which reacts with the active cysteine SH group at the condensation site. Hybridization of the unmodified enzyme with each variant yielded the corresponding (unmodified)(modified) dimers; hybridization of the two variants yielded (chloroacetyl-CoA-modified) (iodoacetamide-modified) dimers. In each case the hybrid dimer was shown to possess up to 50% synthetase activity, indicating that the functional site is comprised of a phosphopantetheine SH group from one subunit and a cysteine SH group from the other subunit as depicted by the "head-to-tail" model proposed by Wakil and co-workers (9).

The phosphopantetheine SH group was highly reactive with DTNB, and the reaction was accompanied by loss of synthetase activity. The reactivity was reduced (and synthetase activity protected) by addition of salt, $NADP^+$, or NADPH, with the nucleotides providing noncompetitive protection. This effect was attributed to a change in the conformational state of the enzyme. The above ligands also reduced subsequent reaction of the thionitrobenzoate derivative with the active cysteine to form cross-linked dimers. On the basis of these results it is proposed that fatty acid synthetase exists in two limiting conformational states: In comformer I the phosphopantetheine and cysteine in adjacent subunits are located proximal to each other to facilitate condensation; the phosphopantetheine arm carrying the ketoacyl product then swings back to sites on the same subunit for subsequent reduction of this product. During the catalyzed reaction the phosphopantetheine arm oscillates seven times between these conformer states for the synthesis of palmitic acid.

References

1. Tian, W. X., Hsu, R. Y., and Wang, Y. S. (1985) *J. Biol. Chem.* 260:11375–11387.
2. Wang, Y. S., Tian, W. X., and Hsu, R. Y. (1984) *J. Biol. Chem.* 259:13644–13647.
3. Stoops, J. K., and Wakil, S. J. (1981) *J. Biol. Chem.* 256:5128–5133.
4. MaCarthy, A. D., and Hardie, D. G. (1982) *FEBS Lett.* 147:256–260.
5. Yun, S. L., and Hsu, R. Y. (1972) *J. Biol. Chem.* 247:2689–2698.
6. Hsu, R. Y., and Wagner, B. J. (1970) *Biochemistry* 9:245–251.
7. Qureshi, A. A., Lonitzo, I. A., Hsu, R. Y., and Porter, J. W. (1976) *Arch. Biochem. Biophys.* 177:379–393.
8. Lindley, H. (1962) *Biochem. J.* 82:418–425.
9. Wakil, S. J., Stoops, J. K., and Joshi, V. C. (1983) *Annu. Rev. Biochem.* 52:537–579.

4
Regulatory Properties of Glucokinase

KENNETH E. NEET, PETER S. TIPPETT,
and ROBERT P. KEENAN

The understanding of abnormal kinetics of even relatively simple enzymes is sometimes a formidable task. Rat liver glucokinase is a monomeric enzyme of only 50,000 daltons (1, 2) but has properties that have stirred considerable debate with regard to its physiological role and its kinetic mechanism. The enzyme catalyzes the phosphorylation of glucose by MgATP and demonstrates positive cooperativity with glucose (3, 4), allosteric inhibition by long-chain acyl-CoA compounds (5, 6), and dependence of its kinetic properties on the oxidation state of its sulfhydryls (7). Thus considerable regulatory potential exists in a small enzyme. Glucokinase is also a good example of the difficulty of analyzing complex kinetic behavior, as three mechanisms that can produce kinetic cooperativity in a monomeric enzyme have been proposed to explain the experimental data (8–10). This chapter analyzes the merits of these mechanisms, presents data on a slow transient under certain assay conditions, and integrates the apparent homotropic cooperativity, the heterotropic allostery, and the hysteresis of glucokinase. The purpose of this study was to examine the mechanistic behavior that underlies the relevant physiological functioning of the enzyme.

Results and Discussion

Physiological Role

Although the intent of this chapter is to discuss the in vitro kinetics of glucokinase, a brief summary of the physiological role of glucokinase as it pertains to these properties is appropriate. Glucokinase is thought to play a role in the uptake of glucose by the liver in response to changing plasma levels (11, 12). A correlation has been made between the apparent cooperativity of glucose uptake in isolated hepatocytes (13) and the in vitro properties of the enzyme; both show a Hill coefficient of about 1.5 to 1.6 and a half-saturation of 5 to 9 m*M* for glucokinase and about 12 to 20 m*M* for glucose uptake. The activity of glucokinase is partially balanced by glucose-6-phosphatase in the substrate cycle in the liver that determines the direction of flux of glucose (11, 12). Thus the positive cooperativity of glucose phosphorylation would provide a sharper response to changing plasma glucose concentrations for its uptake or release. Similar arguments for glucokinase activity and glycolysis in pancreatic islets have been advanced and quantitatively evaluated (14). However, the primary role of glucokinase activity in the liver has been debated in two ways: (a) The suggestion that glucose-6-phosphatase, rather than glucokinase, carries the major burden of phosphorylation in the liver (15) has been contested (12); and (b) the interpretation of metabolic data that the liver does not directly utilize a pulse of administered glucose, i.e., the "glucose paradox," has been proposed (16) and opposed (17, 18).

A second, allosteric regulatory mechanism has been suggested to link lipid metabolism and carbohydrate metabolism in the liver by inhibition of glucokinase at micromolar concentrations of the long-chain acyl-CoA (5, 6, 19). The activity of glucokinase would be partially inhibited under conditions in which hepatic concentrations of acyl-CoA rise, e.g., whole body lipolysis, and thereby promote the preferential utilization of lipids by the liver. This effect is observed at concentrations below the critical micelle concentration of palmitoyl-CoA and in the range of free acyl-CoA in the liver. Indeed this inhibition and the intracellular concentration of palmitoyl-CoA might be responsible for the twofold (13) or tenfold (15) higher saturation measured for glucose uptake or phosphorylation in hepatocytes compared to the kinetics of the pure enzyme (see above). Demonstration of the physiological significance of this response to long-chain acyl-CoA is suggestive (19), but definitive studies are lacking. Nevertheless, the kinetic aspects of the inhibition have been well characterized and bear on the kinetic mechanism addressed below. Thus both allostery and cooperativity appear to be significant contributions to the regulatory role of glucokinase.

Models for Glucokinase Cooperativity

Three models have been proposed for the kinetic cooperativity that glucokinase displays with glucose: (a) the mnemonic mechanism (20) specifically applied to glucokinase by Cornish-Bowden's group (8); (b) the ligand-induced slow transition (LIST) mechanism (21) applied by Cardenas' group (9, 22); and (c) the steady-state random mechanism (23) that was originally considered for glucokinase and discarded by Cardenas et al. (24) but later revived by Pettersson (10). The properties and requirements of these general mechanisms have been discussed (25, 26) and glucokinase kinetics reviewed in detail (27). The first two mechanisms are closely related, as the only significant difference is the number of catalytic cycles; the mnemonic mechanism has only one cycle and two isomeric forms of free enzyme and is therefore a special case of the two full cycles and two interconvertible forms of the LIST mechanism (25). Both of these models are related to the hysteretic properties of enzymes (28) and can be applied to single substrate enzymes and to ordered and random bi–bi mechanisms, whereas the steady-state random mechanism obviously cannot be utilized with an ordered reaction mechanism. Evidence suggests that glucokinase is predominantly ordered, but there is sufficient flux (5%) through the alternate pathway to allow consideration of the influence of the random mechanism on cooperativity (29).

The main experimental observation that has been fit by all three mechanisms is the decrease in cooperativity with a decrease in ATP concentration. The Hill coefficient decreases from about 1.5 to near 1.0 as the ATP is decreased from 5 m*M* to less than 50 μM (4, 24). On this basis alone there is little to choose among the potential mechanisms. The mnemonic and LIST mechanisms attribute this dependency to the ability of the E : Glc to have sufficient time to equilibrate with the isomeric form at low ATP concentrations, where the second-order reaction would be slower. At higher nucleotide concentrations the ATP "traps" the glucose-bound forms in a steady-state distribution without allowing equilibration, thereby allowing kinetic cooperativity. The random mechanism attributes the ATP dependency of the cooperativity to the amount of flux through each limb of the pathway that can occur.

Product inhibition has been proposed as a means of ascertaining the appropriate cooperative mechanism (30, 31). Unfortunately, complex mechanisms are not readily resolved by this approach. MgADP has been shown to be a mixed inhibitor with respect to either MgATP or glucose (8), consistent with the first product released, and to decrease the Hill coefficient for glucose at low MgATP concentrations (8). The latter behavior is predicted by the mnemonic and LIST mechanisms but requires an additional abortive complex with the random mechanism. The nonlinearity of the mixed inhibition by MgADP with MgATP (8) is most easily explained by the LIST mechanism, which also allows "cooperativity" of each product, without invoking additional dead-end inhibition steps. Although the dependence of different measures of cooperativity (e.g., the Hill coefficient and the initial slope of the double reciprocal plot, Γ) on substrate and product concentration has been suggested as a means of distinguishing various mechanisms (31), a more rigorous approach would be to utilize the same data for fitting or simulating the results to each mechanism.

Inhibition and Cooperativity

Two types of inhibition of glucokinase kinetics have been reported that have mechanistic implications. *N*-Acetylglucosamine (GlcNAc) is a linear competitive inhibitor that eliminates the kinetic cooperativity with glucose (9, 17). This observation has been used to suggest that GlcNAc binds rapidly to the glucose-binding conformer of the enzyme, maintains it in a single state, and thus does not allow the slow isomerization necessary for the generation of cooperativity. This type of inhibition does not discriminate between the mnemonic and the slow transition mechanisms because either would allow preferential binding of GlcNAc to one form of glucokinase and maintain the enzyme in predominantly one conformation without significant isomerization. Successful competition with glucose would, of course, still reduce the amount of productive enzyme–substrate complex and cause the competitive inhibition. The GlcNAc inhibition pattern has not been interpreted for the random mechanism (10) but presumably would require the preferential binding to the slow glucose binding form and thus eliminate the small percentage of enzyme tied up in the alternative pathway. The paradox would then arise that the glucose analogue GlcNAc would bind readily to the glucokinase–ATP complex that bound glucose only slowly and would bind poorly to the free enzyme that bound glucose rapidly.

Inhibition by palmitoyl-CoA is quite different. The data demonstrate partial competitive inhibition and have been interpreted in terms of binding at an allosteric site (5, 6). This conclusion was reached based on the observations that the inhibitor had the same partial competitive K_i with respect to both glucose and MgATP, the half-saturation constant ($S_{0.5}$) for glucose was increased but at any given glucose concentration the activity was not zero at the highest palmitoyl-CoA tested, and the measured critical micelle concentration under the same conditions was higher than the apparent K_i. Furthermore, the cooperativity with glucose was not changed with acyl-CoA concentrations (1.5 ± 0.05). The basic interpretation of the palmitoyl-CoA inhibition is the same for all three potential mechanisms; that is, binding at an allosteric site shifts all forms of the enzyme over to a new set that have different activities, i.e., are still active with a poorer affinity for glucose and MgATP, but have the same relation among rate constants to maintain the positive cooperativity with a Hill coefficient of about 1.5. We have demonstrated that this behavior is consistent with the LIST mechanism by model simulations (K. E. Neet, unpublished observations). Less than tenfold changes in the on and off rate constants for glucose binding result in shifts of the $S_{0.5}$ of more than 20-fold without significant change in the Hill coefficient. In contrast, we have been unable to show comparable effects with the mnemonic mechanism; that is, the on and off rates appear to affect the Hill coefficient more in this mechanism because of the limitations consequent to only a single catalytic cycle. The inhibition observed with palmitoyl-CoA itself was not linear, i.e., was hyperbolic or cooperative (6), but we have not yet determined whether this cooperative effect is due to two (or more) acyl-CoA binding sites or the nonlinearity is generated by the slow transition. These contrasting effects of acyl-CoA and GlcNAc on cooperativity emphasize the kinetic effects that are obtained from binding to all forms of the enzyme (palmitoyl-CoA) or preferentially to only some forms (GlcNAc).

Slow Transitions in Glycerol

The basis of hysteretic enzymes (28) or the ligand-induced slow transition (21) lies in the presence of a transition in the enzyme that is slow relative to the rate of catalysis; however, the rate may still be faster than is easily observable

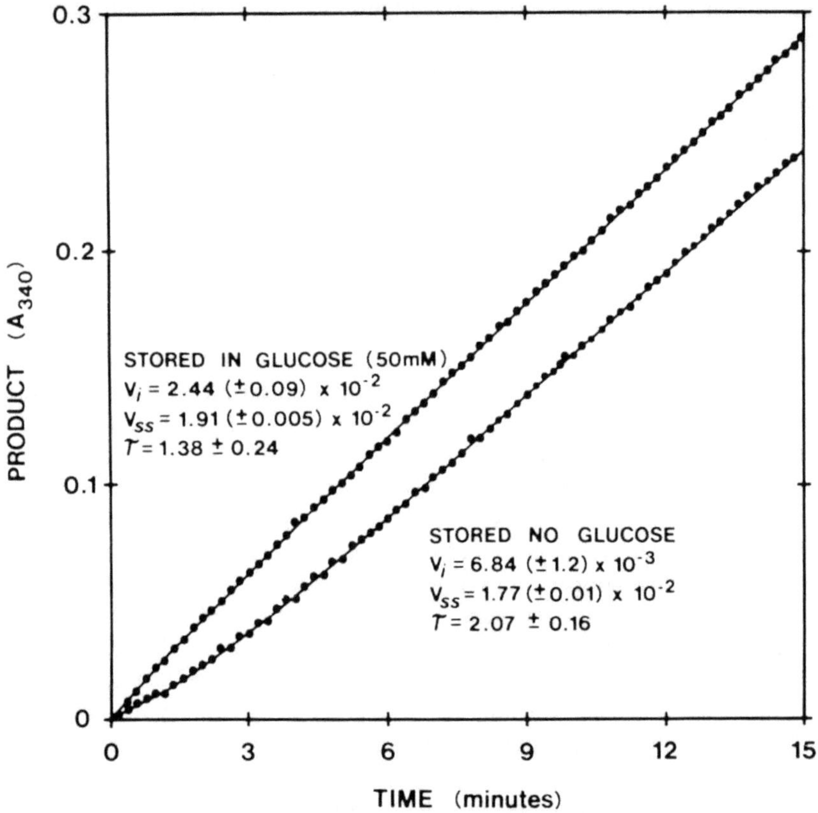

FIG. 4.1. The assay transient in 30% glycerol. Glucokinase was assayed in the standard glycerol assay after preincubation either with (top) or without (bottom) 50 m*M* glucose. The lines are the best fit lines drawn with the parameters indicated. The points shown were acquired digitally at 12-second intervals on a microcomputer attached to the spectrophotometer.

by normal means under normal conditions. We reasoned that altering the conditions of the reaction to increase the viscosity might slow critical enzymatic steps that could lead to observable hysteresis. The perturbation of the glucokinase reaction by glycerol has been utilized (32) to help delineate the mechanism. The kinetics that we have observed have been generally similar to those reported by others (32), except that we have also noted the presence of a slow transient in the assay progress curve. The reason we can readily observe this process appears to be due partially to the handling of the enzyme during its preparation. The curve that is typically seen when glucokinase is stored in high glucose concentration is the "burst" of Fig. 4.1; if the glucose is removed from the enzyme before assay, the "lag" curve is seen. The data presented demonstrate several essential characteristics of the process. The enzyme can exist in two states depending on the presence of glucose in the preincubation, which results in a high (top curve) or a low (bottom curve) initial velocity. The enzyme can be reversibly interchanged between the two states by addition or removal of glucose. The same steady state is attained regardless of the initial state of the enzyme; that is, the slopes of the steady states (Fig. 4.1) are identical. These properties are exactly those that would be expected for a ligand-induced slow transition that generates cooperativity in a monomeric enzyme (21, 25, 28). Studies reported in detail elsewhere (33) document that this behavior is not an artifact of the system and describe the dependence of the rate constant for the transition as a function of several parameters. The apparent transition rate constant decreases with glucose and glycerol concentration, increases with temperature, and is independent of enzyme concentration. Unfortunately, the transition data are not of high enough quality to allow extensive modeling and parameter estimation. The initial and steady-state velocities are sufficiently similar that the transition rate is not well determined and quantitatively reproducible, even with a sophisticated fitting algorithm. We are mainly concerned here with the comparison of initial and steady-state reaction velocities (which are well determined) with normal cooperative velocities in the absence of glycerol.

We interpret the observations of a transient in glycerol to indicate a conformational change induced by glucose that results in two forms of the enzyme with differing kinetic properties. The effect of glycerol on stabilizing protein structure is well known (34), and effects of microviscosity on protein conformational changes are possible. The viscosity of 30% glycerol is about 2 centipoise at 25°C. A macroscopic effect of viscosity on the initial bimolecular collision of glucose and glucokinase would only slow the reaction by about two fold and be insufficient to explain the much slower transition rate (> 100-fold) in glycerol. We propose that the glycerol is simply slowing a preexisting ligand-induced transition. Forms of glucokinase that are interconvertible in the presence or absence of glucose and that are distinguishable by chromatographic properties have been described for rat liver glucokinase by chromatographic behavior on DEAE-Cibracon agarose with magnesium gradients (35) and for pig liver glucokinase on Mono-Q with salt gradients (W. Vogel and K. E. Neet, unpublished observations). Preliminary temperature-dependency studies of intrinsic fluorescence have provided evidence for two (or more) states of rat liver glucokinase with different environments of the tryptophan residues (M. L. Cardenas, personal communication). Similar conformational states have been studied for the related hexokinase from yeast by several methods including crystallography and fluorescence (36, 37).

Cooperativity of Glucose and MgATP in Glycerol

A comparison of the Hill coefficient, as a measure of cooperativity, in the presence and absence of glycerol has suggested that the steady-state velocity in glycerol has properties similar to those of the linear velocity in normal assay conditions and that the initial velocity behaves in a Michaelean manner.

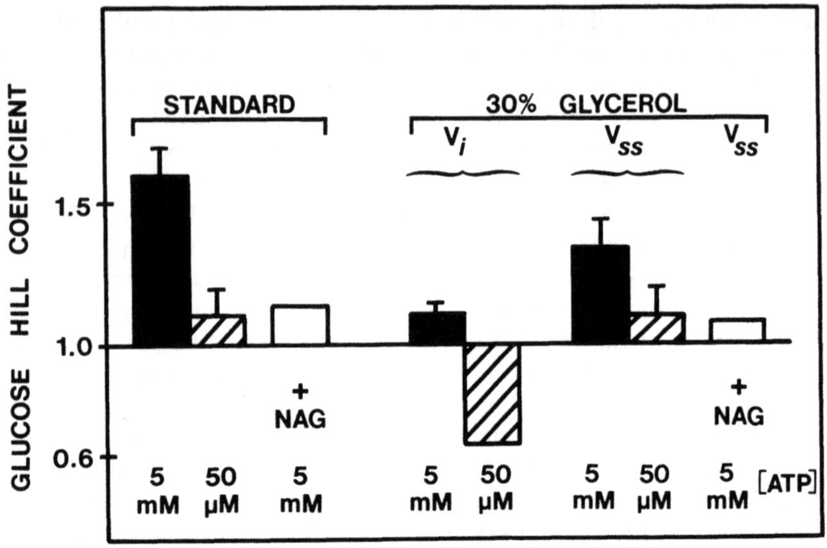

FIG. 4.2. Cooperativity (Hill coefficient) of glucokinase with glucose in the absence (left) or presence (right) of 30% glycerol at high (solid bars) or low (cross-hatched bars) MgATP concentration or with 10 m*M* GlcNAc (NAG, open bars). Initial and steady-state velocities in glycerol are as indicated. The Hill coefficient was determined by a nonlinear least-square fit of the original velocity data to a second order rate equation.

However, the maximal velocity in glycerol, as noted by others (32), is about 50% of that in its absence, and the Km is actually lowered about twofold. We believe that the studies by others are comparable to the initial velocity measured in our laboratory. Comparisons of the cooperativity are shown in Fig. 4.2. The Hill coefficient is known to be dependent on the MgATP concentration and approaches unity at low MgATP concentrations (4, 24). This property holds true for the steady-state velocity in the 30% glycerol (Fig. 4.2). The apparent negative cooperativity of the initial velocity in 0.05 m*M* MgATP is what would be expected for two conformational states that have different kinetic properties and that have not equilibrated at the zero time extrapolation. The behavior of the apparent $S_{0.5}$ and the maximal velocity are not as diagnostic for comparisons of conditions but are consistent with this interpretation.

We have also examined the effects of GlcNAc on the inhibition of the kinetics of glucose phosphorylation in glycerol. Because GlcNAc eliminated the cooperativity in normal assays (9, 17), we reasoned that the effects on the steady-state velocity would be a further test of the nature of the slow transition in glucokinase. Glucokinase was preincubated with GlcNAc and then assayed at a final concentration of 10 m*M* inhibitor. The steady-state velocity was inhibited in a competitive fashion with an apparent K_i of 62 μM compared to a value of 184 μM in the absence of glycerol. No curvature was observed in these experiments, presumably because the GlcNAc keeps the enzyme in only a single kinetic state, as postulated for its effect on decreasing the cooperativity in normal assays. The presence of GlcNAc eliminated the glucose cooperativity in the steady-state velocity in glycerol, just as it does in the standard assays (Fig. 4.2). Palmitoyl-CoA, on the other hand, inhibits in 30% glycerol, again with maintenance of glucose cooperativity, but our experiment with this inhibitor to date have not been extensive in this system. These simple inhibition studies confirm the interpretation that the steady state observed in glycerol is the same functional state as the linear velocities that generate kinetic cooperativity under standard conditions. Furthermore, the elimination of the assay transient in glycerol by GlcNAc is consistent with the interpretation that the observable transition is generated by the same molecular mechanism that is affected by GlcNAc in the standard assay.

When the other substrate, MgATP, is assayed under standard conditions, normal saturation kinetics are observed with no cooperativity. Because the conformational changes of the enzyme are changed in glycerol, however, the probability was that the MgATP kinetics might also be dramatically altered. The kinetics with respect to MgATP concentration were studied in glycerol at low glucose concentrations. When the reactions were initiated with glucokinase that had been stored in 50 m*M* glucose, a transient of about 1 to 3 min occurred, as expected, and the initial velocity was noncooperative, as predicted for recognition of a single enzyme form by MgATP. The steady-state velocity was, however, negatively cooperative with a Hill coefficient of 0.8 or less. At high glucose concentrations this MgATP cooperativity is eliminated, in a manner reminiscent of the effect of MgATP concentration on the glucose cooperativity and interpretable in the same fashion. Thus it appears that the alteration of rate constants in the LIST mechanism has promoted a situation in which the relations in glycerol now generate kinetic negative cooperativity with respect to MgATP concentration. No clear and large effect of MgATP on the apparent rate of the transition was discernible.

Alternative Substrates in Glycerol

Several reports (9, 38) have indicated that glucokinase is not as specific as once thought for glucose and that several other carbohydrate substrates show differing kinetic properties and cooperativity. We have compared several of these substrates in the presence or absence of glycerol with rather interesting results. Examples of sugar substrates with three types of behavior are summarized in Fig. 4.3. Glucose (or mannose) is, as already described, linear in the absence and nonlinear in the presence of glycerol. In the normal assay 2-deoxyglucose is nearly linear (i.e., there is only a hint of a fast, small transient) and shows no cooperativity in the resultant steady state. In the presence of glycerol a slow transition is now readily observable but still with no cooperativity. This behavior suggests that the kinetic states of glucokinase recognize deoxyglucose differently and produce the temporal effect, but that

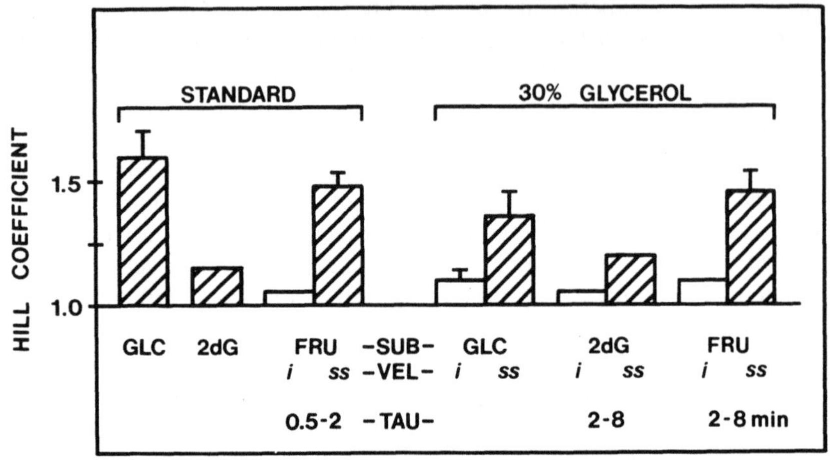

FIG. 4.3. Cooperativity (Hill coefficient) of glucokinase with alternative substrates in the absence (left) or presence (right) of 30% glycerol. The initial and steady-state velocities for 2-deoxyglucose (2dG) and fructose (FRU) were used for the determination of cooperativity. The approximate range of tau values (reciprocal of the rate constant) for the transition is given at the bottom.

the individual rate constants in the two catalytic cycles do not satisfy the requirement for generating cooperativity (25). Fructose in glycerol displays positive cooperativity in the steady state (at high MgATP only) and not in the initial velocity, similar to glucose but with a much higher half-saturation value. As we performed these studies with careful concern for the observation of kinetic transients, we were also able to observe a slow transition of glucokinase with fructose in the standard assay (Fig. 4.3). Thus fructose, which has properties similar to those of glucose when they are compared in the presence of glycerol, also demonstrates a ligand-induced transition with glucokinase in a simple assay buffer (without glycerol) that is still slow enough to see in our hands. These comparisons with alternate substrates emphasize that the occurrence of the observed kinetic transient is a complex function of numerous rate constants (21, 25, 28), even if the same underlying conformational change ($E \Leftrightarrow E'$) is the driving force. The observed transition with fructose or 2-deoxyglucose is slower in glycerol than in simple buffers, supporting the interpretation that this additive is primarily slowing a preexisting enzyme isomerization.

Conclusions

Returning to the question of the appropriate mechanism for glucokinase cooperativity based on the observable data, the following conclusions may be made. (a) Two states of glucokinase have now been described in several ways. The assay transients described here have (Fig. 4.1) demonstrated two *kinetic* states that can slowly isomerize and thus corroborate reports from chromatographic and spectroscopic experiments. Therefore the argument advanced for the random mechanism that only one conformational state has been observed (10) is not pertinent or necessary. (b) Either the mnemonic or the LIST mechanisms can easily contain, if not expect, the assay transient observed in glycerol; this observation is unrelated to requirements of the random mechanism. The observation of either a "burst" or a "lag," depending on the preincubation conditions, adds an additional constraint on the rate constants that must be quantitatively evaluated within the mnemonic or the LIST mechanisms. (c) The main evidence for a mechanism with slow transitions between two catalytic cycles is the observation of Hill coefficients less than unity in several instances. The catalytic heterogeneity with glucose in the initial velocities at low MgATP concentrations in 30% glycerol (Fig. 4.2) and in the linear normal assays at low MgATP concentrations reported by Cardenas et al. (24; personal communication) are most easily explained with a full ligand-induced slow transition mechanism. The mnemonic mechanism does not predict such behavior because of the lack of a second catalytic cycle for the "heterogeneous" reaction to proceed through. The kinetic negative cooperativity of MgATP in the steady-state velocities at low glucose concentrations in 30% glycerol (reported here) are readily explained by the LIST mechanism with the second-order terms in the second substrate that are lacking in the basic mnemonic mechanism. Negative cooperativity has also been observed in the steady-state velocities in ^2H_2O and adequately explained by the mnemonic mechanism (39). The random mechanism does not address these observations of additional "negative cooperativity" at all.

In summary, we believe that the current results for glucokinase require a ligand-induced slow transition mechanism (21) of the type proposed by Cardenas et al. (9, 22), and we propose that the other mechanisms suggested for the enzyme are not sufficient to support all of the available experimental data. Whether the mechanism is totally ordered or partially random is not particularly relevant for future considerations of cooperativity, and we believe that it is satisfactory to represent it as (predominantly) ordered for purposes of modeling and parameter estimation. Subsequent work with the enzyme requires quantitative data fitting all situations in order to provide a consistent overall model. Physicochemical studies of conformational states and transitions of glucokinase are necessary to correlate with the kinetic data already obtained. Glucokinase provides an interesting and important system for evaluating cooperativity in a monomeric enzyme and one that has significant relevance for the role of such positive cooperativity in the physiological function of glucose phosphorylation in the liver.

Acknowledgment. The work described here was supported by USPHS grant AM 12881 to K.E.N.

References

1. Holyroyde, M. J., Allen, M. B., Storer, A. C., Warsy, A. S., Chesher, J. M. E., Trayer, I. P., Cornish-Bowden, A., and Walker, D. G. (1976) *Biochem. J.* 153:363–373.
2. Cardenas, M. L., Rabajille, E., and Niemeyer, H. (1978) *Arch. Biochem. Biophys.* 190:142–148.
3. Niemeyer, H., Cardenas, M. L., Rabajille, E., Ureta, T., Clark-Turri, L., and Penaranda, J. (1975) *Enzyme* 20:321–333.
4. Storer, A. C., and Cornish-Bowden, A. (1976) *Biochem. J.* 159:7–14.
5. Tippett, P. S., and Neet, K. E. (1982) *J. Biol. Chem.* 257:12839–12845.
6. Tippett, P. S., and Neet, K. E. (1982) *J. Biol. Chem.* 257:12846–12852.
7. Tippett, P. S., and Neet, K. E. (1983) *Arch. Biochem. Biophys.* 222:285–289.
8. Storer, A. C., and Cornish-Bowden, A. (1977) *Biochem. J.* 165:61–69.
9. Cardenas, M. L., Rabajille, E., and Niemeyer, H. (1984) *Eur. J. Biochem.* 145:163–171.
10. Pettersson, G. (1986) *Biochem. J.* 233:347–350.
11. Newsholme, E. A., and Start, C. (1973) *Regulation in Metabolism.* Wiley, London.
12. Hue, L. (1981) *Adv. Enzymol.* 52:247–331.
13. Bontemps, F., Hue, L., and Hers, H-G. (1978) *Biochem. J.* 174:603–611.
14. Garfinkel, D., Garfinkel, L., Meglasson, M. D., and Matschinsky, F. M. (1984) *Am. J. Physiol.* 247:R527–R536.
15. Nordlie, R. C., and Sukalski, K. A. (1984) In *The Enzymes of Biological Membranes*, 2nd ed., edited by A. N. Martonosi, pp. 349–398. Plenum Press, New York.
16. Katz, J., and McGarry, J. D. (1984) *J. Clin. Invest.* 74:1901–1909.
17. Scofield, R. F., Kosugi, K., Schumann, W. C., Kumaran, K., and Landau, B. R. (1985) *J. Biol. Chem.* 260:8777–8782.
18. Spence, J. T., and Koudelka, A. P. (1985) *J. Biol. Chem.* 260:1521–1526.
19. Seitz, H. J., Muller, M. J., Krone, W., and Tarnowski, W. (1977) *Arch. Biochem. Biophys.* 183:647–663.
20. Richard, J., Meunier, J., and Buc, J. (1974) *Eur. J. Biochem.* 49:195–208.
21. Ainslie, G. R., Shill, J. P., and Neet, K. E. (1972) *J. Biol. Chem.* 247:7088–7096.
22. Olavarria, J. M., Cardenas, M. L., and Niemeyer, H. (1982) *Arch. Biol. Med. Exp.* 15:365–369.
23. Wells, B. D., Stewart, T. A., and Fisher, J. R. (1976) *J. Theor. Biol.* 60:209–221.
24. Cardenas, M. L., Rabajille, E., and Niemeyer, H. (1979) *Arch. Biol. Med. Exp.* 12:571–580.
25. Neet, K. E., and Ainslie, G. R. (1980) *Methods Enzymol.* 64:192–226.
26. Neet, K. E. (1980) *Methods Enzymol.* 64:139–192.
27. Pollard-Knight, D., and Cornish-Bowden, A. (1982) *Mol. Cell. Biochem.* 44:71–80.
28. Frieden, C. (1970) *J. Biol. Chem.* 245:5788–5799.
29. Gregoriou, M., Trayer, I. P., and Cornish-Bowden, A. (1981) *Biochemistry* 20:499–506.
30. Meunier, J-C., Buc, J., Navarro, A., and Ricard, J. (1974) *Eur. J. Biochem.* 49:209–223.
31. Richard, J., and Noat, G. (1985) *Eur. J. Biochem.* 152:557–564.

32. Pollard-Knight, D., Connolly, B. A., Cornish-Bowden, A., and Trayer, I. P. (1985) *Arch. Biochem. Biophys.* 237:328–334.
33. Tippett, P. S., Keenan, R. P., and Neet, K. E. (1987) Submitted for publication.
34. Gekko, K., and Timasheff, S. N. (1981) *Biochemistry* 20:4677–4686.
35. Meglasson, M. D., Burch, P. T., Hoenig, M., Chick, W. L., and Matschinsky, F. M. (1983) *J. Biol. Chem.* 258:2094–2097.
36. Bennett, W. S., and Steitz, T. A. (1978) *Proc. Natl. Acad. Sci. USA* 74: 4848–4852.
37. Ohning, G. V., and Neet, K. E. (1983) *Biochemistry* 22:2986–2995.
38. Cardenas, M. L., Rabajille, E., and Niemeyer, H. (1984) *Biochem. J.* 222:363–370.
39. Pollard-Knight, D., and Cornish-Bowden, A. (1984) *Eur. J. Biochem.* 141:157–163.

5
Mechanism of Activation of Calmodulin-Dependent Phosphatase by Divalent Metal Ions

CHARLES Y. HUANG, MARINA LANCIOTTI, and AILE ZHANG

Although the precise physiological role of calmodulin (CaM)-dependent phosphoprotein phosphatase (CPP) is not yet known, the significance of this enzyme can be readily appreciated from its participation in regulations involving Ca^{2+}–CaM and phosphorylation/dephosphorylation, its abundance and ubiquity in various tissues, and its action on diverse substrates (see ref. 1 for review), which include phosphatase inhibitor I, the regulatory subunit of cAMP-dependent protein kinase, several brain-specific phosphoproteins (2), and the dihydropyridine-sensitive Ca^{2+} channel (3). The phosphatase, also known as calcineurin, is a hetero dimer consisting of a 61,000-dalton catalytic α subunit that combines with CaM and a 19,000-dalton β subunit that binds four Ca^{2+} ions. It has been shown to be a Zn^{2+}- and Fe^{3+}-containing metalloenzyme that can exist in a phosphorylated form (4, 5). In addition, its activity is enhanced to varying degrees by the presence of divalent metal ions such as Ni^{2+}, Mn^{2+}, Co^{2+}, Mg^{2+}, etc. (6–8). In the absence of these divalent metal ions, activation of CPP by Ca^{2+}–CaM is accompanied by a concomitant time-dependent deactivation process that, when completed, cannot be reversed by dissociation of CaM (4). The deactivated CPP, however, can be reactivated by addition of the above-mentioned cations (4). Thus this divalent metal ion is vital for maintaining the structural stability and full catalytic activity of CPP. Several laboratories have investigated the effect of divalent metal ions on the activation and regulation of this enzyme (9–14). The present investigation on the mechanism of interaction of these ions, particularly the Ni^{2+} ion, is aimed at gaining some insight into their roles in the regulation and catalysis of CPP. Studies on the Ni^{2+} ion are emphasized because it is the most potent activator with a number of protein and nonprotein substrates (7, 8). The synthetic substrate *p*-nitrophenylphosphate (PNPP) was used so that the time course of reactions could be continuously monitored to permit detailed analysis.

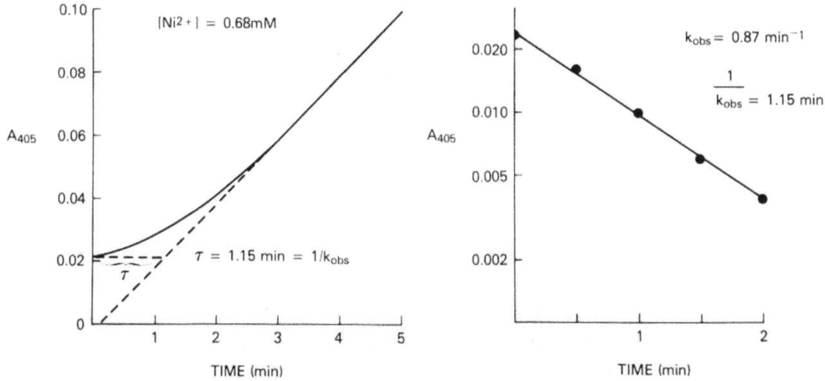

FIG. 5.1. Example of activation of CPP by a low level of Ni^{2+}. CPP (4.6 nM) was assayed at 30°C at a saturating concentration of CaM (~0.1 μM) in 50 mM Hepes buffer pH 7.6 containing 50 mM KC1, 10 mM PNPP, 0.2 mM Ca^{2+}, and 20 μM EDTA. CPP was allowed to reach the totally deactivated state, and Ni^{2+} (0.68 mM in the example) was then added to start the activation reaction. Product formation was monitored continuously at 405 nm. The lag phase was analyzed by plotting the differences between the product formation curve (left panel, solid line) and the extrapolated, fully activated phase (left panel, dashed line) in a first-order plot (right panel); k_{obs} was obtained from the slope of the semilogarithm plot. The lag time (τ) was determined as the distance between the zero time point and the extrapolated line in the left panel.

Results

Activation of the CaM–CPP Complex at Low Ni^{2+} Concentrations

The CPP samples used in the activation studies were preincubated with Ca^{2+}, CaM, and PNPP for at least 45 min to obtain the completely deactivated form. This preincubation ensured that the enzyme existed as a single species at the onset of the activation experiments. As shown in Fig. 5.1, left panel, the time course of activation typically has an initial lag phase that follows first-order reaction kinetics (Fig. 5.1, right panel). In the range of Ni^{2+} concentrations studied, 0.17 to 1.70 mM, the lag time (τ) decreases with increasing Ni^{2+} concentration, but the final extent of activation is unchanged. These observations suggest that the activation of CPP involves a slow conformational rearrangement step and the binding of Ni^{2+} is extremely tight. The tight binding of the Ni^{2+} ion is consistent with the fact that quenching of the tryptophan fluorescence of CPP on binding Ni^{2+} cannot be reversed by the addition of chelators such as EDTA (7). When the concentration of PNPP was varied, the lag time remained constant, indicating that Ni^{2+} binding to the enzyme is unaffected by the presence of this substrate. Consequently, the activation of CPP by Ni^{2+} can be described by the following scheme:

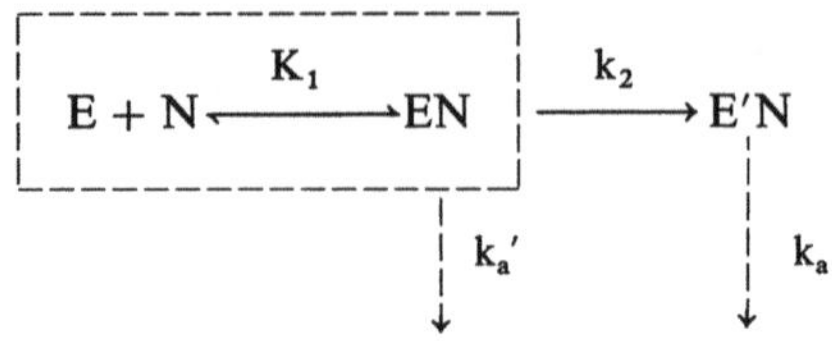

Scheme 1

where E = the deactivated enzyme–CaM complex
N = Ni^{2+}
E′N = the activated EN complex
K_1 = dissociation constant for the EN complex
k_2 = rate constant for the activation process
k_a and k_a' = catalytic constants for the Ni^{2+}-bound enzyme forms at a given substrate concentration.

The area enclosed by the dashed lines denotes the rapid equilibrium segment. The initial binding of Ni^{2+} is assumed to be rapid because only a single first-order process is observed.

Based on scheme 1, the formation of the product (P) as a function of time is given by:

$$P = k_a E_o t + (k_a - k_a' f_{EN}) E_o (e^{-k_2 f_{EN} t} - 1)/k_2 f_{EN} \quad [1]$$

where E_o = total enzyme concentration
f_{EN} = fraction of EN in the rapid-equilibrium segment and is equivalent to $N/(K_1 + N)$

The apparent first-order rate constant (k_{obs}) is equal to $k_2 f_{EN}$, whereas the lag time is defined as:

$$\tau = (1/k_{obs})(1 - k_a' f_{EN}/k_a) \quad [2]$$

Equation 2 shows that, if the EN complex is activated to an appreciable extent (i.e, k_a' is not negligible) as soon as Ni^{2+} is bound, τ should be consistently smaller than $1/k_{obs}$. In all of our experiments, however, the values of τ obtained at different Ni^{2+} concentrations are always similar to those of $1/k_{obs}$, suggesting that the EN complex is inactive and:

$$\tau = 1/k_2 f_{EN} = (1/k_2)(1 + K_1/N) \quad [3]$$

When τ is plotted versus $1/N$ according to Eq. 3, a linear relation can be shown (Fig. 5.2) that is in agreement with the mechanism proposed in scheme 1. From Fig. 5.2, K_1 is estimated to be ~ 2 m*M*, and k_2 is $\sim 4\ \text{min}^{-1}$. These figures reveal that the lag phase is due largely to the initial weak binding of one Ni^{2+} ion to CPP.

Deactivation of CPP by the Binding of a Second Ni^{2+} Ion

After reaching the fully activated steady-state phase, the Ni^{2+}-activated CPP eventually undergoes a very slow reaction that converts the enzyme to a much

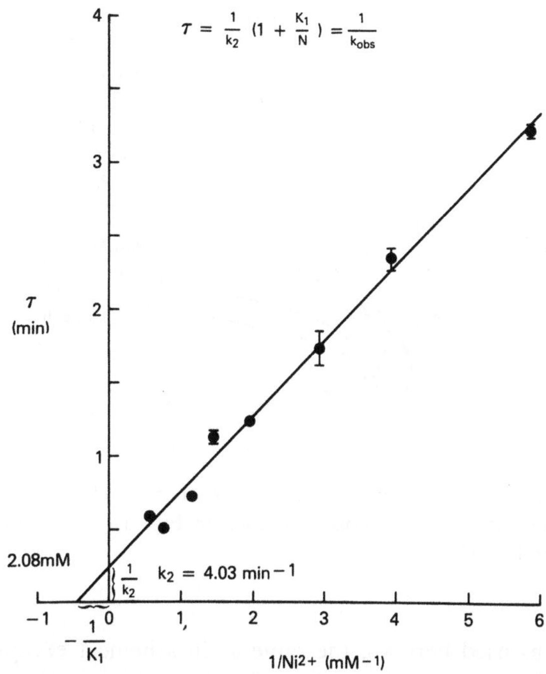

FIG. 5.2. Lag time (τ) in the activation of CPP versus reciprocal Ni^{2+} concentration. The Ni^{2+} concentration range is 0.17 to 1.70 m*M*. Lag times were determined as described in Fig. 5.1. K_1 and k_2 were obtained by linear regression using Eq. 3.

less active form. At high concentrations of Ni^{2+} the initial activation is rapidly attained, and the deactivating process becomes the primary reaction. Figure 5.3 shows a series of such deactivation processes at high Ni^{2+} concentrations (5–33 m*M*). Upon analysis, these processes can be shown to be first-order reactions. Because the rate of inactivation increases with increasing Ni^{2+} concentration and the enzyme seems to attain the same final activity, the deactivation reaction must be due to the irreversible binding of additional Ni^{2+} ions. When the observed first-order rate constants were plotted against $1/Ni^{2+}$ (cf. Eq. 5, below), a linear relation could be demonstrated, further indicating that only one Ni^{2+} ion was involved. Thus deactivation of the Ni^{2+}-activated CPP by the binding of the second Ni^{2+} ion can be depicted by the following scheme:

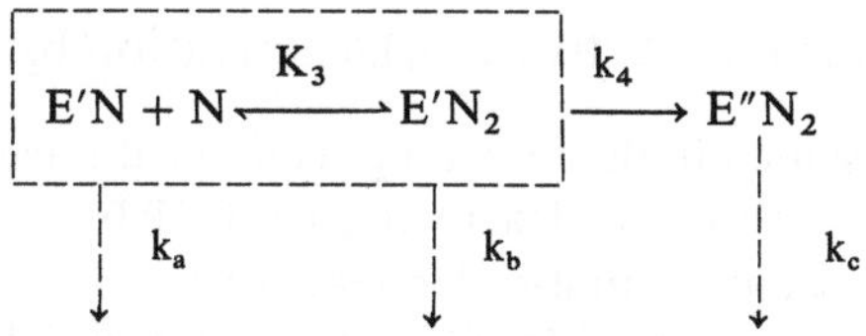

Scheme 2

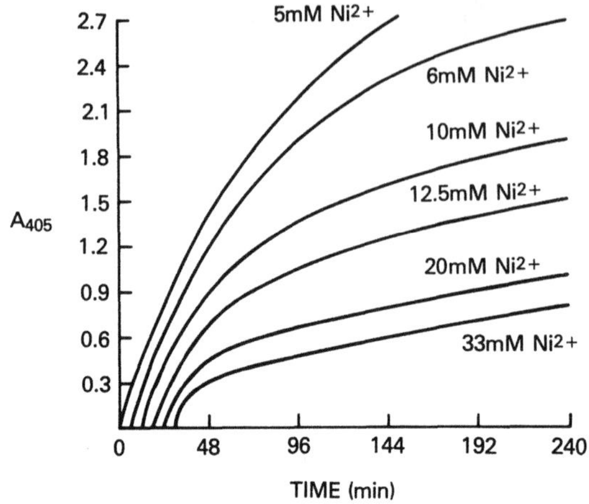

FIG. 5.3. Inactivation of CPP by high levels of Ni^{2+} (5–33 m*M*). Experimental conditions were the same as those described in Fig. 5.1 except that the CPP concentration was 12 μM.

The notations used here are the same as in scheme 1 except that E″ is the deactivated form. From scheme 2, product formation as a function of time has the following expression:

$$P = k_c E_o t - (k_a f_{E'N} + k_b f_{E'N_2} - k_c) E_o (e^{-k_4 f_{E'N_2} t} - 1)/k_4 f_{E'N_2} \qquad [4]$$

where $f_{E'N}$ and $f_{E'N_2}$ represent the fractions of E′N and $E'N_2$ in the rapid-equilibrium segment. The observed first-order rate constant is defined by the reciprocal relation:

$$1/k_{obs} = (1/k_4)(1 + K_3/N) \qquad [5]$$

The values of K_3 and k_4 were determined to be 21 m*M* and 0.078 min^{-1}, respectively. The mechanism proposed in scheme 2 was further verified by a different plot. It we defined the amplitude of each product formation curve, denoted P_A here, as the intercept on the ordinate cut by the line extrapolated from the final, deactivated phase, we can derive from Eq. 4 that:

$$P_A = (E_o/k_4)[(k_b - k_c) + (k_a - k_c)K_3/N] \qquad [6]$$

We were able to obtain a linear P_A versus 1/N plot as predicted by Eq. 6.

Overall Mechanism of Activation/Deactivation by Ni^{2+} Ions

From results presented in the foregoing sections, it is clear that activation of CPP by the first Ni^{2+} ion and deactivation of CPP by the second Ni^{2+} ion are treatable as two separate first-order reactions at low and high Ni^{2+} concentrations owing to the tenfold difference in the initial binding affinities of the Ni^{2+} ions ($K_1 = 2\ mM$ versus $K_3 = 21\ mM$) and the 50-fold difference in the rate constants for the conformational changes involved in the activation and deactivation ($k_2 = 4\ min^{-1}$ and $k_4 = 0.078\ min^{-1}$). Thus the overall mechanism of Ni^{2+} activation/deactivation of CPP can be described by the apparent ordered binding of the two Ni^{2+} ions:

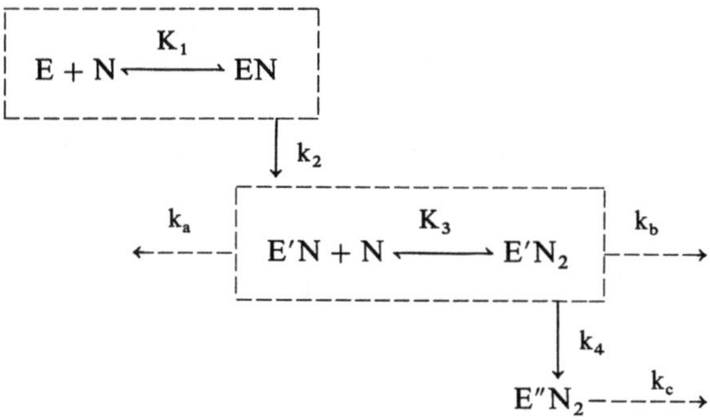

Scheme 3

Formation of P is given by:

$$P = k_cE_ot + \frac{[(k_af_{E'N} + k_bf_{E'N_2}) - (k_ck_4f_{E'N_2}/k_2f_{EN})]E_o(e^{-k_2f_{EN}t} - 1)}{k_2f_{EN_2} - k_4f_{E'N}} + \frac{k_2f_{EN}[(k_af_{E'N} + k_bf_{E'N_2}) - k_c]E_o(e^{-k_4f_{E'N}t} - 1)}{k_4f_{E'N_2}(k_4f_{E'N_2} - k_2f_{EN})} \quad [7]$$

From Eq. 7 it can be shown that the amplitude (P_A) remains a linear function of 1/N:

$$P_A = E_o\{[(k_b - k_c)/k_4 - k_c/k_2] + [(k_a - k_c)K_3/k_4 - k_cK_1/k_2]/N\} \quad [8]$$

The mechanism presented in scheme 3 is supported by the following observations: (a) P_A values from progress curves obtained at intermediate levels of Ni^{2+} (2–5 mM) are linearly proportional to $1/Ni^{2+}$ as predicted by Eq. 8. (b) When equimolar amounts of EDTA were added to remove unbound Ni^{2+} during the early phase of deactivation, CPP remained activated without undergoing the deactivation process. Likewise, when EDTA was added during the early phase of deactivation, no further deactivation occurred. Therefore binding of the second Ni^{2+} is responsible for converting the Ni^{2+}-activated CPP–CaM complex to a deactivated form, and the binding of these two Ni^{2+} ions to CPP is essentially ordered. (c) By substituting experimentally determined constants into Eq. 7, we were able to demonstrate that it reduced to Eqs. 1 and 4, respectively, at low and high Ni^{2+} concentrations. (d) The irreversible combination of two Ni^{2+} ions to the CPP–CaM complex have been confirmed by direct Ni^{2+} binding studies.

Irreversible Binding of Two Ni^{2+} Ions to CPP

Binding studies were carried out by incubating CPP with 5 m*M* $^{63}Ni^{2+}$ in the absence or presence of Ca^{2+}/CaM. EDTA was added to a final concentration of 5 m*M* to terminate the binding after 2 or 60 min of incubation. The samples were exhaustively dialyzed against EDTA-containing buffer and then counted to determine the amount of irreversibly bound Ni^{2+}. The presence of CaM did not appear to affect the outcome. The amount of bound Ni^{2+} was about 0.8 mol/mol of enzyme after 2 min of incubation and about 1.8 mol/mol enzyme after 60 min. The results are consistent with the concept that 2 mol of Ni^{2+}/mol of enzyme are irreversibly bound, and the amount bound depends on the length of reaction time.

Ni^{2+} Binding and CPP Subunit Interaction

The α subunit of CPP, prepared by separation of the α and β subunits on Ultrogel in 6 *M* urea, displayed Trp-quenching on binding Ni^{2+}, which was similar to that observed with the intact enzyme. $^{63}Ni^{2+}$ binding studies carried out with α and β subunits also showed that the primary sites for Ni^{2+} were on the α subunit. In a separate experiment the presence of Ni^{2+} on CPP was found to greatly hinder dissociation of this enzyme by urea, indicating a strong effect of the divalent metal ion on the subunit interaction. Furthermore, activation of the catalytic α subunit by Ni^{2+} seemed to depend on the binding of Ca^{2+} to the β subunit. When large excess (instead of an equimolar amount) of EDTA was added to remove unbound Ni^{2+} during the fully activated phase of the activation reaction, CPP became rapidly and totally inactivated. Because the Ni^{2+} ions that were already bound, as well as the Zn^{2+} and Fe^{3+} ions on CPP, could not be dissociated by addition of chelators, the inactivation by EDTA most likely was due to the removal of Ca^{2+} from the β subunit.

Mechanism of Activation of CPP by Mn^{2+}

The mechanism of Mn^{2+} activation of CPP is similar to that of Ni^{2+} except that the process is not irreversible. The initial binding of the first Mn^{2+} is tighter ($K_d \cong 0.2$ m*M*) compared with the first Ni^{2+}, but the rate constant for the conformational rearrangement is slower ($\sim 0.25\ min^{-1}$). However, binding of the second Mn^{2+} results only in a slight decrease of enzymatic activity.

Discussion

The kinetic and binding experiments presented here are consistent with the mechanism proposed in scheme 3 in which two Ni^{2+} ions bind to the CPP–CaM complex in an ordered fashion, each step involving an initial weak combination followed by a conformational rearrangement. Studies with Mn^{2+} ions reveal a similar mechanism. This finding is not unexpected as

Mn^{2+} and Ni^{2+} appear to compete for the same sites (9). The mechanism may be a general one applicable to the activation of CPP by various divalent metal ions.

There have been conflicting results reported by other laboratories regarding the number of Ni^{2+} ions bound to CPP and the lack of activating effect of Ni^{2+}. In view of the extremely slow binding of the second Ni^{2+} ion (Fig. 5.3), it is not surprising that different laboratories may obtain different numbers of irreversibly bound Ni^{2+}, depending on the concentration of Ni^{2+} used, the presence or absence of CaM, and the incubation time of Ni^{2+} with the enzyme. Failure of Gupta et al. (11) to observe activation of CPP by Ni^{2+} may be explained by the exclusion of Ca^{2+} in their experiments. Our work with chelating agents suggests that binding of Ca^{2+} to the β subunit is required for Ni^{2+} to exert its activating effect. This concept is further supported by the finding that Ca^{2+} protects the Ni^{2+}-stimulated CPP activity against inactivation by *N*-ethylmaleimide (15).

A physiological role for Ni^{2+} or Mn^{2+} seems unlikely based on the mechanism described in this report. Because of the low concentrations of these ions in vivo, the activation process would be too slow to be metabolically significant unless there exists a carrier protein to facilitate the transfer and removal of the cations (4). Mg^{2+} may be a better candidate for serving a regulatory role in CPP activity (4, 10), especially if one considers the effect of Mg^{2+} in reducing the required number of Ca^{2+} ions bound to CaM in the activation of this phosphatase (14).

References

1. Pallen, C. J., and Wang, J. H. (1985) *Arch. Biochem. Biophys.* 237:281–291.
2. King, M. M., Huang, C. Y., Chock, P. B., Nairn, A. C., Hemmings, H. C., Jr., Chan, K-F., and Greengard, P. (1984) *J. Biol. Chem.* 259:8080–8083.
3. Hosey, M. M., Borsotto, M., and Lazdunski, M. (1986) *Proc. Natl. Acad. Sci. USA* 83:3733–3737.
4. King, M. M., and Huang, C. Y. (1984) *J. Biol. Chem.* 259:8847–8856.
5. Patel, J., Lanciotti, M., and Huang, C. Y. (1986) *Fed. Proc.* 45:1884.
6. Stewart, A. A., Ingebritsen, T. S., Manalan, A., Klee, C. B., and Cohen, P. (1982) *FEBS Lett.* 137:80–84.
7. King, M. M., and Huang, C. Y. (1983) *Biochem. Biophys. Res. Commun.* 114:955–961.
8. Pallen, C. J., and Wang, J. H. (1983) *J. Biol. Chem.* 258:8550–8553.
9. Pallen, C. J., and Wang, J. H. (1984) *J. Biol. Chem.* 259:6134–6141.
10. Li, H-C. (1984) *J. Biol. Chem.* 259:8801–8807.
11. Gupta, R. C., Khandelwal, R. L., and Sulakhe, P. V. (1984) *FEBS Lett.* 169:251–255.
12. Wolff, D. J., and Sved, D. W. (1985) *J. Biol. Chem.* 260:4195–4202.
13. King, M. M., Lynn, K. K., and Huang, C. Y. (1985) In *Progress in Nickel Toxicology*, edited by S. S. Brown and F. W. Sunderman Jr., pp. 117–120. Blackwell, Oxford.
14. Huang, C. Y., and King, M. M. (1985) *Curr. Top. Cell Regul.* 27:437–446.
15. King, M. M. (1986) *J. Biol. Chem.* 261:4081–4084.

6
Studies on the Mechanism and Molecular Mode of Regulation of Fructose-1,6-bisphosphatase

JULIE E. SCHEFFLER and HERBERT J. FROMM

Hydrolysis of fructose-1,6-bisphosphate (FBP) to form fructose-6-phosphate (F6P) and inorganic orthophosphate (P_i) is catalyzed by fructose-1,6-bisphosphatase (FBPase) (EC 3.1.3.11). FBPase and phosphofructokinase constitute a highly regulated system for the futile cycle of FBP synthesis and degradation, a focal point in the antagonistic control of glycolysis and gluconeogenesis in the liver. The hydrolysis of FBP by FBPase is inhibited by AMP and fructose 2,6-bisphosphate (2,6-FBP), both of which are activators of phosphofructokinase. The complex hormonal and metabolic controls for 2,6-FBP levels in vivo have been reviewed in detail (1). This chapter focuses on studies from our laboratory on the mechanism and mode of regulation at the molecular level of bovine and rabbit liver FBPases.

Mammalian liver FBPase is a tetramer of MW 140,000 that requires the presence of divalent metal ions (Zn^{2+}, Mn^{2+}, Mg^{2+}, or Co^{2+}) for activity (2). In the absence of substrates, rabbit liver FBPase binds one structural metal ion per subunit (3), whereas the metal stoichiometry for the enzyme from rat liver is two per subunit (4). For both enzymes the presence of F6P or the substrate analogue α-methylfructose-1,6-bisphosphate (α-Me-FBP) induces the appearance of an additional four equivalent metal sites per molecule of FBPase (3,4). The latter sites have been termed catalytic, as the dissociation constants for Mn^{2+} and Zn^{2+} binding roughly approximate their Km values for catalysis (2). Although direct evidence for the participation of a divalent metal ion in catalysis is lacking, it appears that this requirement results from a direct interaction with the enzyme and does not reflect the formation of a Me^{2+}–FBP complex as the true substrate (5).

The inhibition of FBPase by AMP and 2,6-FBP is synergistic (6,7). AMP clearly binds at an allosteric site (2). Inhibition by 2,6-FBP has been described as allosteric because 2,6-FBP alters the FBP kinetics from hyperbolic to sigmoidal (6, 8). Binding studies on the interaction of 2,6-FBP with FBPase (9) are in agreement with the hypothesis that the mode of 2,6-FBP inhibition is competitive. Kinetic, nuclear magnetic resonance (NMR) and fluorescence studies from this laboratory are consistent with an active-site-mediated inhibition by 2,6-FBP (10, 11).

Kinetic Studies on the Mechanism of FBPase

The results of initial rate studies, undertaken in 1978 to determine the kinetic mechanism for bovine liver FBPase, showed that F6P is a linear noncompetitive inhibitor of the substrate, whereas phosphate is a linear competitive inhibitor of FBP (12). A large number of kinetic mechanisms are consistent with these findings when assumptions are made allowing for the formation of abortive complexes (12). Initial-rate experiments yield little information on the mechanism of one-substrate systems; therefore experiments in the nonphysiological direction were carried out by coupling the reaction to aldolase, triosephosphate isomerase, and glycerolphosphate dehydrogenase (12).

The initial-rate results suggested a sequential kinetic mechanism. Four mechanisms, including one involving a phosphoryl-enzyme intermediate, in accord with our findings are shown in Fig. 6.1.

To limit the number of possible mechanisms, we undertook experiments with compounds that act as linear competitive dead-end inhibitors of the substrates (13). Phosphite and 2-deoxyribose-5-phosphate replaced P_i and F6P, respectively. The results eliminated schemes 1 and 2 (Fig. 6.1) from consideration.

To choose between the ordered mechanism of scheme 3 and the random mechanism depicted in scheme 4, inhibition by the multisubstrate analogue 2,5-anhydromannitol-1,6-bisphosphate was studied (13). This approach exploited the fact that subsites exist for both substrates within the active site in the random mechanism only. Furthermore, in scheme 3 F6P binds at an allosteric site, whereas P_i binds at the active site. The linear competitive inhibition observed with 2,5-anhydromannitol-1,6-bisphosphate toward substrates in the forward and reverse directions eliminated the ordered mechanism (13). However, we had also obtained data at variance with the remaining rapid-equilibrium random pathway (scheme 4) (13). First, a small but significant amount of $[^{32}P]P_i$ was trapped by FBPase in a pulse-chase experiment. Second, if the kinetic mechanism is truly rapid equilibrium random, one would have to postulate an abortive complex of the type FBPase · FBP · F6P but not allow for the formation of the analogous abortive FBPase · FBP · Pi. In light of the kinetic results, we conclude that the kinetic mechanism for bovine liver FBPase is probably steady-state random uni–bi (13). Without computer simulation studies of the type carried out with yeast hexokinase (14), we have made the assumption that the rate constants k_3, k_7, and k_8 for the mechanism in scheme 4 approach zero under conditions of the forward reaction; that is, the reaction is essentially irreversible and P_i binds weakly to FBPase. The bovine enzyme is known to exhibit both of these properties (13).

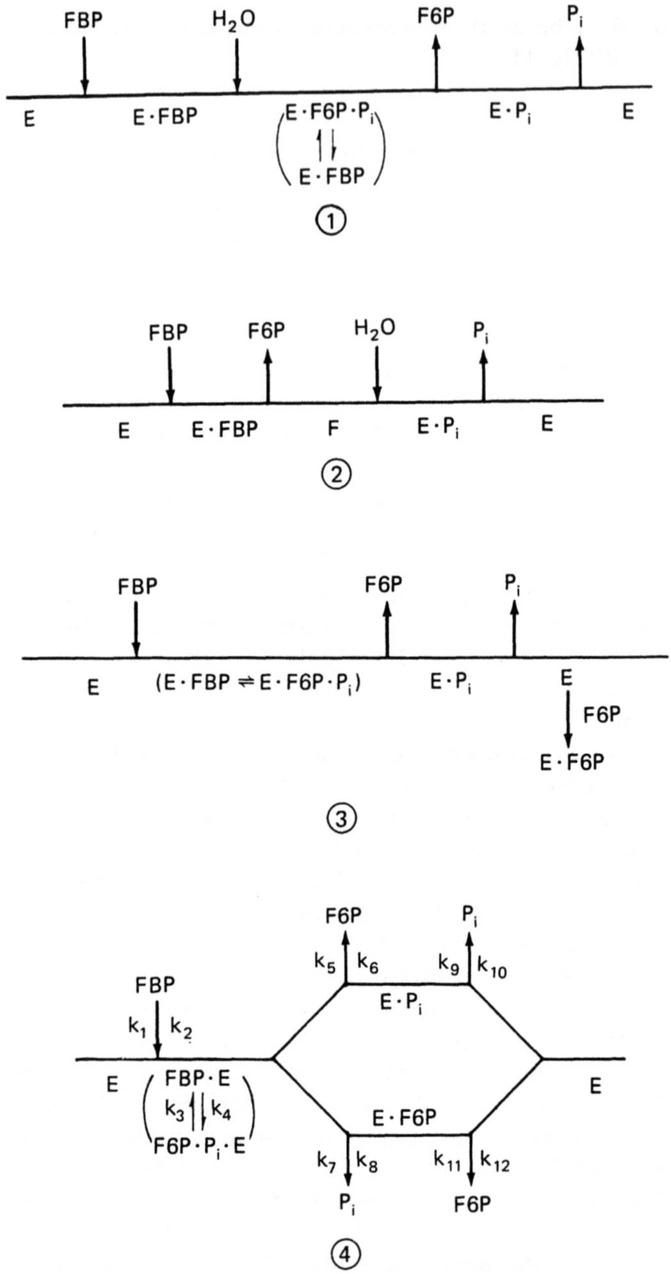

FIG. 6.1. FBPase kinetic mechanisms. Scheme 1: Ordered uni-bi sequential mechanism. Scheme 2: Ping-pong uni-bi mechanism. Scheme 3: Ordered rapid-equilibrium mechanism. Scheme 4: Rapid-equilibrium random bi-uni mechanism.

NMR Studies

Initial-rate experiments with 2,6-FBP (10) show it is a linear competitive inhibitor of the forward and reverse reaction substrates. This observation does not exclude allosteric binding by 2,6-FBP, where the inhibitor and substrates are mutually exclusive binding ligands. The site of 2,6-FBP binding was investigated by taking advantage of the fact that FBPase binds Mn^{2+} tightly, and that this ion can enhance the relaxation rate of diamagnetic nuclei in NMR experiments (11). The effect of paramagnetic Mn^{2+} at the structural site on the relaxation rates of the phosphorus nuclei of F6P, P_i, α-Me-FBP, and 2,6-FBP was measured and the distances from the divalent cation to the phosphoryl groups of these ligands obtained. The distances between Mn^{2+} and the phosphorus atoms at C6 of F6P, α-Me-FBP, and 2,6-FBP are identical (5 Å) within experimental error. Similarly, the Mn^{2+} to phosphorus distance for P_i, the C1 phosphate of α-Me-FBP, and 2,6-FBP all agree within 0.5 Å. The C2 and C6 phosphoryl groups of 2,6-FBP are equidistant from the divalent cation site as are those at C1 and C6 of α-Me-FBP. Cunningham et al. (15) measured Mn^{2+} to phosphorus distances for F6P, P_i, and AMP with the rabbit liver enzyme. For each ligand the calculated distance is approximately 1 Å shorter than that of the bovine liver enzyme. However, the relative positions of the three ligands are similar for the two enzyme species, and again the F6P and P_i phosphoryl groups are equidistant from Mn^{2+} For the bovine enzyme, the distance from Mn^{2+} to AMP is 4.1 Å and is increased to 6.0 Å in the presence of 2,6-FBP (11).

Using the rabbit liver enzyme, Cortes and Fromm (*unpublished results*) observed that the distances from the structural metal site to the P atoms of F6P, FBP, fructose-1-P, and 2,6-FBP were essentially the same. These findings suggest that the structural metal is not directly coordinated to the substrate.

The aromatic region of the ^1H NMR spectrum of rabbit liver FBPase is shown in Fig. 6.2, spectrum A. The variation in chemical shift of resonances 1 and 2 of Fig. 6.2 with pH indicates that these are the imidazole C2 protons of two histidine residues with pKa values of 7.1 and 6.1, respectively. Resolution enhancement of the ^1H NMR spectrum of rabbit liver FBPase allows the observation of approximately eight resolved resonances in the aromatic region of the spectrum when the contribution from broad resonances in this region is subtracted using the convolution difference technique (Fig. 6.2, spectrum B). We have investigated the effects of FBP, 2,6-FBP, and AMP on these observable resonances. The ^1H NMR spectrum of the enzyme · FBP complex (Fig. 6.2, spectrum C) is nearly identical to that of the enzyme · 2,6-FBP complex (Fig. 6.2, spectrum D). The spectral perturbations in each case consist in: (a) a downfield shift of the C2 proton resonance of His 1 by 0.03 ppm; (b) a broadening and 0.03 ppm downfield shift of the His 2 C2 proton resonance; and (c) a decrease in the intensity of several resonances in the Phe/Tyr region of the spectrum (6.6–7.2 ppm). The observed shifts of the

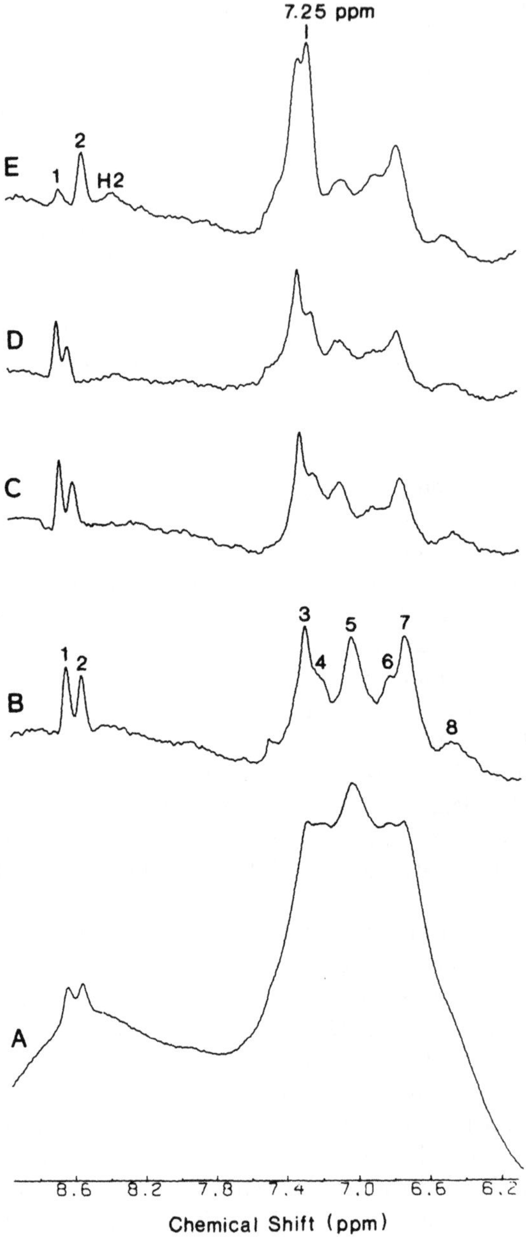

FIG. 6.2. Aromatic region of the 300 mHz ^{1}H NMR spectrum of rabbit liver FBPase at 22°C, 0.8 to 1.2 m*M* FBPase (pH 5.25–5.40), 1 m*M* Pipes, 0.15 *M* KCl. (A) FBPase (B) spectrum A resolution-enhanced by the convolution difference method; (C) + 4.2 m*M* FBP; (D) + 4.1 m*M* 2,6-FBP; (E) + 4.8 m*M* AMP. The shift of resonances 1 and 2 in spectrum E compared with spectrum B is due to a slight difference in pH. H2 refers to the adenine C2 proton of AMP.

imidazole C2 protons result from increases in the pKa values of these residues. In addition, both fructose bisphosphates altered the exchange of AMP from fast/intermediate in the binary enzyme · AMP complex to slow in the ternary complexes. AMP produces similar changes in the Phe/Tyr region of the spectrum (Fig. 6.2, spectrum E). Unlike the fructose bisphosphates, AMP binding produced (a) significant broadening of the C2-H of His 1, (b) no effect on His 2, and (c) enhancement of the intensity of a resonance at 7.25 ppm (pH 5.3).

Experiments involving nuclear Overhauser effects suggest that AMP bound to FBPase is in the syn conformation.

Fluorescence Studies

Fluorescence studies on the interaction of AMP, 2,6-FBP, and metal ions with rabbit liver FBPase are in progress in our laboratory. The fluorescent AMP analogue formycin 5′-monophosphate (FMP) inhibits rabbit liver FBPase. The concentration required for 50% inhibition is 17 μM compared to 13 μM for AMP under identical conditions. Inhibition by FMP is not significantly cooperative (Hill coefficient 1.2), and in this respect FMP differs from AMP (Hill coefficient 2.3). The binding of FMP to FBPase can be monitored by measuring the enhanced fluorescence intensity or polarization of FMP in the presence of FBPase. In this manner the K_D for FMP has been determined to be 6.6 μM. The addition of AMP reduces the fluorescence spectrum of the enzyme · FMP complex to that of free FMP, indicating that FMP most likely binds at the allosteric site for AMP. In the presence of F6P, FBP, α-Me-FBP, or 2,6-FBP, the fluorescence of enzyme-bound FMP is increased by about 10%. The dissociation constant for FMP shows little or no change in response to ligand binding at the active site.

The effect of metal ions on FMP binding has also been examined. Divalent metal ions reduce both the fluorescence intensity and the polarization of enzyme-bound FMP. The order of effectiveness in displacing FMP from the enzyme is $Zn^{2+} = Mn^{2+} > Mg^{2+}$, although concentrations exceeding 50 μM are required for each of these ions. In the presence of active-site binding ligands, i.e., F6P and α-Me-FBP, the concentration of metal ions required to displace FMP from the ternary complexes is significantly reduced. The K_D for Mn^{2+} binding to the enzyme · α-Me-FBP complex determined by FMP displacement is about 14 μM. Because this value is similar to the binding constant for Mn^{2+} at the catalytic metal ion site (4), and the metal ion-mediated displacement of FMP is enhanced by α-Me-FBP and F6P, it appears that binding of the catalytic metal ion and FMP may be competitive. Competition between metal ions and AMP has also been inferred from kinetic and binding experiments (16). In contrast, millimolar concentrations of Zn^{2+}, Mn^{2+}, or Mg^{2+} did not displace FMP from the enzyme · 2,6-FBP · FMP complex. Kitajima and Uyeda (9) observed decreased 2,6-FBP binding to rabbit liver

FBPase in the presence of Mn^{2+}, Co^{2+}, and to a lesser extent Mg^{2+}. It seems that this observation represents a major difference between the interaction of 2,6-FBP and the substrate with FBPase.

The kinetic, NMR, and fluorescence results are consistent with active-site binding of 2,6-FBP. In addition, using space-filling models, we find that α-FBP and β-2,6-FBP have very similar confirmations.

In summary: (a) Initial-rate kinetic studies show that FBP and 2,6-FBP binding is mutually exclusive. (b) The distance from the structural metal ion site to the phosphoryl groups of F6P, P_i, α-Me-FBP, and 2,6-FBP are the same. (c) FBP and 2,6-FBP perturb the same observable aromatic resonances of the ^1H NMR spectrum of FBPase. (d) FBP and 2,6-FBP enhance AMP binding. However, 2,6-FBP does not promote metal ion-mediated displacement of FMP from FBPase and in this respect differs from F6P and α-Me-FBP. Further studies are required to determine if binding of the catalytic metal ion and 2,6-FBP is truly competitive.

Acknowledgment. This research was supported in part by NIH grant NS-10546 and NSF grant DMB-8502211.

References

1. Hers, H. G., and Hue, L. (1983) *Annu. Rev. Biochem.* 52:617–653.
2. Benkovic, S. J., and de Maine, M. M. (1982) *Adv. Enzymol.* 53:45–82.
3. Benkovic, P. A., Caperelli, C. A., de Maine, M. M., and Benkovic, S. J. (1978) *Proc. Natl. Acad. Sci. USA* 75:2185–2189.
4. Pedrosa, F. O., Pontremoli, S., and Horecker, B. L. (1977) *Proc. Natl. Acad. Sci. USA* 74:2742–2745.
5. Marcus, C. J., Geller, A. M., and Byrne, W. L. (1973) *J. Biol. Chem.* 248:8567–8573.
6. Van Schaftingen, E., and Hers, H. G. (1981) *Proc. Natl. Acad. Sci. USA* 78:2861–2863.
7. Pilkis, S. J., El-Maghrabi, M. R., Pilkis, J., and Claus, T. (1981) *J. Biol. Chem.* 256:3619–3622.
8. Hue, L., and Bartrons, R. (1985) In *Regulation in Carbohydrate Metabolism*, edited by R. Beitner, pp. 29–44. CRC Press, Boca Raton, FL.
9. Kitajima, S., and Uyeda, K. (1983) *J. Biol. Chem.* 258:7352–7357.
10. Ganson, N. J., and Fromm, H. J. (1982) *Biochem. Biophys. Res. Commun.* 108:233–239.
11. Ganson, N. J., and Fromm, H. J. (1985) *J. Biol. Chem.* 260:2837–2843.
12. Cassazza, J. P., Stone, S. R., and Fromm, H. J. (1979) *J. Biol. Chem.* 254:4661–4665.
13. Stone, S. R., and Fromm, H. J. (1980) *J. Biol. Chem.* 255:3454–3460.
14. Rudolph, F. B., and Fromm, H. J. (1971) *J. Biol. Chem.* 246:6611–6619.
15. Cunningham, B. A., Raushel, F. M., Villafranca, J. J., and Benkovic, S. J. (1981) *Biochemistry* 20:359–362.
16. Pontremoli, S., Grazi, E., and Accorsi, A. (1968) *Biochemistry* 7:3628–3633.

7
Phosphorylation and dAMP Inhibition of Snake Muscle Fructose-1,6-bisphosphatase

GEN-JUN XU, GUO-FU HU, FU-KUN ZHAO, and QI-CHANG XIA

Fructose-1,6-bisphosphatase (EC 3.1.3.11; $FruP_2$ase) catalyzes the hydrolysis of D-fructose-1,6-bisphosphate ($FruP_2$) to D-fructose-6-phosphate (Fru6P) and inorganic orthophosphate (P_i). Two comprehensive reviews have summarized its mechanism of action and regulation (1, 2). It was found that P_i was a competitive inhibitor with respect to $FruP_2$, whereas the inhibition by Fru6P was S-linear, I-linear noncompetitive (3). The results implied that an ordered sequential or a ping-pong mechanism might be involved in the action of bovine liver $FruP_2$ase. Using 2,5-anhydromannitol-1,6-bisphosphate as an inhibitor of the reverse reaction, it was found that this product analogue was a competitive inhibitor with respect to both Fru6P and P_i (4). 2-Deoxyribose-5-phosphate, as a dead-end inhibitor of Fru6P, was a noncompetitive rather than an uncompetitive inhibitor with respect to P_i (5). Therefore a random mechanism was suggested. The inconsistency of all these results may be due to the complexities of the action of $FruP_2$ase. Because no phosphoryl $FruP_2$ase had been found by labeling the enzyme with both $(^{32}P)P_i$ and (^{32}P)1-phosphate $FruP_2$ (3, 6), the ping-pong mechanism was less emphasized. This chapter presents the results of the formation of a phosphoryl snake muscle $FruP_2$ase and the inhibition of the enzyme by dAMP.

Phosphorylation of $FruP_2$ase

Snake muscle $FruP_2$ase was purified from the muscle of the snake *Zaocys dhumnades* as previously described (7). The purified enzyme was homogeneous and could be crystallized in PEG Tris buffer. Using the time-stop assay to determine the release of inorganic phosphate by the malachite green method (8) and the release of Fru6P by an enzymic method (9), it was found that the rates determined by the two assays were different in 0.5 *M* Tris buffer pH 7.5 or 0.4 *M* TEA-DEA buffer pH 7.5, but almost the same in borate buffer pH 7.5.

Wilson et al. (10) reported that the formation of *p*-nitrophenol exceeded that of phosphate when the assay of alkaline phosphatase with *p*-nitrophenyl-

TABLE 7.1. Incorporation of [^{32}P]P_i into snake muscle $FruP_2$ase.

	^{32}P incorporated	
$FruP_2$ase (nmol)	Mol ^{32}P/mol enzyme recovered by monoclonal antibody	Mol ^{32}P/mol enzyme recovered by TCA
0.22	1.84	
0.32	2.34	
0.64	2.05	
1.32		1.57
1.54		1.60
1.60		1.75
1.72		1.53
3.21		1.95
3.62		1.79

phosphate substrate was made in the presence of Tris or ethanolamine. They suggested that a phosphoryl enzyme intermediate was formed, and the transphosphorylation of the phosphoryl group to a phosphate acceptor such as Tris or ethanolamine would produce the difference of the rates. Because similar results were obtained for snake muscle $FruP_2$ase, it is interesting to study if phosphoryl-enzyme is involved in the catalytic reaction of snake muscle $FruP_2$ase.

Snake muscle $FruP_2$ase 12 μM and [^{32}P]orthophosphate 10 mM (3.2 mCi) were incubated in 25 mM of MOPS buffer pH 6.5 in the absence of Mg^{2+} for 45 min. The reaction mixture was passed through a Sephadex G-25 column to remove most of the unreacted [^{32}P]P_i and then precipitated with 10% TCA. The precipitate was collected on a Millipore filter, washed thoroughly, and counted. Alternatively, instead of TCA, a specific monoclonal antibody of snake muscle $FruP_2$ase was used as the precipitant and the same results were obtained (Table 7.1). There was a linear relation between the [^{32}P]phosphate incorporated into $FruP_2$ase and the concentration of the enzyme. The number of moles of [^{32}P]phosphate incorporated per mole of enzyme was calculated based on a molecular weight of 144,000 for the enzyme. In the absence of Mg^{2+} a maximum of about 2 mol/mol enzyme was incorporated, and in the presence of Mg^{2+} the maximum incorporation of [^{32}P]phosphate was 4 mol phosphate/mol enzyme, showing a "half-of-the-site" reaction in the absence of Mg^{2+} (Fig. 7.1). A molecular weight of 144,000 for [^{32}P]$FruP_2$ase was estimated by gel filtration on Sephadex G-150.

In order to rule out the possibility that the [^{32}P]phosphate found was firmly but noncovalently bound to the enzyme or was simply a contamination of [^{32}P]P_i, a solution containing 1.3 mg of ^{32}P-labeled $FruP_2$ase in 25 mM MOPS buffer pH 7.5 was treated with trypsin (the $FruP_2$ase/trypsin ratio being 30:1). The tryptic digests were separated by reverse-phase HPLC on a C-18 column. The radioactivity was recovered mainly from a peptide peak whose amino acid composition was Gly-Ala-Gly-Ser-Arg. After treating the

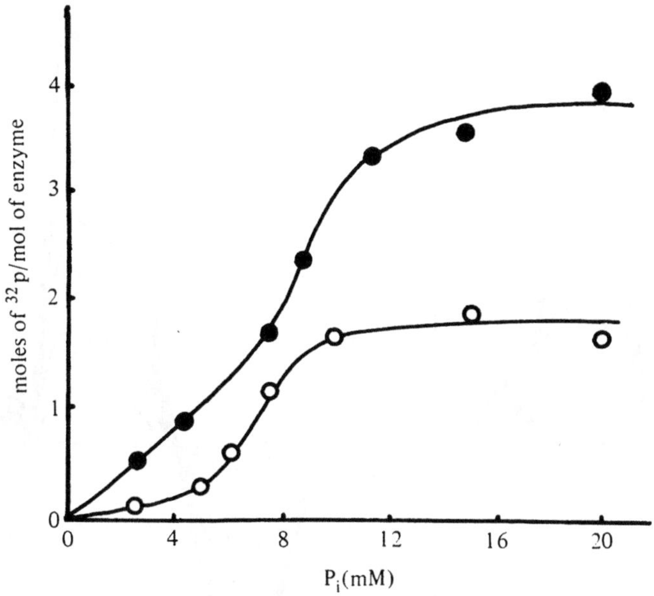

FIG. 7.1. Effects of Mg^{2+} on the phosphorylation of $FruP_2$ase. Phosphorylation was carried out in 25 m*M* MOPS, pH 6.5, containing various concentrations of $(^{32}P)P_i$ as indicated (specific radioactivity was 0.2 mCi/mmol) and 10 μM $FruP_2$ase. Closed circles = presence of 5 m*M* Mg^{2+}; open circles = absence of Mg^{2+}. The phosphorylated enzyme was recovered by TCA precipitation.

^{32}P-labeled peptide with alkaline hydroxyamine, rechromatography yielded a peptide peak with the same amino acid composition but with no radioactivity. All the radioactivity was recovered in the first peak before the gradient was applied. The results suggest that snake muscle $FruP_2$ase can be phosphorylated at a serine residue by $[^{32}P]$orthophosphate. The differences in results obtained by our laboratory and those reported by Pontremoli's (6) and Fromm's (3) laboratories may be due to the difference in enzyme species or experimental conditions.

Dephosphorylation of $FruP_2$ase

The phosphoryl $FruP_2$ase had almost the same specific activity as the native enzyme. Dialysis of the phosphoryl enzyme in the presence of Mg^{2+} removed all the radioactivity, whereas in the absence of Mg^{2+} at pH 6.5 the phosphoryl enzyme was stable. The first-order rate constant of dephosphorylation of the enzyme (2 mol phosphate/mol enzyme) was 0.07 s^{-1}. Casazza et al. (3) also reported that for bovine liver $FruP_2$ase the off-rate constant for P_i dissociation was much less than the maximum velocity of the forward reaction (V_1),

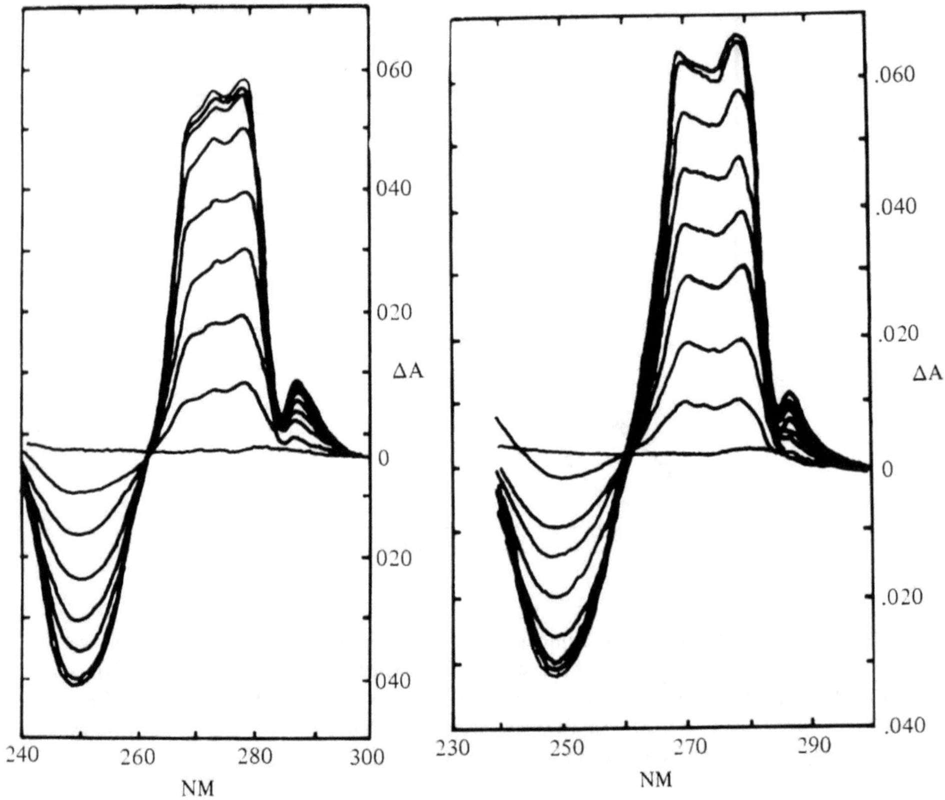

UV Difference spectra of $FruP_2ase$ combining with dAMP

UV Difference spectra of $FruP_2ase$ combining with AMP

FIG. 7.2. UV difference spectra of snake muscle $FruP_2ase$ binding AMP or dAMP. The difference spectra were carried out in 20 m*M* TEA-DEA buffer, pH 7.5, containing 17 m*M* KCl, 1 m*M* EDTA, 10 m*M* Mg^{2+}, and 7.2 μM $FruP_2ase$. *A*: The concentrations of AMP in curves designated in the region of positive peaks, from bottom to top, were 0, 4.18, 8.36, 12.54, 16.72, 20.90, 25.08, 29.26, 33.44, and 37.62 μM, respectively. *B*: The concentrations of dAMP were 0, 4.8, 9.8, 14.7, 19.6, 24.5, 29.4, 34.3, and 39.2 μM, respectively.

and they suggested that isomerization might be involved in the catalytic process.

dAMP Inhibition

We have reported that the inhibition behavior of snake muscle $FruP_2ase$ by dAMP was different from that by AMP. Both AMP and dAMP were strong allosteric inhibitors for snake muscle $FruP_2ase$ (11). The inhibition constants calculated from the concentrations of the nucleotides at 50% inhibition of the enzyme were 1.5 and 3.0 μM for AMP and dAMP, respectively. It is consistent with the inhibition of rat liver $FruP_2$ase by AMP reported by Taketa and Pogell (12). The maximum inhibition by dAMP was about 60 to 70% depending on the temperature of the reaction, whereas for AMP the maximum inhibition approached 100%.

AMP and dAMP generated similar ultraviolet (UV) difference spectra, with three positive peaks at 270, 280, and 288 nm and a broad negative peak at about 250 nm (Fig. 7.2). In addition, the difference spectrum of dAMP had another unique peak at 275 nm (13).

Proteolysis of $FruP_2$ase with trypsin decreased the difference absorbance of these peaks with a concomitant decrease in the AMP inhibition of the enzyme. The binding constants calculated from the difference spectra were 2.0×10^{-6} M and 1.2×10^{-6} M for AMP and dAMP, respectively. The binding constants for AMP and dAMP were similar, and we believe that the unique behavior of dAMP inhibition of the enzyme is likely due to the lack of interaction between the 2′-OH on the ribose moiety and the allosteric site of the enzyme.

We have synthesized a photoaffinity labeling reagent 2′(3′)-O-[*N*-(4-azido-2-nitrophenyl)aminopropyl]adenosine-5′-monophosphate(ANA-AMP) (14). The reagent used was a mixture of 2′ and 3′-ANA-AMP. Similar to AMP, ANA-AMP inhibited snake muscle $FruP_2$ase in the dark with a sigmoidal response with increasing concentrations. The inhibition constant was 9×10^{-6} M compared with 6×10^{-7} M for AMP. The activity of the inhibited snake muscle $FruP_2$ase could be completely recovered by treating the incubation mixture with charcoal or using an AMP-removing system in the enzyme assay. The inhibition seemed to be reversible with no chemical modification taking place.

ANA-AMP also showed a binding difference spectrum with three positive peaks and a broad negative peak at 250 nm for the enzyme. The inhibition by ANA-AMP could be partially overcome by PLP (Fig. 7.3). We have reported that the reversal of the inhibition of rabbit liver $FruP_2$ase was due to PLP competition with AMP for the allosteric site (15, 16). The results suggested that the inhibition by ANA-AMP was on the allosteric site of the enzyme. A mixture of 1 mg snake muscle $FruP_2$ase and 40 mM ANA-AMP was irradiated with a hydrogen lamp at a distance of 5 cm with stirring. After irradiation the mixture was treated with 1 M NH_4HCO_3 and was dialyzed against 20 mM TEA-DEA buffer pH 7.5. Although the activity of the photoaffinity labeled $FruP_2$ase was only slightly affected about 80% of the native enzyme), the labeled enzyme no longer distinguished the inhibition behavior between AMP and dAMP as did the native enzyme, suggesting that a residue on the enzyme molecule that interacts with the 2′-OH on the ribose moiety of AMP has been modified. Appa·ently the modification of this residue has only a limited effect on the activity of the enzyme. The above results support our previous suggestion that this residue is located at the allosteric site and may

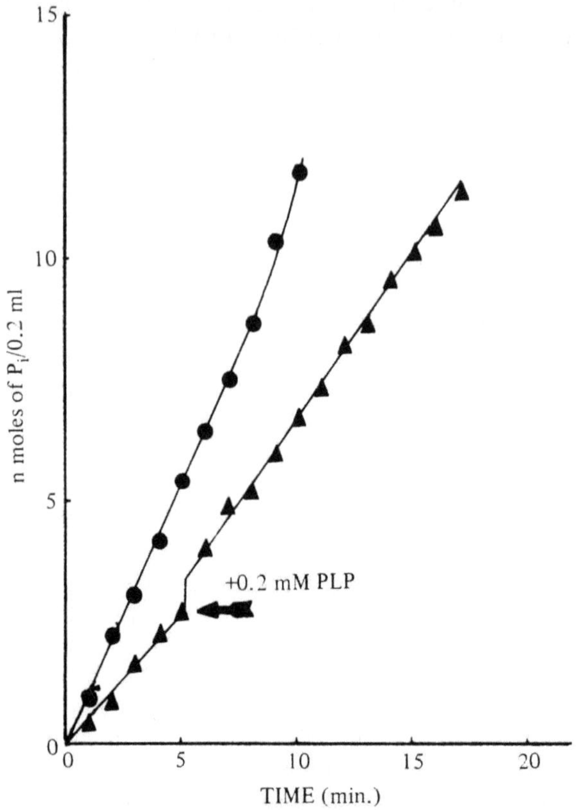

FIG. 7.3. Reversibility of ANA-AMP inhibition by PLP. The assays were carried out in 20 m*M* TEA-DEA buffer, pH 7.5, containing 8 m*M* Mg^{2+}, 0.1 m*M* EDTA, and 0.1 m*M* $FruP_2$ in a total volume of 4 ml. $FruP_2$ase, 4 μg, was added to initiate the reaction. An aliquot of 0.2 ml was removed after 1 min and transferred to a tube containing 0.5 ml water and 0.2 ml color-developing reagents. After 20 min of incubation at room temperature, the absorbance was measured at 650 nm. Circles = control experiment; triangles = with 16 μM of ANA-AMP, PLP being added to a final concentration of 0.2 m*M* at the time indicated.

have a function of signal transmission from the allosteric site to the catalytic site of the enzyme.

Acknowledgment. This research was supported by the Scientific Fund Committee of Academia Sinica. We also thank Dr. R. J. Guillory for helping us synthesize ANA–AMP.

References

1. Benkovic, S. J., and deMaine, M. M. (1982) *Adv. Enzymol.* 53:45–82.
2. Tejwani, G. A. (1983) *Adv. Enzymol.* 54:121–194.

3. Casazza, J. P., Stone, S. R., and Fromm, H. J. (1979) *J. Biol. Chem.* 254:4661–4665.
4. Marcus, C. J. (1976) *J. Biol. Chem.* 251:2963–2966.
5. Stone, S. R., and Fromm, H. J. (1980) *J. Biol. Chem.* 255:3454–3460.
6. Pontremoli, S., Traniello, S., Luppis, B., and Wood, W. A. (1965) *J. Biol. Chem.* 240:3459–3463.
7. Xu, G. J., Shi, J. P., and Wang, Y. L. (1982) *Methods Enzymol.* 90:349–351.
8. Tashima, Y., and Yoshimura, N. (1975) *J. Biochem.* 78:1161–1169.
9. Hohorst, H. J. In *Methods of Enzymatic Analysis*, edited by H. U. Bergmeyer, pp. 134–138, Verlag Chemie, GmbH, Weinheim/Bergstr and Academic Press, New York and London.
10. Wilson, I. B., Dayan, J., and Cyr, K. (1964) *J. Biol. Chem.* 239:4182–4185.
11. Zhao, F. K., Shi, J. P., and Xu, G. J. (1984) *Acta Biochim. Biophys. Sin.* 16:565–570.
12. Taketa, K., and Pogell, B. M. (1965) *J. Biol. Chem.* 240:651–662.
13. Zhao, F. K., and Xu, G. J. (1986) *Sci. Sin.* [*B*] 5:514–523.
14. Jeng, S. J., and Guillory, R. J. (1975) *J. Supramol. Struct.* 3:448–468.
15. Kido, H., Vita, A., Xu, G. J., Pontremoli, S., and Horecker, B. L. (1980) *Biochem. Int.* 1:91–97.
16. Xu, G. J., Datta, A. G., Singh, V. N., Suda, H., Pontremoli, S., and Horecker, B. L. (1981) *Arch. Biochem. Biophys.* 210:98–103.

8
Identification and Characterization of Intermediates in the Mechanism of Enzyme Action

BERT L. VALLEE and DAVID S. AULD

The three-dimensional structure and conformation of enzymes are critical to catalytic potential. Numerous experimental approaches can elucidate structure–function relationships, but few can detect subtle conformational changes that might identify both substrate binding and catalytic groups and particularly their changes during catalytic reactions. Such alterations when visualized at rates at least as fast as catalysis would allow the direct observation of these important dynamic events which are essential to the verification of enzyme mechanisms.

Ideally, structural and functional data are determined on the same material, in the same physical state, and during identical time intervals—objectives not readily met concurrently. Such circumstances cannot be achieved easily either in standard kinetic experiments on the one hand or in thermodynamic experiments on the other: kinetics call for excellent substrates hydrolyzed as rapidly as possible, and thermodynamic studies often demand inhibitors or pseudosubstrates that result in slow reactions or no reaction at all. Consequently, views of the characteristics of a given enzyme · substrate or inhibitor complex derived from either one of the two approaches can diverge widely. At present the structural basis of enzymatic activity is generally deduced from time-averaged procedures at thermodynamic equilibrium, whereas functional conclusions often depend on pre-steady-state or steady-state kinetic conditions. The delineation of the structure and function of the enzyme · substrate complex at the moment of catalysis has proved difficult because of the vastly different time frames when structure and function are studied at conventional temperatures.

Subzero temperatures, which decrease the rates of enzyme-catalyzed reactions, combined with rapid mixing and rapid scanning instrumentation, allow acquisition of the kinetic and electronic absorption spectra of transients in enzyme catalysis (1–3). Carboxypeptidase A has served as our model system (4). The rates of formation and breakdown of intermediates have been monitored directly through fluorescence generated by radiationless energy transfer (RET) between enzyme tryptophans and the dansyl group of enzyme-bound, rapidly hydrolyzed substrates (5).

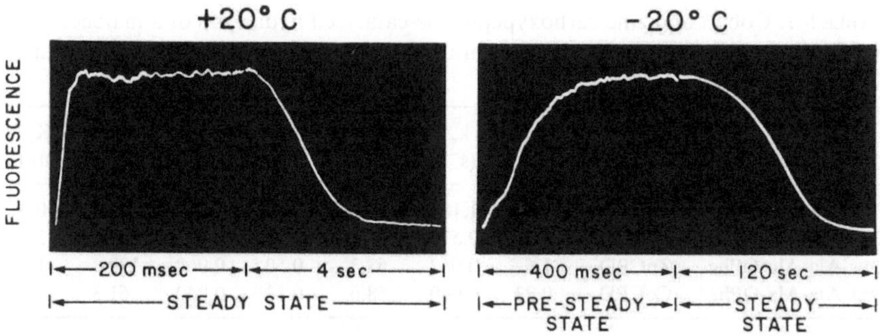

FIG. 8.1. Stopped-flow RET measurement of the binding and hydrolysis of 50 μM Dns-$(Ala)_2$-Phe by 1 μM carboxypeptidase A in 4.5 *M* NaCl, 10 m*M* Hepes, pH 7.5, at +20°C and −20°C (6).

Cryokinetics of Carboxypeptidase A

N-Dansylated oligopeptides and their ester analogues exhibit Michaelis-Menten kinetics over the temperature range −20° to +20°C with k_{cat}/K_m values of 0.3×10^7 to 3.0×10^7 M^{-1} s^{-1} at +20°C, pH 7.5 (6). At substrate concentrations ranging from 0.1 to 100 μM, excellent RET signals are obtained so that assays are convenient at $[S] < [E] < K_m$ (5, 6). Figure 8.1 is a typical example of reactions in which Dns-$(Ala)_2$-Phe is the substrate. At +20°C the initial, rapid increase in dansyl fluorescence to its maximum value coincides with rapid attainment of the steady-state concentration of enzyme · substrate (ES) complexes within the dead time of the stopped-flow instrument (Fig. 8.1A).

At subzero temperatures the pre-steady-state time interval for all substrates is biphasic, indicating the presence of multiple ES and/or enzyme · product (EP) complexes, as shown at −20°C for Dns-$(Ala)_2$-Phe (Fig. 8.1B). When substrate is mixed with enzyme, an initial, rapid rise (<15 ms) in dansyl fluorescence is within the dead time of the instrument, corresponding to formation of the Michaelis complex (ES_1). A slower exponential rise thereafter is due to formation of a second intermediate, ES_2, with a first-order rate constant (k_{obs}). The rates of formation and breakdown of the intermediates have been determined as a function of substrate concentration to quantitate all the individual rate and equilibrium constants of scheme 1.

$$E + S \underset{}{\overset{K_s}{\rightleftharpoons}} ES_1 \underset{k_{-2}}{\overset{k_2}{\rightleftharpoons}} ES_2 \xrightarrow{k_3} E + P$$

Scheme 1

Table 8.1 shows the rate and equilibrium constants for the individual steps in the zinc- and cobalt-catalyzed hydrolysis of Dns-Ala-Ala-Phe and Dns-

TABLE 8.1. Cobalt and zinc carboxypeptidase-catalyzed hydrolysis of a matched peptide-ester pair: Michaelis-Menten parameters and rate and equilibrium constants for elementary steps.

Substrate	Enzyme	K_m (μM)	k_{cat} (s^{-1})	k_2 (s^{-1})	k_{-2} (s^{-1})	k_3 (s^{-1})	K_s (μM)	K'_m[a] (μM)
Dns-Ala-Ala-Phe	ZnCPD	13.5	1.18	40.0	3.48	1.32	102	10.9
Dns-Ala-Ala-Phe	CoCPD	2.77	0.57	36.1	0.17	0.59	154	3.17
Dns-Ala-Ala-OPhe	ZnCPD	1.6	0.062	53.3	0.50	0.062	129	1.3
Dns-Ala-Ala-OPhe	CoCPD	0.23	0.040	58.6	0.13	0.040	61.3	0.18

Conditions of assays were −20°C for peptide and −10°C for ester in 4.5 NaCl/10 mM Hepes, pH 7.5 (6, 7). The kinetic parameters were determined by RET kinetics (5). OPhe, L-β-phenyllactate.
[a] Calculated $K'_m = K_s(k_{-2} + k_3)/(k_{-2} + k_2 + k_3)$.

Ala-Ala-OPhe (6, 7). In every case K'_m calculated from these constants is in close agreement with that determined from Lineweaver-Burk plots, K_m.

The interconversion of ES_1 and ES_2 reflected by two first-order constants, k_2 and k_{-2}, indicates that ES_2 does *not* represent the conventional acyl intermediate observed in serine proteases. In such an acyl enzyme intermediate mechanism, reversal of the acyl complex would occur in a second-order rate process ($k'_{-2} = k_{-2}$ [P_1], and hence k'_{-2} would essentially be *zero* in the absence of added product. Furthermore, product P_1 would be a *mixed* inhibitor, and its addition would *increase* the pre-steady-state rate constant (k_{obs}) for such a mechanism (8, 9).

$$k_{obs} = k_3 + k_{-2}\,[P_1] + k_2/(K_s/[S] + 1)$$

However, carboxypeptidase A fulfills none of these expectations. For all substrates studied k_{-2} is *greater than zero*, in fact usually larger than k_3 (Table 8.1). The P_1 products Phe and OPhe are *competitive* inhibitors, and k_{obs} *decreases* upon addition of P_1. Minimally, these results demonstrate that the C-terminal product is not liberated prior to the rate-limiting step and hence, by inference, that deacylation cannot be rate-limiting.

Furthermore, in enzyme mechanisms known to operate through acyl intermediates, addition of nucleophiles traps the acyl moiety, even if deacylation is not rate-limiting (9). In this regard it is of special relevance that yeast seryl carboxypeptidase behaves in accord with predictions based on the acyl mechanism of chymotrypsin (6): When hydrolysis is carried out in the presence of methanol, the substrate Dns-$(Gly)_3$-OPhe is readily converted to Dns-$(Gly)_3$-OMe. However, when carboxypeptidase A acts on this same substrate, there is no evidence that the product is trapped under any of the conditions investigated, in complete accord with previous studies of other substrates (10). Methanol, in particular, has proved to be an effective nucleophile in reactions of other proteases, as its intrinsic nucleophilicity is much greater than that of water. Yet methanol concentrations of up to 50% (v/v) fail to trap any intermediate during hydrolysis by carboxypeptidase A (6, 11).

Evidence has been presented for an anhydride intermediate in the hydrolysis of the ester *p*-chlorocinnamoyl-OPhe (12), although it has been questioned (13). Studies have also been made on an α-(acylamino)cinnamoyl-OPhe substrate (14), which should readily form chromophoric (300–400 nm) oxazolinones if an anhydride intermediate occurs during catalysis (15, and references therein). No such intermediate was detected either during the reaction or in the products using a variety of quenching conditions with and without enzyme-denaturing agents (13).

Cryospectroscopy of Cobalt Carboxypeptidase A

The decreased rates at subzero temperatures lengthen the lifetimes of intermediates sufficiently to permit both their structural and kinetic characterization (3, 6, 7). At the concentrations of peptides and depsipeptides employed for subzero spectroscopy, the peptide and ester ES_2 intermediates form maximally within 200 ms, at which time they constitute >95% of the total enzyme present (7). The subsequent breakdown of the intermediates is much slower, 25 to 250 s, depending on the substrate examined. The spectroscopic properties of the ES_2 intermediates can therefore be recorded readily at subzero temperatures with a rapid mixing and spectral scanning instrument (1-3).

Thus at −17°C the spectra of the cobalt enzyme during catalysis allow direct visualization of the formation and breakdown of the intermediates, exemplified here for Dns-Ala-Ala-Phe (Fig. 8.2). The figure insets show the time course of the changes in cobalt absorbance at 510 nm for the peptide intermediate. The decrease in absorbance at this wavelength over the first 200-ms interval reflects the rapid formation of the ES_2 intermediate (Fig. 8.2A). The increase in absorbance in the course of the next several seconds (inset, Fig. 8.2B) signals the reduction in the concentration of the intermediate due to its conversion to enzyme plus products, Dns-Ala-Ala and Phe.

The ES_2 intermediates generated with a large number of matched peptide and ester pairs have been examined in this manner. The cobalt spectra of both peptide and ester intermediates identify catalysis-related changes in the active site metal coordination sphere characteristic of each (3, 7). The electronic absorption spectra of the ES_2 intermediates generated with all peptides and analogous esters examined so far fall into two distinct catagories, clearly representative of the two substrate types (Fig. 8.3). The ES_2 intermediates of peptides display an absorption maximum at 570 ±3 nm and a band of low intensity at 473 ±6 nm, whereas those of the ester analogues exhibit two maxima of unequal intensity at 598 ±3 nm and 551 ±2 nm as well as a prominent shoulder at 523 ±3 nm. These two classes of spectra correlate directly and solely with the substrate type, i.e., peptide versus ester (7). Thus the spectral features of the peptide ES_2 intermediates (Fig. 8.3) that differ in length (di, tri, tetra), internal amino acids and their sequences (Gly, Ala, Val,

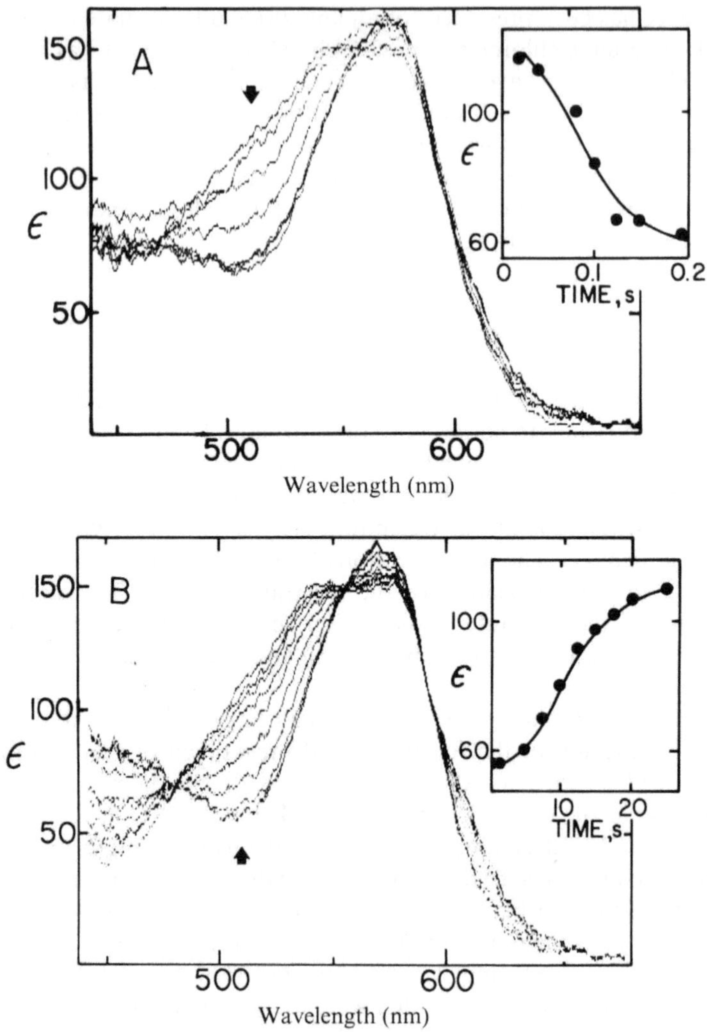

FIG. 8.2. Absorption spectra of the formation (A) and breakdown (B) of the ES_2 intermediate formed in the reaction of cobalt carboxypeptidase, 0.1 m*M*, with Dns-Ala-Ala-Phe, 0.2 m*M*, at −17°C (7). Individual spectra were recorded in 16.48 ms. *Insets*: Absorbance change at 510 nm, indicated by the arrows, as a function of time after mixing.

or Glu), blocking group (Dns, Cbz, Bz), and C-terminal residue (Phe or Leu) are all the same and exemplify those of a large number of peptides. Similarly, the spectral features of the ES_2 intermediates derived from structurally different esters are identical (Fig. 8.3).

The corresponding electron paramagnetic resonance (EPR) spectra support and further amplify the deductions drawn from the electonic spectra of

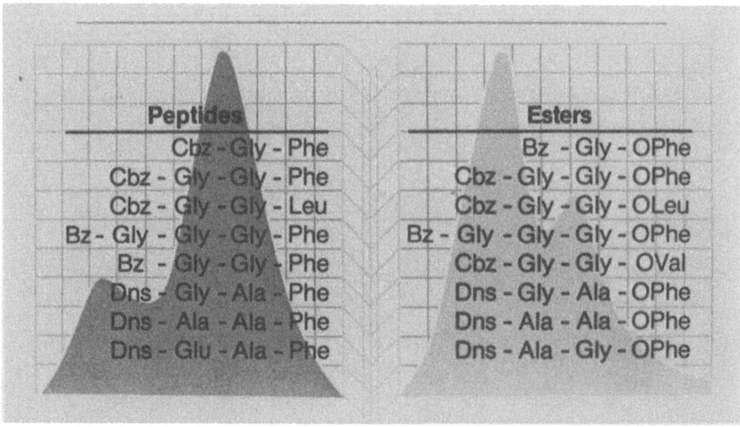

FIG. 8.3. Schematic representation of the cobalt absorption spectra of the ES_2 intermediates generated with peptides (left) and esters (right) of differing lengths, amino acid sequence, COOH-terminal residue, and NH_2-terminal blocking group (7).

the intermediates. Immediately prior to the rate-determining step, both peptides and esters disturb the coordination sphere and the symmetry of the metal in a manner characteristic for each substrate (3, 7).

Hyperfine structure is resolved on the low field resonance in the EPR spectra of the ES_2 peptide and ester intermediates, but that induced by any given peptide and its matching ester differ significantly. The effective hyperfine coupling in g_1 for the ester intermediates is larger by 20×10^{-4} to 40×10^{-4} cm^{-1} than that for the corresponding peptide (3, 7). Thus both the absorption and EPR spectra of the reaction intermediates consistently demonstrate (a) formation of transient metal complexes in the second step of their hydrolysis, (b) differences between the effects induced by peptide and by ester substrates, and (c) identities between those induced by all peptides on one hand and by all esters on the other.

Cryoquench Studies

Because the ES_2 intermediate is the predominant enzyme species present at steady state, its chemical nature is deducible from subzero chemical quench studies done after steady state is established. Extrapolation of the product concentration to zero time, $[P_0]$, measures the concentration of the enzyme species in which bond cleavage has occurred (11).

For the peptides Dns-Gly-Phe in 4.5 *M* NaCl and Dns-Ala-Ala-Phe in 50% methanol, 0.25 *M* NaCl, the extrapolated $[P_0]$ values are zero, indicating that no product is generated prior to turnover and therefore that the steady-state intermediate ES_2 involves a complex between the enzyme and intact peptide substrate (11). These results are confirmed by studying the reaction from the

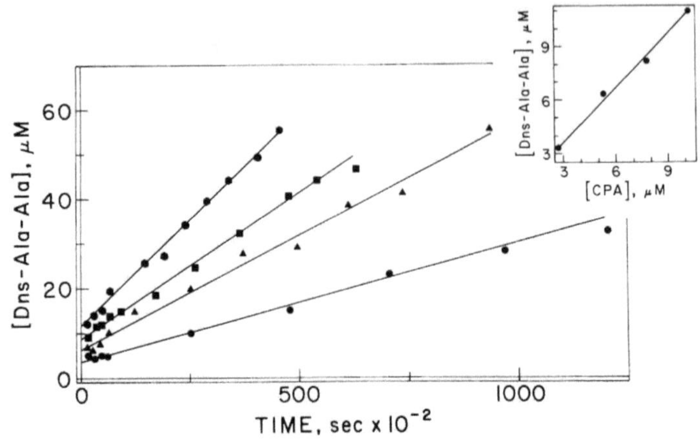

FIG. 8.4. Time courses, determined by HPLC analyses, for the hydrolysis of 102.6 μM Dns-Ala-Ala-OPhe by carboxypeptidase A in 4.5 M NaCl, 10 mM Hepes pH 7.5 at −19.2°C (11). Enzyme concentrations: (⬢) 10.2 μM; (■) 7.8 μM; (▲) 5.3 μM; (●) 2.7 μM. *Inset*: Plot of the *y*-intercepts ([Dns-Ala-Ala] at t = 0) versus enzyme concentration. This plot gives a slope of 1.10 mol of Dns-Ala-Ala/mol of enzyme.

reverse direction, i.e., peptide synthesis. After equilibrium is established for the reverse direction, the predominant enzyme species present has visible absorption and EPR spectra identical with those of the ES_2 intermediate formed in the hydrolytic direction (16). Under these conditions, chemical analyses clearly demonstrate that ES_2 represents a complex of intact peptide with the enzyme.

On the other hand, for the depsipeptide Dns-Ala-Ala-OPhe in both 4.5 M NaCl (Fig. 8.4) and a ternary coorganic cryosolvent, $[P_0]$ extrapolates to 1 mol equivalent of product/mol of enzyme over the entire temperature range from −20° to −50°C (11), and $[P_0]$ is linearly proportional to $[E_T]$ (Fig. 8.4). These results indicate that in depsipeptide hydrolysis the susceptible ester bond is already cleaved in intermediate ES_2 so that this species is more properly denoted by EP_1P_2 in which P_1 and P_2 are the products of the forward reaction (or the substrates for the reverse reaction). The rate-limiting step for depsipeptides thus involves release of products, which may occur directly or through a mandatory conformational change followed by rapid product release.

The marked difference in chemical state of the intermediates of peptide and ester hydrolysis suggests a plausible explanation for a puzzling feature of the zinc protease-catalyzed hydrolysis of amide and ester bonds. The best studied zinc proteases—carboxypeptidase (17), thermolysin (18), and angiotensin-converting enzyme (19)—all display an equal or greater catalytic efficiency toward peptides than toward the corresponding ester analogues. This enzymatic leveling of activity is not observed with serine proteases nor is it expected, as an ester bond is much more labile than an amide bond (20). The results can be rationalized, however, if product release is rate-limiting for the esters but not for amides. In this case the hydrolytic step for catalysis of esters would occur earlier in the pathway and thus could still be faster than cleavage of the amide bond, as it is easier to hydrolyze an ester bond than an amide bond.

The change in rate-determining step observed here for a metalloprotease differs completely in type from that observed for serine proteases. In the latter case it is generally believed that the rate-limiting step for peptides is formation of an acyl intermediate, and the rate-limiting step for esters is deacylation of such an intermediate. Hydrolysis of peptides by carboxypeptidase A does not proceed through nucleophilic catalysis resulting in a covalent acyl species as is observed in serine or cysteine proteases but, more likely, through general base catalysis of a metal-bound intermediate. This mechanism may be a general feature of zinc proteases, as studies on thermolysin (21) and β-lactamase (22, 23) also indicate that peptide hydrolysis involves a metal-bound intermediate and general base catalysis.

Implications of Cryospectrokinetic Studies

This cryospectrokinetic approach greatly expands the range of mechanistic questions that can be answered experimentally. Future studies of intermediates such as those observed here can examine directly a number of important issues regarding the detailed structure of transients of enzymes which contain suitable probe characteristics due either to intrinsic inorganic or organic constituents or to the introduction of extrinsic ones by chemical modification of the protein and/or substrate. The coupling of this approach to molecular biology techniques should ultimately allow identification of which amino acid side chain groups interact with the substrate at each step of catalysis.

References

1. Auld, D. S. (1979) *Methods Enzymol.* 61:318–335.
2. Hanahan, D., and Auld, D. S. (1980) *Anal. Biochem.* 108:86–95.
3. Geoghegan, K. F., Galdes, A., Martinelli, R. A., Auld, D. S., and Vallee, B. L. (1983) *Biochemistry* 22:2255–2262.
4. Vallee, B. L., Galdes, A., Auld, D. S., and Riordan J. F. (1984) In *Metal Ions in Biology: Zinc Enzymes, vol. 5*, edited by T. G. Spiro, pp. 25–75. Wiley, New York.
5. Lobb, R. R., and Auld, D. S. (1984) *Experientia* 40:1197–1206.
6. Galdes, A., Auld, D. S., and Vallee, B. L. (1983) *Biochemistry* 22:1888–1893.
7. Auld, D. S., Galdes, A., Geoghegan, K. F., Holmquist, B., Martinelli, R. A., and Vallee, B. L. (1984) *Proc. Natl. Acad. Sci. USA* 82:5041–5045.
8. Lobb, R. R., and Auld, D. S. (1979) *Proc. Natl. Acad. Sci. USA* 76:2684–2688.
9. Fersht, A. (1985) In *Enzyme Structure and Mechanism*, 2nd Ed., pp. 193–220, W. H. Freeman, San Francisco.

10. Breslow, R., and Wernick, D. L. (1977) *Proc. Natl. Acad. Sci. USA* 74:1303–1307.
11. Galdes, A., Auld, D. S., and Vallee, B. L. (1986) *Biochemistry* 25:646–651.
12. Makinen, M. K., Wells, G. B., and Kang, S. (1984) *Adv. Inorg. Biochem.* 6:1–101.
13. Hoffman, S. J., Chu, S. S-T., Li, H., Kaiser, E. T., and Corey, P. J. (1983) *J. Am. Chem. Soc.* 105:105–107.
14. Suh, J., Cho, W., and Chung, S. (1985) *J. Am. Chem. Soc.* 107:4530–4535.
15. Kogan, R. L., and Fife, T. H. (1985) *J. Org. Chem.* 49:5229–5232.
16. Geoghegan, K. F., Galdes, A., Hanson, G., Holmquist, B., Auld, D. S., and Vallee, B. L. (1986) *Biochemistry* 25:4669–4674.
17. Auld, D. S., and Holmquist, B. (1974) *Biochemistry* 13:4355–4361.
18. Holmquist, B., and Vallee, B. L. (1976) *Biochemistry* 15:101–107.
19. Keung, W. M., Holmquist, B., and Riordan, J. F. (1980) *Biochem. Biophys. Res. Commun.* 96:506–513.
20. Bruice, T. C., and Benkovic, S. J. (1966) *Bioorganic Mechanisms*, pp. 1–119. W. A. Benjamin, New York.
21. Hangauer, D. G., Monzingo, A. F., and Matthews, B. W. (1984) *Biochemistry* 23:5730–5741.
22. Bicknell, R., and Waley, S. G. (1985) *Biochemistry* 24:6870–6887.
23. Bicknell, R., Schäffer, A., Waley, S. G., and Auld, D. S. (1986) *Biochemistry* 25:7208–7215.

9
Conformational Dynamics in RNA–Protein Interactions: Immobilization of the Functional Domains in tRNA^{fMet} and Methionyl-tRNA Synthetase

DAVID C. H. YANG and BLAIR Q. FERGUSON

RNA–protein interactions have been studied extensively. Conformational changes in both nucleic acids and proteins have been demonstrated and may play critical roles in the recognition processes. Conformational flexibility in proteins and its functional significance in some proteins have been elucidated (1). Relatively little is known about the segmental movement in RNA. Specific interactions of tRNA and cognate synthetases have provided model systems that have contributed to our understanding of the mechanisms in protein–RNA interactions (2). One of the particularly advantageous systems is the interaction of tRNA^{fMet} and methionyl-tRNA synthetase (MetRS), as the crystal structures of both tRNA^{fMet} (3) and an active fragment of MetRS (4) have been determined. Although the crystal structure of the tRNA–synthetase complex is yet to be resolved, their interaction has been carefully examined using a variety of approaches including photo-cross-linking (5), chemical modification (6), base substitution (7), fluorescence spectroscopy (8), neutron scattering (9), and affinity chemical modification (10). These studies have provided extensive information on the recognition sites in the tRNA molecule. Furthermore, these studies suggest that conformational changes in interacting tRNA and synthetase are essential in the recognition process. Little is known about the nature of the conformational changes in tRNA and the cognate synthetase upon their association. In this chapter we summarize the results of our studies on the conformational changes in tRNA^{fMet} and MetRS upon their association.

Immobilization of the 3′-Terminus in tRNA^{fMet} by MetRS

In the nucleotide sequence of tRNA^{fMet}, the 3′-terminus, 4-thiouridine (S8), and dihydrouridine (D20) can be specifically modified with fluorescent probes (11). Dansyl hydrazine and fluoresceinyl thiosemicarbazide react specifically with the periodate-oxidized $tRNA^{fMet}$ at the 3′-terminus, and the resultant single labeled $tRNA^{fMet}$ are referred to as $tRNA^{fMet}$ (3′-DNS) and $tRNA^{fMet}$ (3′-Flc), respectively. Proflavin (PF) and ethidium (Etd) specifically replace D20 in $tRNA^{fMet}$ and give $tRNA^{fMet}$ (D-PF) and $tRNA^{fMet}$ (D-Etd), respectively. Irradiation of $tRNA^{fMet}$ at 335 nm followed by $NaBH_4$ reduction yields a highly fluorescent photo-cross-linked product at S8 and C13 in $tRNA^{fMet}$, and this labeled $tRNA^{fMet}$ is designated $tRNA^{fMet}$ (8–13). The rotational freedom of the fluorophores at these locations in $tRNA^{fMet}$ was determined from the slope of the corresponding Perrin plots (1/anisotropy versus 1/viscosity) and respective fluorescence lifetimes. $tRNA^{fMet}$ (D-PF), $tRNA^{fMet}$ (D-Etd), and $tRNA^{fMet}$ (8–13) show rotational correlation times close to that of the entire tRNA molecule (26 ns), indicating little rotational freedom of the fluorophores at these locations. In contrast, the fluorophores at the 3′-terminus in $tRNA^{fMet}$ (3′-DNS) and $tRNA^{fMet}$ (3′-Flc) show high mobility. The fluorophore in $tRNA^{fMet}$ (3′-Flc) is completely immobilized upon binding of MetRS. $tRNA^{fMet}$ (3′-DNS) also shows a similar drastic increase in the steady-state anisotropy of 3′-DNS in the tRNA–synthetase complex. Furthermore, the immobilization of the 3′-terminus is accompanied by a drastic decrease in its solvent accessibility and an increase in the hydrophobicity after the binding of MetRS. Only slight changes in the fluorescence properties of the fluorophores at the 3′-terminus were observed when $tRNA^{Phe}$ (3′-Flc), $tRNA^{Phe}$ (3′-DNS), or $tRNA^{Glu}$ (3′-Flc) was used.

Significant but relatively minor changes in the rotational freedom, solvent accessibility, and polarity were observed for the fluorophores at D20 and 8–13 in the fluorescent-labeled $tRNA^{fMet}$ upon binding of MetRS, suggesting that MetRS induced delocalized conformational changes in the tRNA molecule (11).

Bending of the 3′-Terminal Arm in $tRNA^{fMet}$

Singlet–singlet energy transfer provides a sensitive spectroscopic ruler for studies of structures of biological macromolecules (14). Three species of $tRNA^{fMet}$, each labeled with an energy donor and an energy acceptor, were prepared by sequential chemical modifications (13). The MetRS binding affinity of these double-labeled $tRNA^{fMet}$ species is similar to those of unmodified $tRNA^{fMet}$. The fluorescence properties of the individual fluorophore in these $tRNA^{fMet}$ species, including excitation and emission spectra, anisotropy, and fluorescence quenching by MetRS are similar to those of single-labeled $tRNA^{fMet}$. The transfer efficiencies of the double-labeled $tRNA^{fMet}$ species, as determined by both donor quenching and sensitized emission, showed efficient energy transfer in all cases. Assuming random relative orientation of the donor and the acceptor, the corresponding apparent distances are 25 Å between 8–13 and D-20, 44 Å between 8–13 and the 3′-terminus, and 49 Å between the 3′-terminus and D20, respectively, in free $tRNA^{fMet}$. Upon binding of MetRS, the apparent distances are 25 Å between 8–13 and D-20, 45 Å between 8–13 and 3′-terminus, and 54 Å between 3′-terminus and D-20, respectively. It appears that the overall conformation of the bulk $tRNA^{fMet}$ is similar before and after the binding of MetRS, except for an appreciable increase in the distance between 3′-Flc and D-20. Comparison of these fluorescence-derived apparent distances to the crystallographic structure of $tRNA^{Phe}$ (15, 16) suggests that the 3′-terminus in the tRNA–synthetase complex is oriented away from D-20 without changing its distances to 8–13 and bends toward the anticondon loop, in contrast to the continuing helix in the crystal structure of free $tRNA^{Phe}$. The 3′-terminus in $tRNA^{fMet}$ is thus shielded from the solvent molecules and immobilized in a conformation that is bent toward the anticondon loop upon binding of MetRS.

Cibacron Blue Displaces Synthetase-Bound $tRNA^{fMet}$

The fluorescent groups at the 3′-terminus in the $tRNA^{fMet}$–MetRS complex are likely surrounded by a hydrophobic pocket near the active site of MetRS as suggested by the changes in the fluorescence properties. This hydrophobic site may be comprised within the dinucleotide binding fold (Rossman fold) observed in the crystal structure of trypsin-modified MetRS (4). Part of the dinucleotide binding fold has been tentatively assigned to be the binding site of 8-Br-ATP (4). Cibacron blue F3GA is known to bind with additional hydrophobic interactions to the dinucleotide binding fold of many proteins (17). Cibacron blue binds MetRS in the same manner as $tRNA^{fMet}$, as monitored by the quenching of intrinsic tryptophan or extrinsic AEDANS fluorescence of MetRS (see below). The binding of not more than two Cibacron blue molecules to MetRS is sufficient to displace MetRS-bound $tRNA^{fMet}$ (3′-Flc), as shown by the complete recovery of the quenched fluorescence of fluorescein in $tRNA^{fMet}$ (3′-Flc). The displaced $tRNA^{fMet}$ (3′-Flc) showed the same anisotropy as that of free $tRNA^{fMet}$ (3′-Flc) and did not show any energy transfer from the tryptophan residues in the synthetase to 3′-Flc. These results indicate that $tRNA^{fMet}$ (3′-Flc) is completely displaced from MetRS by Cibacron blue alone. These observations suggest an overlapping site for ATP and $tRNA^{fMet}$ with Cibacron blue in MetRS, and that the binding of $tRNA^{fMet}$ to MetRS likely requires the proper binding of the 3′-terminus of the tRNA molecule.

Conformational Transition in MetRS

Upon binding of $tRNA^{fMet}$, the intrinsic fluorescence of MetRS is quenched nearly 50% (8). Neutron scattering studies showed that the radii of gyration of free and tRNA-bound MetRS are 44.7 and 43.9 Å, respectively

(9). Further conformational studies were carried out using fluorescence-labeled MetRS (12). IAEDANS reacts specifically at a site within the intersubunit domain at the C-terminus of MetRS (130 amino acid residues). The fluorescence-labeled MetRS is fully active in the aminoacylation of tRNA. The emission properties of AEDANS-labeled MetRS are consistent with the following model. AEDANS in the tRNA-binding subunit in AEDANS-MetRS is statically quenched upon binding of $tRNA^{fMet}$, and the residual AEDANS fluorescence from the tRNA-free subunit showed an appreciable decrease in solvent accessibility. The unsymmetrical conformational changes in the two subunits are in accord with the known negative cooperativity in the tRNA binding by MetRS. Because removal of the intersubunit domain converts MetRS to monomeric enzyme and does not impair the aminoacylation activity (18), the observed conformational changes at the AEDANS site likely result indirectly from the binding of tRNA to MetRS. Further conformational changes were demonstrated by anisotropy measurements. The free and tRNA-bound MetRS showed a rotational correlation time of 62 and 80 ns, respectively. The increase in the rotational correlation time, in contrast to the decrease in the radius of gyration in MetRS binding to tRNA, can be best interpreted as tightening of a flexible hinge at the intersubunit domain in MetRS upon association of $tRNA^{fMet}$ (13).

Discussion

Nucleic acid–protein recognition, such as that exhibited by tRNA and cognate aminoacyl-tRNA synthetase, likely involves precisely controlled steps, as illustrated by the various conformational changes we have detected in interacting $tRNA^{fMet}$ and MetRS. Figure 9.1 summarizes the observed changes. In $tRNA^{fMet}$ the 3′-terminus of the amino acid acceptor stem is

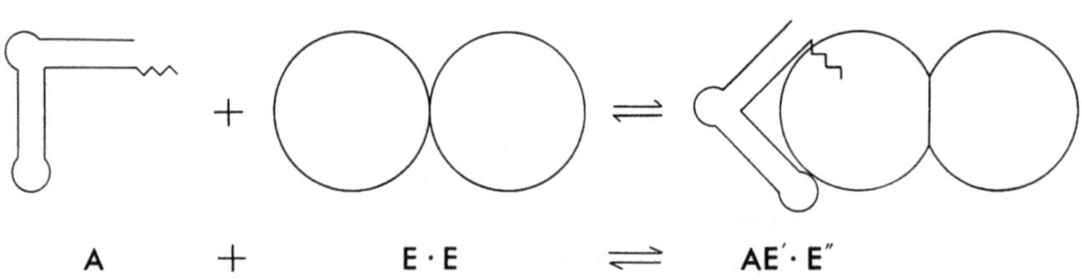

FIG. 9.1. Schematic representation of the conformational changes in the interaction of $tRNA^{fMet}$ and methionyl-tRNA synthetase. *A* denotes the tRNA molecule, and *E*, *E*′, and *E*″ represent different conformational states of the subunits of methionyl-tRNA synthetase. The inner loop of tRNA is oriented toward the enzyme, and the 3′-arm bends toward the hydrophobic pocket in the active site of methionyl-tRNA synthetase. Methionyl-tRNA synthetase shows strong negative cooperativity in tRNA binding and exhibits unsymmetric conformational changes in the two subunits.

immobilized and bent; it binds to a hydrophobic pocket near the active site in MetRS, accompanied by delocalized conformational changes in $tRNA^{fMet}$, although the overall conformation is maintained. The displacement of the synthetase-bound $tRNA^{fMet}$ by Cibacron blue F3GA suggests that the stable association of tRNA and synthetase requires the binding of the 3′-terminus of $tRNA^{fMet}$. It follows that either the bulk of the tRNA molecule or the 3′-terminus of $tRNA^{fMet}$ triggers the conformational changes in MetRS, and the new conformation subsequently binds the rest of the tRNA molecule to form a stable complex with concomitant conformational changes in $tRNA^{fMet}$. In MetRS the two subunits are joined more tightly together and show unsymmetrical conformational changes upon binding of $tRNA^{fMet}$. These conformational changes are likely to be part of the structural mechanisms for the recognition of tRNA by the synthetase and for the observed negative cooperativity in tRNA binding.

In addition to the possible role in the recognition process, the conformational flexibility in RNA and protein may have other functional roles. It is conceivable that flexibility in single-stranded RNA may effectively enhance the size of the target site in the nucleic acid–protein association by reducing the orientational constraints and may thus enhance the rate constants in diffusion-limited reactions. Furthermore, the immobilization of the flexible domain in RNA upon binding of an enzyme may favor subsequent formation of the transition-state complex due to the entropy loss in the association reaction.

The observed conformational changes in $tRNA^{fMet}$ and MetRS may be a general characteristic in tRNA–synthetase interactions. Further elucidation of the mechanism of the mutual adaptation and the recognition process should provide the underlying principles in nucleic acid–protein interactions.

Acknowledgment. This work was supported by NIH grant GM-25848 and NSF grant 81-10818.

References

1. Hubber, R. (1979) *Trends Biochem. Sci.* 4:271–276.
2. Schimmel, P. R., and Soll, D. G. (1979) *Annu. Rev. Biochem.* 48:601–648.
3. Woo, N. H., Roe, B. A., and Rich, A. (1980) *Nature* 286:346–351.
4. Risler, J. L., Zelwer, C., and Brunie, S. (1981) *Nature* 292:384–386.
5. Ackerman, E. J., Joachimiak, A., Klinghofer, V., and Sigler, P. (1985) *J. Mol. Biol.* 181:93–102.
6. Schulman, L. H., and Pelka, H. (1977) *Biochemistry* 16:4256–4265.
7. Uemura, H., Imai, M., Ohtsuka, E., Ikahara, M., and Soll, D. (1982) *Nucleic Acids Res.* 10:6531–6539.
8. Blanquet, S., Iwatsubo, M., and Waller, J. P. (1973) *Eur. J. Biochem.* 36:213–226.

9. Dessen, P., Fayat, G., Zaccai, G., and Blanquet, S. (1982) *J. Mol. Biol.* 154: 603–613.
10. Hountondji, C., Blanquet, S., and Lederer, F. (1985) *Biochemistry* 24:1175–1180.
11. Ferguson, B. Q., and Yang, D. C. H. (1986) *Biochemistry* 25:529–539.
12. Ferguson, B. Q., and Yang, D. C. H. (1986) *Biochemistry* 25:2743–2748.
13. Ferguson, B. Q., and Yang, D. C. H. (1986) *Biochemistry* 25:6572–6578.
14. Stryer, L. (1978) *Annu. Rev. Biochem.* 47:819–846.
15. Kim, S. H., Suddath, F. L., Quigley, G. J., McPherson, A., Susmann, J. L., Wang, A., Seeman, N. C., and Rich, A. (1974) *Science* 185:435–440.
16. Robertus, J. D., Ladner, J. E., Finch, J. T., Rhodes, D., Brown, R. S., Clark, B. F. C., and Klug, A. (1974) *Nature* 250:546–551.
17. Beissner, R. S., Quiocho, F. A., and Rudolph, F. B. (1979) *J. Mol. Biol.* 134: 851–853.
18. Cassio, D., and Waller, J. P. (1971) *Eur. J. Biochem.* 20:283–300.

10
Structural and Kinetic Studies of *Escherichia coli* Glutamine Synthetase

J. J. VILLAFRANCA, C. D. EADS, R. LOBRUTTO, F. C. WEDLER, and J. COLANDUONI

Glutamine synthetase catalyzes the synthesis of glutamine from glutamate, ammonia, and ATP. The enzyme consists of 12 identical subunits, each of MW ~50,000, arranged as two hexameric rings stacked face to face to form the catalytically active enzyme (1). The enzyme has an absolute requirement for two divalent metal ions per subunit for catalysis. The two types of metal ion binding site can be distinguished based on their metal ion affinities, their effects on the properties of the enzyme, and their interactions with substrates. The high-affinity site, designated the n_1 site, is near the substrate glutamate binding site (2, 3). The low-affinity site, designated the n_2 site, is involved in binding the metal–ATP complex (4).

The enzymatic reaction is thought to proceed by a two-step mechanism (5–8). In the first step, nucleophilic attack on the terminal phosphate of ATP by glutamate produces γ-glutamylphosphate (9). In the second step, displacement of the phosphate by ammonia to yield the product glutamine proceeds via formation of a tetrahedral adduct. Methionine sulfoximine (MSOX), which is tetrahedral at the critical position, behaves as a transition state analogue (10). MSOX has been used in numerous biophysical investigations aimed at discerning the role of the n_1 site-bound metal ion in the catalytic mechanism (2, 3, 11–16). The L,S-isomer of MSOX reacts with ATP to form methionine sulfoximine phosphate (11, 17). The enzyme–metal–MSOX phosphate–ADP adduct thus formed is inactive, as these molecules do not dissociate from the enzyme, even after extensive dialysis (17, 18). Only when the pH is lowered can MSOX-P and ADP be released from the enzyme, resulting in reactivation of the enzyme (18).

Using electron paramagnetic resonance (EPR) spectroscopy, it has been established that the binding of racemic MSOX to glutamine synthetase (with Mn^{2+} at the n_1 site) results in an enzyme complex with Mn^{2+} in an environment of nearly octahedral symmetry (3). Water proton relaxation rate studies indicate that MSOX binding reduces the solvent water accessibility of the metal ion (19), and methionine sulfoximine phosphate formation entirely precludes the free exchange of water molecules from the n_1 site (3, 16). These changes in the metal ion environment suggest that the metal ion may be directly involved in the catalytic mechanism. However, the nature of this involvement has not been established.

Studies of the role of the n_1 site-bound metal ion were extended by examining in detail the mode of interaction of solvent water and of the resolved L,S- and L,R-isomers of MSOX with Mn^{2+} bound to this site. Continuous-wave Mn^{2+} EPR spectroscopy was used to determine qualitative differences in the interactions of these isomers with the metal ion. Electron spin echo envelope modulation spectroscopy was used to determine the proximity of MSOX to the metal ion. The results of these experiments gave detailed information on the environment of the n_1 site-bound metal ion and suggested a role for this ion in catalysis. Also, kinetic studies with the transition-state analogue phosphinothricin (PPT) demonstrate that this compound inactivates the enzyme in a manner similar to that of MSOX.

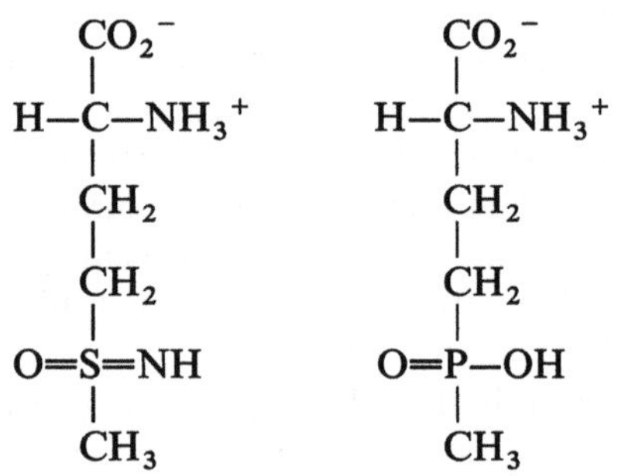

MSOX **PTT**

Materials and Methods

Glutamine synthetase in a low state of adenylylation (1.1–2.5 adenylylated subunits per dodecamer) was prepared by the method of Miller et al. (20). Enzyme subunit concentration and state of adenylylation were determined spectrophotometrically (21). The buffer used for all experiments was 10 m*M* Hepes, 100 m*M* KCl, pH 7.0. Metal-free enzyme was prepared by dialysis against buffer containing 3 m*M* EDTA followed by extensive dialysis against buffer alone.

L-Methionine-(S,R)-sulfoximine and L-methionine sulfone was obtained from Sigma. Phosphinothricin was a gift from Hoechst Chemical Co. All other reagents were obtained from standard chemical supply houses and were used without further purification.

Continuous-wave EPR spectra were recorded using a Varian E-12 Spectrometer at 9.1 gHz. Pulsed EPR spectroscopy was carried out at the University of Pennsylvania.

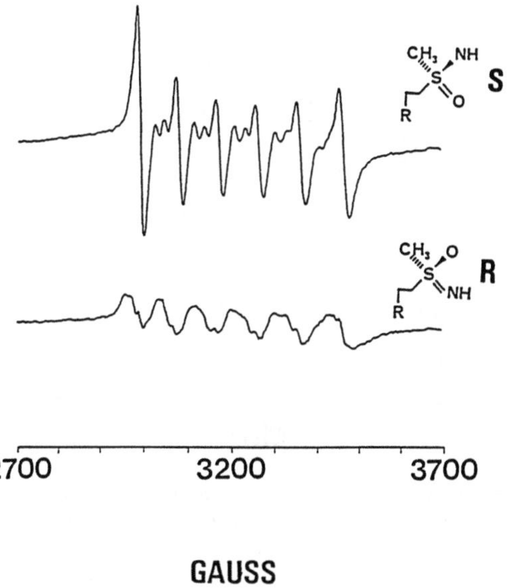

FIG. 10.1. EPR spectra of glutamine synthetase–Mn^{2+} complex with pure L,S-MSOX and L,R-MSOX. Data were obtained at 9 gHz.

Results and Discussion

Continuous Wave EPR Experiments

The continuous wave EPR spectra of glutamine synthetase in the presence of L,S-MSOX and L,R-MSOX at room temperature are given in Fig. 10.1. The spectra obtained with the L,S-MSOX isomer are identical to the spectra obtained using racemic mixtures of the isomers (3), demonstrating that the L,S-MSOX isomer binds tightly enough to exclude the other isomer under these conditions. Therefore, as discussed previously for spectra obtained using racemic MSOX (3), the spectra of the L,S-MSOX isomer are characteristic of an Mn^{2+} environment of high symmetry and low solvent accessibility, resulting in sharp lines caused by a small zero field splitting (ZFS).

The spectrum found with the L,R-MSOX isomer can be rationalized in terms of an environment of lower symmetry than for L,S-MSOX that also may have altered solvent accessibility. This situation results in higher values of the ZFS and "broadened" lines. Based on the appearance of these spectra it is clear that the two isomers of MSOX interact with the metal ion differently.

Spin-Echo EPR Experiments

Electron–nuclear couplings that are much less than the EPR line width are difficult to detect using conventional EPR techniques. However, such weak

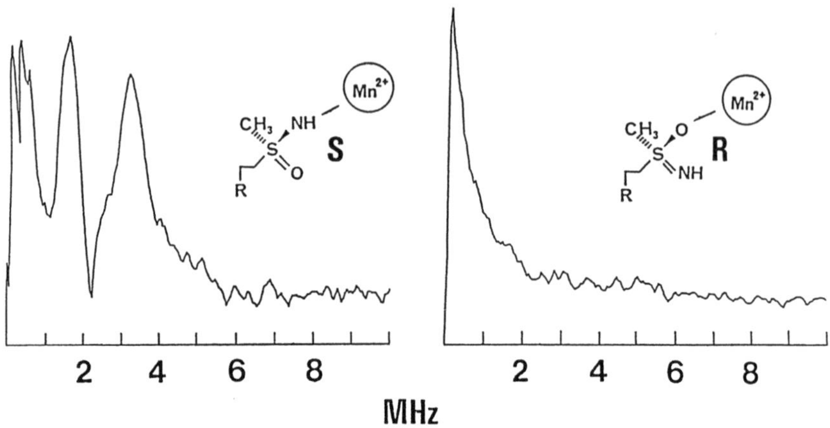

FIG. 10.2. Fourier-transformed electron spin-echo modulation spectra of glutamine synthetase–Mn^{2+} complex with pure L,S-MSOX and L,R-MSOX.

coupling can be detected by the electron spin echo envelope modulation technique (22, 23). With this technique a series of microwave pulses are applied to a sample at low (liquid helium) temperature in order to induce a long-lived electron spin-echo. A plot of the echo intensity as a function of the time between pulses is called the electron spin-echo envelope, which decays exponentially. Often the electron spin-echo envelope is modulated by one or more frequencies related to the frequencies of coupled nuclear spins, thereby giving structural information.

As Fig. 10.2 shows, the Fourier power transforms of the spin-echo spectra of glutamine synthetase-bound Mn^{2+} in the presence of resolved, L,S-MSOX and L,R-MSOX, respectively. The peaks at < 1 mHz are not reproducible and are probably artifacts of the data analysis. The peaks in Fig. 10.2 at 1.7 and 3.3 mHz can be assigned to superhyperfine coupling with the imine nitrogen of MSOX. To account for the possibility of the peaks arising from coupling to ligands donated by the protein, a control was run using L-methionine sulfone, which differs from the sulfoximine by replacement of the imine nitrogen by oxygen. No peaks at 1.7 or 3.3 mHz were detected with this sample or with another transition state analogue, 2-amino-4-phosphonobutyric acid. The instrument used for this experiment is limited to detection of coupling frequencies in the approximate range of 1 to 10 mHz by its pulse power and resolution. Therefore the lack of coupling seen in the data for the L,R-MSOX isomer could indicate either an absence of coupling or coupling outside the range of detectability.

For uncharacterized systems such as the present one, structural interpretation of these data is best carried out with reference to model complexes. LoBrutto and Reed (personal communication) have carried out electron-spin-echo studies on systems believed to have nitrogen ligands directly coordinated to Mn^{2+}, including thiocyanide, certain complexes of creatine kinase, and complexes of concanavalin A (ConA). In these cases peaks in the Fourier transforms of the electron spin-echo envelopes are detected in the range found for the glutamine synthetase–Mn^{2+}–(L,S)-MSOX complex reported herein. Thus the detection of superhyperfine coupling for the L,S-MSOX isomer can be taken as evidence that the imine nitrogen of this transition-state analogue resides in the inner coordination sphere of the enzyme-bound metal ion.

With regard to the ligation scheme of the metal ion in the presence of the L,R-MSOX isomer, at least two possibilities exist. First, this isomer may bind to the Mn^{2+} with oxygen as a ligand, or it may bind to the enzyme in the outer coordination sphere of the metal ion. This situation has been observed for the glutamate analogue 2-amino-4-phosphonobutyric acid. The tetrahedral phosphorus atom of this compound was found to be 5 to 8 Å distant from the metal ion as determined by nuclear magnetic resonance (NMR) relaxation measurements (19).

Regardless of the identity of the ligand to the metal ion, the observation that L,S-MSOX lies in the inner coordination sphere of enzyme-bound Mn^{2+} has an interesting implication concerning the role of the n_1-site-bound metal ion in the catalytic mechanism of glutamine synthetase. To the extent that MSOX is a good analogue of the putative tetrahedral transition state, the above result suggests that one of the catalytic roles of the metal ion is to stabilize the negative charge that forms on the carbonyl oxygen during the attack of ammonia on γ-glutamyl phosphate to produce the tetrahedral adduct. Thus the glutamyl phosphate intermediate would be coordinated to the metal ion, resulting in polarization of the carbonyl oxygen, facilitating nucleophilic attack by ammonia. The tetrahedral adduct is shown in Fig. 10.3. The major difference in the initial phosphorylation step when MSOX is present versus the normal substrate glutamate is that coordination of the imine nitrogen of MSOX to the metal ion as shown in Fig. 10.3 would facilitate its attack on the γ-P of ATP. This coordination may not be essential for the sp^2 hybridized γ-carboxyl of glutamate in the "normal" catalytic mechanism.

Inactivation of Glutamine Synthetase by PPT and MSOX

When glutamine synthetase was incubated with MSOX and ATP, a time-dependent inhibition was observed. The inactivation followed saturation kinetics, and the data were fit to the following model.

$$E + I \underset{}{\overset{K_a}{\rightleftharpoons}} EI \xrightarrow{k_i} EI^*; \quad 1/k_{obs} = (K_a/k_i)(1/[I]) + 1/k_i$$

The k_i was 0.012 s^{-1} at pH 7.0 and $K_a = 1.0\ \mu M$. With PPT and ATP together, a time-dependent inactivation of the enzyme was observed. The rate of inactivation was followed at extremely low PPT concentrations ($<20\ \mu M$), as the inactivation was extremely fast. With 10 μM PPT and 1.3 μM glutamine synthetase, the observed rate constant for inactivation, $k_{obs'}$, was 0.025 s^{-1}.

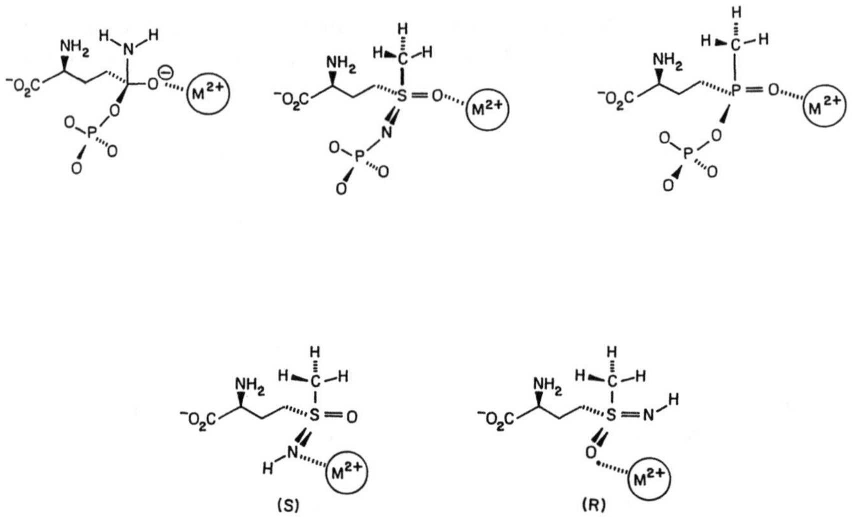

FIG. 10.3. Structures for various complexes found by glutamine synthetase, substrates, or substrate analogues. *Top left*: Tetrahedral adduct in the "normal" reaction. *Top middle*: Phosphorylated MSOX adduct. *Top right*: Phosphorylated PPT adduct. *Bottom left*: L,S-MSOX–Mn^{2+} complex. *Bottom right*: L,R-MSOX–Mn^{2+} complex.

The same mechanism as above was used to analyze the data. Assuming the K_{IS} value is equivalent to the K_a value, a k_i value of 0.10 s^{-1} was determined. Thus PPT is approximately an eightfold better inhibitor of glutamine synthetase at saturating levels of inhibitor. In control experiments, PPT alone does not inhibit glutamine synthetase, and phosphinothricin plus AMPPCP showed no inactivation of the enzyme.

Summary

The inactivation of *Escherichia coli* glutamine synthetase by MSOX and PPT proceeds via the formation of a reversible Michaelis complex prior to the formation of the "irreversible" inactive complex. The latter process requires ATP and the phosphorylation of MSOX or PPT. The binding constants for MSOX and PPT are approximately 1 and 18 μM, respectively, from steady-state experiments. Under similar conditions, glutamate binds to the enzyme with a K_D of approximately 3 mM in the presence of ATP. Because PPT is an amino acid with a tetrahedral phosphorus atom at the γ position, tight binding of this molecule to the enzyme is attributed to it being a mimic of the tetrahedral adduct formed in the normal reaction with glutamate, ATP, and NH_3 (Fig. 10.3). More interesting is the fact that this inhibitor behaves like MSOX in that it forms an inactive complex with the enzyme and ATP.

Acknowledgment. This research was supported in part by NIH grants GM 23529 (J. J. V.) and GM 26582 (F. C. W.), and NSF grant PCM-8409737 (J. J. V.).

References

1. Stadtman, E. R., and Ginsburg, A. (1974) In *The Enzymes*, Vol. 10, 3rd ed., edited by P. D. Boyer, pp. 755–807.
2. Villafranca, J. J., Ash, D. E., and Wedler, F. C. (1976) *Biochemistry* 15:536–543.
3. Villafranca, J. J., Ash, D. E., and Wedler, F. C. (1976) *Biochemistry* 15:544–553.
4. Hunt, J. B., Symyrnoitis, P. Z., Ginsburg, A., and Stadtman, E. R. (1975) *Arch. Biochem. Biophys.* 166:102–124.
5. Clark, D. D., and Villafranca, J. J. (1985) *Biochemistry* 24:5147–5152.
6. Meek, T. D., Johnson, K. A., and Villafranca, J. J. (1982) *Biochemistry* 21:2158–2167.
7. Midelfort, C. F., and Rose, I. A. (1976) *J. Biol. Chem.* 251:5881–5887.
8. Rhee, S. G., and Chock, P. B. (1976) *Proc. Natl. Acad. Sci. USA* 73:476–480.
9. Krishnaswamy, P. R., Pamiljans, V., and Meister, A. (1962) *J. Biol. Chem.* 237:2932–2940.
10. Manning, J. M., Moore, S., Rowe, W. B., and Meister, A. (1969) *Biochemistry* 8:2681–2685.
11. Shrake, A., Whitley, E. J., Jr., and Ginsburg, A. (1980) *J. Biol. Chem.* 255:581–589.
12. Gibbs, E. J., Ransom, S. C., Cuppett, S., and Villafranca, J. J. (1984) *Biochem. Biophys. Res. Commun.* 120:939–945.
13. Wedler, F. C., Sugiyama, Y., and Fisher, K. E. (1982) *Biochemistry* 21:2168–2177.
14. Hunt, J. B., and Ginsburg, A. (1980) *J. Biol. Chem.* 255:590–594.
15. Balakrishnan, M. S., and Villafranca, J. J. (1978) *Biochemistry* 17:3531–3538.
16. Ransom, S. C. (1984) PhD thesis, Pennsylvania State University.
17. Rowe, W. B., Ronzio, R. A., and Meister, A. (1969) *Biochemistry* 8:2674–2680.
18. Maurizi, M. R., and Ginsburg, A. (1982) *J. Biol. Chem.* 257:4271–4278.
19. Eads, C. D. (1985) PhD thesis, Pennsylvania State University.
20. Miller, R. E., Shelton, E., and Stadtman, E. R. (1974) *Arch. Biochem. Biophys.* 163:155–171.
21. Shapiro, B. M., and Stadtman, E. R. (1970) *Methods Enzymol.* 17A:910–922.
22. Rowan, L. G., Hahn, E. C., and Mims, W. B. (1965) *Phys. Rev.* 137:A61.
23. Kevan, L. (1979) In *Time Domain Electron Spin Resonance*, edited by L. Kevan and R. N. Schwarts, p. 279. Wiley, New York.

11
Cu,Zn Superoxide Dismutase: A Case of Metalloenzyme Catalysis in Which the Protein Moiety Plays a Major Role

ADELIO RIGO, LILIA CALABRESE, and GIUSEPPE ROTILIO

The structure and catalytic mechanism of Cu,Zn superoxide dismutase (SOD) are known from studies of the bovine enzyme (for review see refs. 1, 2). This enzyme consists of two identical subunits of 16 kD, each containing a catalytically active copper ion and a zinc ion, which shares a common ligand—the imidazole of His 61—with the copper. It has been demonstrated (3, 4) that the enzyme-catalyzed superoxide ion dismutation proceeds by a cyclic oxidation-reduction mechanism involving the copper ion:

$$\mathrm{E\,Cu(II)} + \mathrm{O_2}^- \xrightarrow{k_1} \mathrm{E\,Cu(I)} + \mathrm{O_2} \qquad [1]$$

$$\mathrm{E\,Cu(I)} + \mathrm{O_2}^- + 2\mathrm{H}^+ \xrightarrow{k_2} \mathrm{E\,Cu(II)} + \mathrm{H_2O_2} \qquad [2]$$

Our previous pulse-radiolysis work (4) has demonstrated that the reactions between O_2^- and either reduced or oxidized forms of the enzyme are diffusion-limited and occur at the same rate of $\sim 2 \times 10^9\ M^{-1}\ \mathrm{s}^{-1}$ for each Cu site. No rate-limiting first-order process has been observed at the superoxide concentration accessible to pulse radiolysis. A Km $\simeq 0.4\ \mathrm{m}M$ could be measured by a polarographic measurement (5). However, all detailed kinetic studies are performed at $[O_2^-] \ll$ Km because of technical limitations. Under these conditions the rate-limiting process is the binding of substrate to the enzyme active site. Previous results showed that the binding rates are independent of pH in the range between 5.0 and 9.5 (6, 7) and are unaffected by D_2O (8), indicating that proton transfer is not a part of the binding process. This finding is in line with the identical, diffusion-controlled values of the rate constants measured for the individual redox reactions 1 and 2 (Eqs. 1 and 2). Under saturating conditions, which cannot be explored by current kinetic studies, proton transfer in Eq. 2 could become rate-limiting.

Most aspects of the Cu,Zn superoxide dismutase catalysis can be rationalized in terms of displacement by O_2^- of a copper-bound water molecule followed by rapid electron exchange between O_2^- and the copper. The metal ion would alternate between a five-coordinate Cu(II) state and four-coordinate, near to tetrahedral Cu(I) state, with little ligand rearrangement (1, 2). The proton source for the formation of the peroxide product in the reoxidation step would be the imidazolate nucleus of His 61 bridging the copper and the zinc ions (9). These details fit in well with a refined crystallographic model (10). However, some peculiar features of the enzymic reaction are not easily explained with a model based only on the chemical properties of the active site copper. First, why is O_2^- so selective in driving the redox cycle? Secondly, why does O_2^- bind at the same rate to both the oxidized and reduced forms of the enzyme although the structures of the Cu(II) and Cu(I) states of the enzyme are different? Finally, some work (11–15) has indicated that electrostatic interaction of O_2^- with positively charged amino acid residues near the active site may play an important role in superoxide dismutase catalysis. The present chapter describes in more quantitative terms such interactions and identifies the residue(s) believed to be responsible for catalysis.

Materials and Methods

The enzyme was isolated from bovine erythrocytes as previously described (16). The catalytic constant was determined at $O_2^- \simeq 2 \times 10^{-4}$ *M* by a polarographic method that allows work with a wide pH range (17). The steady-state concentration of reduced and oxidized enzyme was assayed by EPR as previously described (18). Room temperature liquid solution EPR spectra were recorded with an E-9 Varian instrument in the flat cell configuration.

Results

Anion Binding to Bovine Cu,Zn SOD

Anions are known to be inhibitors of Cu,Zn SOD (1,2). Several singly charged anions display a competitive pattern of inhibition (19). They are likely to compete with O_2^- for either the metal or some activity-linked positively charged side chains, or both. To differentiate between these two possibilities, whether a given anion affects the spectroscopic properties of the copper ion bound to the enzyme was tested. For example, the liquid solution EPR spectrum of the metal is typically modified into a more axial line shape by inhibitory anions entering the first coordination sphere of the enzyme-bound Cu(II) (20). A stability constant for E-Cu(II)–anion complexes was calculated from this parameter and compared with the apparent binding constant as determined by activity measurements (Table 11.1). In all cases the inhibition constant was higher than the constant of the binding process to E-Cu(II). The reasons for the discrepancy are explored further below.

TABLE 11.1. Affinity constants for binding of various anions to bovine Cu,Zn SOD.

	$K(M^{-1})$ at 20°C	
Anion	Activity ($[O_2^-] \ll Km$)	EPR
CN^{-a}	2.9×10^5	1×10^5
OH^{-b}	6×10^3	3×10^{2c}
ClO_4^{-a}	3	0^d
F^{-a}	3	0^d

[a] pH 9.8, 0.02 *M* borate.
[b] Different pH values, 0.0 2 *M* borate.
[c] The value is approximate because of the uncertainty to the endpoint due to protein denaturation beginning at pH ≃ 12.3.
[d] Frozen solution EPR spectra display axial transitions for these ions, which turn out to be an artifact of freezing (20).

TABLE 11.2. Oxidation state of the active site of 1×10^{-4} *M* bovine Cu,Zn SOD.[a]

Condition	[E-Cu(II)] / [E-Cu(II) + E-Cu(I)]	Activity (%)
Control	0.44	100
4×10^{-7} *M* OH^{-b}	0.48	100
3×10^{-3} *M* OH^{-c}	0.49	25
0.1 *M* F^{-b}	0.53	98
0.5 *M* F^{-b}	0.56	95
0.05 *M* ClO_4^{-b}	0.44	75
0.25 *M* ClO_4^{b}	0.43	30
Carbamoylated enzyme[b,d]	0.43	45

[a] Under conditions of turnover in the presence of a stationary flux of O_2^- produced by 5×10^{-3} *M* NADH + 2×10^{-5} *M* phenazine methosulfate (18).
[b] Measurement in 0.05 *M* morpholine-ethane-sulfonate buffer, pH 7.4.
[c] Measurements in 0.02 *M* borate buffer.
[d] See Table 11.3.

Effect of OH^- and ClO_4^- on the E-Cu(I)/E-Cu(II) Ratio at Steady-State

In view of the results reported in Table 11.1, it appears that OH^-, F^-, and ClO_4^- are particularly suited for studying those factors that influence the SOD activity in addition to the binding of O_2^- to the protein-bound Cu(II). Both F^- and ClO_4^- inhibit the activity even though they had no detectable effect on the spectral properties of E-Cu(II). Among the ions that can be detected to interact with Cu(II), OH^- produced the largest difference between the two constants as it affected the activity at a concentration much lower than the one that produces the spectral effects. Because the Cu(I) form of the enzyme cannot be directly investigated, a suitable approach to test possible interactions of anions with the reduced metal is to assay the percent fraction of E-Cu(II) under steady-state conditions. If binding to the lower oxidation form of the enzyme determined the inhibition constant in the presence of a given anion, the concentration of the anion required to obtain a 50% inhibition of the enzyme maybe different from that required to half-saturate the E-Cu(II). Table 11.2 shows that the E-Cu(II)/E-Cu(I) ratio remains constant with increasing either $[OH^-]$ or $[ClO_4^-]$. These data suggest that the anion is likely bound to group(s) of the protein moiety that are not affected by the oxidation state of the enzyme.

Dependence of k_{cat} at $[O_2^-] < Km$ on $[OH^-]$ and $[NaClO_4]$

Plots of k_{cat} versus $[OH^-]$ or $[NaClO_4]$ are shown in Figs. 11.1 and 11.2. In the former case (Fig. 11.1) the data can be fitted to a superimposing titration curve from two activity-linked basic side chains according to the equation

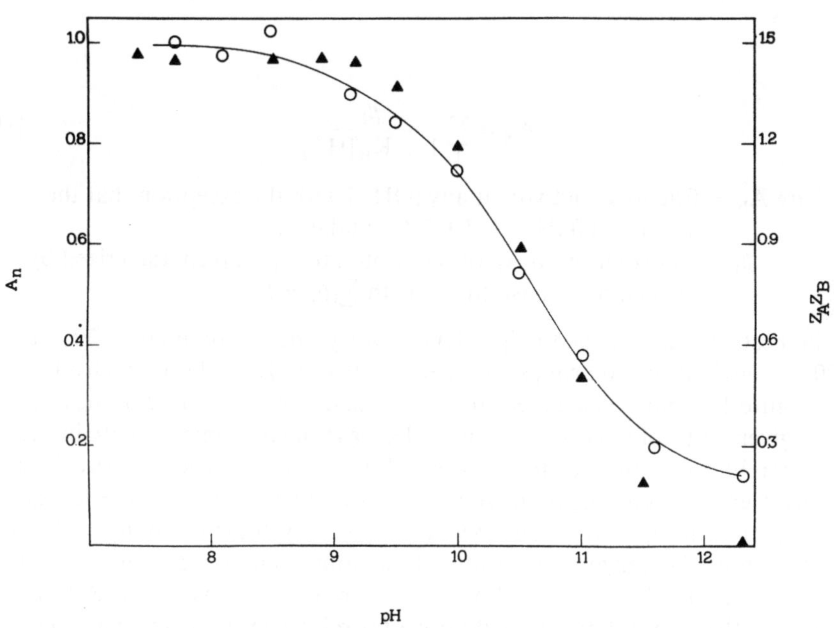

FIG. 11.1. The k_{cat} of bovine Cu,Zn SOD between pH 7.5 and 12.5. (○) 0.02 *M* Tris, borate, or *n*-propylamine buffer, 0.02 *M* perchlorate. Best fit of the points (continuous line) according to Eq. 3. (▲) Average protein charge, calculated as described in the text. A_n = normalized value of k_{cat}.

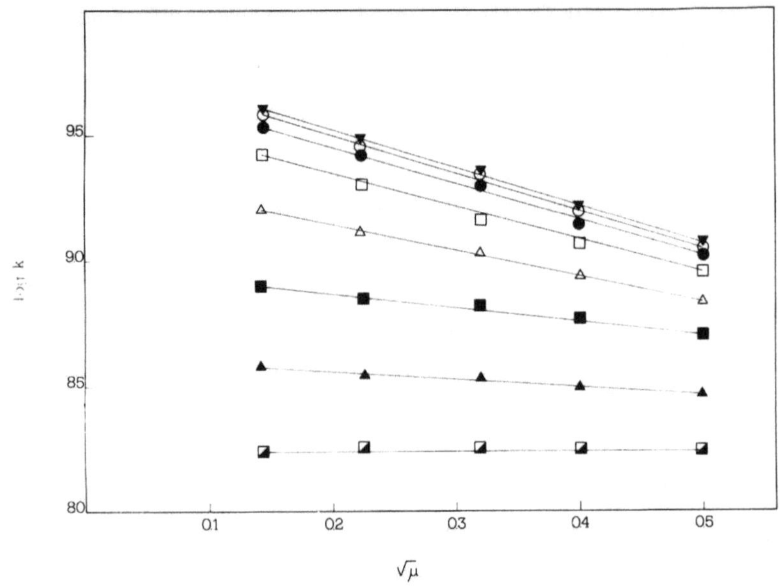

FIG. 11.2. Dependence of SOD activity on ionic strength ($NaClO_4^-$). (◪) pH 12.3. (▲) pH 11.5. (■) pH 11.0. (△) pH 10.5. (□) pH 10.0. (●) pH 9.5. (○) pH 9.0. (▼) pH 7.5, 8.0, 8.5.

$$A_N = \sum_i \frac{\beta_i}{1 + K_i/[H^+]} \qquad [3]$$

where A_N = fractional activity at any $[OH^-]$ with the exception that the activity at $[OH^-] \leqslant 10^{-6}$ *M* is taken as 1

β_i = contribution to A_N of the protonated group characterized by a dissociation constant K_i, with $\sum_i \beta_i = 1$

The best fit was obtained (Fig. 11.1) by assigning approximately 30% and 70% contributions to groups with $pK_1 \simeq 10$ and $pK_2 \simeq 11$, respectively.

Figure 11.2 plots, calculated from the data of Argese et al. (21), show that, at a given value of $[OH^-]$, the activity decreases logarithmically with increasing perchlorate concentration expressed as the square root of perchlorate ionic strength, according to the Brϕnsted-Bjerrum equation: $\log k = \log k_0 + Z_A Z_B \sqrt{\mu}$. This equation, which is valid only in very dilute solution, was found (22) to hold for enzyme–substrate interactions even at high ionic strength. The results of Fig. 11.2 can thus be interpreted in terms of the Brϕnsted-Bjerrum theory as follows. (a) Addition of perchlorate has no secondary effect on the SOD activity, as the values measured at $[OH^-] \geqslant 10^{-3}$ *M* are nearly independent of $[NaClO_4]$ and approach the values measured at very low $[OH^-]$ and highest $[NaClO_4]$. (b) The phenomena described above are in line with ā reaction between opposite charges in the presence of a salt that has the expected marked inhibitory effect on the rate without binding to the active site metal. (c) The experimentally determined k value allows calculation, from the Brφnsted-Bjerrum equation, of the average charge (Z_A) actually seen by the O_2^- electric charge ($Z_B = -1$) when approaching the enzyme surface. The plot of the calculated $Z_A Z_B$ value versus pH overlaps the activity titration curve (Fig. 11.1) and has a maximum value approaching -1.5 at pH 10.

Identification of the Residues Responsible for Electrostatic Interaction

Table 11.3 shows that neutralization of the charge of all lysine residues by chemical modification tends to abolish the crossed pH–perchlorate effect and causes the decrease of the inhibition constant of ions that interact directly with the copper, whereas it scarcely affects the spectroscopic constant. Figure 11.3 shows that selective modification of a single lysine residue, i.e., Lys 134, renders the enzyme activity much less sensitive to large changes of either $[OH^-]$ or $[NaClO_4]$, thus mimicking the native protein at high pH or high ionic strength.

Discussion

The results presented here support the hypothesis that the rate-limiting step of O_2^- dismutation by Cu,Zn SOD at low $[O_2^-]$ is the encounter of the substrate with surface positive side chains that have pK values in the pH range 9.5 to 11.0. The activity–pH profile shown in Fig. 11.1 can be interpreted in terms of two overlapping titration curves with $pK_1 \simeq 10$ and $pK_2 \simeq 11$. At $[OH^-] \leqslant 10^{-4}$ *M* the reaction is strongly dependent on $[NaClO_4]$, even though ClO_4^- does not bind to the copper. This dependence obeys the Brφnsted-Bjerrum

TABLE 11.3. Stability constant for ion binding to lysine-modified Cu,Zn SOD.

Ion	K_N[a] Activity	K_N[a] EPR
N_3^-	0.41	0.7
OH^-	0.05	Unchanged
ClO_4^-	0.60	Unchanged

All lysine side chains were carbamoylated by reaction with NCO^-, as previously described (23).
[a] K_N is the normalized value of the stability constant with reference to the value of the enzyme in the absence of anions taken as 1.

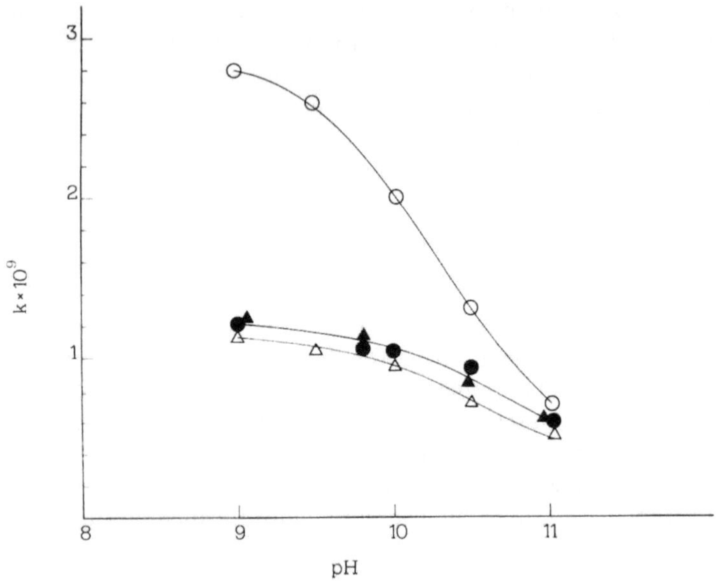

FIG. 11.3. The k_{cat} of bovine Cu,Zn SOD in native and Lys 134-modified SOD as a function of pH and [$NaClO_4$]. (○) Native SOD, 0.02 *M* perchlorate. (●) Native SOD, 0.25 *M* perchlorate. (△) Lys 134-modified SOD, 0.02 *M* perchlorate. (▲) Lys 134-modified SOD, 0.25 *M* perchlorate. Lys 134 was modified by carbamoylation with pyriodoxal-5′-phosphate, and the selectively labeled protein was purified according to a procedure to be published in detail elsewhere.

type of relation to ionic strength and allows one to calculate the average charge on the protein that the incoming O_2^- encounters. The overlapping of the experimental and calculated curves (Fig. 11.1) is strong evidence for the control of the enzyme activity by electrostatic interactions. This conclusion is further confirmed by the identity of the catalytic constant measured at pH values around pH 12 and at lower pH in the presence of high [$NaClO_4$].

The results of this work strongly suggest that the side chains of two lysine residues are crucial for enzyme activity. A refined crystallographic model (14) indicates that Lys 120 and Lys 134 are strategically located at the edge of the active site. The experiment described in Fig. 11.3 confirms that Lys 134 is important for the enzyme mechanism. Lys 134 is likely to be the group with the higher pK, as there is a salt bridge present between Lys 134 and the adjacent Glu 131 (24). The residual activity of the lysine-neutralized enzyme, which appears to be nearly insensitive to salt effects (Fig. 11.2 and 11.3) may be due to a positive electrostatic potential more centered on the metal-binding region of the protein.

In conclusion, Cu,Zn SOD represents an interesting example of a metal-dependent oxidoreductase in which the protein moiety plays a major role in determining the rate-limiting step of the electron transfer reaction. Work is in progress to extend and refine this model. We are pursuing a site directed mutagenesis approach and the comparative study of superoxide dismutases of various species with different surface charge.

References

1. Rotilio, G., Calabrese, L., and Rigo, A. (1978) In *Frontiers in Physicochemical Biology*, edited by B. Pullman, pp. 357–368. Academic Press, New York.
2. Fielden, F. M., and Rotilio, G. (1984) In *Copper Proteins and Copper Enzymes*, Vol. 2, edited by R. Lontie, pp. 27–61. CRC Press, Boca Raton.
3. Klug-Roth, D., Fridovich, I., and Rabani, J. (1973) *J. Am. Chem. Soc.* 95:2786–2790.
4. Fielden, E. M., Roberts, P. B., Bray, R. C., Lowe, D. J., Mautner, G. N., Rotilio, G., and Calabrese, L. (1974) *Biochem. J.* 139:49–60.
5. Rigo, A., Viglino, P., and Rotilio, G. (1975) *Biochem. Biophys. Res. Commun.* 63:1013–1013.
6. Rotilio, G., Bray, R. C., and Fielden, F. M. (1972) *Biochim. Biophys. Acta* 268:605–609.
7. Klug, D., Rabani, J., and Fridovich, I. (1972) *Biochim. Biophys. Acta* 268:605–609.
8. Hodgson, E. K., and Fridovich, I. (1975) *Biochemistry* 14:5294–5299.
9. McAdam, M. E., Fielden, F. M., Lavelle, F., Calabrese, L., Cocco, D., and Rotilio, G. (1977) *Biochem. J.* 167:271–274.
10. Tainer, J. A., Getzoff, E. D., Richardson, J. S., and Richardson, D. C. (1983) *Nature* 306:284–287.
11. Koppenol, W. H. (1981) In *Oxygen and Oxy-Radicals in Chemistry and Biology*, edited by M. A. J. Rodgers and E. L. Powers, pp. 671–674. Academic Press, New York.
12. Marmocchi, F., Mavelli, I., Rigo, A., Stevanato, R., Bossa, F., and Rotilio, G. (1982) *Biochemistry* 21:2853–2856.
13. Cudd, A., and Fridovich, I. (1982) *J. Biol. Chem.* 257:1144–11447.
14. Getzoff, E. D., Tainer, J. A., Weiner, P. K., Kollman, D. A., Richardson, J. S., and Richardson, D. C. (1983) *Nature* 306:287–290.
15. Mota de Freitas, D., and Valentine, J. S. (1984) *Biochemistry* 23:2079–2082.
16. McCord, J. M., and Fridovich, I. (1969) *J. Biol. Chem.* 244:6049–6055.
17. Argese, E., De Carli, B., Orsega, E., Rigo, A., and Rotilio, G. (1983) *Anal. Biochem.* 132:110–114.
18. Viglino, P., Rigo, A., Argese, F., Calabrese, L., Cocco, D., and Rotilio, G. (1981) *Biochim. Biophys. Res. Commun.* 100:125–130.
19. Rigo, A., Stevanato, P., Viglino, P., and Rotilio, G. (1977) *Biochem. Biophys. Res. Commun* 79:776–783.
20. Calabrese, L., Cocco, D., and Rotilio, G. (1983) In *Oxy-Radicals and Their Scavenger Enzymes*, edited by G. Cohen, and R. A., Greenwald, pp. 177–186. Elsevier, New York.
21. Alberty, R. A., and Hammes, G. G. (1958) *J. Phys. Chem.* 62:154–159.
22. Argese, E., Viglino, P., Rotilio, G., Scaipa, M., and Rigo, A. (1987) *Biochemistry* 26:3224–3228.
23. Cocco, D., Rossi, L., Barra, D., Bossa, F., and Rotilio, G. (1982) *FEBS Lett.* 150:303–305.
24. Tainer, J. A., Getzoff, E. D., Beem, K. M., Richardson, J. S., and Richardson, D. C. (1982) *J. Mol. Biol.* 160:181–217.

12
Dynamic Participation of Protein Domains in Catalysis by 2-Oxo Acid Dehydrogenase Multienzyme Complexes

RICHARD N. PERHAM

The 2-oxo acid dehydrogenase multienzyme complexes (reviewed in refs. 1–3) consist of multiple copies of each of three enzymes which catalyze successive steps in the oxidative decarboxylation of the relevant 2-oxo acids, notably pyruvate, 2-oxoglutarate, and branched-chain 2-oxo acids derived from transamination of isoleucine, leucine, and valine. The overall reaction scheme is shown in Fig. 12.1.

The three constituent enzymes of the pyruvate dehydrogenase complex are pyruvate decarboxylase (E1; EC 1.2.4.1), lipoate acetyltransferase (E2; EC 2.3.1.12), and dihydrolipoamide dehydrogenase (E3; EC 1.8.1.4). The corresponding enzymes of the 2-oxoglutarate dehydrogenase complex are 2-oxoglutarate decarboxylase (E1; EC 1.2.4.2), lipoate succinyltransferase (E2; EC 2.3.1.61), and dihydrolipoamide dehydrogenase (E3; EC 1.8.1.4, formerly EC 1.6.4.3). The three short branched-chain 2-oxo acids are oxidatively decarboxylated by a single complex of comparable structure (4, 5). In all these complexes substrate specificity resides in the E1 and E2 components, whereas, as is evident from Fig. 12.1, E3 is required only to bring about the reoxidation of the intermediate dihydrolipoamide. This feature is reflected in the fact that the E3 polypeptide chain is normally identical in the pyruvate, 2-oxoglutarate, and branched-chain 2-oxo acid dehydrogenase complexes (reviewed in ref. 5), a good example of the principle of parsimony in the generation of biologically successful structures (6). An interesting exception to this rule, the reason for which remains obscure, is the existence of different E3 polypeptide chains in the pyruvate and branched-chain 2-oxo acid dehydrogenase complexes of *Pseudomonas putida* (7).

The 2-oxo acid dehydrogenase complexes also represent one of the classic examples of a "swinging arm" mechanism (1–3), whereby the substrate is carried in thioester linkage to lipoic acid residues, which are in turn in amide linkage with the N^6-amino groups of lysine residues of the E2 component (Fig. 12.1). This permits, in principle, rotation of the substrate between the three active sites in three enzymes whose participation is required in the overall catalysis. Spin-label experiments with the *Escherichia coli* pyruvate dehydrogenase complex indicate that the correlation time for the dithiolane

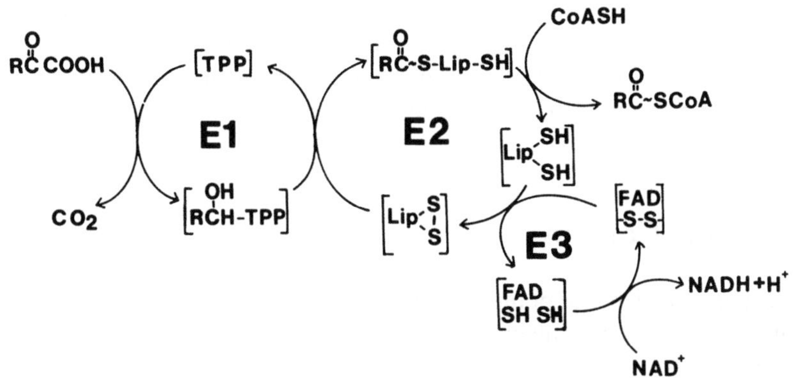

Net Reaction

$$RC(=O)COOH + NAD^{+} + CoASH \longrightarrow RC(=O){\sim}SCoA + NADH + H^{+} + CO_2$$

FIG. 12.1. Reaction mechanism of the 2-oxo acid dehydrogenase multienzyme complexes. TPP = thiamine pyrophosphate; Lip = lipoic acid.

ring at the end of this swinging arm ($\tau_c \simeq 10^{-9}$ s) is consistent with this idea (8, 9).

Two Classes of Symmetry

The 2-oxo acid dehydrogenase complexes have a high M_r (5×10^6 to 10×10^6) and are seen as particles of 30 to 40 nm in the electron microscope (1, 10–12). Thus they are significantly larger than a ribosome. The E2 components of the complexes perform two functions: they contribute the acyltransferase activity responsible for transfer of the relevant acyl group from its thioester linkage with the lipoic acid residue in E2 to the thiol group of coenzyme A (Fig. 12.1), and they constitute the structural cores of the complexes to which the E1 and E3 components bind tightly but noncovalently.

In the pyruvate and 2-oxoglutarate dehydrogenase complexes of *E. coli* this core consists of 24 copies of the E2 chain arranged with octahedral symmetry (1, 10); whereas in the pyruvate dehydrogenase complexes of mammals (1), *Bacillus stearothermophilus* (11), and yeast (13) the core comprises 60 E2 chains arranged with icosahedral symmetry. Despite their location as core enzymes, the acyltransferases of all 2-oxo acid dehydrogenase complexes are highly susceptible to limited proteolysis: fragments of the E2 chains carrying the lipoyl groups are readily cleaved off, suggesting that they protrude physically from an inner core, interdigitating between E1 and E3 subunits. After such proteolysis, the inner core still carries the acyltransferase activity and remains assembled, and it is this residual fragment of the E2 chain that evidently determines the underlying symmetry of the overall complex (3, 14, 15).

Active Site Coupling

Several lines of evidence point to properties of unusual interest in the mechanism of active site coupling in the 2-oxo acid dehydrogenase complexes. It has been known for some time that the three types of active site are not represented in equal numbers in the enzyme particles. For example, the pyruvate dehydrogenase complex of *E. coli* has been reported to have an E1 : E2 : E3 chain ratio of 1.0 : 1.0 : 0.5 (1, 2, 16), but there is also evidence that these chain ratios may not be strictly adhered to (3, 17, 18) and that the complex may exist as a population of particles with limited heterogeneity (19, 20). Similar considerations apply to other 2-oxo acid dehydrogenase complexes. Thus for the 2-oxoglutarate dehydrogenase complex of *E. coli*, more than 100,000 potential quaternary structures may exist, generated by different permutations of E1 and E3 subunit packing during assembly around the E2 core (21). For the pyruvate dehydrogenase complexes of icosahedral E2 symmetry, the E3 : E2 chain ratios can be significantly <0.5, and once again there is evidence of some variability of chain ratios (1, 11). It is clear that these complexes do not conform entirely to the simple design principles expected of oligomeric systems (6).

A further remarkable property of all 2-oxo acid dehydrogenase complexes is that a single E1 subunit can bring about the reductive acylation of all the lipoic acid residues in the E2 core and not just the lipoyl group(s) of the E2 subunit to which that E1 subunit is bound (22–24). Lipoic acid residues on neighboring E2 subunits appear to be able to transfer acyl groups (and perhaps electron pairs) from one to another by means of an extended network of intramolecular transacylation reactions (Fig. 12.2).

The curiosities do not end there; for when the pyruvate dehydrogenase complexes of *E. coli* (14) or *B. stearothermophilus* (15) are subjected to limited proteolysis, fragments of the E2 chains carrying about half of the lipoyl groups of the E2 core can be removed without much loss of overall catalytic activity. Similarly, the lipoic acid residues of the *E. coli* complex can be chemically modified (25) or enzymically removed (25, 26) more rapidly than pyruvate dehydrogenase complex activity is lost. Thus what have hitherto been regarded as catalytically essential groups in the mechanism can be excised or modified without a corresponding loss of catalytic activity. Clearly we must seek a molecular model that does not require a simple direct transfer of substrate among three active sites organized with 1 : 1 : 1 stoichiometry in a strict geometric arrangement.

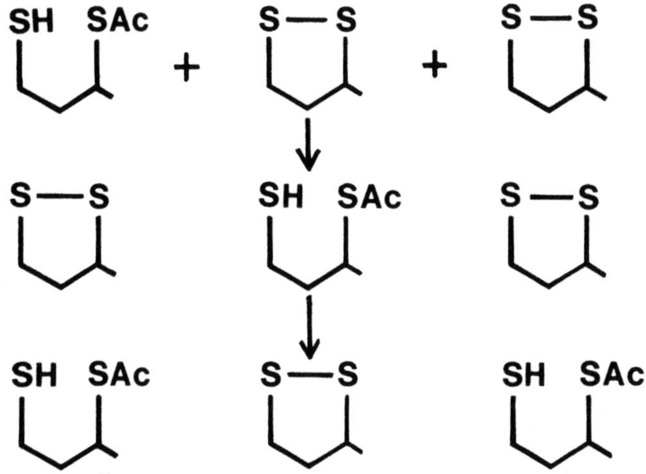

FIG. 12.2. Transfer of acyl groups between lipoic acid residues on neighboring E2 polypeptide chains.

^{1}H-NMR Spectroscopy and Domain Structure of the E2 Chains

As we have seen above, the E2 chains of all 2-oxo acid dehydrogenase complexes are divisible into at least two parts, one the inner core-forming domain containing the acyltransferase active site and the other an outer part or domain containing the lipoic acid residues. Given the fuzzy appearance of the E2 cores in the electron microscope and their susceptibility to nicking with proteinases, it was suggested that the lipoyl domains might enjoy sufficient flexibility to facilitate the necessary visits of lipoic acid residues to the various active sites in E1, E2, and E3 (25, 26).

Direct evidence for conformational flexibility in the E2 chains of all 2-oxo acid dehydrogenase complexes has come from ^{1}H-nuclear magnetic resonance (NMR) spectroscopy (reviewed in ref. 27). The calculated line width of a methylene proton resonance from structures as large as these ($\tau_c \simeq 10^{-5}$ s) is about 8 kHz, and, as expected, a broad component of this type dominates the spectra (15, 28, 29). However, superimposed on the broad component is a collection of relatively sharp resonances between -0.6 and $+3.0$ ppm from dioxan, with line widths of only about 50 Hz, implying that some regions of the polypeptide chains in the complexes enjoy significant conformational mobility with respect to the rest of the structure (Fig. 12.3). Resonances from aromatic side chains, normally found at 2 to 6 ppm, are notably absent. On the other hand, in the spectrum of the *E. coli* pyruvate dehydrogenase complex, the prominent peak at -2.3 ppm suggests that the flexible regions may be rich in alanine (or threonine) residues (27, 28).

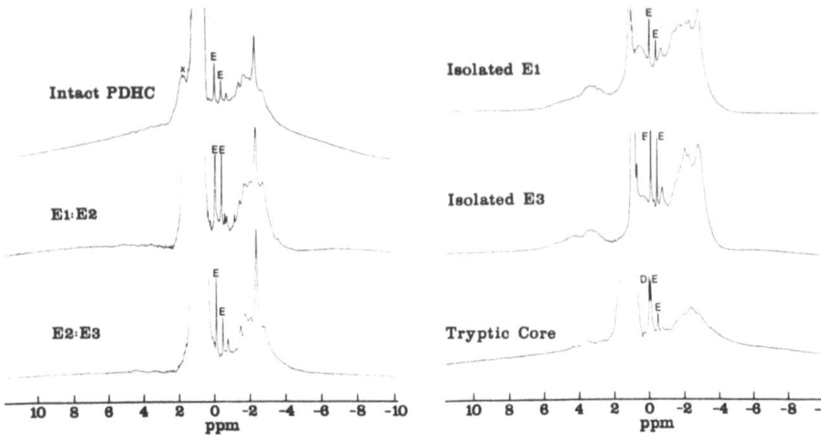

FIG. 12.3. ^{1}H-NMR spectra (270 mHz) of *E. coli* pyruvate dehydrogenase complex and its various subcomplexes and components. E = resonances from EDTA; D = dioxan. Chemical shifts are measured from dioxan.

A more detailed analysis of the E2 chain of the *E. coli* pyruvate dehydrogenase complex has been made possible by determination of the DNA sequence of the *aceF* gene encoding it (30). This E2 chain is oddly complicated in that it appears to contain up to three lipoyl-lysine residues in repeating, identical amino acid sequences (31). The lipoylated section of the protein in fact consists of three highly homologous segments of about 75 to 80 amino acids, each containing a potential lipoyl-lysine residue, tandemly repeated to form the N-terminal half of the polypeptide chain; these segments in turn are separated from each other and then linked to the C-terminal half of the protein by regions of about 25 amino acids rich in proline and alanine residues (Fig. 12.4). Treatment of the intact complex with *Staphylococcus aureus* V8 proteinase readily produces selective cleavages in the three (alanine + proline)-rich sequences (Fig. 12.4), releasing the three lipoylated segments as individually folded domains. All three domains retain their ability to function as substrates for reductive acetylation (17).

Similar treatment of the complex with trypsin, which releases all three lipoyl domains by cleavage of the E2 chain at lysine 316 (17), causes a significant fall in the sharp resonances in the 270 MHz ^{1}H-NMR spectrum, especially those tentatively attributed to the methyl protons of alanine (or threonine) residues (Fig. 12.3). The three (alanine + proline)-rich sequences contain no aromatic residues, sharp resonances from which are of course lacking in the spectra. Given, in addition, their susceptibility to selective cleavage with proteinases (Fig. 12.4), it is likely that these sequences are those that, on the basis of ^{1}H-NMR spectroscopy, enjoy unusual conformational flexibility in the complex.

The E2 chains of the icosahedral pyruvate dehydrogenase complexes of *B. stearothermophilus* and mammalian mitochondria and of the octahedral

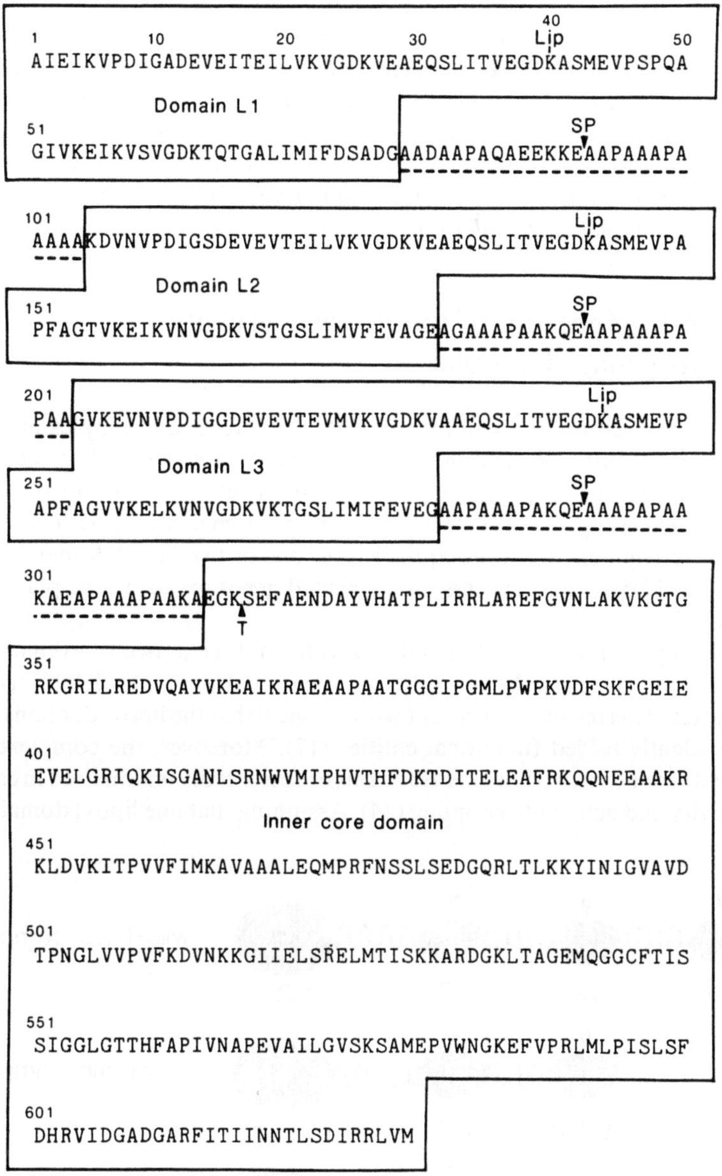

FIG. 12.4. Primary structure of the E2 chain of the *E. coli* pyruvate dehydrogenase complex. Lip = lipoyl group; SP = principal cleavage sites with *S. aureus* V8 proteinase; T = principal cleavage site with trypsin. (----) regions rich in alanine, proline, and charged amino acids, assigned high conformational flexibility. The demarcations of folded lipoyl domains and the inner core domain are still uncertain, and those shown are for illustrative purposes only.

2-oxoglutarate dehydrogenase complex of *E. coli* conform to the same underlying structural principle but are distinctly simpler. They contain only one lipoyl domain, but it again is linked to the inner core (catalytic and subunit-binding) domain by a region of polypeptide chain that, to judge from ^{1}H-NMR spectroscopy, is conformationally flexible (29, 32). There is considerable sequence homology between the one lipoyl domain of the 2-oxoglutarate dehydrogenase complex and the three lipoyl domains of the pyruvate dehydrogenase complex of *E. coli* (33).

Protein Engineering of the *E. coli* Pyruvate Dehydrogenase Complex

An obvious question to ask of the pyruvate dehydrogenase complex of *E. coli* is whether all three lipoyl domains per E2 chain are essential for catalytic activity. Given the cloned *aceF* gene and the repetitive nature of the DNA sequence encoding the three successive lipoyl domains (30), it has proved possible to engineer deletions equivalent to one or two lipoyl domains in the gene (Fig. 12.5) and to express the truncated genes on a suitable plasmid in *E. coli* (34, 35).

Lowering the number of lipoyl domains from three to two or even one per E2 chain still permits the assembly of functional pyruvate dehydrogenase complexes. This result is consistent with the view that the lipoyl domains form independently folded functional entities (17). Moreover, the complexes assembled around such truncated E2 chains are essentially unimpaired in catalytic activity and active site coupling (34). Assuming that one lipoyl domain per

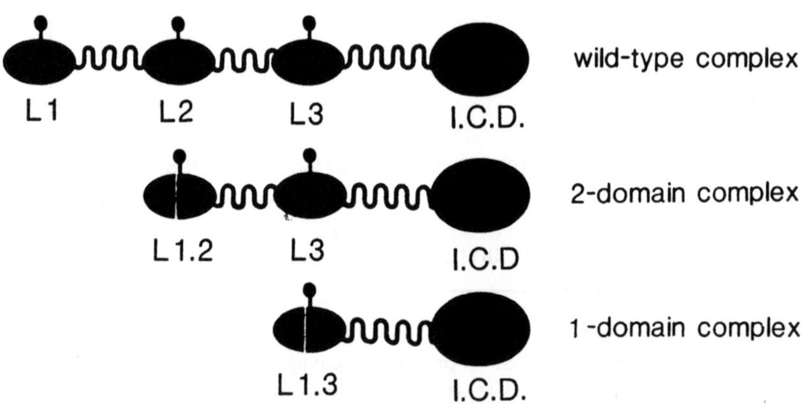

FIG. 12.5. Structure of the E2 subunit in the wild-type and genetically engineered pyruvate dehydrogenase complexes. L1, L2, L3 = native lipoyl domains; L1.2, L1.3 = hybrid lipoyl domains generated by genetic deletion; ICD = inner core domain. (〰) flexible linker regions.

E2 chain therefore represents the minimal structural requirement for an effective acetyltransferase component, it is difficult to see why *E. coli* has elaborated two additional lipoyl domains in the N-terminal half of the E2 chain of its pyruvate dehydrogenase complex. Highly homologous and tandemly repeated sequences in the *aceF* gene would seem to be good candidates for recombinational deletion unless some selective advantage were conferred by their retention. Perhaps a more stringent kinetic analysis of the restructured complexes will reveal what that advantage might be.

Dynamic Participation of Lipoyl Domains in Catalysis

Our growing knowledge of the segmental structure of the E2 chains and of the presence and location of flexible amino acid sequences between the folded domains permits a plausible answer to the kinetic paradoxes posed above in the description of active site coupling in the 2-oxo acid dehydrogenase complexes.

If E1 catalyzes the rate-determining step in the mechanism, as it does in the *E. coli* pyruvate dehydrogenase complex (36), no loss of catalytic activity need ensue if the function of any lipoyl group that has been modified or excised is taken over by one that remains (25, 26, 28). This effect can be achieved by a given E1 active site being accessible to several different lipoyl groups (domains) from neighboring E2 chains (28, 37), which is permitted by the segmental flexibility in the E2 chains (Fig. 12.6). To these properties we can

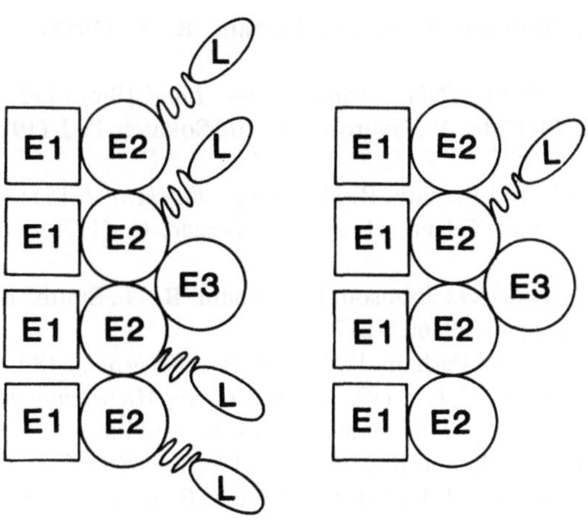

FIG. 12.6. Section of a 2-oxo acid dehydrogenase complex before and after partial proteolysis. A given E2 lipoyl domain can service more than one E1 active site, and a single E3 active site can service more than one E2 lipoyl domain either directly or by intramolecular transacylation reactions (Fig. 12.2).

add the system of intramolecular transacylation reactions, which permits acyl transfer between lipoyl groups in neighboring domains, again facilitated perhaps by segmental flexibility in the E2 chains. Catalysis is therefore no longer restricted to successive participation of three juxtaposed active sites, thereby reducing the problems of substrate diffusion, especially at low substrate concentrations. Thus the complexes are relieved of any necessity to be organized in a strictly ordered geometric arrangement, with 1 : 1 : 1 stoichiometry of subunits. In this regard, it is of particular interest that the E3 component, which fulfills a service function in these complexes, can be used so sparingly. The 2-oxo acid dehydrogenase complexes have clearly evolved as a highly effective solution to the problem of catalysis of a multistep reaction.

Acknowledgments. I am grateful to many colleagues, notably Dr. L. C. Packman, Professor G. C. K. Roberts, and Professor J. R. Guest, who have made valuable contributions to the work reviewed above. I thank the Science and Engineering Research Council for financial support.

References

1. Reed, L. J. (1974) *Acc. Chem. Res.* 7:40–46.
2. Hammes, G. G. (1981) *Biochem. Soc. Symp.* 46:73–90.
3. Perham, R. N. (1983) In: *Frontiers in Biochemical and Biophysical Studies of Proteins and Membranes*, edited by T-Y. Liu, S. Sakakibara, A. N. Schechter, K. Yagi, H. Yajima, and K. Yasunobu, pp. 249–258. Elsevier, New York.
4. Pettit, F. H., Yeaman, S. J., and Reed, L. J. (1978) *Proc. Natl. Acad. Sci. USA* 75:4881–4885.
5. Lowe, P. N., Hodgson, J. A., and Perham, R. N. (1983) *Biochem. J.* 215: 133–140.
6. Perham, R. N. (1975) *Philos. Trans. R. Soc. Lond.* [Biol.] 272:123–136.
7. Sokatch, J. R., McCully, V., Gebrosky, J., and Sokatch, D. J. (1981) *J. Bacteriol.* 148:639–646.
8. Ambrose, M. C., and Perham, R. N. (1976) *Biochem. J.* 155:429–432.
9. Grande, H. J., Van Telgen, H. J., and Veeger, C. (1976) *Eur. J. Biochem.* 71:509–518.
10. Danson, M. J., Hale, G., Johnson, P., Perham, R. N., Smith, J., and Spragg, P. (1979) *J. Mol. Biol.* 129:603–617.
11. Henderson, C. E., and Perham, R. N. (1980) *Biochem. J.* 189:161–172.
12. Oliver, R. M., and Reed, L. J. (1982) In *Electron Microscopy of Proteins*, Vol. 1, edited by R. Harris, pp. 1–48. Academic Press, London.
13. Keha, E. E., Ronft, H., and Kresze, G-B. (1982) *FEBS Lett.* 145:289–292.
14. Bleile, D. M., Hackert, M. L., Pettit, F. H., and Reed, L. J. (1981) *J. Biol. Chem.* 256:514–519.
15. Duckworth, H. W., Jaenicke, R., Perham, R. N., Wilkie, A. O. M., Finch, J., and Roberts, G. C. K. (1982) *Eur. J. Biochem.* 124:63–69.
16. Yang, H., Hainfeld, J. F., Wall, J. S., and Frey, P. A. (1985) *J. Biol. Chem.* 260:16049–16051.

17. Packman, L. C., Hale, G., and Perham, R. N. (1984) *EMBO J.* 3:1315–1319.
18. De Kok, A., and Westphal, A. H. (1985) *Eur. J. Biochem.* 152:35–41.
19. Schmitt, B., and Cohen, R. (1980) *Biochem. Biophys. Res. Commun.* 93:709–712.
20. Gilbert, G. A., and Gilbert, L. M. (1980) *J. Mol. Biol.* 144:405–408.
21. Wagenknecht, T., Frances, N., and DeRosier, D. (1986) *Biochem. Biophys. Res. Commun.* 135:802–807.
22. Bates, D. L., Danson, M. J., Hale, G., Hooper, E. A., and Perham, R. N. (1977) *Nature* 268:313–316.
23. Collins, J. H., and Reed, L. J. (1977) *Proc. Natl. Acad. Sci. USA* 76:4385–4389.
24. Packman, L. C., Stanley, C. J., and Perham, R. N. (1983) *Biochem. J.* 213: 331–338.
25. Berman, J. N., Chen, G.-X., Hale, G., and Perham, R. N. (1981) *Biochem. J.* 199:513–520.
26. Stepp, L. R., Bleile, D. M., McRorie, D. K., Pettit, F. H., and Reed, L. J. (1981) *Biochemistry* 20:4555–4560.
27. Roberts, G. C. K., Duckworth, H. W., Packman, L. C., and Perham, R. N. (1983) *Ciba Found. Symp.* 93:47–62.
28. Perham, R. N., Duckworth, H. W., and Roberts, G. C. K. (1981) *Nature* 292:474–477.
29. Packman, L. C., Perham, R. N., and Roberts, G. C. K. (1984) *Biochem. J.* 217:219–227.
30. Stephens, P. E., Darlison, M. G., Lewis, H. M., and Guest, J. R. (1983) *Eur. J. Biochem.* 133:481–489.
31. Hale, G., and Perham, R. N. (1980) *Biochem. J.* 187:905–908.
32. Perham, R. N., and Roberts, G. C. K. (1981) *Biochem. J.* 199:733–740.
33. Spencer, M., Darlison, M. G., Stephens, P. E., Duckenfield, I. K., and Guest, J. R. (1984) *Eur. J. Biochem.* 414:361–374.
34. Guest, J. R., Lewis, H. M., Graham, L. D., Packman, L. C., and Perham, R. N. (1985) *J. Mol. Biol.* 185:743–754.
35. Graham, L. D., Guest, J. R., Lewis, H. K., Miles, J. S., Packman, L. C., Perham, R. N., and Radford, S. E. (1986) *Philos. Trans. R. Soc. Lond.* [*A*] 317:391–404.
36. Danson, M. J., Fersht, A. R., and Perham, R. N. (1978) *Proc. Natl. Acad. Sci. USA* 75:5390.
37. Hackert, M. L., Oliver, R. M., and Reed, L. J. (1983) *Proc. Natl. Acad. Sci.* USA 80:2907–2911.

13
Membrane-Bound GTP-Transducin Efficiently Activates Retinal cGMP Phosphodiesterase

THEODORE G. WENSEL and LUBERT STRYER

Transducin (T, M_r 85,000) is the GTP-binding protein of retinal rod outer segments (ROS) that receives the excitation signal from photoisomerized rhodopsin (R*) in the first amplified step of the cyclic GMP cascade of vision (1, 2). Transducin, in turn, activates a phosphodiesterase (PDE) that rapidly hydrolyzes cGMP, leading to the closure of cGMP-gated channels in the plasma membrane (for review, see refs. 3–5). R* induces the release of GDP by transducin, thus catalyzing the uptake of GTP by its α subunit (T_α-GTP; M_r 39,000); T_α-GTP then activates the PDE (2). When it binds GTP, T_α dissociates from the β (M_r 35,000) and γ (M_r 6,000) subunits of transducin ($T_{\beta\gamma}$) as well as from R*.

Transducin, a peripheral membrane protein, is bound in the $T_{\alpha\beta\gamma}$-GDP form to disk membranes in the dark at physiological ionic strength. Under moderate salt conditions ($I \sim 0.1$ *M*), the binding of GTP solubilizes T_α, whereas PDE remains tightly bound (6). Purified transducin, extracted from bleached membranes with GTP, can fully activate the phosphodiesterase in the presence of R* when GTP or a GTP analogue is present. The α subunit, isolated in the GTP form by using guanosine-5′-O-(3-thiotriphosphate) (GTPγS) or another hydrolysis-resistant analogue, can also activate PDE. In reconstitution experiments the fraction of PDE activated is strongly dependent on transducin concentration; half-maximal activation requires about 10^{-6} *M* transducin (7–10). In contrast, the fraction of PDE activated in suspensions of whole ROS was found to be independent of concentration from 2 to 14 μM R* (11). Analysis of the kinetics of activation suggested that rapid activation of PDE depends on transducin–PDE interactions on the surface of the membrane (12). These results raised the possibility that activation of PDE is mediated by a membrane-bound form of transducin. We therefore examined the effects on PDE activation of removing solubilized transducin from ROS.

Experimental Procedures

Reagents were obtained from the following sources: [^{35}S]GTPγS from New England Nuclear; [8-^{3}H]cGMP and 5′[U-^{14}C]GMP from Amersham; GTPγS and GTP from Boehringer Mannheim; TPCK-treated trypsin from Worthington; soybean trypsin inhibitor and *Ophiophagus hannah* snake venom from Sigma; hexyl agarose from Miles; and frozen bovine retinas from J. A. Lawson (Lincoln, NB).

Rod outer segments were isolated from frozen bovine retinas as described (2). Transducin (7), PDE (10), and urea-stripped ROS membranes (13) were prepared as described.

Membranes were washed with buffer A (containing 20 m*M* 3-[*N*-morpholino]propanesulfonic acid (MOPS) pH 7.4, 120 m*M* NaCl, 30 m*M* KCl, 2 m*M* $MgCl_2$, 1 m*M* dithiothreitol) or buffer B (containing 20 m*M* MOPS pH 8.0, 0.15 *M* KCl, and 2 m*M* $MgCl_2$).

Phosphodiesterase assays were performed by either a pH recording method (14) or following the hydrolysis of [^{3}H]cGMP (15, 16). Reconstitution of transducin-stimulated PDE activity in urea-stripped ROS membranes and PDE assays were carried out as described (10). In GTPγS uptake assays (2, 13) membranes (0.6 μM R*) were incubated at 23°C for 10 min in buffer A containing 0.1 μM [^{35}S]GTPγS (56 Ci/mmol), diluted 30-fold with buffer A containing 0.67 μM GTPγS, and filtered through nitrocellulose. Filters were washed twice with 3 ml buffer A and the bound radioactivity determined by scintillation counting.

Results

PDE Is Activated More Efficiently by Endogenous Transducin in ROS than by Added Transducin in Reconstituted Membranes

Figure 13.1 shows that in a reconstituted system consisting of PDE bound to urea-stripped ROS membranes, micromolar concentrations of transducin are needed to activate PDE to a high level. Similar results were obtained previously when T_α-GppNHp [guanosine-5′-(β,γ-imino)triphosphate] was added to unbleached ROS (7) and when transducin was added to PDE bound to phospholipid vesicles (8). We reported elsewhere a detailed study of the dependence of PDE activity on the concentrations of transducin and the γ subunit of PDE (10).

In contrast, when PDE is assayed in ROS, activity is linear with total membrane concentration (Fig. 13.2). Because PDE and transducin are present in fixed proportion to R* in these membranes, this result indicates that a constant fraction of PDE is activated in each experiment even though the solution concentration of transducin varies from ~0.5 to 3.5 μM (Fig. 13.2A) or from ~0.08 to 0.76 μM (Fig. 13.2B). A similar result had been obtained

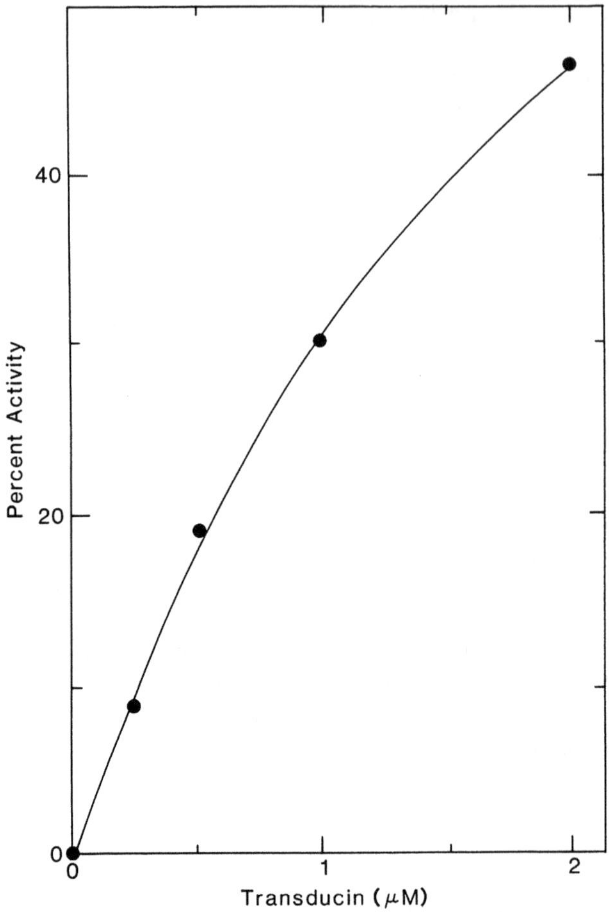

FIG. 13.1. PDE activation in reconstituted membranes requires micromolar concentrations of transducin. PDE activity was measured in a system reconstituted from the following components: urea-stripped disk membranes (57 nM R*), purified PDE (0.56 nM), GTPγS (10 μM), and the indicated concentrations of purified transducin. The 100% level of PDE activity was determined by trypsin activation.

previously (11) over a narrower concentration range. These results point to activation by a membrane-bound transducin (T_m) rather than a soluble transducin (T_s).

Removal of T_s Does not Eliminate PDE Activation

As shown in Fig. 13.3, solubilizing T_α with GTPγS and removing it by centrifugation has almost no effect on the PDE activity stimulated in bleached RO$\bar{S}$ by GTPγS. This result could be due to either (a) the activator remaining

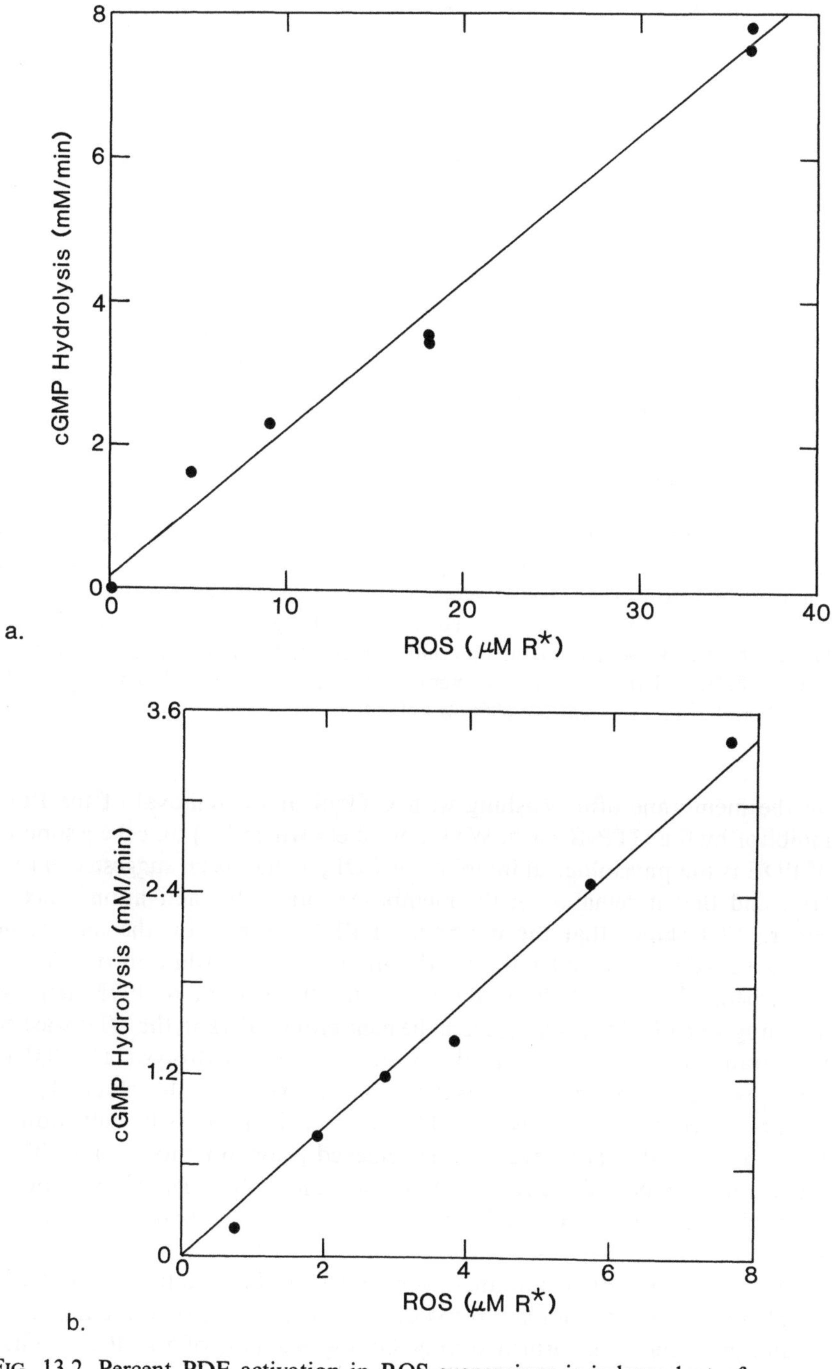

FIG. 13.2. Percent PDE activation in ROS suspensions is independent of concentration. *A*: The formation of [^{3}H]5′-GMP from [^{3}H]cGMP was followed by ion-exchange chromatography. *B*: Using a different preparation of ROS, PDE activity was measured by determining the initial velocity of proton generation (resulting from cGMP hydrolysis) using a pH electrode. The concentration of GTPγS was 10 μM in each assay.

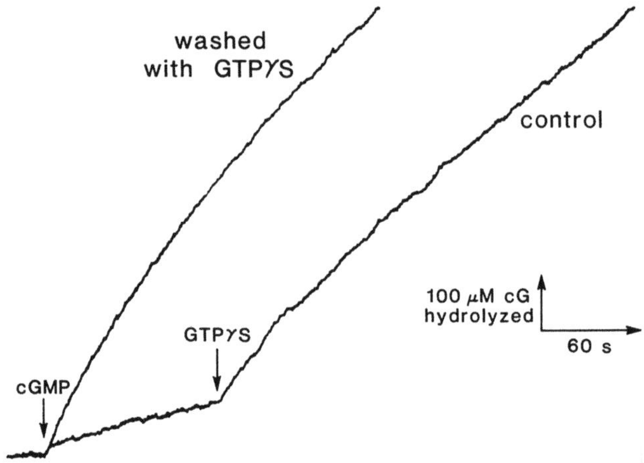

FIG. 13.3. Activator (T_m) remains on membranes washed with GTPγS. When GTPγS is added to ROS membranes, most of the transducin becomes soluble (T_s) and can be removed by centrifugation. The pH electrode traces follow the hydrolysis of cGMP in bleached ROS membranes resuspended to 1.9 *μM* R* after washing. Control membranes were washed with moderate salt buffer lacking GTPγS (bottom trace) and their level of PDE activity compared to membranes washed once at 76 *μM* R*/36 *μM* GTPγS and again at 25 *μM* R*/600 *μM* GTPγS.

on the membrane after washing with GTPγS or (b) removal of the PDE inhibitor by the GTPγS wash. We reported elsewhere (10) that the γ subunit of PDE is the physiological inhibitor of PDE, as had been suggested earlier (16), and that it remains on the membrane during the activation process. Figure 13.4 shows that the inhibitor of PDE remains on the membrane following activation under the conditions used here. Addition of GTP to unwashed, bleached ROS results in strong stimulation of PDE activity. Washing with GTP removes T_s from the membranes. PDE in the GTP-washed membranes returns to the inactive state following hydrolysis of GTP to GDP by membrane-bound transducin. Then, even though soluble T_α has been removed by the GTP wash, addition of GTPγS results in activation of PDE to nearly the same level as was achieved prior to removal of T_s. Thus the membrane-bound activator shows all the hallmarks of transducin: R*-catalyzed GTP-GDP exchange, GTP hydrolysis, and activation of PDE.

The persistence of membrane-associated transducin activity is not due simply to trapping of solubilized α subunit in sealed vesicles or multilamellar membrane structures formed during homogenization of the ROS. When membranes are sonicated during GTPγS washing, the same persistent activation of PDE in the pellet is obtained in either the presence or absence of the supernatant fraction containing T_s (Fig. 13.5).

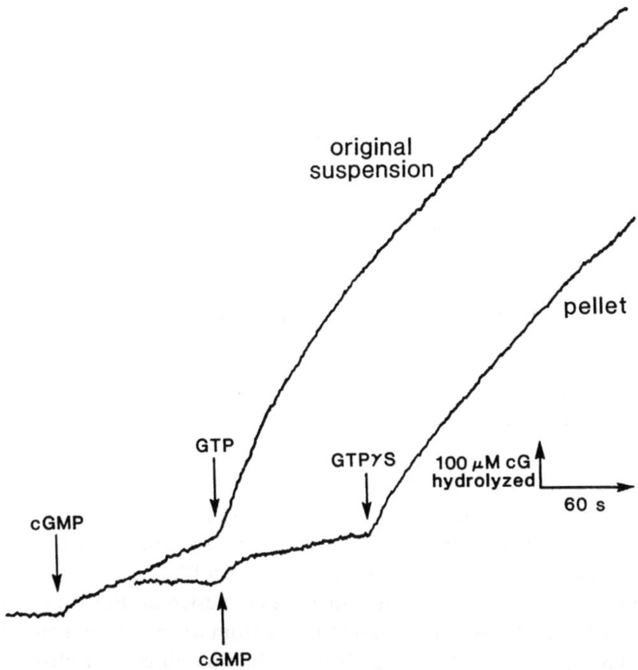

FIG. 13.4. Inhibitor and activator remain on the membrane. To determine whether the persistent activation of PDE illustrated in Fig. 13.3 was due to removal of inhibitor (γ subunit of PDE) or retention of activator Tα-GTP-γS), ROS (19 μM R*) was washed with GTP (0.1 mM) to remove T_s. The traces show cGMP hydrolysis monitored by a pH electrode, as in Fig. 13.3, at 1.9 μM R*. The top trace shows the level of PDE activity in these membranes in the presence of GTP before washing. After 1 h of incubation to allow hydrolysis of residual GTP by T_m, the GTP-washed membranes were assayed for PDE activity before and after addition of 1 μM GTPγS (bottom trace). Little activity was observed before addition of GTPγS, indicating that the inhibitor had remained on the membranes and that PDE had returned to its inhibited (dark) state. Addition of GTPγS resulted in virtually full recovery of GTP-dependent activity, indicating that the activator (T_m) remained on the membranes.

Quantitation of T_s Removal and PDE Activation

In order to determine the relation between release of T_s and loss of PDE activation, ROS were washed repeatedly with GTP. After each washing step, the pellet was assayed for PDE activity and GTPγS uptake activity (Fig. 13.6). The first wash removed 57% of the GTPγS uptake activity. PDE activity, however, stayed at ~80% of its initial value through this first and the subsequent washing steps. Although removal of T_s appears to be virtually complete after the four washing steps, some of the nucleotide uptake activity remaining after the fourth wash may be due to nonspecific retention of

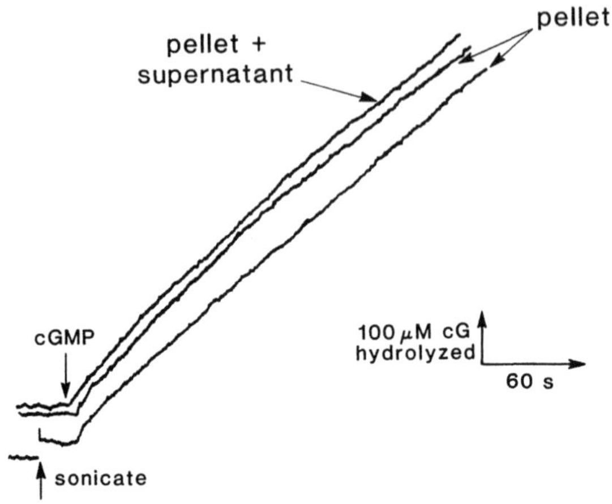

FIG. 13.5. Sonication does not release T_m. In order to determine if T_m is simply T_s trapped in multilamellar membrane structures, sonication was employed during GTPγS washing of ROS membranes (18.5 *μM* R*) before assaying at 1.9 *μM* R*. PDE activity in the pellet (middle trace) was compared with PDE activity in the combined pellet and supernatant (top trace). The same level of persistent PDE activation due to T_m was observed in each case. As shown in the bottom trace, further sonication in the assay vessel just prior to addition of cGMP also had no observable effect.

radioactivity by the membranes rather than binding to T_m. Consequently, the 27% of initial GTPγS uptake activity remaining after the four washes reflects an upper limit on the amount of T_m. Based on this upper limit, the PDE activation observed in these assays was stimulated by T_m at a solution concentration of <16 n*M* and a surface density on the membranes of $<800\ \mu m^{-2}$.

PDE is not saturated with T_α in dilute ROS suspensions. Figure 13.7 shows that PDE activity in ROS can be stimulated beyond the level induced by saturating R*, GTPγS, and endogenous transducin simply by adding more transducin. This result indicates that most of the PDE present in suspensions of ROS under these conditions is in the inactive state and capable of being activated by transducin.

Discussion

These experiments show that there are two populations of transducin activity, referred to here as T_s and T_m. Although both exhibit R*-catalyzed GDP-GTP exchange, GTP hydrolysis, and PDE activation, they are strikingly different in their affinities for the membrane and in their abilities to activate PDE. T_s is easily removed from membranes in moderate salt once GTP or GTPγS is bound. Under the same conditions, T_m activity in the membranes remains undiminished after repeated washing. Moreover, there is at least a 60-fold

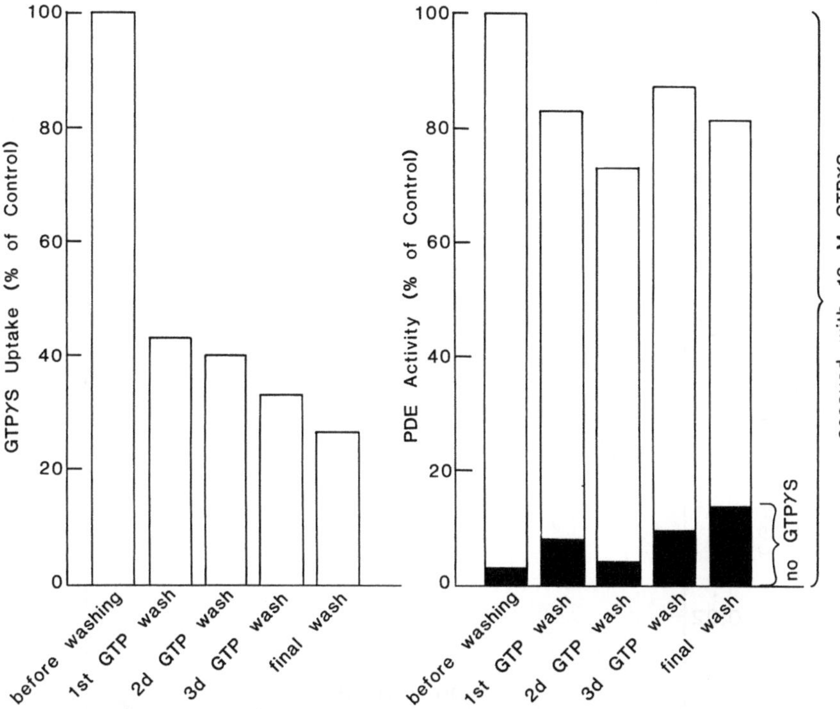

FIG. 13.6. PDE remains activated after T_s is removed. To quantitate the amount of membrane-bound transducin, both GTPγS uptake and PDE activity were measured in ROS membranes subjected to extensive washing with GTP (0.1 mM). Membranes were diluted to 6 μM R* during washing and 0.6 μM during assays. PDE activity was assayed using [^{3}H]cGMP, and GTPγS uptake was measured by determining the amount of [^{35}S]GTPγS bound to nitrocellulose filters.

difference in the strength of their interactions with PDE; <16 nM T_m is required for levels of PDE activation similar to those stimulated by micromolar T_s. T_m accounted for 80% of the PDE activation observed in these experiments, and the present results imply that, in previous studies of PDE activation in dilute suspensions of ROS, T_m was responsible for most of the stimulation of PDE.

We have not yet determined the chemical basis for these striking functional differences. One possibility is that T_s and T_m are intrinsically the same but differ because of the existence of a membrane-docking site that is present in lower concentrations than is transducin. For example, PDE or its γ subunit (PDE_γ) might serve as a docking site. It has been found (17) that PDE_γ forms a membrane-bound complex with T_α and that addition of PDE to ROS membranes increases the proportion of membrane-bound T_α. However, only a fraction of PDE is activated by T_m (Fig. 13.7). This finding indicates that

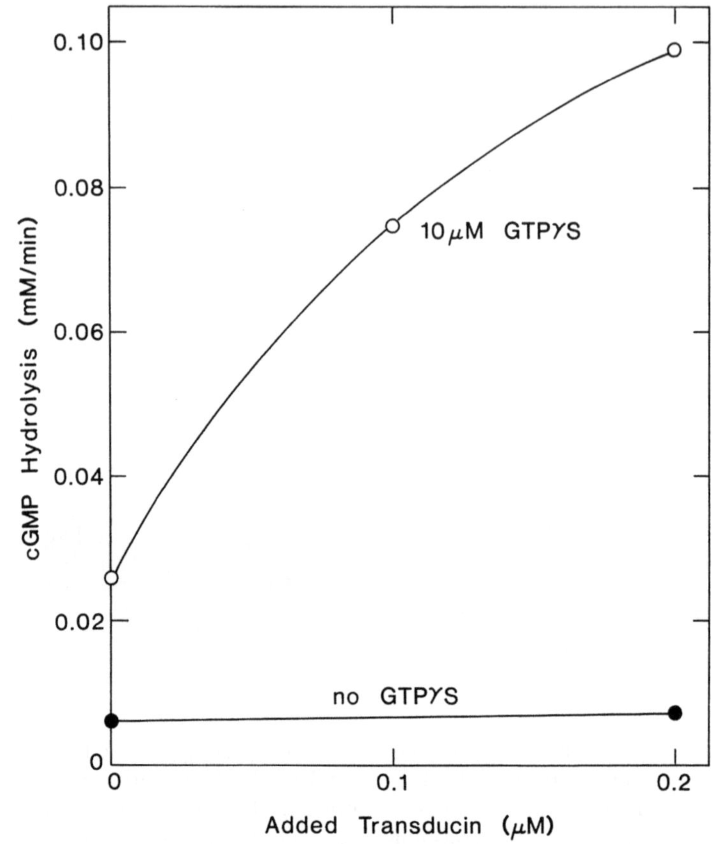

FIG. 13.7. ROS PDE activity stimulated by endogenous transducin can be further stimulated by addition of exogenous transducin. PDE activity was monitored by following the hydrolysis of [^{3}H]cGMP in a suspension of ROS (0.3 *μM* R*) in the presence (open circles) or absence (closed circles) of 10 *μM* GTPγS, along with the indicated concentrations of added transducin.

there is more PDE_γ than T_m. Hence the binding of T_α to PDE_γ cannot account for the finding that there are two distinct populations of transducin.

Another possibility is that the two populations of transducin differ in amino acid sequence or covalent modification. A reversible covalent modification could serve a regulatory function in excitation, recovery, or adaptation. For example, ADP-ribosylation of a cysteine residue by pertussis toxin (18–20) or of an arginine (21) by cholera toxin has profound effects on the properties of T_α; it is possible that one or both of these residues could serve as a target site for covalent modification. Fatty acylation of a cysteine residue in mammalian *ras* proteins (22, 23) is required for their binding to membranes. The finding that of eight cysteines coded for by T_α cDNA (24–26) only three could be chemically modified under denaturing conditions (27) suggests that one or more of the remaining cysteines may be involved in such a modification and also raises the possibility of different patterns of disulfide bonds that could give rise to functional differences. Another potential site for differential modification is the terminal amino group. This group is blocked in T_α (28); and acylation by myristic acid, as is found in a number of proteins at this position (29), could anchor T_α to the membrane. Other possible modifications include phosphorylation and partial proteolysis. Distinguishing between these alternatives is important for a full understanding of the mechanism of PDE activation.

Summary

The light-triggered activation of the cGMP PDE of ROS is mediated by the signal-coupling protein transducin. Photoexcited rhodopsin catalyzes the conversion of transducin from the inactive $T_{\alpha\beta\gamma}$-GDP form to the active T_α-GTP form, which stimulates the PDE (7).

We have found that there are two distinct populations of transducin in ROS. The major species, T_s, is released from the membranes on binding GTP. Micromolar concentrations of T_s are required for half-maximal activation of PDE. A minor species, T_m, which accounts for less than 30% of the total GTPγS uptake activity, remains membrane-bound on binding GTP under physiological salt conditions. T_m is a potent activator of PDE; 80% of the PDE activity observed in ROS (0.6 μM R*) in the presence of both T_s and T_m remained when T_s was removed and the total concentration of T_m was <16 nM. Moreover, the fraction of PDE activated by endogenous trandcucin in ROS was found to be independent of ROS concentration over the range 0.8 to 36.0 μM R*. Thus in dilute suspensions of ROS, PDE activation is primarily due to T_m.

Acknowledgments. This research was supported by a grant from the National Eye Institute (EY-02005) to L. S. and a National Research Service Award (EY-05815) to T. G. W.

References

1. Stryer, L., Hurley, J. B., and Fung, B. K-K. (1981) *Curr. Top. Membr. Transport* 15:93–108.
2. Fung, B. K-K., and Stryer, L. (1981) *Proc. Natl. Acad. Sci. USA* 77:2500–2504.
3. Stryer, L. (1986) *Annu. Rev. Neurosci.* 9:87–119.
4. Chabre, M. (1985) *Annu. Rev. Biophys. Biophys. Chem.* 14:331–360.
5. Yau, K-W., Haynes, L. W., and Nakatani, K. (1986) *Fortschritte der Zoologie* (*Prog. Zool.*) 33:343–366.
6. Kühn, H. (1981) *Curr. Top. Memb. Transplant.* 15:171–201.

7. Fung, B. K-K., Hurley, J. B., and Stryer, L. (1981) *Proc. Natl. Acad. Sci. USA* 78:152–156.
8. Tymininski, P. N., and O'Brien, D. F. (1984) *Biochemistry* 23:3986–3993.
9. Roof, D. J., Applebury, M. L., and Sternweis, P. C. (1985) *J. Biol. Chem.* 260:16242–16249.
10. Wensel, T. G., and Stryer, L. (1986) *Proteins* 1:90–99.
11. Yee, R., and Liebman, P. A. (1978) *J. Biol. Chem.* 253:8902–8909.
12. Liebman, P. A., and Sitaramayya, A. (1984) *Adv. Cyclic Nucleotide Protein Phosphorylation Res.* 17:215–225.
13. Yamanaka, G., Eckstein, F., and Stryer, L. (1985) *Biochemistry* 24:8094–8101.
14. Liebman, P. A., and Evanczuk, T. (1982) *Methods Enzymol.* 81:532–542.
15. Thompson, W. J., and Appleman, M. M. (1971) *Biochemistry* 10:311–316.
16. Hurley, J. B., and Stryer, L. (1982) *J. Biol. Chem.* 257:11094–11099.
17. Deterre, P., Bigay, J., Robert, M. Pfister, C., Kühn, H., and Chabre, M. (1986) *Proteins* 1:188–193.
18. Manning, D. R., Fraser, B. A., Kahn, R. A., and Gilman, A. G. (1984) *J. Biol. Chem.* 259:749–756.
19. West, R. E., Jr., Moss, J., Vaughn, M., Liu, T., and Liu, T-Y. (1985) *J. Cell Biol.* 101:354a.
20. Van Dop, C., Yamanaka, G., Steinberg, F., Sekura, R. D., Manclark, C. R., Stryer, L., and Bourne, H. R. (1984) *J. Biol. Chem.* 259:23–26.
21. Van Dop, C., Tsubokawa, M., Bourne, H. R., and Ramachandran, J. (1984) *J. Biol. Chem.* 259:696–698.
22. Sefton, B. M., Trowbridge, I. S., Cooper, J. A., and Scolnick, E. M. (1982) *Cell* 31:465–474.
23. Willumsen, B. M., Norris, K., Papageorge, A. G., Hubbert, N. L., and Lowry, D. R. (1984) *EMBO J.* 3:2581–2585.
24. Medynski, D. C., Sullivan, K., Smith, D., Van Dop, C., Chang, F. H., Fung, B. K-K., Seeburg, P. H., and Bourne, H. R. (1985) *Proc. Natl. Acad. Sci. USA* 82:4311–4315.
25. Yatsunami, K., and Khorana, H. G. (1985) *Proc. Natl. Acad. Sci. USA* 82: 4316–4320.
26. Tanabe, T., Nukada, T., Nishikawa, Y., Sugimoto, K., Suzuki, H., Takahashi, H., Noda, M., Haga, T., Ichiyama, A., Kangawa, K., Minamino, N., Matsuo, H., and Numa, S. (1985) *Nature* 315:242–245.
27. Ho, Y-K., and Fung, B. K-K. (1984) *J. Biol. Chem.* 259:6694–6699.
28. Hurley, J. B., Simon, M. I., Teplow, D. B., Robishaw, J. D., and Gilman, A. G. (1984) *Science* 226:860–862.
29. Magee, A. I., and Courtneidge, S. A. (1985) *EMBO J.* 4:1137–1144.

14
Signal Transduction in the β-Adrenoceptor-Dependent Adenylate Cyclase

ALEXANDER LEVITZKI

Receptors to hormones and neurotransmitters are transmembrane protein molecules that couple with specific biochemical effector systems. The mode of interaction between specific receptors and their biochemical effector system is a key element in the regulation of cell function. One such system is the hormonally regulated adenylate cyclase. Over the past 20 years, the key features of this multicomponent regulatory system have been unravelled, and very recently the components have been purified and reconstituted into a functionally active adenylate cyclase.

The Universal Structure of Hormonally Regulated Adenylate Cyclase

The hormonally regulated adenylate cyclase is a multicomponent system composed of five functional units (Fig. 14.1): (a) The stimulatory receptor R_s, which binds the stimulatory ligand. We shall discuss in detail the β-adrenoceptor, which is a member of this family. This receptor is a transmembrane glycoprotein, and it is likely that other stimulatory receptors are also transmembrane proteins. (b) The G_s (or N_s) protein, which binds GTP and is tightly associated with the catalytic unit. G_s is composed of three subunits: α_s, which possesses the GTP binding site and is the target for ADP-ribosylation by NAD^+ and cholera toxin, the β-subunit, and the γ-subunit, both of which are tightly associated with each other. The heterotrimetric G_s protein is embedded in the inner portion of the plasma membrane and is much less hydrophobic than either the β-receptor or the catalytic unit C. (c) The catalytic unit is a single transmembrane polypeptide carrying its catalytic site at the inner surface of the membrane. (d) An inhibitory receptor R_i. (e) The inhibitory GTP-binding protein G_i (or N_i); like G_s, it is a heterotrimer composed of three subunits: α_i, β, and γ, where the βγ complex is identical to the one found in G_s. α_i is homologous to α_s and is a substrate for ADP-ribosylation catalyzed by the toxin of *Bordetella pertussis*, but not by cholera toxin. The homology is, as expected, high in the sequences involved in the

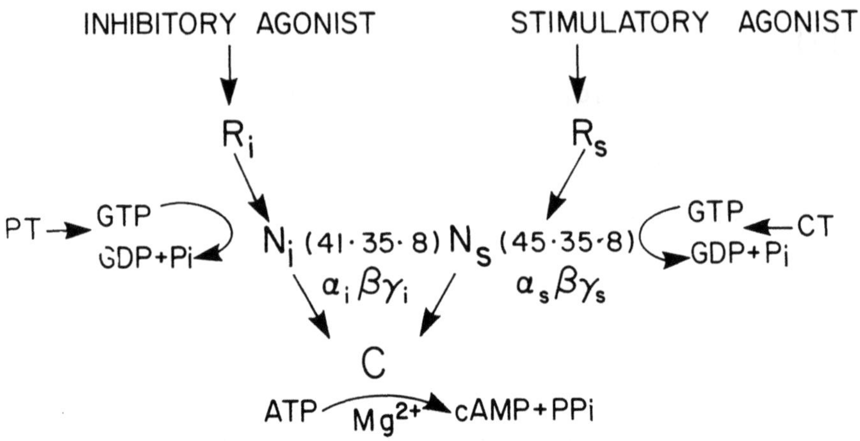

FIG. 14.1. The components of adenylate cyclase. R_s = stimulatory receptor; R_i = inhibitory receptor; N_s = the stimulating GTP regulatory protein; N_i = the inhibiting GTP regulatory protein; the β-subunits of both G-proteins are identical and presumably γ_s and γ_i are identical; CT = cholera toxin; PT = pertussis toxin—both toxins catalyse the ADR-ribosylation of the α subunits of the respective G-proteins.

binding of GTP. The GTP binding domain is composed of a number of conserved sequences including those found in yeast RAS proteins and protein synthesis elongation factors.

Universality of the System

It has been shown that stimulatory receptors from one cell type can hybridize with a heterologous adenylate cyclase system when transplanted in another cell type. For example, β-adrenoceptors from turkey erythrocytes and glucagon receptors from rat liver can be transferred into Friend erythroleukemia cells and activate its adenylate cyclase (1, 2). Kinetic experiments demonstrated independently that different receptors in the same cell couple to a single pool of adenylate cyclase: A_2 adenosine receptors and β_1-adrenoceptors in turkey erythrocytes couple to the same pool of adenylate cyclase molecules (3, 4). These experiments suggest that receptors that differ markedly in their pharmacologic specificity possess common structural domains that participate in the interaction between the stimulatory receptor and G_s. Recent cloning and sequencing studies on the mammalian β-2 adrenoceptors support this hypothesis. It was found (5) that certain sequences are homologous to bovine rhodopsin sequences, which also interact with a G protein (transducin). This finding supports the view that receptors that possess different pharmacologic specificities may have common structural domains.

The GTP stimulatory protein G_s is also a universal protein since it can couple to a variety of C units from different sources. For example, restoration of a functional β-receptor-dependent adenylate cyclase in G_s-deficient S49AC-T-lymphoma cells can be achieved by G_s from various sources: wild type S49 lymphoma cells, rabbit liver, turkey erythrocytes, and human erythrocytes (for review, see reference 6). Similar findings are now emerging for the inhibitory binding protein G_i. Recently, we have shown (7) that the catalytic unit of adenylate cyclase from either bovine brain or rabbit heart can couple equipotently with G_s from either turkey erythrocytes or rabbit liver. This result indicates that the catalytic unit C is also a highly conserved molecule.

Mode of Hormonal Control

Hormone-dependent activation of adenylate cyclase is summarized in Fig. 14.2. Hormone binding to R_s and GTP binding to G_s are both required to induce activation of the catalytic unit C to the cAMP-producing state. The active state of the system decays concomitantly with the hydrolysis of GTP to GDP and Pi at the G_s regulatory site. Replenishment of G_s with GTP and the continued presence of hormone at the receptor assure the ability of the system to reacquire its active cAMP-producing state. This type of "on-off" GTPase cycle accounts well for the properties of hormone-dependent adenylate cyclase (8):

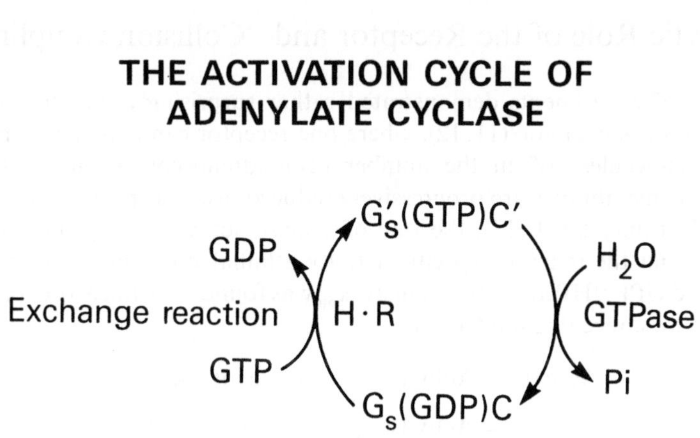

FIG. 14.2. The "on-off" cycle of adenylate cyclase activation. The steady-state level of active adenylate cyclase is given by:

$$[\text{Active enzyme}] = \frac{\text{Total enzyme}}{1 + k_{on}/k_{off}}.$$

As derived in reference 11, k_{on} is a linear function of the total receptor concentration.

1. Hormone-dependent adenylate cyclases also possess hormone dependent GTPase activity. This activity has since been directly demonstrated in a glucagon-dependent cyclase, a pancreozymin-dependent cyclase, and a prostaglandin E_1-dependent cyclase (for review, see reference 9).
2. Nonhydrolyzable GTP analogs increase adenylate cyclase activity. Indeed, GPPNHP, $GPPCH_2P$, and GTPγS generate a high activity form of cyclase since the enzyme is "trapped" in a persistently active form due to the complete blockade of the "turn-off" GTPase reaction.
3. Inhibition of the GTPase reaction by ADP-ribosylation catalyzed by cholera toxin activates the cyclase due to the slow-down of the "off" GTPase reaction. This covalent interconversion of the G_s unit is due to the ADP ribosylation of the α_s subunit, which is the GTP binding site. Recently, the purified G_s protein was shown to acquire GTPase activity when coreconstituted with purified 1-adrenoceptor in phospholipid vesicles and challenged with β-agonists plus GTP.
4. From the quantitative and independent measurements of the rate of enzyme activation by hormone and guanyl nucleotide (the "on" reaction) and the decay of the cAMP-producing state to its basal state (the "off" reaction), one can compute the fraction of the total pool of cyclase, which is in the active state in the presence of GTP. Such measurements have indeed been conducted and have demonstrated that the two-state model suggested early in 1977 (10) and presented in Fig. 14.2 is sufficient to account for the "on-off" cycle.

Catalytic Role of the Receptor and "Collision Coupling"

A series of experiments demonstrated rather surprisingly that the role of the receptor is catalytic (11, 12), where one receptor can activate numerous cyclase molecules. When the number of β-adrenoceptors on the turkey erthrocyte membranes are progressively reduced, using a specific β-receptor-directed affinity label (13), the rate of cyclase activation is proportionally reduced, but the maximal specific activity attainable is retained when hormone and GPPNHP are used. Namely, k_{on} was found to be linearly dependent on receptor concentration (eq. 1).

$$\begin{aligned} \text{Activity} &= \text{Activity}_{max} \times \{1 - \exp(-k_{on})t\} \\ &= \text{Activity}_{max} \times \{1 - \exp(-k[R])t\} \end{aligned} \quad [1]$$

This discovery led to the formulation of the "collision coupling" model (11).

$$HR_s + G_s \cdot (GDP)C \rightleftharpoons HR_s \cdot G_s(GDP) \cdot C$$

$$HR_s \cdot G_s \cdot (GDP)C + GTP \rightleftharpoons HR_s \cdot G_s(GTP) \cdot C + GDP$$

$$HR_s \cdot G_s(GTP) \cdot C \rightleftharpoons HR_s \cdot G_s'(GTP)C'$$

$$HR_s \cdot G_s'(GTP) \cdot C' \rightleftharpoons HR_s + G_s'(GTP) \cdot C'$$

$$G_s' \cdot (GTP) \cdot C' + H_2O \rightarrow G_s(GDP) \cdot C + Pi$$

where the intermediate $HR_s \cdot G_s \cdot (GTP) \cdot C$, does not accumulate.

This formulation predicts: (a) strict first-order kinetics of adenylate cyclase activation by hormone and GPPNHP; (b) a linear dependence of the rate constant of activation k_{on} on receptor concentration; and (c) a Michaelian (noncooperative) dependence of k_{on} on the hormone concentration, yielding the same hormone-receptor dissociation constant as binding experiments. Similar results were obtained for the activation of adenylate cyclase by glucagon and GPPNHP in the rat liver system (14).

A number of investigators suggested alternative molecular models (15, 16), which postulate that the G_s unit, being the "transducer" element, physically shuttles between the receptor and the catalytic unit.

The suggestion that G_s physically shuttles between the β-receptor and the catalytic unit is based on the following arguments: (a) activation of G_s by the β-receptor can be performed in the absence of the catalytic unit (see above); (b) G_s can be shown to associate with the β-adrenoceptor in the presence of β-agonists, but in the absence of GTP or its analogs; and (c) GTP and its analogs are known to reduce β-receptor affinity towards agonists (for review, see reference 8). The essence of this model is that G_s in the presence of GTP is first activated by the β-adrenoceptor, then physically separates from the receptor, migrates away from the receptor, and associates with C to form the active enzyme $G'_s(GTP) \cdot C'$. Upon hydrolysis of GTP to G_s (GDP) plus P_i, G_s (GDP) separates from C and reassociates with the β-receptor. In view of these suggestions (14, 15), we have examined in detail the kinetic consequences of "G_s shuttle" models and tested experimentally the feasibility of such models to explain the mode of coupling between β-receptors and adenylate cyclase. Rigorous analysis showed (17) that "shuttle" models predict nonfirst-order kinetics of adenylate cyclase activation, complex kinetics of activation by hormone and GPPNHP, as well as complex dependence of the rate of activation on receptor concentration.

Experiments on native membranes (18) and current studies on systems reconstituted from purified components (7, 19) establish that activation of the C unit is first order and that the β-receptor acts catalytically, while the rate of adenylate cyclase activation depends linearly on receptor concentration. These findings definitively exclude all shuttle models because none of them can account for these kinetic features. Acceptance of the "collision coupling" mechanism also means that G_s is associated with C at all times, namely during the *entire* course of the C to C′ transition. It can be independently demonstrated (20) that the G_s unit and the catalytic unit C are permanently associated in the detergent-solulized state. Furthermore, in the presence of phospholipid the stoichiometric complex $G_s \cdot C$ can be purified 250-fold both in the GDP-basal state and in the GPPNHP preactivated state. These results support the kinetic findings.

G_s Subunit Dissociation Model of Hormonal Activation and Inhibition

Exposure of G_s to the nonhydrolyzable GTP analog GTPγS, high Mg^{2+} in lubrol induces the irreversible dissociation of G_s:

$$G_s(\text{GDP}) + \text{GTP}\gamma\text{S} \xrightarrow[\text{lubrol PX}]{Mg^{2+}} \alpha_s(\text{GTP}\gamma\text{S}) + \beta\gamma$$

This finding led to a model (6) that suggests that hormonal stimulation in the presence of the natural ligand GTP leads to the reversible dissociation of G_s to α'_s (GTP), which becomes associated with the catalytic unit C. Upon hydrolysis of GTP to GDP, α_s (GDP) reassociates with $\beta\gamma$ (Fig. 14.3). The $\beta\gamma$-subunits complex, according to this scheme, competes with C for α_s. Hence, elevation of the level of the $\beta\gamma$-subunits causes α_s to be less accessible to C and, therefore, induces inhibition. According to this scheme, an agonist bound to an inhibitory receptor R_i facilitates the dissociation of G_i to α_i (GTP) and $\beta\gamma$, and the latter scavenging α_s and thus inhibiting C (Fig. 14.3). The key feature of this model is that it ties in the mode of hormonal inhibition of adenylate cyclase. This model also accounts for the failure to demonstrate a G_i to C interaction in reconstituted systems, and provides a mechanism by which G_i confers inhibition of C without a direct physical interaction.

This scheme implies that the α_s subunit shuttles between G_s and the catalytic unit C, and that G_s and C are physically separate entities within the membrane bilayer. Any mechanism, however, that implies $G_s \cdot C$ dissociation cannot be compatible with the kinetic data obtained for native membranes as well as for reconstituted systems (19), which clearly demonstrate that the activation of C is strictly first order, as previously explained. A modified G_s dissociation model (21, 22), however, can accommodate with the "collision coupling" model. The modified G_s dissociation scheme suggests that both species G_s (GTP) $\cdot$ C and α_s (GTP) $\cdot$ C represent active cyclase molecules, where the G_s (GTP)C species is the less active species. An important constraint on the system is that the nucleotide exchange reaction at α_s requires the presence of $\beta\gamma$ subunits. This has been directly demonstrated in reconstitution experiments (19). This constraint forces one to conclude that the stimulatory receptor interacts with the intact complex $G_s \cdot C(\beta\alpha_s C)$ to bring about the GDP to GTP exchange. If, as suggested, G_s (GTP)C readily dissociates to the active α'_s (GTP) $\cdot$ C and $\beta\gamma$, it must also readily recombine with the $\beta\gamma$ subunits once GTP is hydrolyzed, allowing another round of activation (Fig. 14.3). The G_s dissociation models assume that α_i has no direct effect on the catalytic unit C and its function is only to bind or release the $\beta\gamma$ subunits. These suggestions do not agree with the data of Hildebrandt et al. (23), who showed that within the S49 AC^- membranes, G_i interacts directly with C. Furthermore, there is evidence that G_s does not dissociate in the presence of the physiological ligand GTP (24). It is likely that G_s subunits never come apart during adenylate cyclase activation. If this is indeed the case, the mode of action of G_i needs to

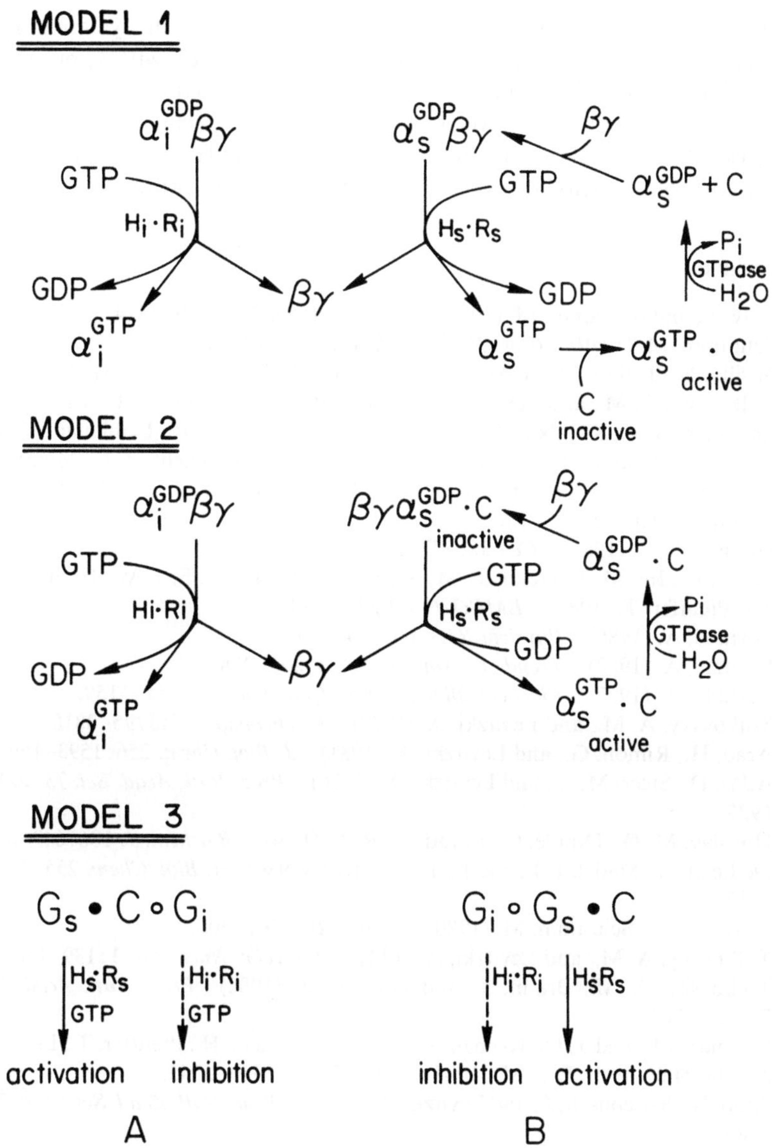

FIG. 14.3. Mode of regulation of adenylate cyclase by hormones and G-proteins. *Model 1* is essentially the original G-dissociation model, which assumes that the G-protein is a separate component and that the α_s subunit "shuttles" between G_s and C (6). *Model 2* is a modification of model 1 and assumes that α_s remains associated with C at all times. Still, like in model 1, the $\beta\gamma$ subunits play a key role in the regulation of the enzyme. *Model 3A* assumes a direct G_i to C interaction, which is much weaker than G_s to C interaction. *Model 3B* assumes that G_i attenuates adenylate cyclase by interacting with G_s.

be defined since the $\beta\gamma$ subunits would not mediate adenylate cyclase inhibition. The possibilities are (a) that G_i does interact directly with C but more weakly than G_s, and (b) G_i interacts with G_s, thus attenuating its stimulatory activity. These alternatives have yet to be examined experimentally.

Figure 14.3 summarizes the different molecular models that may account for the control of adenylate cyclase by hormones and G-proteins.

References

1. Orly, J., and Schramm, M. (1976) *Proc. Natl. Sci.* 73:4410–4414.
2. Schramm, M. (1979) *Proc. Natl. Acad. Sci.* 76:1174–1178.
3. Sevilla, N., Tolkovsky, A. M., & Levitzki, A. (1977) *FEBS Lett.* 81:339–341.
4. Tolkovsky, A. M., and Levitzki, A. (1978) *Biochemistry* 17:3811–3817.
5. Dixon, R. A. F., Kobilka, B. K., Strader, D. J., Benovic, J. R., Dohlanan, M. G., Frielle, T., Bolanowski, M. A., Bennet, C. D., Rands, E., Diehl, R. E., Mumford, R. A., Slater, E. E., Sigal, I. S., Caron, M. G., Lefkowitz, R. J., and Strader, C. D. (1986) *Nature* 321:75–79.
6. Gilman, A. G. (1984) *Cell* 36:577–579.
7. Feder, D., Im, M. J., Hekman, M., Klein, H., Helmreich, E. J. M., Levitzki, A., and Pfeuffer, T. (1986) *EMBO J.* 5:1509–1514.
8. Levitzki, A. (1986) *Physical Revs.* 66:819–842.
9. Levitzki, A. (1982) *Trends Pharmocol. Sci.* 3:203–208.
10. Levitzki, A. (1977) *Biochem. Biophys. Res. Commun.* 74:1154–1159.
11. Tolkovsky, A. M., and Levitzki, A. (1978) *Biochemistry* 17:3795–3810.
12. Arad, H., Rimon, G., and Levitzki, A. (1981) *J. Biol. Chem.* 256:1593–1597.
13. Atlas, D., Steer, M. L., and Levitzki, A. (1976) *Proc. Natl. Acad. Sci.* 73:1921–1925.
14. Houslay, M. D., Dipple, C., Elliott, K. R. F. (1980) *Biochem. J.* 186:649–658.
15. De Lean, A., Stadel, J. J., and Lefrowitz, R. J. (1980) *J. Biol. Chem.* 255:7108–7117.
16. Citri, Y., and Schramm, M. (1980) *Nature* 287:297–300.
17. Tolkovsky, A. M., and Levitzki, A. (1981) *J. Cyclic. Nucl. Res.* 1:139–150.
18. Tolkovsky, A. M., Braun, S., and Levitzki, A. (1982) *Proc. Natl. Acad. Sci.* 79:213–217.
19. Hekman, M., Feder, D., Keenan, A. K., Gal, A., Klein, H., Pfeuffer, T., Levitzki, A., and Helmreich, E. J. M. (1984) *EMBO J.* 3:3339–3345.
20. Arad, H., Rosebusch, J., and Levitzki, A. (1984) *Proc. Natl. Acad. Sci.* 81:6579–6583.
21. Levitzki, A. (1984) *J. Rec. Res.* 4(1–6):339–409.
22. Levitzki, A. (1987) *FEBS. Lett.* 211:113–118.
23. Hildebrandt, J. D., Codina, J., and Birnbaumer, L. (1984) *J. Biol. Chem.* 259:13178–13185.
24. Huff, R. M., and Neer, E. J. (1986) *J. Biol. Chem.* 26:1105–1110.

15
Acetyl-CoA Carboxylase: Correlation of Phosphorylation State with Allosteric Properties and Physiological State

HARIS JAMIL and NEIL B. MADSEN

Acetyl-CoA carboxylase (EC 6.4.1.2) may be considered to be the enzyme subjected to the most important regulation in the synthesis of long-chain fatty acids. It has been the subject of several reviews, which may be consulted for more details (1–4). In addition to variations in the amount of enzyme protein, there is an acute control that responds quickly to hormonal and nutritional conditions. The acute control has been attributed to both allosteric mechanisms and covalent modification by phosphorylation. Inhibition by acyl-CoA or activation by citrate may play a role, but consideration of citrate levels in the liver cell suggested that reversible phosphorylation, changing the sensitivity to citrate, may be the major mechanism for the acute control. Thus using a partially purified rat liver carboxylase containing phosphatase(s) and kinase(s), it was shown that dephosphorylation decreased the Km for citrate activation and the sensitivity to inhibition by acyl-CoA and ATP. It was suggested that only the dephosphorylated form would be active under in vivo conditions (4). Although carboxylase contains up to 8 mol of phosphate per subunit of 250,000 M_r, only one, at a specific serine residue, appears to be associated with a change in enzymic properties in the case of the mammary gland enzyme (5, 6). Phosphorylation of the liver carboxylase by a specific kinase showed a decrease in activity until the limit of approximately 1 mol P/subunit was incorporated (7).

Determination of the in vivo status of other enzymes that undergo reversible phosphorylation has been facilitated by enzymic tests that distinguish between the two forms, as exemplified by the sensitivity of glucose-6-phosphate activation of glycogen synthase to the latter's level of phosphorylation (8). In this chapter we report on the development of such a test for phosphorylation of liver carboxylase based on the enzymic characteristics of the citrate activation.

Materials and Methods

Details will be published elsewhere, and only a brief summary of the materials and methods is given here. Carboxylase was purified from rat livers by a method slightly modified from a previous report (9) and yielded a homogeneous preparation giving a single band on SDS-PAGE. A kinase that associates with it was purified to homogeneity by adding an FPLC chromatographic step to a previous method (7). Protein phosphatase was purified from rabbit liver using a published method (10) and contains the catalytic activities of protein phosphatases 1 and 2. Carboxylase activity was measured by the fixation of ^{14}C-labeled bicarbonate (11). Phosphorylation of the carboxylase was measured by the incorporation of ^{32}P from [γ-^{32}P]ATP into trichloracetic acid precipitates or a filter disk method, these methods yielding similar results.

For measuring enzymic activities in vivo, liver samples were removed from anesthetized rats and immediately frozen between aluminum blocks at liquid nitrogen temperatures before being pulverized at Dry Ice temperatures, as previously detailed (12). The frozen powdered liver was extracted with a stopping buffer containing 50m*M* Tris-HCl pH 7.5, 2 m*M* EDTA, 1 m*M* EGTA, 100 m*M* NaF, and six protease inhibitors. For comparison, livers were homogenized in the stopping buffer at 0°C.

Results and Discussion

Acetyl-CoA carboxylase was incubated with either protein phosphatase or the carboxylase kinase, and activities were measured at timed intervals in the presence of 10m*M* citrate; ^{32}P incorporation by the kinase was also determined. As shown in Figure 15.1, phosphatase treatment increased the activity by 1.4-fold, suggesting that the enzyme as isolated contained some phosphate, the removal of which caused an activity increase. The kinase, on the other hand, caused a 65% decrease in the enzymic activity accompanied by the incorporation of 0.45 mol P/mol subunit. The data of Fig. 15.1 are replotted in Fig. 15.2 as percent activity versus incorporation of ^{32}P per mole of subunit, assigning 100% activity to that obtained after exhaustive phoshatase treatment. This figure suggests that the enzyme as isolated has only 72% of its potential activity and contains approximately 0.3 mol P/mol subunit that affects activity, whereas kinase raises the level of phosphate, which affects activity to 0.75 mol P/subunit, resulting in an enzyme that has only 24% of its potential activity. The inverse linear relation between phosphate incorporated by the kinase and the carboxylase activity, as shown in Fig. 15.2, was confirmed in a separate experiment in which carboxylase was treated exhaustively by the phosphatase, then reisolated on an avidin column before incubation with the kinase under conditions similar to those used for the lower part of Fig. 15.1. In this case the kinase caused a 85% loss of activity accompanied by incorporation of 0.71 mol P/subunit, and the linear inverse relation between the two measured parameters had a correlation coefficient of -0.98 (data not shown, to be published elsewhere). The extra phosphate incorporated into the phosphatase-treated carboxylase ($0.71 - 0.45 = 0.26$) also agrees with the amount of "functional" phosphate

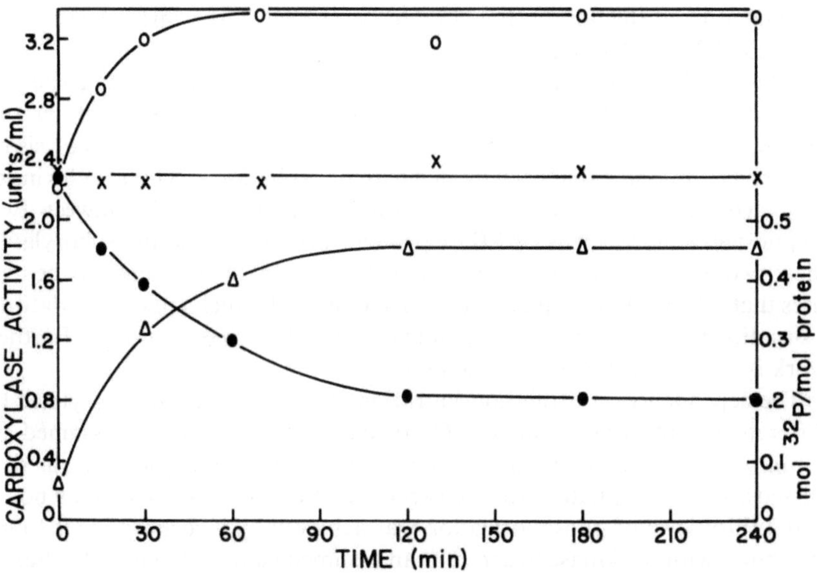

FIG. 15.1. Effect of phosphorylation and dephosphorylation on acetyl-CoA carboxylase activity. Carboxylase was incubated with the kinase and ATP (●), the phosphatase (○), or neither (×), and activities were measured with 10 m*M* citrate. Phosphate incorporation by the kinase was measured using ^{32}P-ATP (△).

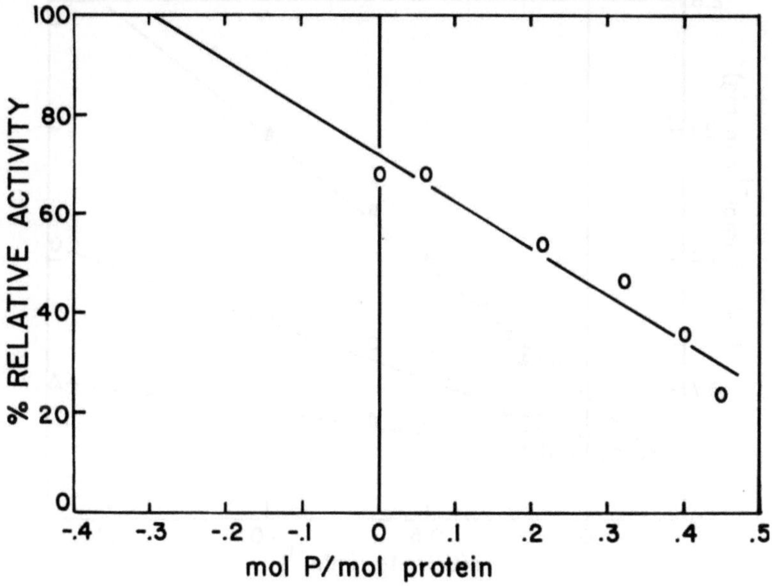

FIG. 15.2. Inverse relation between phosphate-incorporated and enzyme activity. The data from Fig. 15.1 are replotted, assigning 100% activity to the enzyme treated exhaustively with the phosphatase.

calculated from the data in Fig. 15.1 to be present in the carboxylase when isolated (0.3 mol P/subunit).

Whereas the results presented above suggested that only a single site per carboxylase subunit was affecting activity when phosphorylated by the kinase, as was found for the enzyme from mammary gland (5, 6), analysis of carboxylase that had first been treated with phosphatase and then with kinase indicated that radioactive phosphate was distributed equally between two tryptic peptides, whereas cAMP-dependent protein kinase phosphorylated these two peptides and one other (results not shown). Thus there may be two sites that affect activity equally but which limit each other's phosphorylation, a situation that also appears to occur in the case of glycogen synthase. Further work is needed to clarify this situation.

The dependence on citrate concentration of carboxylase phosphorylated to different extents was examined. Carboxylase as isolated was assumed to contain approximately 0.3 mol P/subunit on sites affecting activity, as discussed above; phosphatase-treated carboxylase was assumed to contain none; and a third level of phosphorylation was achieved by incorporating 0.3 mol P/subunit with the kinase to arrive at an assumed value of 0.6 mol P/subunit. The effects of citrate concentration on these three preparations are shown in the form of reciprocal plots in Fig. 15.3. For the carboxylase preparations containing 0, 0.3, and 0.6 mol of "functional" phosphate per subunit, the Ka

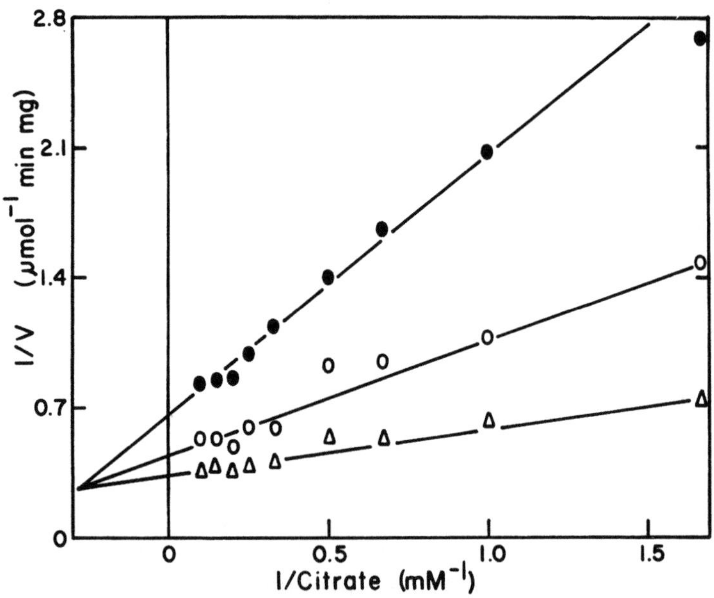

FIG. 15.3. Reciprocal plot of the effect of citrate concentration on activities of carboxylase preparations containing various levels of phosphate, as defined in the text. Moles of P per subunit were zero (▵), 0.3 (○), and 0.6 (●).

values for citrate activation were 0.8, 1.4, and 2.1 m*M*, respectively, and the corresponding maximal velocities were 3.0, 2.2, and 1.5 μmol/min/mg. It may be observed that although the ratio of maximal velocities varied only twofold with the extent of phosphorylation, the ratio was >10 at low or zero citrate concentrations. This finding suggested a basis for a sensitive enzymic test that would give a measure of the extent of phosphorylation at the sites affecting activity.

Carboxylase that had been treated with protein phosphatase and then reisolated, as described above, was incubated with the carboxylase kinase and ^{32}P-ATP for various times before stopping the reaction so that a series of carboxylase preparations with increasing amounts of incorporated phosphate were obtained. They were then examined for enzymatic activities at several citrate concentrations. Whereas activities at all citrate concentrations were inversely proportional to the amount of phosphate incorporated, the ratio of activities at two citrate concentrations was considered to be the best measure as it provided an internal control when used with crude extracts of tissues. The ratio of activities at 0 and 2 m*M* citrate was especially promising, and the data in Table 15.1 show its relation to the amount of phosphate incorporated in each subunit by the kinase. The equation relating the two parameters is $R = 0.50 - 0.64\,P$, where R is the ratio of activities at 0 and 2 m*M* citrate and P is the moles of phosphate incorporated into each subunit. The coefficient of correlation is -0.95.

This activity ratio method was used to estimate the occupancy of phosphorylatable sites in the carboxylase in the livers of rats subjected to different nutritional conditions that are known to affect carboxylase activity. Starvation for 48 h, normal feeding, and starvation followed by refeeding with a fat-free diet yielded average values of 0.44, 0.26, and 0 mol P/subunit of carboxylase when samples were frozen to liquid nitrogen temperatures within 1 s

TABLE 15.1. Effect of phosphate incorporated into acetyl-CoA carboxylase by its kinase on the ratio of enzymic activities at 0 and 2 m*M* citrate.

Phosphate (mol/mol subunit)	Activity Ratio at 0 and 2 m*M* Citrate
0	0.49
0.07	0.46
0.19	0.37
0.37	0.39
0.50	0.16
0.51	0.16
0.54	0.18
0.61	0.07
0.65	0.08

of excision followed by extraction with stopping buffer at temperatures below freezing. If more conventional methods of homogenizing liver samples with the stopping buffer at 0°C was followed, the activity ratio method applied to the crude homogenate yielded 0.65, 0.52, and 0.28 mol P/subunit, respectively, for the three nutritional conditions. We interpret the increased phosphate contents to indicate that the inhibitors in the stopping buffers did not stop in kinase activity rapidly enough, a result similar to that found for glycogen muscle phosphorylase by Danforth et al. (13). For acetyl-CoA carboxylase, as for other metabolically interconvertible enzymes, its state in tissues must be estimated on samples that have been frozen rapidly, then pulverized and extracted at temperatures well below 0°C.

The effects of nutritional state on the extent of phosphorylation of the carboxylase, as estimated by the citrate activity ratio method, was confirmed by independent means, including isolation of the carboxylase and the use of both the kinase and the phosphatase to estimate occupancy of the sites affecting activity. Although absolute figures varied, the trend noted above was maintained.

Returning to the effect of phosphate content on the enzymic activity of the carboxylase, the effects of phosphatase and kinase action (Fig. 15.1) agree with earlier work (4, 14), but the linear inverse relation between phosphate incorporated by the kinase was not clear from the previous work. Our kinetic study of the effect of citrate on the activity of carboxylase containing various amounts of phosphate agree in general with that of Carlson and Kim (15), although these authors observed a sigmoidal response of the phosphorylated form. It may be noted that the covalently bound phosphate exhibits a noncompetitive inhibition of the citrate activation, with an intersection point at 3.6 m*M* citrate. This finding suggests that citrate can activate the fully phosphorylated enzyme, but the interaction constant of 4.5 (3.6 m*M* versus 0.8 m*M* for the dephosphorylated enzyme) suggests a strong allosteric effect of the covalently bound phosphate on the binding of citrate to its separate site.

Additional details and applications of this research have now been published (16, 17).

Acknowledgments. We thank Mr. M. Monanu for the gift of protein phosphatase and Mrs. S. Shechosky and Mrs. Rita Langford for excellent technical assistance.

This investigation was supported by grant MT-1414 from The Medical Research Council of Canada. H. J. is a Fellow of the Alberta Heritage Foundation for Medical Research.

References

1. Lane, M. D., Moss, J., and Palakis, S. E. (1974) *Curr. Top. Cell. Regul.* 8: 139–195.
2. Block, K., and Vance, D. (1977) *Annu. Rev. Biochem.* 46:263–298.

3. Geelen, M. J. H., and Beynen, A. C. (1981) In *Short-term Regulation of Liver Metabolism*, edited by L. Hue and G. Van de Werve, pp. 231–246. Elsevier, Amsterdam.
4. Kim, K-H. (1983) *Curr. Top. Cell. Regul.* 22:143–176.
5. Hardie, D. G. (1980) In *Recently Discovered Systems of Enzyme Regulation by Reversible Phosphorylation*, edited by P. Cohen, pp. 33–62. Elsevier, Amsterdam.
6. Hardie, D. G., and Holland, R. (1983) In *15th FEBS Meeting Abstracts*, p. 34.
7. Lent, B. A., and Kim, K-H. (1982) *J. Biol. Chem.* 257:1897–1901.
8. Guinovart, J. J., Salavert, A., Massaque, J., Ciudad, C. J., Salsas, E., and Itarte, E. (1979) *FEBS Lett.* 106:284–288.
9. Tipper, J. P., and Witters, L. A. (1982) *Biochim. Biophys. Acta* 715:924–929.
10. Brandt, H., Capulong, Z. L., and Lee, E. Y. C. (1975) *J. Biol. Chem.* 250: 8038–8044.
11. Majerus, P. W., Jacobs, R., Smith, M. B., and Morris, H. P. (1968) *J. Biol. Chem.* 243:3588–3595.
12. Maddaiah, V. T., and Madsen, N. B. (1966) *Biochim. Biophys. Acta* 121: 261–268.
13. Danforth, W. H., Helmreich, E., and Cori, C. F. (1962) *Proc. Natl. Acad. Sci. USA* 48:1191–1199.
14. Thampy, K. G., and Wakil, S. J. (1985) *J. Biol. Chem.* 260:6318–6323.
15. Carlson, C. A., and Kim, K-H. (1974) *Arch. Biochem. Biophys.* 164:478–489.
16. Jamil, H., and Madsen, N. B. (1987) *J. Biol. Chem.* 262:630–637.
17. Jamil, H., and Madsen, N. B. (1987) *J. Biol. Chem.* 262:638–642.

16
Role of Cyclic Cascades in Metabolic Regulation

P. BOON CHOCK, STEWART R. JURGENSEN, SUE GOO RHEE, EARL R. STADTMAN, and JACKIE R. VANDENHEEDE

Covalent interconversion of enzymes was first observed in glycogen phosphorylase from rabbit skeletal muscle (1–4). The interconversion involved the phosphorylation of the phosphorylase that converted the enzyme to its active form, and dephosphorylation returned the enzyme to its relatively inactive form. Since then, the number of enzymes/proteins reported to undergo covalent interconversion has mushroomed (5–7). Many of these interconvertible enzymes/proteins occupy key positions in metabolic networks, and their enzymic activities vary as a function of the fractional modification of the enzyme. Reversible covalent modifications can be grouped into seven classes based on the nature of the modification (Table 16.1). Of this list, the reversibility of tyrosine sulfation has yet to be established. Nevertheless, this process is likely to be reversible because protein sulfation has been reported to be involved in some essential regulatory pathways, e.g., tyrosine sulfation of interleukin-2 receptor (8), and reduced levels of tyrosine sulfate were found in retrovirus-induced transformed cells (9). Each of the modifications and demodifications listed in Table 16.1 is catalyzed by specific converter enzymes; e.g., phosphorylation and dephosphorylation are catalyzed by protein kinases and phosphatases, respectively. This action of one enzyme on another constitutes a cascade, and the coupling of two opposing cascades forms a cyclic cascade system (Fig. 16.1). The physiological significance of the cyclic cascade mechanism in regulating cellular functions has been extensively analyzed (6, 7, 10–12). The results reveal that cyclic cascades are endowed with unique properties for well coordinated cellular regulation. Therefore it is not surprising that a broad array of cellular pathways such as protein, carbohydrate, lipid, and nucleic acid metabolism, muscle contraction, viral oncogenesis, and hormonal signal transduction are modulated through cyclic covalent modification of enzymes/proteins.

As reported earlier (10–12), in metabolic regulation cyclic cascades can function as signal amplifiers that allow the system to respond effectively to the presence of low effector concentrations, well below the dissociation constant of the effector–enzyme complex involved. Thus they provide a means for sensing low effector concentrations without the normal requirement for tight

TABLE 16.1. Reversible covalent modifications of proteins.[a]

Modification	Donor Molecule	Modified Amino Residues
Phosphorylation	ATP, GTP	Ser, Thr, Tyr, LysOH
ADP ribosylation	NAD^+	Arg, Glu, Lys, diphthamide
Nucleotidylylation (adenylylation and uridylylation)	ATP, UTP	Tyr, Ser
Methylation	*S*-Adenosylmethionine	Asp, Glu, Lys, His, Gln
Acetylation	Acetyl-CoA	Lys
Tyrosylation	Tyrosine	Carboxyl terminus
Sulfation	3-Phosphoadenosine-5-phosphosulfate	Tyr

[a] For original references, see ref. 7.

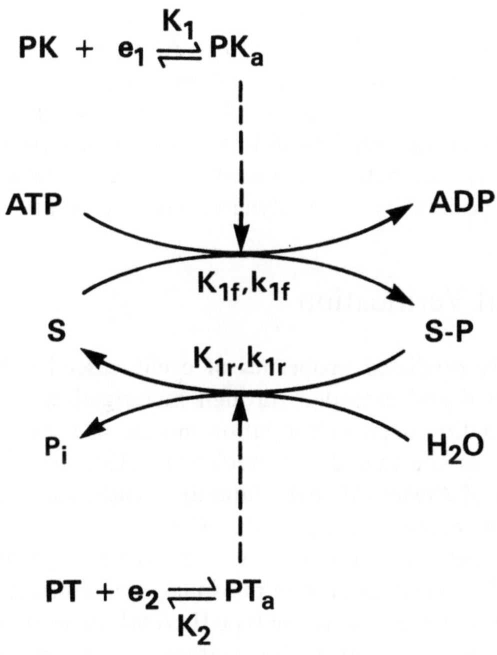

FIG. 16.1. Monocyclic cascade. PK and PT are inactive converter enzymes that can be converted to their active forms, PK_a and PT_a, by allosteric effectors, e_1 and e_2, respectively. S and S-P are unmodified and modified forms of the interconvertible substrate. K(s) and k(s) are dissociation and rate constants for the reaction designated.

binding that would reduce the temporal efficiency of the control process. Cyclic cascades can also enhance the sensitivity of fractional modification of the interconvertible enzyme(s) to changes in the concentrations of effectors by generating an apparent positive cooperative response for fractional modification of the interconvertible enzyme to increasing effector concentration, modulate the amplitude of the maximal response obtainable at a saturating level of effector concentration, and serve as rate amplifiers. Because of the facts: that a minimum of three enzymes, an interconvertible enzyme and two converter enzymes (Fig. 16.1), is required to constitute a monocycle; that the number of converter and interconvertible enzymes increases as the number of cycles increases in the cyclic cascade system; and that each of these enzymes can serve as targets for one or more allosteric effectors; cyclic cascades can sense simultaneous fluctuations in concentrations of numerous metabolites and adjust accordingly the specific activity of the targeted interconvertible enzymes. They can therefore serve as effective biological integrators. The price the cell must pay for such an effective means of regulation is the consumption of energy-rich donor molecules (Table 16.1). As depicted in Fig. 16.1, for each cycle of a phosporylation/dephosphorylation cycle, one equivalent of ATP is consumed. This ATP flux provides the free energy required to maintain the steady-state level of fractional modification of the interconvertible enzyme, which reflects its specific activity, at metabolite-specified levels that are different from those specified by thermodynamic consideration.

Experimental Verification

The theoretically predicted properties of cyclic cascades described earlier have been verified and extended through investigation of (a) a simple in vitro phosphorylation/dephosphorylation monocyclic cascade developed to test the validity of the theoretical predictions (13), (b) the adenylylation/deadenylylation of *Escherichia coli* glutamine synthetase with both purified protein and permeabilized cells (14–16); Rhee et al., this volume), and (c) a synchronous cascade consisting in type II cAMP-dependent protein kinase and Mg(II)-ATP-dependent protein phosphatase as converter enzymes (17).

The in vitro system (13) consists in type II cAMP-dependent protein kinase and a 38-kD, type 2A protein phosphatase as converter enzymes and a nanopeptide, Leu-Arg-Arg-Ala-Ser-Val-Ala-Gln-Leu, as interconvertible substrate. The results obtained from this simple model show that at constant ATP levels a steady state is established for the fractional phosphorylation of the nanopeptide. At steady state, ATP hydrolysis continues at a constant rate, which is a measure of the interconversion rate of the nanopeptide between its phosphorylated and dephosphorylated forms. Furthermore, as predicted by the theoretical analysis, both signal amplification and the capacity to generate a cooperative response to increasing effector concentrations were observed. Because of signal amplification, only one-tenth of the cAMP concentration required to half-activate the kinase is needed to obtain 50% phosphorylation of the nanopeptide. Apparent cooperativity, derived from tight binding between the nanopeptide and the kinase, was observed for the fractional phosphorylation of the nanopeptide as a function of cAMP concentration. This increase in sensitivity leads to the observation that phosphorylation of the nanopeptide responds more sharply to increasing concentrations of cAMP than does activation of the cAMP-dependent protein kinase.

In the case of *E. coli* glutamine synthetase, its activity in catalyzing the biosynthesis of glutamine under physiological conditions is regulated by the adenylylation/deadenylylation of a specific tyrosine residue in each subunit of the dodecamer (6, 12). The unadenylylated enzyme is active, whereas the adenylylated enzyme is inactive. Studies have been carried out using purified proteins for the monocyclic (14) and bicyclic (15; Rhee et al., this volume) cascades of *E. coli* glutamine synthetase, as well as for permeabilized *E. coli* cells (16). The results are significantly more complex than those observed with the simple in vitro system because the effectors, glutamine and α-ketoglutarate, each exhibits dual roles in the adenylylation/deadenylylation reactions; α-ketoglutarate also plays dual roles in the uridylylation/deuridylylation reactions, whereas glutamine inhibits only the deuridylylation reaction. In addition, more than a single converter enzyme–effector complex is catalytically active. Nevertheless, the data confirmed the predicted properties of cyclic cascades. For more details of this study, see Rhee et al. (this volume).

In search of a well synchronized regulatory mechanism involving a cyclic cascade system, we investigated the activation mechanism and the regulatory properties of rabbit skeletal muscle type I protein phosphatase. This enzyme consists of two types of subunits, a catalytic subunit that migrates as a 38-kD protein in SDS-gel electrophoresis and the modulator subunit (also known as heat-stable inhibitor-2) that migrates as a 31-kD protein. The enzyme complex is inactive as isolated. it can be activated by various methods (Vaudenheede et al., this volume), among them, activation by a protein kinase F_A, which is also known as glycogen synthase kinase-3, and MgATP. Therefore we refer to it as Mg(II) · ATP-dependent protein phosphatase. This activation involved the phosphorylation of a modulator subunit that induces a conformational change of the catalytic subunit to its active form (18–20). The finding that the regulatory subunit of type II cAMP-dependent protein kinase (R_2C_2) is a potent inhibitor of Mg(II) · ATP-dependent protein phosphatase (17, 21) allowed one to formulate a highly synchronous cascade (Fig. 16.2) when type II cAMP-dependent protein kinase and the Mg · ATP-dependent protein phosphatase are the converter enzymes. The reaction scheme depicted in Fig. 16.2 was derived from mechanistic studies showing that: (a) addition of type II R_2C_2 inhibits the phosphatase activity only if cAMP is present, and inhibition observed with the regulatory subunit of R_2C_2 can be reversed by adding the catalytic subunit of R_2C_2 in the absence of cAMP; (b) the regulatory subunit

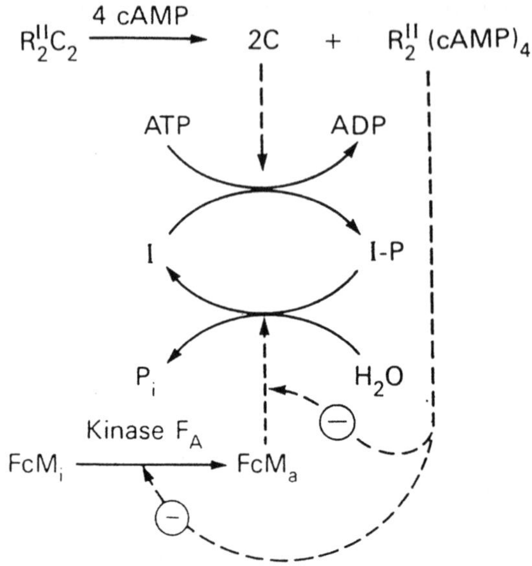

FIG. 16.2. Scheme for the synchronous regulation of cAMP-dependent protein kinase (R_2^{II} C_2) and the Mg(II)·ATP-dependent protein phosphatase ($F_c \cdot M$). $F_c \cdot M_i$ and $F_c \cdot M_a$ = inactive and active forms of $F_c \cdot M$. I and I-P = dephosphorylated and phosphorylated forms of the interconvertible proteins, respectively. C and R_2^{II} (cAMP)$_4$ = catalytic subunit and cAMP-bound regulatory subunit of the cAMP-dependent protein kinase, respectively.

of R_2C_2 inhibits both the activated Mg(II)·ATP-dependent protein phosphatase and the ethanol-treated 38-kD catalytic subunit of the phosphatase; and (c) the regulatory subunit also inhibits the activation process of the Mg(II)·ATP-dependent protein phosphatase as indicated by the observations that the extent of inhibition is enhanced significantly if the regulatory subunit was added before the addition of kinase F_A, and the degree of inhibition increases as one decreases the concentration of kinase F_A used in the activation of the phosphatase. Although these inhibition data indicate that the regulatory subunit of R_2C_2 inhibits both the activated form of the phosphatase and the activation process of the inactive phosphatase by kinase F_A, the two effects can be accounted for by the binding of a single regulatory subunit to the phosphatase; presumably it binds to the catalytic subunit.

Based on the scheme shown in Fig. 16.2, one can calculate the fractional phosphorylation of the interconvertible protein (I) as a function of the cAMP concentration and compare it with the case in which the regulatory subunit is not an inhibitor for the protein phosphatase (Fig. 16.3). In Fig. 16.3 it is assumed that the regulatory subunit causes 60% inhibition of the activated phosphatase and behaves as a competitive inhibitor of the kinase F_A. The calculated results indicate that a synchronous cyclic cascade could further

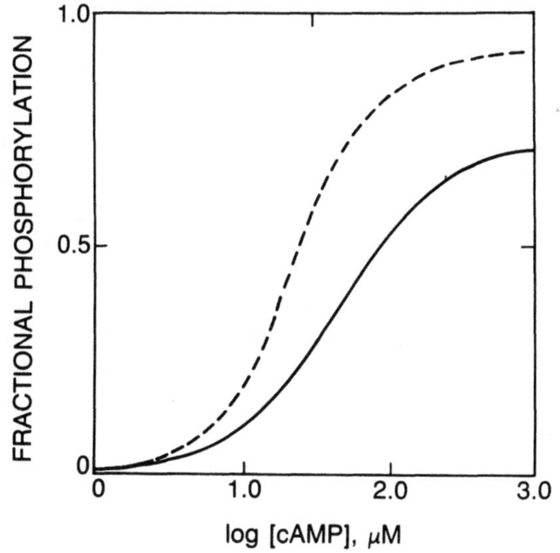

FIG. 16.3. Computer-simulated fractional phosphorylation of I, based on the reaction scheme depicted in Fig. 16.2, as a function of cAMP concentration. (---) Obtained with the synchronous cascade shown in Fig. 16.2 and R_2^{II} $(cAMP)_4$ was assumed to give 60% inhibition of the activated protein phosphatase; it also functions as a competitive inhibitor for kinase F_A. (——) Fractional phosphorylation of I when R_2^{II} $(cAMP)_4$ is not an inhibitor for the phosphatase.

enhance the predicted properties of cyclic cascades. For example, it could lead to an increase in the amplitude of fractional modification of the interconvertible substrate, in sensitivity as reflected by increasing the apparent cooperativity, and in signal amplification. It should also be pointed out that a synchronous cyclic cascade also provides a means for conserving the ATP hydrolysis required for maintaining the cyclic cascades in an operative state.

Concluding Remarks

The cyclic cascade model, derived mainly from detailed investigation on the regulation of *E. coli* glutamine synthetase, is applicable to all covalent interconvertible enzyme systems. Analysis reveals the regulatory advantages of the cyclic cascades, e.g., signal amplification, rate amplification, enhanced sensitivity to changes in metabolic concentrations, and flexibility. In addition, they serve as biological integrators by sensing changes in the concentrations of multiple metabolites and adjust accordingly the specific activity of the target enzymes. The unique properties of cyclic cascades can be attributed to the combined usage of both covalent modification of proteins and allosteric interactions. It has been shown (7) that allosteric interactions alone fail to provide some of the essential properties of cyclic cascades, particularly the temporal efficiency and the catalytic effect of the control process. Although cyclic cascades are suitable for modulating the specific activity of the interconvertible enzyme smoothly and continuously over a wide range of conditions, they can also serve as "on–off" switches for the target enzymes under extreme physiological conditions. The price for the usage of a cyclic cascade mechanism is the consumption of ATP and other energy-rich donor molecules. It has been shown (22) that the amount of ATP required for maintaining a cyclic cascade is negligible compared to the total cellular ATP hydrolysis. However, the ATP consumption can be minimized by using synchronous cyclic cascade systems. In view of the regulatory advantages of cyclic cascades, it is reasonable to expect a large number of key enzymes to be regulated by covalent interconversion mechanisms, and the cyclic cascades involved could be highly synchronized.

References

1. Cori, G. T., and Green, A. A. (1943) *J. Biol. Chem.* 151:31–38.
2. Cori, C. F., and Cori, G. T. (1945) *J. Biol. Chem.* 158:341–345.
3. Sutherland, E. W., Jr., and Wosilait, W. D. (1955) *Nature* 175:169–170.
4. Fischer, E. H., and Krebs, E. G. (1955) *J. Biol. Chem.* 216:121–132.
5. Krebs, E. G., and Beavo, J. A. (1979) *Annu. Rev. Biochem.* 48:923–959.
6. Chock, P. B., Rhee, S. G., and Stadtman, E. R. (1980) *Annu. Rev. Biochem.* 49:813–843.
7. Shacter, E., Chock, P. B., Rhee, S. G., and Stadtman, E. R. (1986) *In The Enzymes*, Vol. 17, edited by P. D. Boyer and E. G. Krebs, pp. 21–42. Academic Press, New York.
8. Leonard, W. J., Depper, J. M., Krönke, M., Robb, R. J., Waldmann, T. A., and Greene, W. C. (1985) *J. Biol. Chem.* 260:1872–1880.
9. Lin, M. C., and Lipmann, F. (1984) *Proc. Natl. Acad. Sci. USA* 81:3695–3698.
10. Stadtman, E. R., and Chock, P. B. (1977) *Proc. Natl. Acad. Sci. USA* 74:2761–2765.
11. Chock, P. B., and Stadtman, E. R. (1977) *Proc. Natl. Acad. Sci. USA* 74:2766–2770.
12. Stadtman, E. R., and Chock, P. B. (1978) *Curr. Top. Cell. Regul.* 13:53–95.
13. Shacter, E., Chock, P. B., and Stadtman, E. R. (1984) *J. Biol. Chem.* 259:12252–12259.
14. Rhee, S. G., Park, R., Chock, P. B., and Stadtman, E. R. (1978) *Proc. Natl. Acad. Sci. USA* 75:3138–3142.
15. Rhee, S. G., Chock, P. B., and Stadtman, E. R. (1985) *In the Enzymology of Post-Translational Modification of Proteins* Vol. 2, edited by R. Freedman, pp. 273–297. Academic Press, New York.
16. Mura, U., Chock, P. B., and Stadtman, E. R. (1981) *J. Biol. Chem.* 256:13022–13029.
17. Jurgensen, S. R., Chock, P. B., Taylor, S., Vandenheede, J. R., and Merlevede, W. (1985) *Proc. Natl. Acad. Sci. USA* 82:7565–7569.
18. Hemmings, B. A., Resink, T. J., and Cohen, P. (1982) *FEBS Lett.* 150:319–324.

19. Jurgensen, S. R., Shacter, E., Huang, C. Y., Chock, P. B., Yang, S-D., Vandenheede, J. R., and Merlevede, W. (1984) *J. Biol. Chem.* 259:5864–5870.
20. Villa-Morrcozzi, E., Ballou, L. M., and Fischer, E. H. (1984) *J. Biol. Chem.* 259:5857–5863.
21. Jurgensen, S. R. (1984) *Fed. Proc.* 43:2817.
22. Shacter, E., Chock, P. B., and Stadtman, E. R. (1984) *J. Biol. Chem.* 259:12260–12264.

17
Regulation of Glutamine Synthetase Activity and Its Biosynthesis in *Escherichia coli*: Mediation by Three Cycles of Covalent Modification

SUE GOO RHEE, WONG GI BANG, JA HYUN KOO, KYUNG HEE MIN, and SANG CHUL PARK

Glutamine synthetase (GS), which catalyzes the first step of nitrogen assimilation in prokaryotes, occupies a central position in the regulation of nitrogen metabolism and is vigorously regulated in response to the availability of the nitrogen source. *Escherichia coli* cells grown under nitrogen limitation contain active GS, and upon addition of excess ammonia to this culture GS is rapidly inactivated. This modulation is achieved by means of covalent modification of GS, which involves two linked nucleotidylation cycles (for review, see refs. 1,2). One involves the adenylylation and deadenylylation of GS (cycle on the upper right of Fig. 17.1); the other involves the cyclic uridylylation and deuridylylation of regulatory protein P_{II} (cycle on the middle left of Fig. 17.1).

The adenylylation of GS involves the transfer of an adenylyl group from ATP to each subunit of GS. The adenylyl group is attached to the enzyme by phosphodiester linkage to the hydroxyl group of a specific tyrosyl residue in each subunit. Because GS is a dodecamer composed of identical subunits, as many as 12 adenylyl groups can be attached to each enzyme molecule. Moreover, because adenylylated subunits are catalytically inactive under most physiological conditions, the specific activity of the enzyme is inversely proportional to the average number of adenylylated subunits per enzyme molecule.

The adenylylation of GS is opposed by enzyme-catalyzed phosphorylysis of the adenylyltyrosyl bonds to regenerate unmodified (active) GS and produce ADP. These two opposing activities reside on a single polypeptide enzyme with an M_r of 110,000 (3), and several lines of evidence (4) indicate that there are two separate catalytic sites on the adenylyltransferase: One catalyzes the adenylylation reaction (AT_a), and the other is for the deadenylylation reactions (AT_d). Because two opposite reactions are catalyzed on the same enzyme, it follows that the two processes are intimately coupled. Moreover, unless the capacity of the adenylyltransferase to catalyze one reaction with respect to the other is strictly regulated, GS undergoes senseless cycling between adenylylated and unadenylylated states, the net result of which is simply phosphorylysis of ATP to form ADP and PP_i. Such futile cycling is

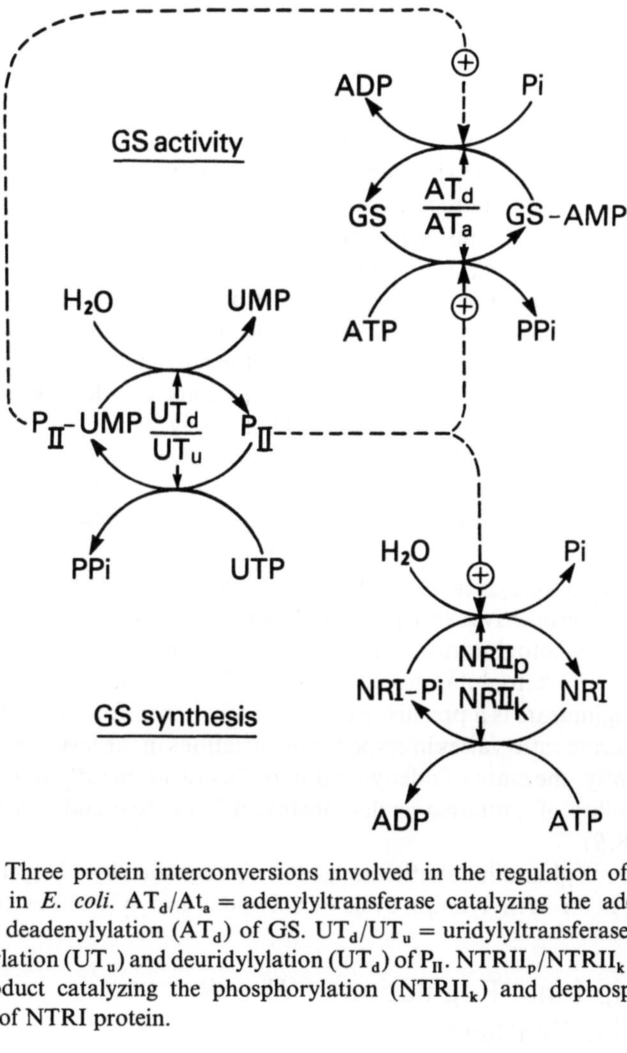

FIG. 17.1. Three protein interconversions involved in the regulation of glutamine synthetase in *E. coli.* AT_d/At_a = adenylyltransferase catalyzing the adenylylation (AT_a) and deadenylylation (AT_d) of GS. UT_d/UT_u = uridylyltransferase catalyzing the uridylylation (UT_u) and deuridylylation (UT_d) of P_{II}. $NTRII_p/NTRII_k$ = the *glnL* (*ntrB*) product catalyzing the phosphorylation ($NTRII_k$) and dephosphorylation ($NTRII_p$) of NTRI protein.

prevented by interaction of the adenylylating cycle with another nucleotidylation cycle that involves the uridylylation and deuridylylation of P_{II} protein (5).

As shown in Fig. 17.1, a specific uridylyltransferase catalyzes transfer of a uridylyl group from UTP to each subunit of P_{II}. Because P_{II} protein is composed of four identical subunits with M_r 11,000, as many as four uridylyl groups can be bound per P_{II} molecule (6). As in the case of GS adenylylation, the uridylyl groups are attached to P_{II} through phosphodiester linkage to the hydroxyl group of a specific tyrosyl residue in each subunit. The uridylylation reaction is opposed by the action of a separate uridylyl-removing enzyme (UT_d) activity. This enzyme catalyzes hydrolysis of the uridylyltyrosyl bond to form UMP and unmodified P_{II}. The uridylyltransferase (UT_u) and deuridylylation (UT_d) activities are due to a single polypeptide protein with an M_r of 95,000 (7).

It is known that linkage of the GS adenylylation cycle with the P_{II} uridylylation cycle occurs because of the fact that the unmodified form of P_{II} stimulates the capacity of AT_a to catalyze the adenylylation of GS, whereas the uridylylated form of P_{II} (P_{II}-UMP) is required to activate the AT_d activity. Ultimately, however, the cyclic interconversions of P_{II} and GS are regulated by the concentrations of various metabolites that influence the activities of the converter enzymes, i.e., AT_a, AT_d, UT_u, and UT_d. Whereas at least 40 metabolites have been shown to affect one or more of these enzymes, two metabolites, α-ketoglutarate and glutamine, play a dominant role in the regulation of GS (4). Glutamine stimulates the adenylylation of GS and the deuridylylation of $P_{II}(UMP)_n$, whereas it inhibits the deadenylylation of $GS(AMP)_n$ and the uridylylation of P_{II}. Conversely, α-ketoglutarate inhibits the adenylylation of GS, but it stimulates the deadenylylation of $GS(AMP)_n$ and the uridylylation of P_{II}. In other words, the AT_a activity that leads to the inactivation of GS is stimulated by glutamine and inhibited by α-ketoglutarate, whereas the activities of AT_d and UT_u that lead to the reactivation of GS are stimulated by α-ketoglutarate and inhibited by glutamine. The UT_d activity that initiates the inactivation cascade of GS is stimulated by glutamine but not inhibited by α-ketoglutarate. The beauty of these reciprocal effects of L-glutamine and α-ketoglutarate can be more fully appreciated if one considers that α-ketoglutarate is a precursor of L-glutamine; therefore the L-glutamine/α-ketoglutarate ratio varies in response to variations in the levels of ammonia. Consequently, the state of adenylylation of GS varies rapidly in response to the availability of ammonia as demonstrated in in vitro and in vivo experiments (4, 8, 9).

Thus *E. coli* glutamine synthetase activity is regulated by a bicyclic cascade consisting in an adenylylation cycle of GS and a uridylylation cycle of P_{II}.

Transcriptional Regulation of *E. coli* Glutamine Synthetase

The synthesis of GS is also regulated in response to the availability of nitrogen and carbon sources. *E. coli* cultures grown on limited nitrogen and excess glucose contain induced levels of deadenylylated GS. In the presence of excess nitrogen in glucose-limited culture, GS is repressed and fully adenylylated. The mechanism of transcriptional regulation became better understood as a consequence of efforts mainly by the two laboratories of Magasanik (10, 11) and Kustu (12, 13). The structural gene for GS, *glnA*, is a member of the glnALG operon, where *glnL* (ntrB) and *glnG* (ntrC) are the structural genes for NRII (NTRB) and NRI (NTRC), respectively (10–13). NRII and NRI are dimers of identical subunits with respective molecular weights of 68,000 and 110,000 (14, 15). This operon has three promoters: glnApl, with a transcriptional start site located 187 bp upstream from the translational start site in *glnA*; glnAp2 with a transcriptional start 73 bp upstream from the translational start site in *glnA*; and glnLp, with a transcriptional site located 256 bp downstream from the translational termination site in *glnA* and 33 bp upstream from the translational start site in *glnL* (16, 17).

The promoters glnApl and glnLp have the nucleotide sequences characteristic of most of the promoters of enteric bacteria in the regions, 10 bp and 35 bp upstream from their respective transcriptional starts sites, and their transcription requires core polymerase with the most abundant σ subunit (σ^{70}). The initiation of transcription from these promoters is negatively regulated by NRI and NRI $\sim$ P_i, irrespectively. In addition, the initiation of transcription at glnApl is stimulated by cAMP and cAMP binding protein (16, 17).

The glnAp2 promoter is responsible for the transcriptional control of the GS synthesis in response to nitrogen. The nucleotide sequence of glnAp2 fits a consensus found in other nitrogen-regulated promoters. In addition, like other nitrogen-regulated promoters, the transcription from glnAp2 requires σ^{60} (σ^{55} in *S. typhimurium*), the product of *glnF* (alternatively designated *ntrA*), rather than the abundant σ^{70} (16, 18). The activation of transcription at glnAp2 has been studied in systems consisting in a DNA template containing glnAp2 and purified protein components (16, 18). It could be shown that the initiation of transcription required core RNA polymerase and σ^{60} and was greatly increased by the addition of NRI and NRII. The addition of only NRI had no effect. Subsequently, it could be shown that, in the absence of P_{II}, NRII catalyzes the conversion of NRI to NRI-P_I (19). This phosphorylated protein activates the initiation of transcription at the σ^{60}-dependent promoter glnAp2 (19).

Although uridylylated P_{II} has effects that are essentially opposite to those of unmodified P_{II}, in the activity modulating the bicyclic cascade uridylylated P_{II} appears to have no effect on the NRII activity in vitro. However, the fact that the synthesis of GS is not fully activated in a mutant lacking both P_{II} and UTase indicates that uridylylated P_{II} might play an activating role for the transcription. Therefore the transcription at glnAp2 is dependent on the P_{II}/uridylylated P_{II} ratio; and the bifunctional enzyme UTase, which changes the ratio in response to the concentration of α-ketoglutarate and glutamine, acts as a principal sensor of nitrogen. These results lead to the conclusion that the transcriptional regulation of GS also involves a bicyclic cascade consisting in the uridylylation cycle of P_{II} and the phosphorylation cycle of NRI.

In Vitro Reconstitution

Although the importance of two converter enzymes, adenylyltransferase (ATase) and uridylyltransferase (UTase), and the regulatory protein P_{II} in the regulation of GS has long been recognized, detailed studies on these proteins have been hampered by difficulties in obtaining homogeneous preparations of catalytically active enzymes. To obtain strains in which these enzymes are enriched, we cloned the *glnD* gene (structural gene for UTase) and the *glnE* gene (structural gene for ATase) on a cloning vector (PKC 30) that had been constructed to achieve highly efficient transcription, and the *glnB* gene (structural gene for P_{II}) on PBR 322 (20). In this way, UTase and ATase were overproduced to the levels approaching 800- and 500-fold, respectively, and P_{II} by 70-fold. Subsequently, these proteins were purified to homogeneity, and antibodies were derived against them. A relative abundance of the proteins involved in the GS activity regulation cascade was measured using these antibodies. The ratio for the polypeptides (subunits) of GS, P_{II} (including uridylylated P_{II}), ATase, and UTase is 411:42:2.6:1.0, respectively, in *E. coli* K12 obtained under a derepressed growth condition of GS, ATase, and UTase.

The GS activity regulation cascade system was reconstituted using protein components purified from the overproducing *E. coli* strains, and its response to the changes in the concentration of glutamine and α-ketoglutarate were studied (W. G. Bang, S. C. Park, and S. G. Rhee, unpublished observations). The concentrations of protein components used in the in vitro experiments were similar to their concentration in *E. coli* cells grown under a nitrogen-limiting condition, which are 16 μM of GS subunits, 1.6 μM of P_{II} subunits, 0.1 μM of ATase, and 0.04 μM of UTase.

When an adenylylation reaction mixutre containing GS_2 (GS containing on the average two adenylylated subunits per dodecameric molecule), a 1:1 mixture of P_{II} and $P_{II}(UMP)_4$ and ATase was incubated with glutamine and α-ketoglutarate at various concentrations (Fig. 17.2), the state of adenylylation ($\bar{n}_A$) of GS changed rapidly (<5 min) and reached a steady-state level. The steady-state values varied from near 0 to 12 depending on the concentrations of the two metabolites. For example, in the presence of 0.1 mM glutamine, the $\bar{n}_A$ values decreased from 12.0 to 10.0 to 0.5 as the concentration of α-ketoglutarate increased from 0 to 0.01 to 0.10 mM. When a uridylylation reaction mixture of P_{II} and UTase was incubated with various levels of glutamine and α-ketoglutarate, the state of uridylylation ($\bar{n}_U$) also reached various steady-state levels (Fig. 17.3). At a fixed concentration of glutamine, the value of $\bar{n}_U$ increases as the concentration of α-ketoglutarate increases. However, unlike the full-scale response of the adenylylation cycle, the $\bar{n}_U$ reaches the maximum value of 4 only when the concentration of glutamine is 0. Otherwise, the state of uridylylation is lower than 4 even at a saturating level of α-ketoglutarate. For example, in the presence of 0.1 mM glutamine, the state of uridylylation rises from 0.4 to 1.3, 1.5, and 1.6 as the concentration of α-ketoglutarate increases from 0 to 0.01, 0.10, and 1.00 mM, respectively.

The different responses of $\bar{n}_A$ and $\bar{n}_U$ to changes in glutamine and α-ketoglutarate could be also seen in the bicyclic cascade that consists in GS_{TT}, P_{II}, ATase, and UTase (Figs. 17.4 and 17.5). In the experiment described in

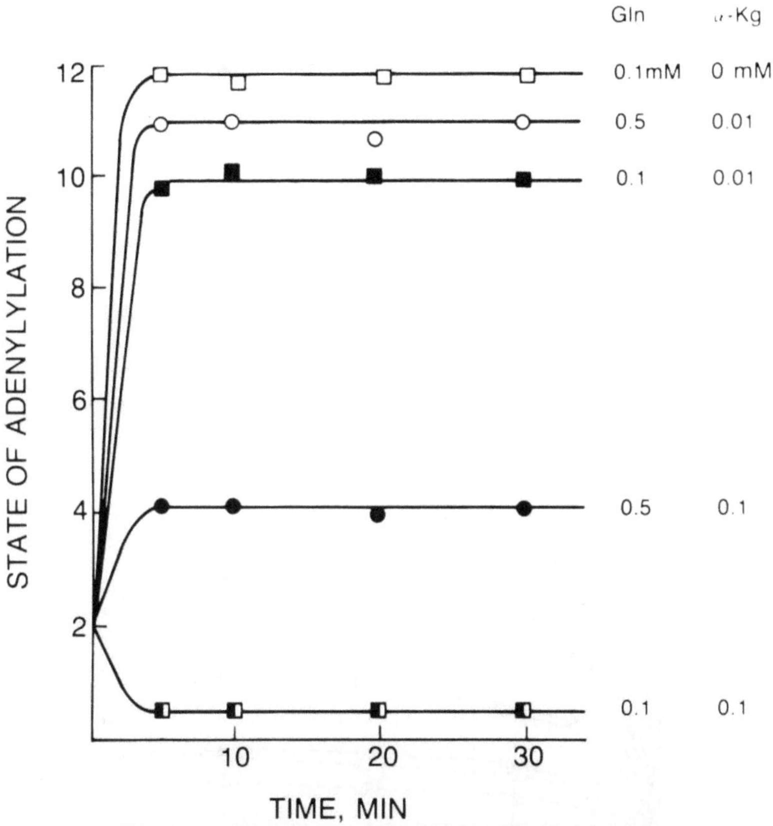

FIG. 17.2. Time-dependent changes in the state of adenylylation of GS at various concentrations of glutamine and α-ketoglutarate. A solution (250 μl) containing 10 μg $GS_{2.0}$, 2.5 μg P_{II}, 2.5 μg $P_{II}(UMP)_4$, 3 μg ATase, 2 m*M* ATP, 20 m*M* P_i, 1 unit pyruvate kinase, 2m*M* phosphoenolpyruvate, 20 m*M* Hepes pH 7.0, 20 m*M* $MgCl_2$, and indicated levels of α-ketoglutarate and glutamine was incubated at 37°C. At the indicated time, the state of adenylylation was measured by γ-glutamyltransferase assay.

Fig. 17.4 the concentration of glutamine was fixed at various levels and the concentration of α-ketoglutarate was varied; in Fig. 17.5, glutamine was varied at several fixed levels of α-ketoglutarate. The values of $\bar{n}_A$ changed in full range from 12 to 0 (Fig. 17.4) or from 0 to 12 (Fig. 17.5) as the effector concentration was increased. However, in Fig. 17.4 the $\bar{n}_U$ value plateaued at the steady state of 1.0 when [glutamine] = 0.2 μ*M*. As the glutamine concentration decreased from 0.20 to 0.10 to 0.05 m*M*, the plateau value increased from 1.0 to 1.5 to 2.0, respectively. On the other hand, the $\bar{n}_U$ value varied in nearly full range in Fig. 17.5.

These different responses of $\bar{n}_A$ and $\bar{n}_U$ to changes in the concentration of

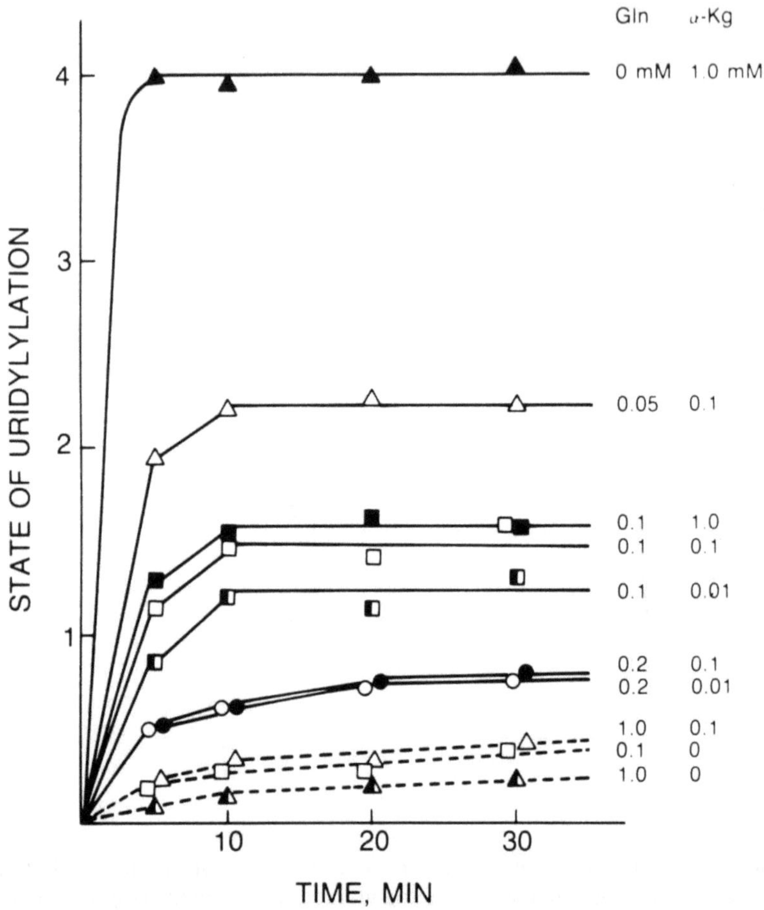

FIG. 17.3. Time-dependent changes in the state of uridylylation of P_{II}. A solution (1 ml) contained 20 μg P_{II}, 4 μg UTase, 2 m*M* ATP, 0.2 m*M* [^{3}H]UTP, 20 m*M* P_i, 20 m*M* Hepes pH 7.0, 20 m*M* $MgCl_2$, 0.2 m*M* $MnCl_2$, and indicated levels of α-ketoglutarate and glutamine. At the indicated time, the state of uridylylation was calculated from the ^{3}H radioactivity in TCA-precipitable protein.

glutamine and α-ketoglutarate are due to the fact that: (a) glutamine and α-ketoglutarate affect both directions of the adenylylation cycle in a reciprocal manner; (b) they antagonize each other's binding to the ATase; (c) in the uridylylation cycle glutamine affects both directions, the deuridylylation reaction as a stimulator and the uridylylation reaction as an inhibitor, whereas α-ketoglutarate affects only the uridylylation reaction as an activator; and (d) glutamine and α-ketoglutarate do not affect each other's binding to UTase.

In other words, glutamine exerts its effect on all four reactions of the bicyclic

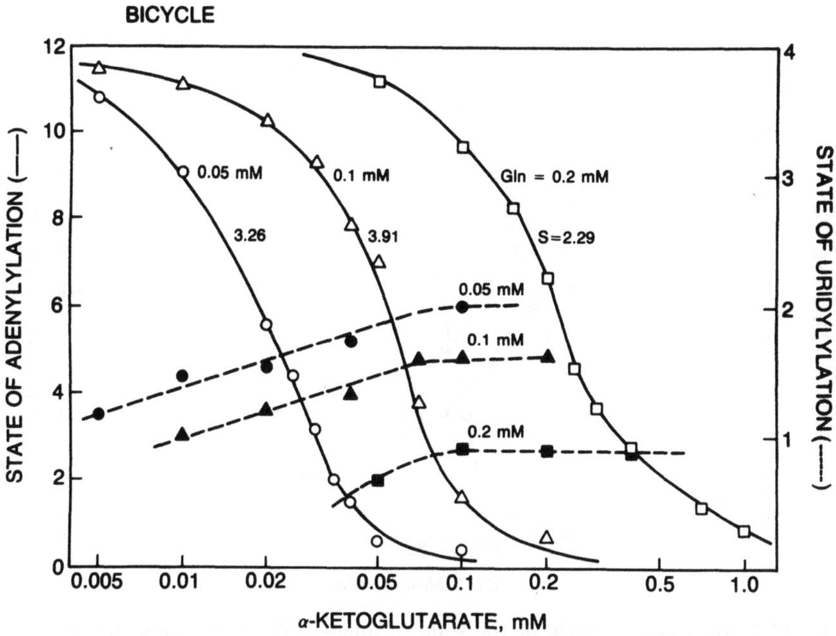

FIG. 17.4. Steady-state levels of the state of adenylylation (-----) and the state of uridylylation (-----) as a function of the α-ketoglutarate concentration at various fixed concentrations of glutamine. A solution (500 μl) containing 200 μg GS_{II}, 10 μg P_{II}, 6 μg ATase, 2 μg UTase, 2 m*M* ATP, 0.2 m*M* [^{3}H]UTP, 20 m*M* P_i, 2 m*M* phosphoenolpyruvate, 1 unit pyruvate kinase, 20 m*M* Hepes pH 7.0, 20 m*M* $MgCl_2$, and 0.2 m*M* $MnCl_2$ was incubated for 10 min at 37°C. The state of adenylylation of GS and the state of uridylylation of P_{II} were measured.

cascade, and α-ketoglutarate modulates three reactions. Consequently, the bicyclic cycle responds with a higher sensitivity to glutamine. The sensitivity index (S) for the interconvertible enzyme cascade was previously defined by Stadtman and Chock (21) and is 1.0 for the interaction of a ligand with an enzyme that exhibits michaelian kinetics. In the GS cascade shown here, the sensitivity index is a function of both glutamine and α-ketoglutarate: The S values with respect to α-ketoglutarate are 3.26, 3.91, and 3.29 when the glutamine concentration is 0.05, 0.10, and 0.20 m*M*, respectively (Fig. 17.4). However, as expected, the S values with respect to glutamine are higher than the values obtained as a function of α-ketoglutarate. They are 4.07 and 5.95 when the α-ketoglutarate concentration is set at 0.02 and 0.10 m*M*, respectively (Fig. 17.5).

Thus adenylyltransferase, P_{II}, and uridylyltransferase are purified from strains capable of overproducing them, and the bicyclic cascade is reconstituted in vitro.

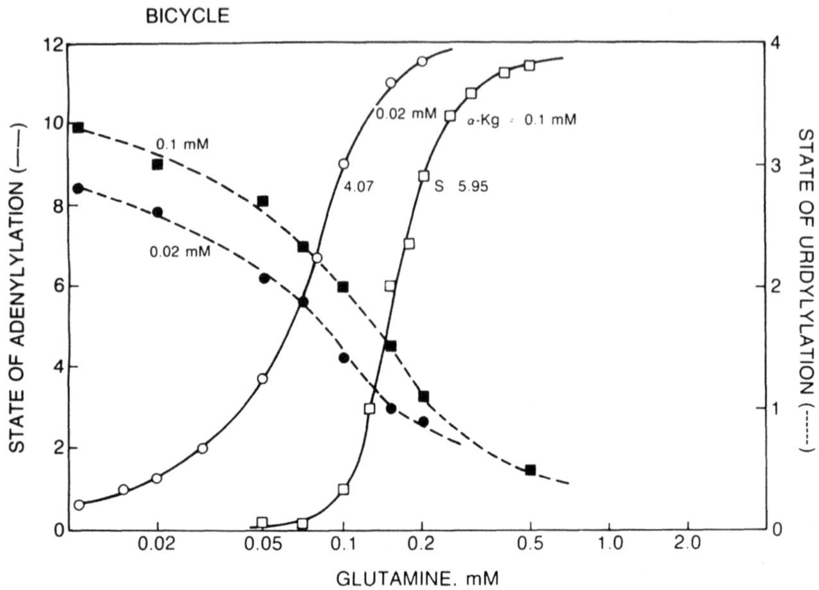

FIG. 17.5. Steady-state levels of the state of adenylylation (-----) and the state of uridylylation (-----) as a function of the glutamine concentration at various fixed concentrations of α-ketoglutarate. All other experimental procedures were as described in Fig. 17.4.

Integrated Cyclic Cascade: Regulation of Enzymatic Activity and Biosynthesis of *E. coli* Glutamine Synthetase

From the work described above, it is clear that the enzymatic activity of GS is modulated by a bicyclic cascade that consists in adenylylation/deadenylylation of GS and uridylylation/deuridylylation of a regulatory protein P_{II}. The adenylylation of GS converts the enzyme to an inactive form under physiological conditions. The converter enzyme for the adenylylation/deadenylylation cycle is adenylyltransferase, a bifunctional enzyme. Similarly, another bifunctional enzyme, uridylyltransferase, catalyzes the interconversion of P_{II}. As shown in Fig. 17.1, the synthesis of GS is regulated by another bicyclic cascade that consists in uridylylation/deuridylylation of P_{II} protein and phosphorylation/dephosphorylation of NRI protein. The phosphorylation/dephosphorylation of NRI is catalyzed by a bifunctional enzyme NRII. The coupling of the uridylylation/deuridylylation cycle of P_{II} and the phosphorylation/dephosphorylation cycle of NRI is derived from the fact that the unmodified form of P_{II} activates the phosphatase activity for the dephosphorylation of phosphorylated NRI. The phosphorylated NRI activates the initiation of transcription at the σ^{60}-dependent promoter glnAp2, which then promotes the synthesis of GS. The two bicyclic cascades, one regulating the enzymatic activity and the other regulating the synthesis of the enzyme, are linked through the sharing of the uridylylation/deuridylylation cycle of P_{II}. As a consequence, UTase occupies a vital position in regulating the specific activity of GS in the cells. In addition, the two bicyclic cascades are well coordinated in such a way that the unmodified P_{II} activates the adenylylation reaction (inactivation) of GS and represses the GS synthesis by enhancing the dephosphorylation of the phosphorylated NRI.

References

1. Stadtman, E. R., and Ginsburg, A. (1974) In *The Enzymes*, Vol. 10, 3rd ed., edited by P. D. Boyer, pp. 755–807. Academic Press, New York.
2. Chock, P. B., Rhee, S. G., and Stadtman, E. R. (1980) *Annu. Rev. Biochem.* 49:813–843.
3. Caban, C. E., and Ginsburg, A. (1976) *Biochemistry* 15:1569–1580.
4. Rhee, S. G., Park, R., Chock, P. B., and Stadtman, E. R. (1978) *Proc. Natl. Acad. Sci. USA* 75:3138–3142.
5. Brown, M. S., Sega, A., and Stadtman, E. R. (1971) *Proc. Natl. Acad. Sci. USA* 68:2949–2953.
6. Adler, S. P., Purich, D., and Stadtman, E. R. (1975) *J. Biol. Chem.* 250:6264–6272.
7. Garcia, E., and Rhee, S. G. (1983) *J. Biol. Chem.* 258:2246–2253.
8. Senior, P. J. (1975) *J. Bacteriol.* 123:407–418.
9. Mura, U., Chock, P. B., and Stadtman, E. R. (1981) *J. Biol. Chem.* 256:13022–13029.
10. Magasanik, B. (1982) *Annu. Rev. Genet.* 16:135–168.
11. Magasanik, B., and Bueno, R. (1985) *Curr. Top. Cell. Regul.* 27:215–220.
12. Kustu, S., Sei, K., and Keener, J. (1986) In *Regulation of Gene Expression*, edited by I. Booth and C. Higgins. Cambridge University Press, New York, pp. 139–154.
13. Keener, J., Wong, P., Popham, D., Wallis, J., and Kustu, S. (1987) In *RNA Polymerase and Regulation of Transcription*, edited by W. R. Rednikoff, R. R. Burgess, J. E. Dahlberg, C. A. Gross, M. P. Reckard, and R. Rickens. Elsevier/North Holland, Amsterdam pp. 159–175.
14. Reitzer, L. J., and Magasanik, B. (1983) *Proc. Natl. Acad. Sci. USA* 80:5554–5558.
15. Ninfa, A. J., Ueno-Nishio, S., Hunt, T. P., Robustell, B., and Magasanik, B. (1986) *J. Bacteriol.* 168:1002–1004.
16. Hirschman, J., Wong, P. K., Sei, K., Artz, S., Keener, J., and Kustu, S. (1985) *Proc. Natl. Acad. Sci. USA* 82:7525–7529.
17. Ueno-Nishio, S., Mango, S., Reitzer, J., and Magasanik, B. (1984) *J. Bacteriol.* 160:379–384.
18. Hunt, T. P., and Magasanik, B. (1985) *Proc. Natl. Acad. Sci. USA* 82:8453–8457.
19. Ninfa, A. J., and Magasanik, B. (1986) *Proc. Natl. Acad. Sci. USA* 83:5909–5913.
20. Rhee, S. G., Park, S. C., and Koo, J. H. (1985) *Curr. Top. Cell. Regul.* 27:221–232.
21. Stadtman, E. R., and Chock, P. B. (1978) *Curr. Top. Cell. Regul.* 13:53–95.

18
Regulation of ATP, Mg-Dependent Protein Phosphatases

JACKIE R. VANDENHEEDE, CARLINE VANDEN ABEELE, and WILFRIED MERLEVEDE

"ATP,Mg-dependent protein phosphatase" designates a family of enzyme forms whose catalytic subunit can interconvert between an active and an inactive conformation through the specific actions of the heat-stable modulator protein (inactivation) and protein kinase F_A (activation). The catalytic subunit of these enzymes has a wide substrate specificity (1, 2) so that the ATP, Mg-dependent phosphatase activity can account for the major dephosphorylation of the key enzymes that regulate glycogen metabolism, i.e., glycogen phosphorylase, glycogen synthase, and phosphorylase kinase (β subunit).

Several forms of ATP,Mg-dependent phosphatase have been isolated from different tissues or cellular localizations. All species contain the same 38-kD protein as catalytic subunit (3–10), which for the sake of simplicity we designate "C" if in the active conformation and "F_C" when present in the inactive (or kinase F_A and ATP,Mg-dependent) form. Most of the information regarding its function and regulation stems from studies of the rabbit skeletal muscle system. The present report mainly deals with the characterization and interrelation of the various forms of the ATP,Mg-dependent protein phosphatase in this tissue.

Inactive Protein Phosphatase

An inactive protein phosphatase can be isolated from rabbit skeletal muscle extract (11) or cytosol (12) whose potential activity can be revealed in two seemingly different ways. One way is to proteolyze the enzyme in the presence of millimolar concentrations of Mn^{2+} (3, 4), and another, perhaps more physiological method is to incubate it with ATP,Mg and protein kinase F_A (11–13). The inactive enzyme has been purified as a two-subunit protein complex [F_CM] where F_C represents the 38-kD inactive catalytic subunit and M the (32 kD) regulatory or "modulator" subunit (3–7, 10). The modulator protein has been identified (14) as the previously characterized heat-stable inhibitor-2 (15, 16) and was found to play a crucial role in the reversible activation-inactivation cycle of the ATP,Mg-dependent protein phosphatase

(1, 10, 17). The kinase F_A-mediated activation of [F_CM] is initiated by phosphorylation of the M subunit (5, 7, 18, 19), which induces conversion of the catalytic subunit F_C to the active C state (5, 19). ATP[γS] can replace ATP in the kinase F_A-mediated activation of the catalytic subunit, although the activated enzyme cannot express its phosphorylase phosphatase activity while in the thiophosphorylated form (19, 20). This observation, together with the fact that there is no good correlation between the amount of phosphate incorporated into [F_CM] and the production of phosphorylase phosphatase activity, has led to the conclusion that the activated enzyme undergoes autodephosphorylation, which is necessary for the accommodation of exogenous substrates.

Results have pointed to the importance of Mg^{2+} ions in the isomerization of the catalytic subunit (20) and in the consequent autode(thio)phosphorylation of the modulator subunit (6). It is therefore conceivable that the net result of the kinase F_A-mediated phosphorylation of the [F_CM] enzyme is Mg^{2+} activation of the phosphatase as originally suggested (1, 12). Moreover, the metal ion dependency constitutes a common element in the two seemingly unrelated mechanisms of activation of the [F_CM] enzyme.

The inactive F_C subunit by itself cannot be activated by kinase F_A as it lacks the phosphorylation site, which is localized on the M subunit. Addition of purified modulator protein restores this activation in a dose-dependent manner, although excess M inhibits the phosphatase activity, probably by binding at a low-affinity inhibitory site on the catalytic subunit (10, 17, 19, 20). Metal ions alone, e.g., Mn^{2+} or Co^{2+} and Mg^{2+}, in combination with low amounts of detergents can bring out more than 70% of the potential phosphatase activity from a "free F_C" enzyme. Addition of M subunit to "free F_C" blocks the metal-induced phosphatase activity; and like the "native" [F_CM], these reconstituted enzymes require proteolytic destruction of the modulator for the metal ion to be effective (3, 4, 20). This finding suggests that the M subunit blocks a metal ion binding site on the catalytic subunit of the inactive phosphatase. The regulatory properties of the modulator protein are discussed later in more detail.

One important aspect of the kinase F_A-mediated activation of the phosphatase is that it is reversible: The enzyme slowly reverts back to its inactive conformation after removal of the activating system. This reversal seems to be an intrinsic property of the enzyme and is controlled by the inactivating properties of the modulator subunit. Limited proteolysis of the activated [F_CM] enzyme destroys the M subunit and produces a stable, active 40-kD phosphatase. Readdition of purified modulator produces a 60- to 70-kD complex and restores the inactivating process to the [F_CM] form (21). An inactive form of the ATP,Mg-dependent phosphatase has also been reconstituted by incubating purified M with active catalytic subunit isolated by the ethanol procedure (7, 18). All known properties of the reconstituted and "native" [F_CM] enzyme were reported to be similar (18).

A major difference between the inactive and the activated enzyme is that the

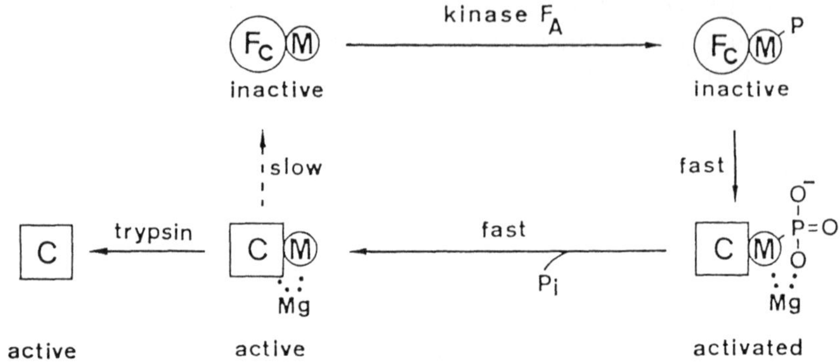

FIG. 18.1. Kinase F_A-mediated activation of the $[F_CM]$ enzyme.

potential activity of the inactive F_C protein is destroyed by limited proteolysis with trypsin, whereas the activity of the C unit is resistant to proteolysis. This observation confirms the proposed isomerization of the phosphatase catalytic subunit between an active and an inactive state (5). The kinase F_A-mediated activation of the $[F_CM]$ enzyme is presented in Fig. 18.1.

Evidence has been presented (22) for synergistic phosphorylation of the modulator protein and the $[F_CM]$ enzyme by the kinase F_A (GSK-3) and casein kinase TS (PC 0.7 or casein kinase II). It was shown that the phosphorylation of M at serine residues by casein kinase TS stimulates the phosphorylation at threonine by kinase F_A. When $[F_CM]$ was used as substrate for the kinases, it was observed that this synergistic phosphorylation also resulted in more effective activation of the inactive phosphatase by kinase F_A. The only explanation for these results, which are in agreement with the above proposed mechanism of activation of the phosphatase, is that the serine phosphorylation of the modulator subunit by casein kinase TS stabilizes the active conformation of the catalytic subunit after activation by kinase F_A and subsequent autodephosphorylation of the threonine site. The synergistic activation of the phosphatase cannot be the direct result of the increased amount of threonine phosphate in the modulator subunit, as also suggested (22), if the autodephosphorylation is a necessary step in activation of the protein phosphatase (6, 19, 20).

Modulator Unit

It has been reported (11, 19) that the phosphorylase phosphatase activity produced after kinase F_A activation is not linear with respect to the concentration of $[F_CM]$ used in the assay. This finding suggests that the $[F_CM]$ enzyme is partially inhibited by an association/dissociation mechanism. It could mean either that the inactive enzyme is better represented as a partially inhibited $[F_CM]_2$ dimer or that it contains more than one M per F_C unit, with the second M bound at a low-affinity inhibitory site on the catalytic subunit (19). The observation that combined kinase F_A-ATP,Mg and trypsin treatment produces more phosphorylase phosphatase activity than kinase F_A activation alone would confirm this hypothesis (19, 20).

Several reports have described two modulator binding sites (*m* and *i*) on the catalytic subunit of the ATP,Mg-dependent protein phosphatase (10, 17, 19, 20). Binding of M to the high-affinity binding site (site *m*) controls the reversible activation-inactivation cycle of the catalytic subunit. The presence of modulator at this site does not have an inhibitory effect on the phosphatase. Modulator bound at the low-affinity binding site (site *i*) causes instantaneous inhibition of phosphorylase phosphatase activity but does not affect activation or inactivation of the [F_CM] enzymes. It seems therefore likely that these two M-binding sites add flexibility to the regulation of the enzyme's catalytic unit by either inhibiting the active catalytic subunit or converting it to the inactive F_C conformation. The site *i* may also be the location where the phosphoinhibitor-1 exerts its inhibitory action on the catalytic subunit.

Physiological Modulator Complexes

We have pointed out (10, 17, 23) that unboiled tissue preparations do not contain a "free" modulator, but that the modulator activity is always associated with higher-molecular-weight complexes. Such a complex is the inactive [F_CM] enzyme itself, but a second modulator-containing protein complex [MX] has been purified and characterized in rabbit skeletal muscle (10). Its subunit composition appears to be identical to that of the [F_CM] enzyme (38-kD and 32-kD subunits on SDS-PAGE); and like the inactive phosphatase, it migrates on sucrose density gradients as a 70-kD protein and on gel filtration as a 140-kD complex. However, [MX] does not exhibit any significant phosphorylase phosphatase activity before or after preincubation with kinase F_A and ATP,Mg. The 38-kD subunit of the preparation has been called [X] until its identity is revealed. A remarkable property of the native [MX] complex is that it exhibits the activation-inactivation characteristics of the "free M" but is not able to inhibit the phosphatase activity of the C catalytic unit. This finding suggests that the intact [MX] complex can bind to the high-affinity modulator subunit site but not at the inhibitory site on the catalytic subunit. Boiling the [MX] complex dissociates the modulator subunit, which now inhibits the active phosphatase just as effectively as the "free M" isolated in the conventional way (24). It indicates that the 38-kD X protein blocks a modulator domain that binds to the inhibitory site on the phosphatase catalytic unit. The part of the modulator that links to the high-affinity binding site seems to be fully exposed in the [MX] complex. If X is identical to the F_C protein, one could visualize the [MX] complex as an inactive [F_CM] enzyme that has the modulator bound at the inhibitory site *i* instead of the usual high-affinity modulator subunit site *m*. These two [F_CM] forms are represented in Fig. 18.2.

FIG. 18.2. [MX] and [F_CM] complexes. X may be identical to F_C. m = Modulator subunit site; i = inhibitory site.

Origin of the Inactive Phosphatase Forms

The origin of the inactive [F_CM] enzyme is still ill-defined. Incubation of fresh rabbit skeletal muscle extracts with kinase F_A and ATP,Mg does not generate a considerable amount of extra phosphorylase phosphatase activity over the already present active phosphatase. This fact suggests that the ATP,Mg-dependent activity isolated as [F_CM] is produced during the purification procedure. Isoelectric precipitation at pH 5.8 of a protein–glycogen complex separates the phosphorylase phosphatase activity into two fractions: (a) a glycogen-associated activity that can be isolated as such and does not spontaneously convert to the inactive, ATP,Mg-dependent state; and (b) a cytosolic phosphatase activity that largely inactivates upon removal of kinase F_A and ATP,Mg in the first step of its purification (11, 12). It seems therefore logical to accept that the [F_CM] activity originates from the cytosolic active phosphatase fraction. It has been reported (25) that the cytosolic phosphorylase phosphatase activity is drastically increased by a trypsin treatment, which suggests the presence of a "latent" or inhibited protein phosphatase whose catalytic subunit is in the active conformation. The exact subunit composition of this cytosolic inhibited enzyme still escapes us, as upon isolation it "spontaneously" converts to an inactive species. It is conceivable that [MX] and [F_CM] are produced from this cytosolic-inhibited phosphatase, either independently or consecutively. It was noted that during their purification from rabbit skeletal muscle the relative amounts of [MX] and [F_CM] varied considerably from preparation to preparation and from one step to the next, which could mean that one is converted to the other. [F_CM] appeared to be the prevalent or more stable form, which is probably produced by migration of the modulator subunit from the inhibitory site i to the higher-affinity modulator subunit site m. We have not been able to control or prove this interconversion. One plausible hypothesis is that the stability of the active phosphatase and the binding of M at the low-affinity inhibitory site is promoted or controlled by the presence of serine phosphate at the casein kinase TS phosphorylation sites on the modulator protein. This possibility has already been mentioned in this chapter and is presently under investigation. Another possibility that cannot be excluded is that the active, glycogen-bound protein phosphatase present in the extract is being converted to an inactive, ATP,Mg-dependent form through the action of [MX] (in its inactive or active conformation). [MX] could then be considered a "physiological" inactivating protein.

Active Protein Phosphatase

About 50% of the total phosphorylase phosphatase activity present in a rabbit skeletal muscle extract is found associated with the glycogen particle (9, 26). This glycogen-bound enzyme has been isolated and characterized as a two-subunit protein: (a) a 38-kD catalytic unit "C"; and (b) a 103-kD regulatory protein that is responsible for the binding of the enzyme to glycogen and was therefore called the G subunit (27). The [GC] enzyme interconverts to an inactive, ATP,Mg-dependent form by the addition of free modulator protein so that one can assume it contains the same 38-kD catalytic unit as the cytosolic ATP,Mg-dependent phosphatase discussed above.

Glycogen-associated protein phosphatases from several tissues have been shown to be rather insensitive to inhibition by the heat-stable inhibitor-1 and modulator. This has been reported for mouse (28) and dog (29, 30) liver as well as for rabbit skeletal muscle (22, 31). It was invariably observed that trypsin treatment greatly increased the sensitivity of these enzymes to inhibition by the heat-stable proteins, which in dog liver was attributed to proteolytic destruction of the "deinhibitor" subunit of the glycogen-bound phosphatase (30). Later it became clear that the protective action of the deinhibitor subunit could also be manipulated in a more physiological way: The cyclic AMP-dependent protein kinase was found to phosphorylate the isolated 10-kD deinhibitor peptide, and this action neutralized its protective effect on the catalytic subunit against the heat-stable inhibitors (32). Dephosphorylation and reactivation of the purified deinhibitor was achieved through the specific action of the high-molecular-weight polycation-stimulated protein phosphatases, which also dephosphorylate the phosphoinhibitor-1 (33, 34).

Similar cyclic AMP-dependent phosphorylation has been shown to control the inhibition of the purified rabbit skeletal muscle [GC] enzyme (27). In this case, C is a 38-kD catalytic subunit and the regulatory unit G is a 103-kD glycogen-binding subunit that protects the catalytic subunit against inhibition by the heat-stable protein inhibitors. It has been suggested that the 10-kD deinhibitor peptide might be a fragment of the larger dog liver G subunit, which has retained the phosphorylation site that regulates its activity (27). The rabbit skeletal muscle glycogen particle does not seem to contain any inhibited or inactive ATP,Mg-dependent phosphatase activity inasmuch as trypsin or kinase F_A ATP,Mg treatment does not noticeably increase its phosphorylase phosphatase activity. It has, however, been shown to harbor some [MX] activity (17).

The glycogen-associated [GC] phosphatase does not contain an M subunit and therefore does not convert to its inactive conformation unless modulator protein is added. The in vitro inactivation of the GC enzyme by added M has been documented to occur without dissociation of the G unit (17). A ternary [GCM] complex was observed that inactivated to an ATP,Mg-dependent form. Such an inactive phosphatase complex, comprising both G and M subunits, has not been identified in any tissue preparation so that in vivo conversion of the glycogen-bound active phosphatase to the inactive ATP,Mg-dependent enzyme form remains to be demonstrated. Also, because there is no "free M" in unboiled tissue extracts, the inactivating molecule may have to be [MX] instead of M, which would produce a different inactive phosphatase form containing a G protein, an M unit, and possibly two C units. Again, such an enzyme form has not been observed. Third, activation of the cytosolic [F_CM] enzyme proceeds without dissociation of the M subunit, so that if [F_CM] and [GC] were to interconvert in vivo, one would have to invoke the dissociation of the two respective regulatory subunits M and G, which has not been reported to occur under physiological conditions.

A more logical extrapolation of the available data is to assume that the physiological regulation of the [GC] enzyme is restricted to the cyclic AMP-mediated inhibition of the phosphatase through phosphorylation of the [G-deinhibitor] subunit and inhibitor-1. The physiological process that reverses this inhibition is less well documented but is likely to involve the dephosphorylation of "G" and inhibitor-1, resulting in dissociation of the dephosphoinhibitor-1 and recovery of the [GC] phosphatase in its active form.

Conclusions

Most if not all tissues contain a multisubstrate protein phosphatase whose catalytic subunit can be made to interconvert between an active "C" and an inactive, ATP,Mg-dependent conformation "F_C." This protein phosphatase can be isolated in different forms, all of which contain the same catalytic subunit but which have different regulatory proteins associated with it. These regulatory subunits are responsible for intricate control of the protein phosphatase activity that involves substrate specificity changes, increased resistance or sensitivity to inhibition by the heat-stable phosphoinhibitor-1, and reversible activation-inactivation of the catalytic subunit. Some active forms of the enzyme contain a stabilizing subunit "G" that protects them from either inhibition or inactivation, whereas the inactive phosphatase forms have a modulator subunit "M," which controls conversion of the catalytic subunit to the active form. The ATP,Mg-dependent phosphatase is regulated by protein kinases. The cyclic AMP-dependent protein kinase blocks the phosphatase activity via phosphorylation of two proteins: inhibitor-1 and the G-regulatory subunit of the active phosphatase. The kinase F_A is the activating enzyme for the inactive, ATP,Mg-dependent phosphatase: It transiently phosphorylates the M subunit of the [F_CM] enzyme and provides the Mg^{2+} ions that are necessary for protein phosphatase activity. It is suggested that the different forms of the ATP,Mg-dependent protein phosphatase may not all be interconverting in vivo.

Acknowledgments. The original work carried out in our laboratory was supported by grants from the National Fund for Scientific Research (Belgium) and from the "Onderzoeksfonds K. U. Leuven." J. R. V. is a Senior Research Associate and C. V. A. a Research Assistant of the National Fund for Scientific Research (Belgium).

References

1. Merlevede, W., Vandenheede, J. R., Goris, J., and Yang, S-D. (1984) *Curr. Top. Cell. Regul.* 23:177–215.
2. Stewart, A. A., Hemmings, B. A., Cohen, P., Goris, J., and Merlevede, W. (1981) *Eur. J. Biochem.* 115:197–205.
3. Ballou, L. M., Brautigan, D. L., and Fischer, E. H. (1983) *Biochemistry* 22: 3393–3399.
4. Villa-Moruzzi, E., Ballou, L. M., and Fischer, E. H. (1984) *J. Biol. Chem.* 259:5857–5863.
5. Jurgensen, S., Shacter, E., Huang, C. Y., Chock, P. B., Yang, S-D., Vandenheede, J. R., and Merlevede, W. (1984) *J. Biol. Chem.* 259:5864–5870.
6. Li, H-C., Price, D. J., and Tabarini, D. (1985) *J. Biol. Chem.* 260:6416–6426.
7. Tung, H. Y. L., and Cohen P. (1984) *Eur. J. Biochem.* 145:57–64.
8. Alemany, S., and Cohen, P. (1986) *FEBS Lett.* 198:194–202.
9. Alemany, S., Pelech, S., Briarley, C. H., and Cohen, P. (1986) *Eur J. Biochem.* 156:101–110.
10. Vandenheede, J. R., Vanden Abeele, C., and Merlevede, W. (1986) *Biochem. Biophys. Res. Commun.* 135:367–373.
11. Yang, S-D., Vandenheede, J. R., Goris, J., and Merlevede, W. (1980) *J. Biol. Chem.* 255:11759–11767.
12. Vandenheede, J. R., Yang, S-D., and Merlevede, W. (1981) *J. Biol. Chem.* 256:5894–5900.
13. Vandenheede, J. R., Yang, S-D., Goris, J., and Merlevede, W. (1980) *J. Biol. Chem.* 255:11768–11774.
14. Yang, S-D., Vandenheede, J. R., and Merlevede, W. (1981) *J. Biol. Chem.* 256:10231–10234.
15. Brandt, H., Killilea, S. D., and Lee, E. Y. C. (1974) *Biochem. Biophys. Res. Commun.* 61:598–604.
16. Huang, F. L., and Glinsmann, W. H. (1975) *Proc. Natl. Acad. Sci. USA* 72:3004–3008.
17. Vandenheede, J. R., and Merlevede, W. (1985) *Adv. Protein Phosphatases* 1:87–103.
18. Hemmings, B., Resink, T. J., and Cohen, P. (1982) *FEBS Lett.* 150:319–324.
19. Vandenheede, J. R., Yang, S-D., Merlevede, W., Jurgensen, S., and Chock, P. B. (1985) *J. Biol. Chem.* 260:10512–10516.
20. Vandenheede, J. R., Vanden Abeele, C., and Merlevede, W. (1986) *Biochem. Biophys. Res. Commun.* 136:16–21.

21. Vandenheede, J. R., Goris, J., Yang, S-D., Camps. T., and Merlevede, W. (1981) *FEBS Lett.* 127:1–4.
22. DePaoli-Roach, A. A. (1984) *J. Biol. Chem.* 259:12144–12152.
23. Vandenheede, J. R., Yang, S-D., and Merlevede, W. (1983) *Biochem. Biophys. Res. Commun.* 113:439–445.
24. Yang, S-D., Vandenheede, J. R., and Merlevede, W. (1981) *FEBS Lett.* 132:293–296.
25. Lee, E. Y. C., Silberman, S. R., Ganapathi, M. K., Petrovic, S., and Paris, H. (1980) *Adv. Cyclic Nucleotide Res.* 13:95–131.
26. Ingebritsen, T. S., Stewart, A. A., and Cohen, P. (1983) *Eur. J. Biochem.* 132:297–307.
27. Stralfors, P., Hiraga, A., and Cohen, P. (1985) *Eur. J. Biochem.* 149:295–303.
28. Laloux, M., and Hers. H-G. (1979) *FEBS Lett.* 105:239–243.
29. Defreyn, G., Goris, J., and Merlevede, W. (1977) *FEBS Lett.* 79:125–128.
30. Goris, J., Waelkens, E., Camps, T., and Merlevede, W. (1984) *Adv. Enzyme Regul.* 22:467–484.
31. Khatra, B. S., and Soderling, T. S. (1983) *Arch. Biochem. Biophys.* 227:39–51.
32. Goris, J., Parker, P., Waelkens, E., and Merlevede, W. (1984) *Biochem. Biophys. Res. Commun.* 120:405–410.
33. Goris, J., Waelkens, E., and Merlevede, W. (1985) *Adv. Protein Phosphatases* 1:347–365.
34. Waelkens, E., Goris, J., and Merlevede, W. (1987) *J. Biol. Chem.* 262:1049–1059.

19
Molecular Mechanisms of Allosteric Regulation in Aspartate Transcarbamylase

Guy Hervé

Escherichia coli aspartate transcarbamylase (ATCase) catalyzes the first reaction of the pyrimidine pathway, i.e., the carbamylation of the amino group of aspartate by carbamylphosphate. This catalysis is feedback inhibited by the end-product CTP and is stimulated by ATP, an antagonism that in the cell tends to balance the production of purines and pyrimidines. ATCase has been extensively studied as a model system for allosteric regulation and its properties thoroughly reviewed (1–3). In terms of structure, this enzyme is composed of two trimers of identical catalytic chains (MW 33,000 daltons each) that are in contact and held together by three dimers of regulatory chains (MW 17,000 daltons each). This association requires the presence of six atoms of zinc (one per regulatory chain), which are linked to the SH groups of cysteine residues clustered in the C-terminal region of the regulatory chains (4). Catalytic trimers (called catalytic subunits) and regulatory dimers (called regulatory subunits) can be easily and reversibly separated under the influence of mercurials (5). The isolated catalytic subunits are fully active but do not exhibit any of the regulatory properties of the native enzyme. The isolated regulatory subunits bear the regulatory sites where the effectors CTP and ATP bind competitively. The amino acid sequences of these two kinds of chain are known (6–9).

Because of the extensive work in crystallography done in Lipscomb's laboratory, the three-dimensional structure of ATCase was resolved with a resolution of 2.6 Å (10), allowing interesting correlations to be made between structure, catalysis, and allosteric regulation.

Homotropic Cooperative Interactions

ATCase shows homotropic cooperative interactions between the catalytic sites for the binding of its substrate aspartate. Based on this observation the six catalytic sites are considered equivalent (11). These homotropic cooperative interactions are explained by a transition from a low-affinity conformation (T state) to a high-affinity conformation (R state). Each of these two extreme conformations can be experimentally stabilized by modification of the protein–solvent interactions through the use of organic co-solvents (12). The simplest interpretation of these results is that the T state is stabilized through reinforcement of the intersubunits' ionic bonds due to the decrease of the dielectric constant of the medium, and the stabilization of the R state results from exposure to the solvent of a larger surface of the protein as a consequence of a substantial decrease in the polarity of the medium. The T and R states differ in pH dependence of the catalytic reaction (13, 14), a phenomenon suggesting that these two extreme states differ by the pK values of some groups involved in substrate binding and/or in the catalytic process. This interesting aspect of the allosteric transition is currently being investigated in this laboratory.

As already mentioned, the three-dimensional structure of ATCase in the absence of ligands (T state) has been determined at a resolution of 2.6 Å (10). The three-dimensional structure of the enzyme saturated with the bisubstrate analogue *N*-phosphonacetyl-L-aspartate (15) has been described at a resolution of 2.9 Å (16), providing a description of the structural differences between the two extreme conformations.

The homotropic cooperative interactions between the catalytic sites cannot be explained by a simple two-state thermodynamic equilibrium, which would be shifted toward the R state through the exclusive binding of aspartate to this form. This conclusion is based on several lines of evidence. Among them is the fact that the variation in pH dependence of the reaction shows that at low aspartate concentration not only is the T state largely predominant but it contributes predominantly to the catalysis (14). In addition, it was found that L-alanosine (Fig. 19.1), a pseudosubstrate for ATCase that can be carbamylated, cannot promote the allosteric transition despite the fact that it shows a higher affinity for the R state than for the T state (17). The same striking observation was made in the case of L-cysteine-sulfinate, another pseudosubstrate for ATCase (18), and in the case of the nonphysiological direction of the reaction, using carbamylaspartate and phosphate as substrates (19).

Recent x-ray solution scattering experiments allowing one to monitor the proportion of T and R states under different experimental conditions led to the same conclusion (20).

Heterotropic Interactions

ATCase shows heterotropic interactions between regulatory and catalytic sites, allowing feedback inhibition of the catalytic activity by CTP and stimulation of this activity by ATP. Evidence that was previously reviewed (21) led to the conclusion that the heterotropic interactions do not operate through the same conformational transition as that involved in the homotropic cooperative interactions between the catalytic sites. Among the evidence is the gener-

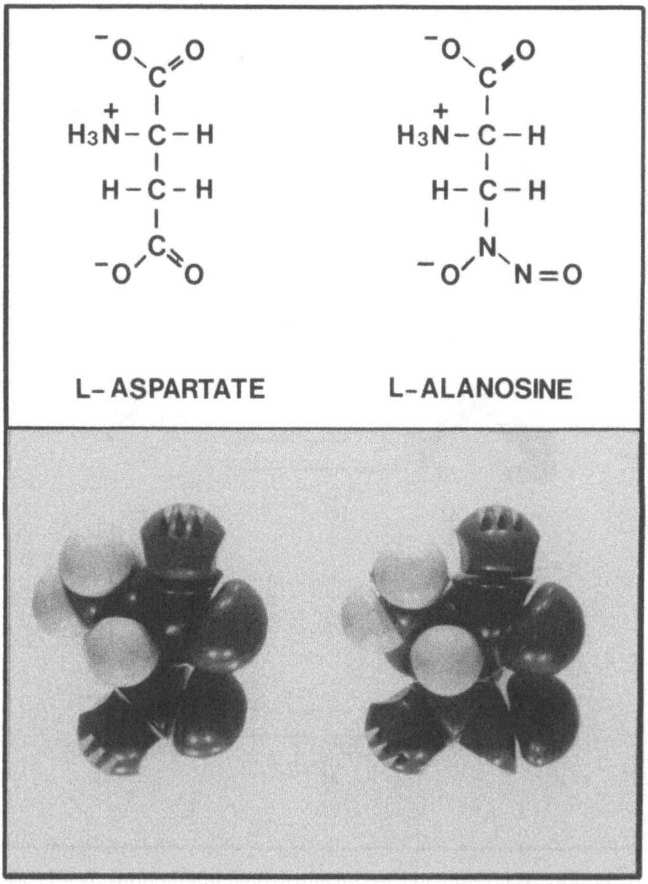

FIG. 19.1. L-Alanosine, a pseudosubstrate for ATCase. (From Baillon, J., Tauc, P., and Hervé, G. [1985] *Biochemistry* 24:7182–7187; with permission.)

ation over the past few years of about 12 species of chemically or genetically modified ATCases in which the homotropic cooperative interactions between the catalytic sites are specifically abolished but which are still sensitive to the effectors (22–25). It was concluded that the effectors act through a "primary effect," which is a local conformational change affecting site by site the affinity of the catalytic sites for aspartate and a "secondary effect" on the T/R ratio, which is in fact mediated by the substrate aspartate (21). This model is shown in a purely descriptive way in Fig. 19.2, although some of its quantitative parameters are already available. When ATP binds to a regulatory site it promotes a local conformational change that increases the affinity of the corresponding catalytic site for aspartate (primary effect). In the presence of a given concentration of aspartate, this enhanced affinity provokes an increase of the degree of occupancy of the catalytic sites, which in turn promotes some

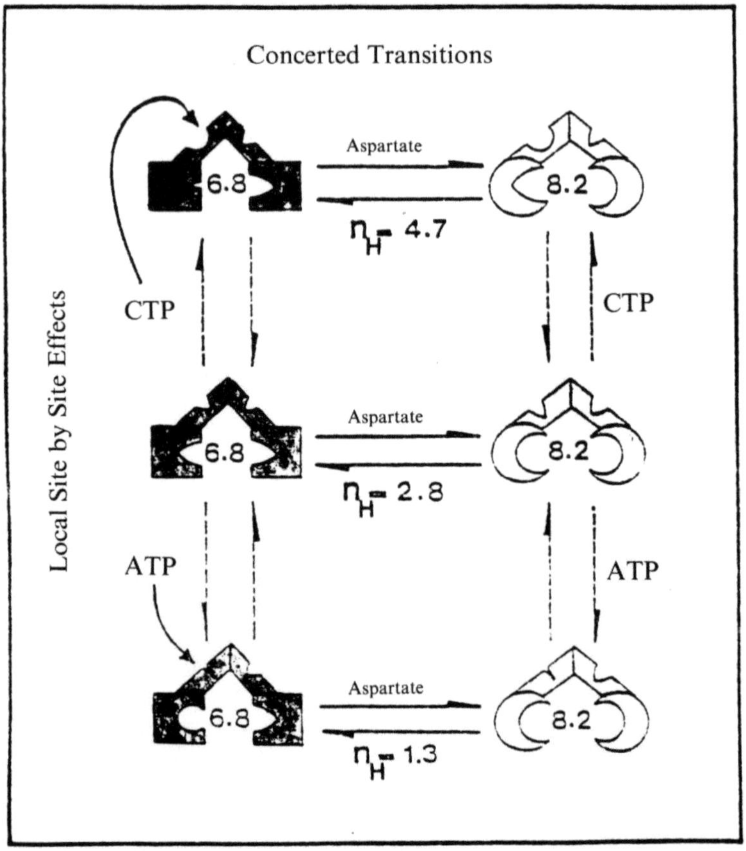

FIG. 19.2. Coupling of homotropic and heterotropic interactions in ATCase. Squares and circles correspond, respectively, to the T and R states. h = Hill coefficient; 6.8 and 8.2 = optimum pH of activity of the T and R states, respectively. (From Tauc, P., Leconte, C., Kerbiriou, D., Thiry, L., and Hervé, G. [1982] *J. Mol. Biol.* 155:155–168; with permission.)

transition from T to R. CTP acts similarly in the reverse direction, although in a slightly more complex way. The primary effect is responsible for maximal 100% stimulation by ATP. The extent of stimulation through the all-or-none process depends on the proportion of T and R forms. In the presence of very low concentrations of aspartate, conditions under which ATCase is essentially in the T state, the maximal stimulation increases 250%. As expected, all the modified forms of ATCase that are "frozen" in R state show only the primary effect and consequently exhibit uniformly maximal stimulation 100% (21, 25). More recently this model, which was primarily based on enzyme kinetics, binding studies, and enzyme modifications, was directly confirmed by x-ray solution scattering experiments, which take advantage of the fact that the T and R states show very different scattering patterns (26).

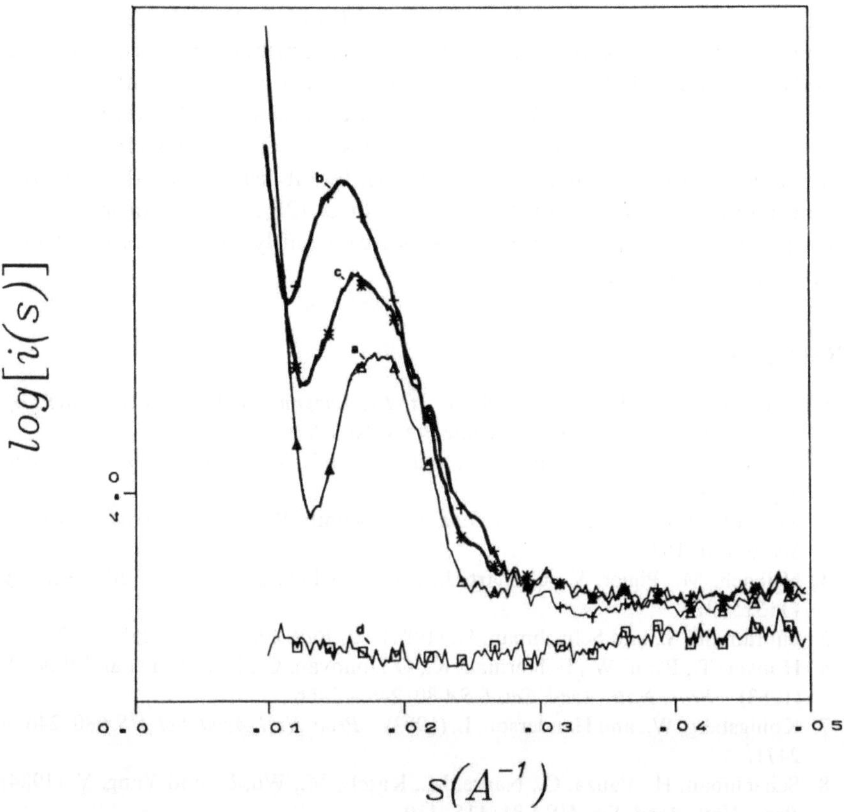

FIG. 19.3. X-ray solution scattering curves of ATCase. a: X-ray scattering curve in absence of any ligand (T state). b: X-ray scattering curve in the presence of saturating *N*-phosphonacetyl-L-aspartate (R state). c: Intermediary curve obtained in the presence of a nonsaturating concentration of succinate in the presence of carbamylphosphate. d: Buffer. (From Hervé, G., Moody, M., Tauc, P., Vachette, P., and Jones, P. [1985] *J. Mol. Biol.* 185: 189–199; with permission.)

These experiments showed that the presence of unligated substrate or substrate analogue is absolutely required for any influence of the effectors on the T/R ratio (20). The scattering spectra of the T and R conformational states are shown in Fig. 19.3.

In terms of a structure–function relation, a peptide substitution at the C-terminus of the regulatory chain of ATCase allowed demonstration of the involvement of that particular region of the molecule in both homotropic and heterotropic cooperative interactions. The generation, of this modified form of enzyme results from the fact that the distal region of the gene coding for the regulatory chain of the enzyme contains a pseudo-site for phage λ attachment (27). In this modified ATCase the last eight amino acids of the regulatory chains are replaced by six amino acids coded for by λ DNA. In this modified form, in which ATCase is frozen in the R state, the primary effects of ATP and CTP are uncoupled. This enzyme is normally sensitive to ATP but is insensitive to CTP despite the fact that this nucleotide binds with a normal affinity to the regulatory sites. The observation suggests that the primary effects of the two effectors involve, at least partially, different features of the catalytic chain/regulatory chain interaction interface (25). The conformational changes involved in these primary effects are currently under investigation by site-directed mutagenesis.

References

1. Jacobson, G., and Stark, G. (1973) In *The Enzymes*, Vol. 9, 3rd ed., edited by P. D. Boyer pp. 225–308. Academic Press, New York.
2. Kantrowitz, E., Pastra-Landis, S., and Lipscomb, W. (1980) *Trends Biochem. Sci.* 5:124–128.
3. Kantrowitz, E., Pastra-Landis, S., and Lipscomb, W. (1980) *Trends Biochem. Sci.* 6:150–153.
4. Nelbach, M., Pigier, V., Gerhart, J., and Schachman, H. (1972) *Biochemistry* 11:315–327.
5. Subramani, S., and Schachman, H. (1981) *J. Biol. Chem.* 256:1255–1262.
6. Hoover, T., Roof, W., Folkerman, K., O'Donovan, G., Bencini, D., and Wild, J. (1983) *Proc. Natl. Acad. Sci. USA* 80:2462–2466.
7. Konigsberg, W., and Henderson, L. (1983) *Proc. Natl. Acad. Sci. USA* 80:2467–2471.
8. Schachman, H., Pauza, C., Navre, M., Karels, M., Wu, L., and Yang, Y. (1984) *Proc. Natl. Acad. Sci. USA* 81:115–119.
9. Weber, K. (1968) *Nature* 218:1116–1119.
10. Honzatko, R., Crawford, J., Monaco, H., Ladner, J., Edwards, B., Evans, D., Warren, S., Willey, D., Ladner, R., and Lipscomb, W. (1982) *J. Mol. Biol.* 160:219–263.
11. Gibbons, I., Ritchey, J., and Schachman, H. (1976) *Biochemistry* 15:1324–1330.
12. Dreyfus, M., Fries, J., Tauc, P., and Hervé, G. (1984) *Biochemistry* 23:4852–4857.
13. Gerhart, J., and Pardee, A. (1964) *Fed. Proc.* 3:727–735.
14. Thiry, L., and Hervé, G. (1978) *J. Mol. Biol.* 125:515–534.
15. Collins, K., and Stark, G. (1971) *J. Biol. Chem.* 246:6599–6605.
16. Krause, K., Volz, K., and Lipscomb, W. (1985) *Proc. Natl. Acad. Sci. USA* 82:1643–1647.
17. Baillon, J., Tauc, P., and Hervé, G. (1985) *Biochemistry* 24:7182–7187.
18. Foote, J., Lauritzen, A., and Lipscomb, W. (1985) *J. Biol. Chem.* 260:9624–9629.
19. Foote, J., and Lipscomb, W. (1981) *J. Biol. Chem.* 256:11428–11433.
20. Hervé, G., Moody, M., Tauc, P., Vachette, P., and Jones, P. (1985) *J. Mol. Biol.* 185:189–199.
21. Tauc, P., Leconte, C., Kerbiriou, D., Thiry, L., and Hervé, G. (1982) *J. Mol. Biol.* 155:155–168.
22. Kerbiriou, D., and Hervé, G. (1973) *J. Mol. Biol.* 78:687–702.

23. Kantrowitz, E., and Lipscomb, W. (1977) *J. Biol. Chem.* 252:2873–2880.
24. Chan, W., and Enns, C. (1979) *Can. J. Biochem.* 57:798–805.
25. Ladjimi, M., Ghélis, C., Feller, A., Cunin, R., Glandsdorff, N., Piérard, A., and Hervé, G. (1985) *J. Mol. Biol.* 186:715–724.
26. Moody, M., Vachette, P., and Foote, A. (1979) *J. Mol. Biol.* 133:517–532.
27. Cunin, R., Jacobs, A., Charlier, D., Crabeel, M., Hervé, G., Glansdorff, N., and Piérard, A. (1985) *J. Mol. Biol.* 186:707–713.

20
Dynamic Interactions of the Second Messenger Systems

CHIAYENG WANG, RAJENDRA K. SHARMA,
and JERRY H. WANG

The response of living cells to changes in cell environment depends on the production and actions of second messenger molecules. The two most extensively studied second messengers are Ca^{2+} and cAMP, each of which has been demonstrated to regulate unique metabolic reactions and pathways. Examination of these regulatory reactions suggests that the two second messenger systems also undergo complex interactions during cell activation that serve to modulate the overall cell response.

Although individual reactions can be demonstrated biochemically in vitro, the sheer complexity of the potential relations among various regulatory reactions during cell stimulation makes them much more difficult to establish. Interactions between these systems can occur on two levels: a metabolic level where the actions of one second messenger affects the cellular concentration of the other, and a functional level where the actions of one second messenger modulates the effects of the other second messenger. Furthermore, these actions can occur at multiple points in a reaction pathway to provide feedback inhibition or stimulation of previous, future, or ongoing reactions.

In this chapter we examine certain cAMP- and Ca^{2+}/calmodulin-regulated reactions that illustrate several mechanisms of interaction of these two second messenger systems. These examples demonstrate the importance of elucidating the temporal relation between various events in order to understand the nature of sequential coordination of second messenger actions. Such coordination enables the cell to maintain sensitive and versatile responses to external stimuli.

General Characteristics of Second Messenger Systems

The stereospecific binding of an environmental stimulus to its membrane receptor elicits sequence of events, one of the earliest of which is the transient elevation of intracellular concentration of second messengers such as Ca^{2+} and cAMP. The second messenger molecules bind to and activate specific intracellular receptors, which in turn modify various target proteins to ultimately result in alterations of cellular activities.

TABLE 20.1. Regulatory proteins in second messenger systems.

Proteins and Enzymes	Mode of Action
Calmodulin (CaM)-dependent adenylate cyclase	Catalyzes Ca^{2+}, CaM-dependent conversion of ATP to cAMP
Calmodulin	Undergoes Ca^{2+}-dependent association with regulated proteins
	Exists as a Ca^{2+}-binding subunit of regulated proteins
cAMP-dependent protein kinase (pKA)	Catalyzes cAMP-dependent phosphorylation of regulated proteins
Phosphorylase kinase	Catalyzes Ca^{2+}-dependent and pKA-stimulated conversions of phosphorylase *b* and glycogen synthetase *a* to phosphorylase *a* and glycogen synthetase *b*, respectively
Myosin light-chain kinase	Catalyzes Ca^{2+}, CaM-dependent phosphorylation of myosin light chains
CaM-dependent cyclic nucleotide phosphodiesterase	Catalyzes Ca^{2+}, CaM-dependent conversion of cAMP to 5′-AMP
CaM-dependent phosphatase (calcineurin)	Catalyzes Ca^{2+}, CaM-dependent dephosphorylation of regulated proteins
Multifunctional CaM-dependent protein kinase	Catalyzes Ca^{2+}, CaM-dependent phosphorylation of regulated proteins

Table 20.1 shows some key proteins and enzymes involved in the complex second messenger-dependent interactions in the signal systems, as well as their possible modes of action. Membrane-bound adenylate cyclase catalyzes the production of cAMP from ATP, and in eukaryotes the only well established cAMP receptors are the cAMP-dependent protein kinases (1). Upon binding cAMP, the previously inactive kinase dissociates into its active catalytic subunits and its inactive regulatory subunits. The catalytic subunits can phosphorylate and thereby modulate the activities of many cellular proteins. The influx of extracellular Ca^{2+} or the release of Ca^{2+} from intracellular stores results in an elevated intracellular Ca^{2+} concentration. In contrast to the case with cAMP, a number of high-affinity Ca^{2+}-binding proteins serve as intracellular receptors of the Ca^{2+} second messenger. Notable among them is calmodulin, which is unique in possessing ubiquitous distribution and the capability of modulating the function of multiple target proteins. The major mechanism of mediation of the Ca^{2+} signal by calmodulin involves a Ca^{2+}-dependent, reversible association of calmodulin with its target proteins, resulting in the alteration of target protein activity (2).

The cAMP and Ca^{2+} second messenger systems interact to affect both the metabolism and the functional actions of these two second messengers. On the metabolic level, both cAMP metabolism and translocation of cell Ca^{2+} are under second messenger-dependent regulation. On the functional level, cellular processes regulated by one second messenger are often further modulated by another second messenger.

Examples of cAMP- and Ca^{2+}-Dependent Regulations

Fast skeletal muscle phosphorylase kinase and smooth muscle myosin light-chain kinase are two examples that involve both cAMP- and Ca^{2+}-dependent regulations. In the case of phosphorylase kinase, it exists in non-activated and activated states, and the enzymes activities at both states depend on Ca^{2+} and calmodulin (3). When nonactivated phosphorylase kinase is phosphorylated by cAMP-dependent kinase the enzyme activity is greatly enhanced, and it can be activated by much lower concentrations of Ca^{2+}, about one order of magnitude lower (4). Smooth muscle myosin light-chain kinase, upon activation by Ca^{2+}–calmodulin complex, can phosphorylate a pair of regulatory light chains of myosin, which is then allowed to interact with actin to cause muscle contraction (5). When myosin light-chain kinase is phosphorylated by cAMP-dependent protein kinase, the myosin kinase activity decreased owing to a decrease in the binding constant of the phosphorylated kinase for the Ca^{2+}–calmodulin complex (6).

Both muscle phosphorylase kinase and the smooth muscle contractile system are regulated by multiple calmodulin actions. Fast skeletal muscle phosphorylase kinase, as shown by Shenolikar and Cohen (7), contains calmodulin as a subunit and can also bind exogenously added calmodulin. The nonactivated phosphorylase kinase requires micromolar concentrations of Ca^{2+} for its activation, and at this Ca^{2+} concentration exogenous calmodulin can further stimulate the enzyme by two- to six fold. When nonactivated phosphorylase kinase is phosphorylated by cAMP-dependent protein kinase, the response of the enzyme activity to exogenously added calmodulin is no longer expressed. These observations suggest that the calmodulin subunit supports regulation of phosphorylase kinase by cAMP, whereas exogenous calmodulin serves to mediate Ca^{2+} activation of the enzyme at elevated cellular Ca^{2+} concentrations. Thus the two modes of regulatory action of calmodulin are used under different stimulating conditions. The observation that cardiac and red skeletal muscle phosphorylase kinases contain calmodulin as a subunit but are not responsive to exogenous calmodulin is consistent with the suggestion (8, 9).

It has been found that when the cytosolic Ca^{2+} concentration is elevated the velocity of shortening in smooth muscle correlates with the phosphorylation state of myosin light chains (LC 20). However, as the Ca^{2+}concentration decreases to a level slightly above that of the resting muscle, LC 20 is dephosphorylated without muscle relaxation. Relaxation occurs only upon further decrease in cytosolic Ca^{2+}. The LC 20 phosphorylation-independent contractile state of smooth muscle, also known as latch state, is believed to be regulated by a secondary Ca^{2+}-dependent mechanism (10) (Fig. 20.1). Although the nature of this secondary Ca^{2+} regulatory mechanism of smooth muscle contraction is not known and it is uncertain if calmodulin is involved, in vitro studies have shown that calmodulin can exert regulation on smooth muscle actomyosin Mg^{2+}-ATPase in addition to that of activation by myosin

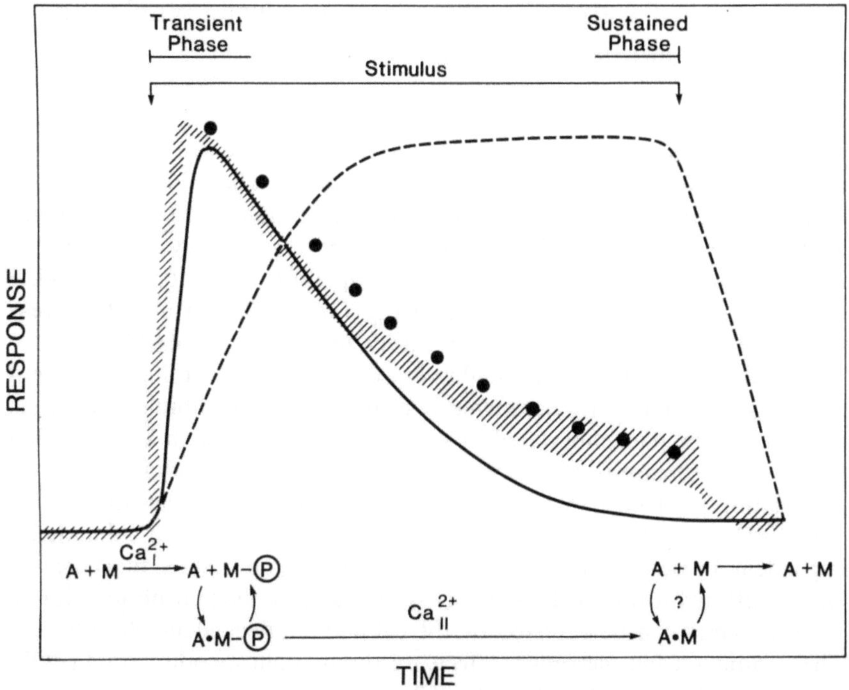

FIG. 20.1. Ca^{2+}-dependent smooth muscle contraction. Two Ca^{2+} concentration-dependent regulatory mechanisms, Ca_I^{2+} and Ca_{II}^{2+}, are involved in different time frames of muscle contraction. Ca_I^{2+} is involved in the initiation of the cyclic interaction of myosin (M) with actin (A) through $Ca_4^{2+} \cdot$calmodulin activation of myosin light-chain kinase. When myosin is phosphorylated, attached cross-bridges (A · M-Ⓟ or A · M) may be rapidly cycling, which results in force development (----) and maximal shortening velocities (•). The cycling process either slows or ceases as dephosphorylation of myosin takes place. Ca_{II}^{2+}, at a concentration slightly above that of resting muscle, is able to maintain the force in the absence of myosin phosphorylation. (From Kamm, K. E., and Stull, J. T. [1985] *Annu. Rev. Pharmacol. Toxicol.* 25:593–620; with permission.)

light-chain kinase. Fodrin, a spectrin-like actin and calmodulin-binding protein, can stimulate the MLCK-supported actomyosin Mg^{2+}-ATPase activity by 50 to 60% in a Ca^{2+}, calmodulin-dependent manner. This stimulation cannot be expressed in the absence of myosin light-chain kinase (11). Thus the fodrin-mediated calmodulin effect on smooth muscle actomyosin Mg^{2+}-ATPase is dependent on the MLCK-dependent calmodulin effect. These two calmodulin effects therefore appear to have a definite temporal relation.

The two examples given here not only demonstrate the interrelation between Ca^{2+} and cAMP second messenger systems on enzyme activities but also the possible regulatory significance of multiple calmodulin regulations.

Regulation of Calmodulin-Dependent Cyclic Nucleotide Phosphodiesterase Isozymes

Bovine brain contains two major calmodulin-dependent cyclic nucleotide phosphodiesterase isozymes that are homodimeric proteins of subunit molecular weights 60,000 and 63,000 daltons, respectively. The two isozymes have been separated by using a 60-kD subunit-specific monoclonal antibody immunoaffinity column (12). Both isozymes catalyze the hydrolysis of cAMP as well as cGMP and have similar kinetic properties, except that the 63-kD subunit-containing isozyme has a higher V_{max} for cGMP than cAMP, whereas the 60-kD subunit-containing isozyme has a higher V_{max} for cAMP (13).

The two bovine brain phosphodiesterase isozymes are differentially regulated by protein phosphorylation mechanisms. The 60-kD subunit-containing isozyme can be phosphorylated by cAMP-dependent protein kinase with a subsequent decrease in the enzyme affinity toward calmodulin. This phosphorylation is blocked by Ca^{2+} and calmodulin (14). On the other hand, the 63-kD subunit-containing isozyme is phosphorylated by a calmodulin-dependent protein kinase, and the phosphorylation results in decreased enzyme affinity toward calmodulin (15). In addition, the phosphorylation of both isozymes can be reversed by the calmodulin-dependent phosphatase. The various second messenger effects on the two phosphodiesterase (PDE) isozymes are summarized in Table 20.2.

The multiple calmodulin actions are in evidence with both isozymes. It is clear that some of the regulatory actions on the individual isozymes have opposing effects. For such regulatory actions to be physiologically significant, it is necessary that they do not occur simultaneously in the cells. One possible mechanism by which different calmodulin actions can be triggered at different times or under different physiological conditions is if the different calmodulin actions exhibit different Ca^{2+} sensitivities.

TABLE 20.2. Regulation of phosphodiesterase isozymes by Ca^{2+} and cAMP.

Isozyme	Regulation
60-kD Subunit-containing isozyme	1. Activation by Ca^{2+} and calmodulin 2. Phosphorylation by cAMP-dependent protein kinase with reduction of PDE affinity towards calmodulin. 3. Blockage of PDE phosphorylation by Ca^{2+} and calmodulin 4. Reversal of PDE phosphorylation by Ca^{2+} and calmodulin-dependent phosphatase
63-kD Subunit-containing isozyme	1. Activation by Ca^{2+} and calmodulin 2. Phosphorylation by calmodulin-dependent protein kinase with consequent decrease in affinity towards calmodulin 3. Reversal of PDE phosphorylation by calmodulin-dependent phosphatase

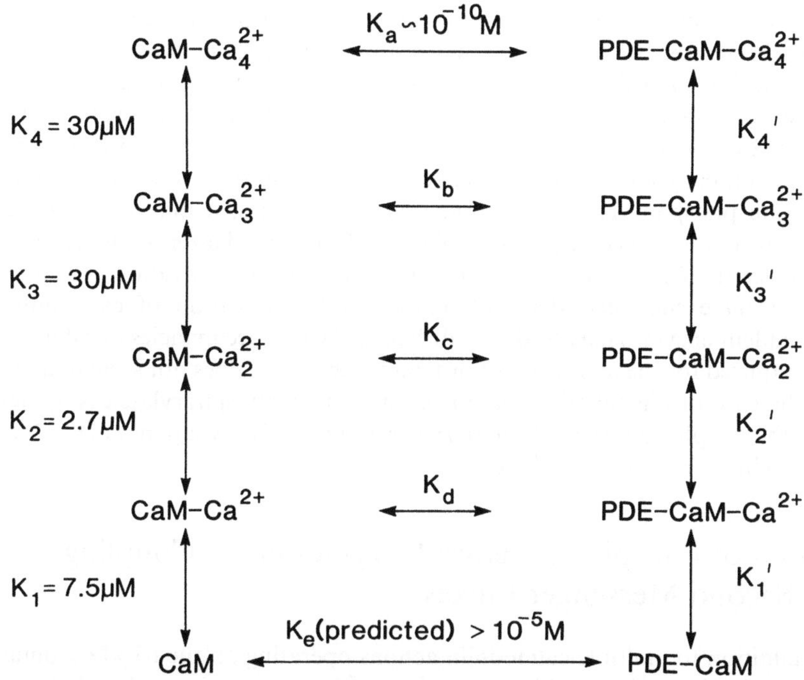

FIG. 20.2. Energy coupling between Ca^{2+} binding of calmodulin (CaM) and the association of calmodulin with phosphodiesterase (PDE). Two pathways for the formation of the fully liganded CaM–PDE complex are illustrated. One pathway (left) is initiated by the stepwise binding of Ca^{2+} to calmodulin followed by the association of the fully liganded calmodulin to the enzyme. The other pathway (right) is via the association of free calmodulin and phosphodiesterase and then the sequential binding of Ca^{2+} to the protein complex. K_1 to K_4 = dissociation constants of Ca^{2+} and calmodulin; K_1' to K_4' = dissociation constants of protein complexes and Ca^{2+}; K_a to K_e = dissociation constants of phosphodiesterase and liganded calmodulin.

Mechanism of Differential Ca^{2+} Sensitivity

For enzymes regulated by Ca^{2+}-dependent and reversible calmodulin binding, the formal mechanism of calmodulin regulation may be depicted as shown in Fig. 20.2 (16). The scheme indicates that the interactions of Ca^{2+} and the target enzymes with calmodulin are energetically coupled. Such energy-coupled processes have been demonstrated by a number of investigators (16–19). The consequence of this energy coupling is that Ca^{2+} and calmodulin could regulate the target enzyme synergistically (16–19); when calmodulin concentration is increased, the Ca^{2+} concentration required for the enzyme modulation decreases.

The mutual dependence of enzyme and Ca^{2+} affinities toward calmodulin (Fig. 20.2) suggests that when a regulated enzyme is modified to change its calmodulin affinity the Ca^{2+} concentration required for the enzyme regulation also changes. The calmodulin-dependent phosphodiesterase isozymes have been used to test such a suggestion. At constant concentration of calmodulin, phosphorylated isozymes were found to require higher Ca^{2+} concentration for activation than the corresponding nonphosphorylated isozymes. Because short-term fluctuation in cellular calmodulin concentration is unlikely and calmodulin concentration in brain is sufficient to fully activate phosphorylated phosphodiesterase isozymes, it may be suggested that the consequence of phosphorylation of phosphodiesterase in vivo is to decrease the Ca^{2+} sensitivity required for the enzyme activation. Although others have suggested that differential Ca^{2+} stimulation of calmodulin-dependent enzymes may be due to differential Ca^{2+} occupancies of calmodulin required for the activation of different enzymes, it does not seem to apply in the case of phosphodiesterase isozymes. Both phosphorylated isozymes, like the nonphosphorylated counterparts, require a full occupancy of Ca^{2+} on calmodulin for enzyme activation.

Roles of Phosphodiesterase Isozymes in the Coupling of Second Messenger Fluxes

In addition to multiple calmodulin actions operating on the 63-kD subunit-containing isozymes at different cellular Ca^{2+} concentration, calmodulin can also activate adenylate cyclase. Figure 20.3 shows a working hypothesis that describes how the different calmodulin-regulatory actions on the phosphodiesterase isozymes are separated temporarily from the activation of brain adenylate cyclase by calmodulin during a transitory increase in cytosolic Ca^{2+} in order to exert meaningful regulation of cAMP metabolisms in cells. It is postulated that the initial increase in cell Ca^{2+} concentration brings about the activation of adenylate cyclase and calmodulin-dependent protein kinase; activation of the latter enzyme results in the phosphorylation and inhibition of the phosphodiesterase isozymes. The two calmodulin-regulatory actions acting in concert cause (or enhance) the elevation in cAMP concentration. As the Ca^{2+} concentration in cells is further increased, the phosphatase is activated by calmodulin to dephosphorylate the phosphodiesterase isozyme, which is then activated by Ca^{2+} and calmodulin. The concerted regulation by these two calmodulin actions results in the rapid decrease in cellular cAMP. The working hypothesis illustrates that a transitory increase in cAMP concentration in cells can be achieved by the time-dependent regulation of phosphodiesterase activity during a transitory increase in Ca^{2+} concentration. Thus the fluxes of the two second messengers can be tightly coupled, and the phosphodiesterase isozyme may play a key role in this coupling. It should be noted that it is yet to be established if the calmodulin-dependent protein kinase can be activated by lower concentrations of Ca^{2+} than the calmodulin-dependent phosphatase and phosphodiesterase. However, brain adenylate cyclase has been shown to be activated by much lower concentrations of Ca^{2+} than the phosphodiesterase (20).

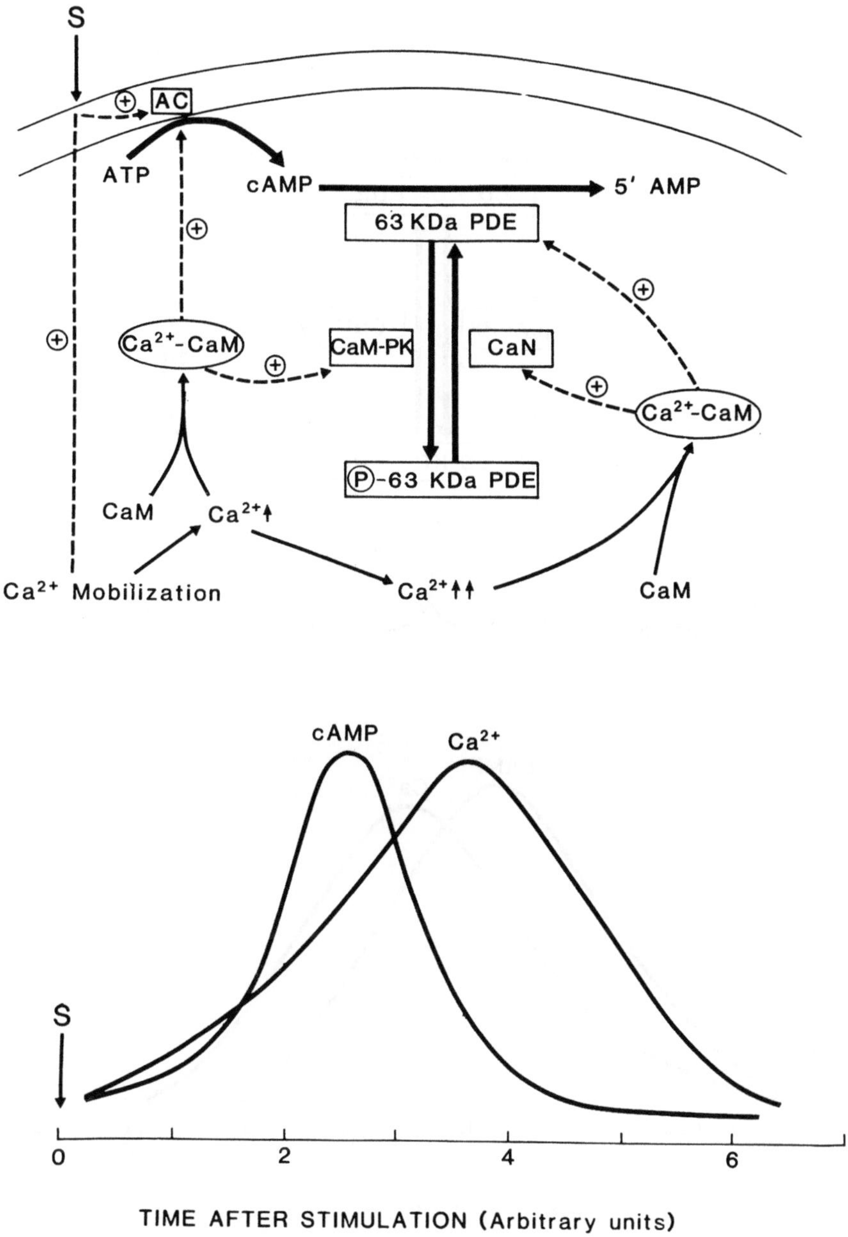

FIG. 20.3. Hypothesis of the temporally separated regulations of 63-kD subunit-containing phosphodiesterase isozyme by Ca^{2+} (top panel) and the phosphodiesterase-mediated coupling between Ca^{2+} and cAMP fluxes (bottom panel). AC = adenylate cyclase; PDE = phosphodiesterase; CaN = calcineurin; CaM-PK = calmodulin-dependent protein kinase; P – = phosphorylated; + = activation; $Ca^{2+}\uparrow$ = small increase in Ca^{2+} concentration; $Ca^{2+}\uparrow\uparrow$ = large increase in Ca^{2+} concentration.

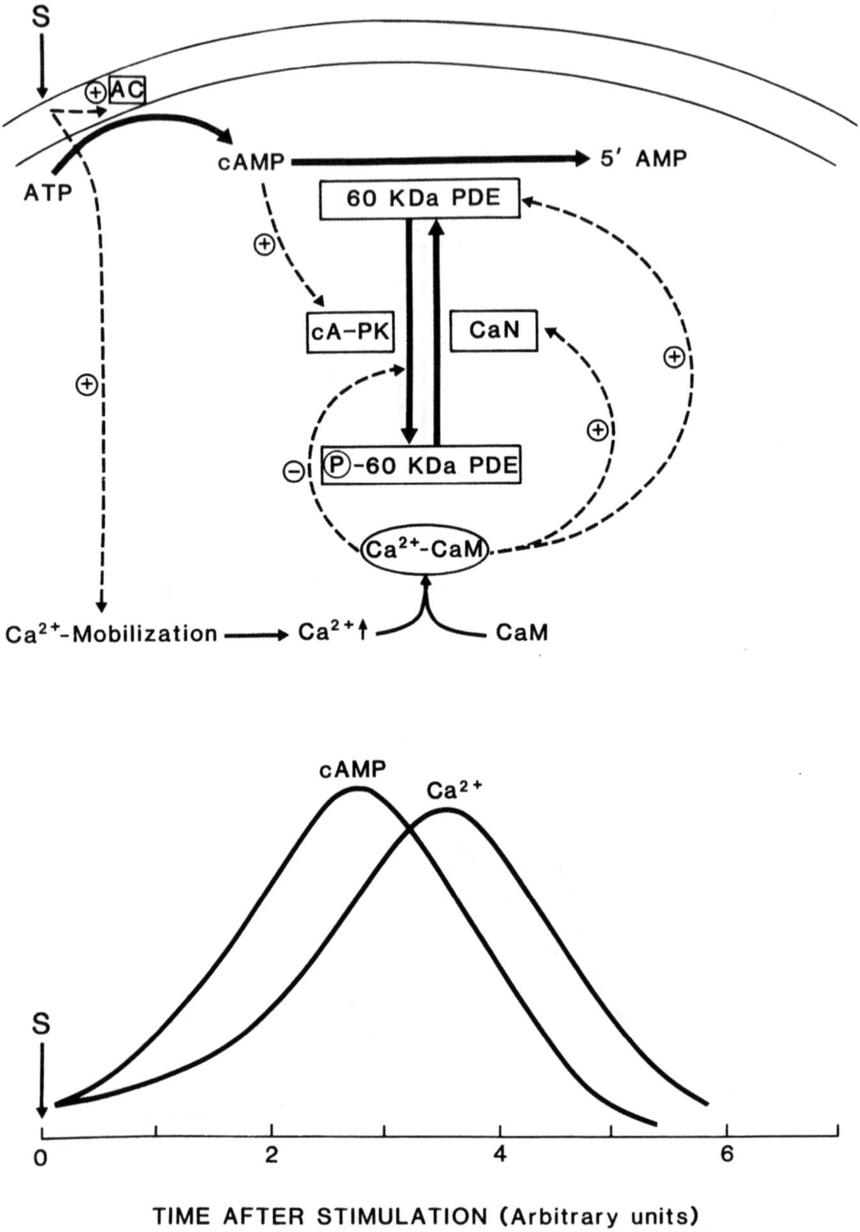

FIG. 20.4. Hypothesis of the temporally separated regulations of the 60-kD subunit-containing phosphodiesterase isozyme by Ca^{2+} and cAMP (top panel) and the phosphodiesterase-mediated coupling between Ca^{2+} and cAMP fluxes (bottom panel). AC = adenylate cyclase; PDE = phosphodiesterase; CaN = calcineurin; cA-pK = cAMP-dependent protein kinase; Ⓟ = phosphorylated; ⊕ = activation; ⊖ = inhibition.

A different coupling between the two messenger fluxes can be brought about by the various regulatory actions of second messengers on the 60-kD subunit-containing isozymes. In this case, cAMP itself also participates in the regulation of cAMP metabolism. A working hypothesis describing the role of this phosphodiesterase isozyme in the coupling between the two messenger fluxes is presented in Fig. 20.4. An initial increase in cAMP concentration during cell activation may bring about phosphorylation of the 60-kD isozyme and thereby prevent it from being activated by the low concentration of Ca^{2+} existing at the early stage of cell activation. The phosphodiesterase reaction is therefore inhibited in coordination with the stimulation of adenylate cyclase (by external signals). This action ensures a sharp rise in intracellular concentration of cAMP. At later stages of cell activation, when the cell Ca^{2+} concentration is increased, the phosphatase reaction is activated to reverse the enzyme phosphorylation, and the phosphodiesterase isozyme is now fully activated by Ca^{2+} and calmodulin. Because the Ca^{2+}–calmodulin complex can block the 60-kD isozyme phosphorylation, the dephosphorylated state of the isozyme is maintained even though cAMP concentration may still be high in cells. The concerted actions of these three regulatory mechanisms on the 60-kD isozyme, i.e., the phosphatase reaction, Ca^{2+}/calmodulin stimulation of the isozyme and Ca^{2+}/calmodulin inhibition of phosphorylation result in a rapid decrease in cAMP concentration.

Role of Calmodulin-Stimulated Phosphatase

For both phosphodiesterase isozymes the calmodulin-stimulated phosphatase reverses their phosphorylation and therefore releases the inhibition of the enzyme. This finding indicates that one of the regulatory actions of the phosphatase is to facilitate the Ca^{2+}-dependent termination of the cAMP signal (21), which is in concert with its action of facilitating the termination of cAMP second messenger effect as suggested by others (22).

Figure 20.5 shows how calmodulin-dependent phosphatase acts to facilitate the abolition of both elevated cell cAMP and its effects. Mammalian tissues contain a protein phosphatase that is active toward many proteins that are phosphorylated by cAMP-dependent protein kinase (23, 24). This phosphatase, designated phosphatase 1, can be inhibited by a specific protein (inhibitor I) that exhibits the inhibitory effect only when it is phosphorylated by cAMP-dependent protein kinase (25). Thus at an elevated cAMP concentration, the phosphorylation of various protein substrates by cAMP-dependent protein kinase is increased, whereas their dephosphorylation is blocked by the phosphorylated inhibitor I (22). Phosphorylated inhibitor I is a substrate of calmodulin-stimulated phosphatase. Therefore when the cell Ca^{2+} level is increased, dephosphorylation of inhibitor I occurs and subsequently reverses the inhibition of phosphatase I. Alternatively, some of the cAMP-dependent protein kinase-phosphorylated proteins can be dephos-

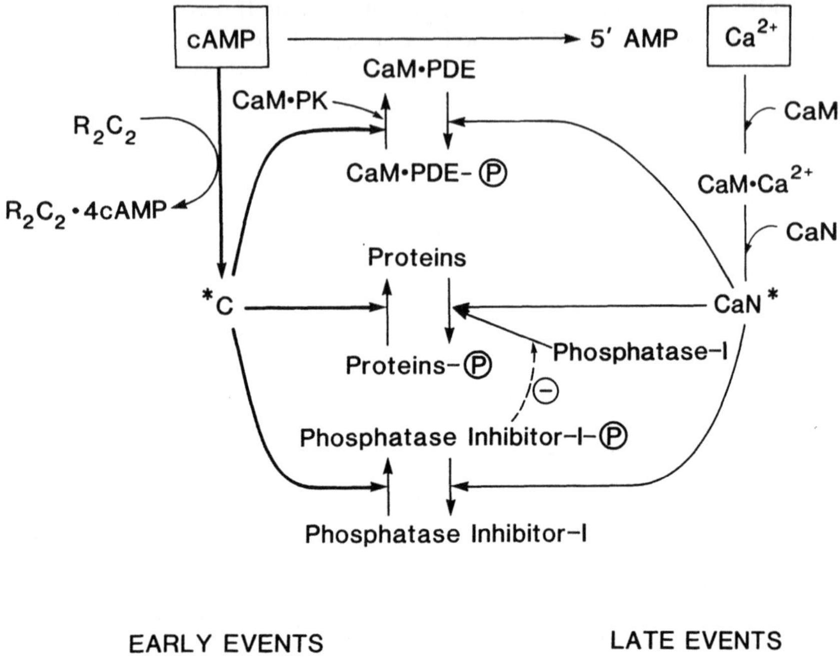

FIG. 20.5. Interactions between Ca^{2+} and cAMP second messenger systems. R_2C_2 = regulatory (R) and catalytic (C) subunits of cAMP-dependent protein kinase; CaM · PK = calmodulin-dependent protein kinase; CaM · PDE = calmodulin-dependent phosphodiesterase; CaN = calcineurin; Ⓟ = phosphorylated; ⊖ = inhibition; * = active form of enzyme.

phorylated directly by calcineurin (e.g., 60-kD subunit of phosphodiesterase isozyme).

The protein kinase-mediated cAMP effect and the calmodulin-stimulated phosphatase-mediated Ca^{2+} effect have to be temporally separated to be regulatorily significant. Thus when the role of the calmodulin-stimulated phosphatase is considered in a broader context and in relation to an overall cAMP effect, the time dependence of the phosphatase action again emerges as an important consideration.

Concluding Remarks

Analysis of various regulatory reactions involving Ca^{2+} and cAMP has revealed the importance of time dependence of these reactions during cell activation. Although no measurement is available for the time of occurrence of specific regulatory reactions, the relative sequence of different reactions during cell activation may be suggested on teleological considerations. The various examples given here illustrate how meaningful regulation may be obtained when certain sequences of regulatory events are postulated. One outcome of the analysis is that the complex regulation of cAMP metabolic enzymes by cAMP and Ca^{2+} may play important roles in coupling the cellular fluxes of Ca^{2+} and cAMP. On the other hand, as both cAMP and Ca^{2+} elevations are transitory, the time dependence of regulatory actions of second messengers appears to be logically inescapable.

Acknowledgement. Work carried out in the authors' laboratory is supported by the Medical Research Council of Canada (MT 2381) and Alberta Heritage Foundation for Medical Research (AHFMR). C. W. is a Medical Research Council Studentship Awardee. J. H. W. is an AHFMR Medical Scientist.

References

1. Krebs, E. G., and Beavo, J. A. (1979) *Annu. Rev. Biochem.* 48:923–959.
2. Klee, C. B., Crouch, J. H., and Richman, P. G. (1980) *Annu. Rev. Biochem.* 49:489–515.
3. Ozawa E. (1973) *J. Neurochem.* 20:1487–1488.
4. Brostrom, C. O., Huang, Y. C., Breckenridge, B. McL., and Wolff, D. J. (1975) *Proc. Natl. Acad. Sci. USA* 72:64–68.
5. Small, J. V., and Sobieszek, A. (1977) *Eur. J. Biochem.* 76:521–530.
6. Conti, M. A., and Adelstein, R. S. (1980) *Fed. Proc.* 39:1569–1573.
7. Shenolikar, S., and Cohen, P. (1978) *FEBS Lett.* 86:92–96.
8. Sharma, R. K., Tam, S. W., Waisman, D. M., and Wang, J. H. (1980) *J. Biol. Chem.* 255:11102–11105.
9. Cooper, R. H., Sul, H. S., and Walsh, D. A. (1981) *J. Biol. Chem.* 256:8030–8038.
10. Kamm, K. E., and Stull, J. T. (1985) *Annu. Rev. Pharmacol. Toxicol.* 25:593–620.
11. Wang, C., Ngai, P. K., Walsh, M. P., and Wang, J. H. (1987) *Biochemistry* 26:1110–1117.
12. Sharma, R. K., Adachi, A., Adachi, K., and Wang, J. H. (1984) *J. Biol. Chem.* 259:9248–9254.
13. Sharma, R. K., Pallen, C. J., and Wang, J. H. (1987) *Brain, Behavior and Education* (in press).
14. Sharma, R. K., and Wang, J. H. (1985) *Proc. Natl. Acad. Sci. USA* 82:2603–2607.
15. Sharma, R. K., and Wang, J. H. (1985) *J. Biol. Chem.* 261:1322–1328.
16. Huang, C. Y., Chau, V., Chock, P. B., Wang, J. H., and Sharma, R. K. (1981) *Proc. Natl. Acad. Sci. USA* 78:871–874.
17. Wolff, D. J., and Brostrom, C. O. (1979) *Adv. Cyclic Nucleotide Res.* 11:27–88.
18. Blumenthal, D. K., and Stull, J. T. (1980) *Biochemistry* 19:5608–5614.
19. LaPorte, D. C., Toscano, W. A., and Storm, D. R. (1979) *Biochemistry* 18:2820–2825.
20. Piascik, M. T., Wisler, P. L., Johnson, C. L., and Potter, J. D. (1980) *J. Biol. Chem.* 255:4176–4182.

21. Pallen, C. J., Sharma, R. K., Matsui, H., and Wang, J. H. (1985) *Adv. Protein Phosphatase* 1:147–162.
22. Stewart, A. A., Ingebritsen, T. S., Manalan, A., and Klee, C. B. (1982) *FEBS Lett.* 137:80–88.
23. Killiea, S. D., Brandt, H., Lee, E. Y. C., and Whelan, W. J. (1976) *J. Biol. Chem.* 251:2363–2368.
24. Khandelwal, R. L., Vandenheede, J. R., and Krebs, E. J. (1976) *J. Biol. Chem.* 251:4850–4858.
25. Huang, F. L., and Glinsmann, W. H. (1976) *Eur. J. Biochem.* 70:419–426.

21 Role of Multienzyme Complexes in the Integration of Cellular Metabolism

B. I. KURGANOV

A large number of degradation processes provide the energetic needs of the living organism and its need of "building materials." The synthesis of biomacromolecules is a result of numerous biosynthetic reactions. The striking peculiarity of the living system lies in the existence of rather fine and effective regulatory mechanisms that bring each metabolic process to conform with the needs of the whole organism. Thus cellular metabolism represents an integrated system of processes coordinated in space and time. The elucidation of the principles of integration of cellular metabolism is one of the important problems of modern biological chemistry.

In this chapter the general properties of enzymes and multienzyme complexes as controllable systems and the role of multienzyme complexes in the integration of cellular metabolism are discussed.

Mechanisms of Integration of Cellular Metabolism

The regulatory mechanisms in living cells may be classified into two types: (a) the regulatory mechanisms providing maintenance of the levels of key metabolites (e.g., coenzyme), i.e., the mechanisms of homeostasis; and (b) the regulatory mechanisms that make it possible to trace any magnitude determined by external conditions, i.e., the mechanisms of tracing. The mechanisms of homeostasis allow existence of the living cell as an integral organized system. So far as the mechanisms of tracing are concerned, it can be said that in multicellular organisms these mechanisms are realized with the participation of second messengers, i.e., Ca^{2+} ions, cyclic adenosine-3′,5′-monophosphate (cAMP), cyclic guanosine-3′,5′-monophosphate (cGMP), inositol-1,4,5-triphosphate, diacylglycerol, and 2′,5′-oligoadenylate. When compared with the mechanisms of homeostasis, the mechanisms of tracing occupy a higher position in the hierarchy of the control of cellular metabolism. One can suppose that the mechanisms of homeostasis and tracing are realized at different levels of material organization of a cell. Therefore it is well to discuss comparatively the mechanisms of the control of functioning of enzymes and multienzyme complexes.

Enzymes as Controllable Systems

The basic regulatory mechanisms for low-molecular-weight catalysts are the mechanisms of isosteric regulation that result from direct action of the modifier. The appearance of high-molecular-weight catalysts of polypeptide nature (enzyme) in the process of biological evolution denotes not only the creation of highly effective catalysts with high substrate specificity. In fact, the enzymes are controllable catalytic systems. The comparatively large dimensions of a protein molecule enable the disposition on its surface of not only working sites (i.e., active sites) but also sites of control. The role of the control sites is played by the so-called allosteric sites (effector sites). The function of regulators is fulfilled by the metabolites (allosteric effectors) with chemical structures different from those of the substrates and the products of the controllable enzyme reaction (1).

When discussing the behavior of the enzymes as controllable systems, two aspects should be taken into account. First, it should be noted that the working site (active site) and the control site (allosteric site) in the enzymes are spatially isolated. Second, it is significant that the regulation is realized with the aid of factors that are "external" to the controllable enzymic process. The regulatory factors are the metabolites, which provide optimum functioning of the enzyme in the biological system at higher levels of complexity, i.e., in definite metabolic pathways.

Assembly of Multienzyme Complexes

The enzymes taking part in the common metabolic pathway tend to form the ordered structures. Mowbray and Moses (2, 3) have discovered the presence of a multienzyme complex of glycolytic enzymes in the soluble fraction of mechanically ruptured spheroplasts of *Escherichia coli*. When investigating the activities of the glycolytic enzymes in the membrane fractions prepared from hemolyzed beef erythrocytes, Green et al. (4) inferred that the glycolytic enzymes adsorbed to the erythrocyte membrane form the ordered multienzyme complex. The structural organization of the complex of the glycolytic enzymes adsorbed to actin-containing filaments has been discussed by Clarke et al. (5).

We proposed the structure of the glycolytic multienzyme complex adsorbed to the erythrocyte membrane (6, 7). The role of the anchor site providing fixation of the complex on the erythrocyte membrane is played by the band 3 protein. The main function of the band 3 protein is anion transport. We have suggested that the glycolytic multienzyme complex is adsorbed to the band 3 dimer and has a symmetry axis of the second order. However, the experimental data obtained by Schubert and his colleagues (8, 9) show that the band 3 protein is able to form a larger aggregated species than the dimer. It should be noted that the type of the association of the band 3 protein is not known.

We assume that the association of the band 3 protein leads to formation of closed structures, i.e., hexamers (the trimer of the dimers). Supporting this assumption is the finding that certain integral membrane-bound proteins possess trimeric structure (10). According to Klingenberg (10) trimeric integral proteins have a symmetry axis of the third order. This symmetry axis is perpendicular to the membrane plane. The central part of such trimeric structures is a channel fulfilling transport functions. For example, the channel formed by the hexameric form of the band 3 protein may participate in water transport. [The possibility of the participation of the band 3 tetramer in water transport was discussed by Benz et al. (11).]

The hypothetical structure of the glycolytic multienzyme complex with symmetry axis of the third order is shown in Fig. 21.1. The first step in the assembly of the multienzyme complex is the adsorption of three molecules of phosphofructokinase to the band 3 hexamer embedded in the erythrocyte membrane. Each molecule of phosphofructokinase has contact with the band 3 dimer. The complex contains a triple set of the glycolytic enzymes and has a molecular mass of 4.5×10^6 daltons. The assembly of the glycolytic multienzyme complex brings about the formation of a microcompartment, with a glycolytic process proceeding without release of the glycolytic intermediates in the medium.

NAD-dependent dehydrogenases occupy the neighboring positions in that complex. Therefore the oxidized form of NAD, which is necessary for the reaction catalyzed by glyceraldehyde-phosphate dehydrogenase (4), may be regenerated because of the reactions catalyzed by lactate dehydrogenase (9) and glycerol-3-phosphate dehydrogenase (11). The contact between phosphofructokinase (2) and pyruvate kinase (8) allows utilization of ATP produced by pyruvate kinase in the reaction catalyzed by phosphofructokinase.

The possibility of assembling the glycolytic enzymes in the structurally ordered complex is predetermined by the existence of the corresponding recognition sites in the enzyme molecules. The final structure is unique because of the participation of the anchor protein in the assembly process. It should be noted that the complexes of individual glycolytic enzymes are rather unstable. In this aspect it is difficult to investigate such complexes in solution. However, in the multienzyme complex each enzyme has contacts with several neighbors. Therefore the state corresponding to the completely assembled complex is thermodynamically favored. In other words, assembly of the glycolytic multienzyme complex represents the process of cooperative association, which proceeds according to the "all or none" principle.

Multienzyme Complexes as Controllable Systems

Owing to the interaction of the enzymes in the multiprotein complex, the complex acts as an integral system able to function in cooperative fashion. The coordinated changes in conformation of the enzymes in the multiprotein

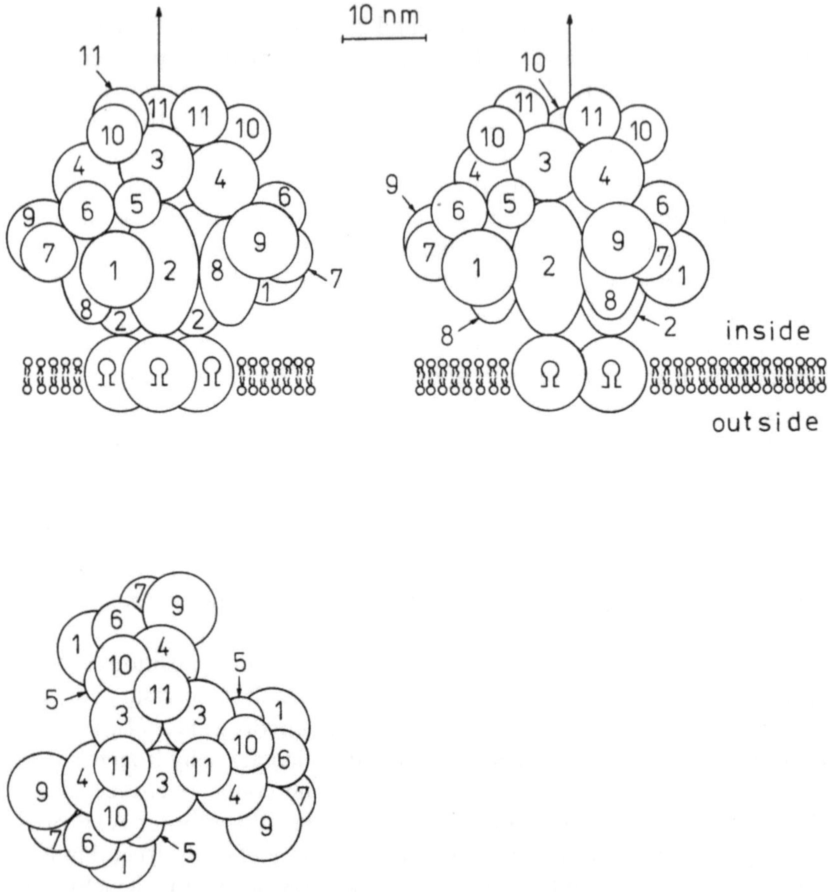

FIG. 21.1. Hypothetical structure of the complex of the glycolytic enzymes adsorbed to the erythrocyte membrane. The numbers refer to: glucose phosphate isomerase (1); 6-phosphofructokinase (2); fructose-bisphosphate aldolase (3); glyceraldehyde-phosphate dehydrogenase (4); phosphoglycerate kinase (5); phosphoglyceromutase (6); enolase (7); pyruvate kinase (8); lactate dehydrogenase (9); triosephosphate isomerase (10); and glycerol-3-phosphate dehydrogenase (11). Ω = band 3 protein dimer.

complex may ensure the effective transfer of the intermediates with the conveyer of active sites in the microcompartment.

When discussing the enzymes as controllable systems we have paid attention to two features of the control of functioning: (a) the spatial isolation of the working sites and the sites of the control in the enzyme molecule; and (b) the presence of the control factors external to the controllable enzyme system. In our opinion, these features of functioning of the enzyme systems are important in the control of functioning of the biological systems at any level of complexity, including the multienzyme complexes. The multienzyme complex as a controllable system must have both the spatially isolated working site and the site of control. It is evident that the working site of the multienzyme complex is a microcompartment where the chemical transformation of entering substrate takes place. The function of the control site is fulfilled by the anchor protein of the support, which participates in assembly of the multienzyme complex (12). For example, the site of the control of glycolytic multienzyme complex adsorbed to the erythrocyte membrane is the band 3 protein.

Kurganov (12) suggested that the general control of functioning of the multienzyme complex (by "switching on/switching off") is exerted with participation of second messengers affecting the state of the anchor protein. In other words, the role of the factors responsible for the control of functioning of the multienzyme complex as a whole is played not by the metabolic intermediates but the external factors (i.e., second messengers), which provide the optimum functioning of the multienzyme complex in the biological system at the higher level of complexity, i.e., in the cell.

It is known that the band 3 protein is able to interact with Ca^{2+} ions (13). The concentration of Ca^{2+} ions in the erythrocytes is 10 to 20 μM (14). The intracellular concentration of Ca^{2+} may increase in response to binding of hormones and neurotransmitters to corresponding receptors incorporated in the erythrocyte membrane. For example, Tang et al. (15) have shown that binding of the cholinergic agonist carbachol to muscarinic cholinergic receptors of the erythrocyte membrane brings about an increase in Ca^{2+} concentration. The maximum uptake (250% in comparison with the initial level of Ca^{2+}) occurred 25 to 30 s after addition of carbachol.

One can expect that the change in the degree of saturation of the anchor protein (band 3 protein) by Ca^{2+} ions results in the conformational changes of the band 3 protein and, further, in the conformational changes of the enzyme molecules that have contact with the anchor protein, i.e., the phosphofructokinase molecules in the case of the glycolytic multienzyme complex, and neighboring enzyme molecules. Thus the binding of Ca^{2+} ions to the band 3 protein may produce the alterations in catalytic functions of the enzymes in the glycolytic multienzyme complex.

The control of functioning of the multienzyme complex by "switching on/switching off' may be represented as follows. The multienzyme complex may have a spatial arrangement of the enzymes and conformation of the enzyme molecules that blocks the entrance of the substrates into the microcompartment. Therefore in this state the complex is functionally inactive. The action of a second messenger on the site of the control induces changes in the conformation of the enzyme molecules (and possibly changes in the spatial arrangement of the enzyme in the complex) that lead to elimination of the steric hindrances, thereby allowing the substrates to enter the microcompartment as well as permitting their chemical transformations.

Conclusion

The main regulatory mechanisms for soluble enzymes are isosteric and allosteric. These regulatory mechanisms stabilize the levels of key metabolites and ensure the self-regulation of metabolic pathways.

The assembly of the complex of the enzymes participating in the common metabolic pathway leads to the appearance of the regulatory mechanism, which occupies the higher position in the hierarchy of the levels of cellular metabolic control. Functioning of the metabolic system as a whole is realized with the participation of second messengers. It represents the tracing mechanism and determines the functioning of the metabolic system in accordance with signals produced by nervous, hormonal, and immune systems of the organism.

The multienzyme complexes adsorbed to subcellular structures are mobile structures. A dynamic equilibrium exists between the complex and the soluble enzymes, which makes possible effective operation of the homeostasis mechanisms as well as the tracing mechanisms.

References

1. Kurganov, B. I. (1982) *Allosteric Enzymes. Kinetic Behaviour*. Wiley, New York.
2. Moses, V. (1978) In *Microenvironments and Metabolic Compartmentation*, edited by P. A. Srere and R. W. Estabrook, p. 169. Academic Press, New York.
3. Mowbray, J., and Moses, V. (1976) *Eur. J. Biochem.* 66:25
4. Green, D. E., Murer, E., Hultin, H. O., Richardson, S. H., Salmon, B., Brierly, C. P., and Baum, H. (1965) *Arch. Biochem. Biophys.* 112:635.
5. Clarke, F. M., Stephan, P., and Morton, D. J. (1985) In *Regulation of Carbohydrate Metabolism*, edited by R. Biethner. CRG Press, New York.
6. Kurganov, B. I., (1984) *J. Theor. Biol.* 111:707.
7. Kurganov, B. I., Sugrobova, N. P., and Mil'man, L. S. (1985) *J. Theor. Biol.* 116:509.
8. Dorst, H-J., and Schubert, D. (1979) *Hoppe Seyler Z. Physiol. Chem.* 360:1605.
9. Pappert, G., and Schubert, D. (1983) *Biochim. Biophys. Acta* 730:32.
10. Klingenberg, M. (1981) *Nature* 290:449.
11. Benz, R., Tosteson, M. T., and Schubert, D. (1984) *Biochim. Biophys. Acta* 775:347.
12. Kurganov, B. I. (1986) *J. Theor. Biol.* 119:445.
13. Passing, R., and Schubert, D. (1983) *Hoppe Seyler Z. Physiol. Chem.* 364:873.
14. Harrison, D. G., and Long, C. (1968) *J. Physiol.* (*Lond.*) 198:367.
15. Tang, L. G., Schomaker, E., and Wiesmann, W. P. (1984) *Biochim. Biophys. Acta* 772:235.

22 Dynamics of Deoxynucleotide Synthesis in Relation to DNA Replication

ROLF ELIASSON, MARC FONTECAVE, and PETER REICHARD

The biosynthesis of the four deoxyribonucleoside triphosphates (dNTPs) required for DNA replication is a highly regulated affair. Not only do cells engaged in DNA synthesis require a steady supply of building blocks, but the relative amounts of the four dNTPs appear to be important and disturbances in the normal balance lead to negative genetic consequences and disease (1).

The enzyme ribonucleotide reductase and its regulation occupy key positions in this context. A single protein catalyzes the reduction of the four common ribonucleotides and is responsible for the production of all dNTPs (2). The size of the four pools is controlled mainly by the complicated allosteric regulation of the reductase, which determines both the substrate specificity of the enzyme and its overall activity (3). The content of the tyrosyl free radical of the enzyme (4, 5), necessary for its catalytic activity, also appears to be of importance. Finally, auxiliary enzyme reactions involving breakdown of nucleotides to nucleosides and the reverse phosphorylation of the nucleosides, integrated as substrate cycles between deoxyribonucleosides/deoxyribonucleotides, contribute to the fine tuning of the levels of dNTP pools in mammalian cells (6).

This chapter is concerned with one of these regulatory aspects: the tyrosyl free radical of the enzyme. First we describe briefly the structural background for the radical. Then in more detail we discuss experiments that led to the discovery and characterization of an enzyme system in *Escherichia coli* (7) that has the capacity to remove one electron from the ring of a specific tyrosine (Tyr-122) of the polypeptide chain in the B2 subunit of the *E. coli* reductase and thereby generates a tyrosyl free radical in the primary structure. The enzyme system transforms an inactive form of the reductase into an active one.

Structure of *E. coli* Ribonucleotide Reductase and Its Tyrosyl Free Radical

The *E. coli* enzyme has the general structure $\alpha_2\beta_2$ (2). For historical reasons α_2 is called protein B1 and β_2 protein B2. The structure of the reductase is shown in Fig. 22.1. Protein B1 has two substrate binding sites, but the catalytic site

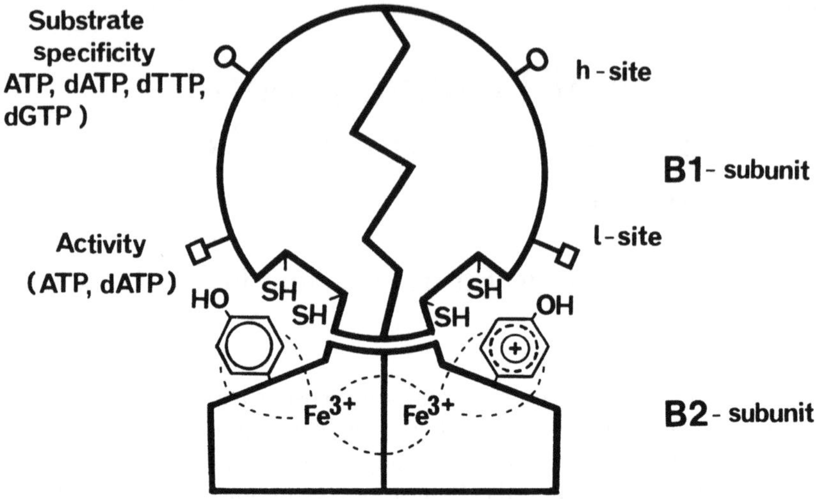

FIG. 22.1. Structure of *E. coli* ribonucleotide reductase.

is made up from both subunits, and protein B2 contributes to the site with a tyrosyl free radical (5). The radical resides on a specific tyrosine residue (Tyr-122) of one of the two polypeptide chains of protein B2 (8). It is an oxidized species, as the tyrosine ring has lost one electron. Protein B2 also contains two irons that stabilize the radical. The irons exist as a pair of unequivalent high-spin Fe(III) ions, antiferromagnetically coupled by a μ-oxo bridge (9, 10). The iron ions are bonded directly to the protein structure and not via sulfur, as in ferredoxin and similar proteins. In this and in many other aspects protein B2 shows a striking similarity to the iron center of hemerythrin, a small oxygen-carrying protein of some invertebrates (11). As shown in Fig. 22.1 only one of the two polypeptide chains of B2 carries the radical.

A close connection exists between the iron center and the free radical (5). When iron is removed by chelation, the metal-free apoprotein B2 (apo B2) also loses the radical and is enzymatically inactive. Both the radical and the enzyme activity can be restored simply by adding ferrous iron to apo B2 in the presence of oxygen (9). This addition results in the simultaneous oxidation of ferrous to ferric iron and Tyr-122 to a tyrosyl free radical.

The iron center is thus a requirement for the stability and formation of the radical. No form of B2 is known that contains the free radical but lacks iron. It is, however, possible to prepare a form that lacks the radical but contains the intact iron center. This form is obtained by treatment of native B2 with hydroxyurea or similar radical scavengers (4) and is called B2/HU. It is completely devoid of enzyme activity. Figure 22.2 shows the interrelations between native B2 and its two inactive forms, apo B2 and B2/HU.

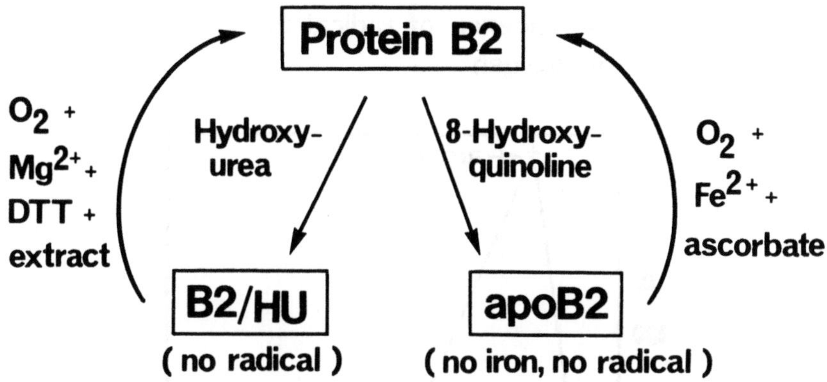

FIG. 22.2. Interrelations between different forms of the B2 subunit of ribonucleotide reductase.

Enzymatic Modification of a Specific Tyrosine Residue of Protein B2 by an *E. coli* Extract

Extracts from *E. coli* were shown to contain an enzyme system capable of introducing the tyrosyl radical into B2/HU (7). The reaction depended on the addition of dithiothreitol or similar dithiols, Mg^{2+}, and the presence of oxygen. The transformation of the enzymatically inactive B2/HU into native B2 could be monitored by the appearance of either enzyme activity or the typical EPR spectrum of the tyrosyl radical.

The activity of the extract was crucially dependent on the presence of oxygen. Figure 22.3 depicts an experiment in which B2/HU was incubated with dithiothreitol, Mg^{2+}, and bacterial extract in the presence or absence of oxygen (7). The ordinate of Fig. 22.3 records the increase in the amplitude of the radical signal with time and demonstrates that this parameter increased rapidly in the sample incubated in air, whereas in argon no such increase was found. Interestingly, the signal disappeared rapidly when air was replaced by argon in the activated sample and reappeared when oxygen was reintroduced. These results not only show that radical formation required oxygen but suggest that the extract contains an activity that removes the radical under anaerobic conditions. The data suggest that *E. coli* contains two opposite activities: one introducing the radical and requiring the presence of oxygen, the other removing the radical. The first activity cannot operate under anaerobic conditions, a condition that allows demonstration of the second one. Because only the radical-containing form of B2 is enzymatically active these two activities might have a regulatory function. This "radicalization" of a specific tyrosine residue is reminiscent of the well known modification of tyrosines of enzymes involving phosphorylation or dephosphorylation (12).

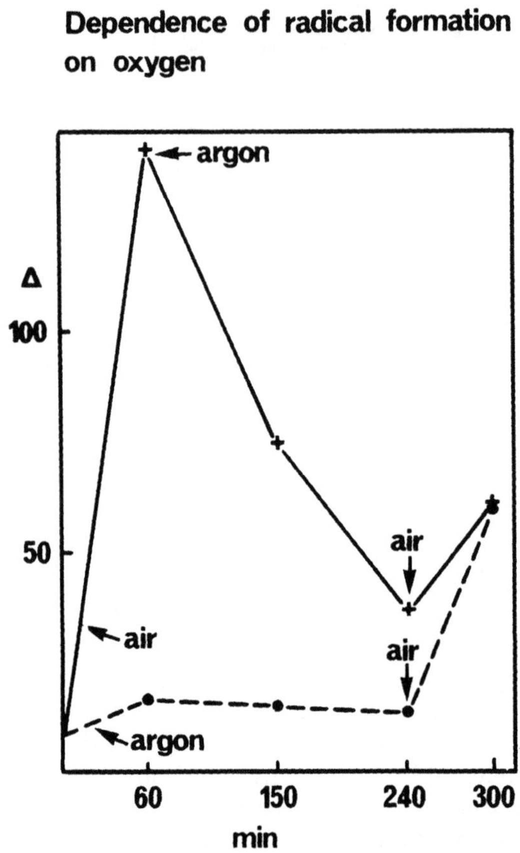

FIG. 22.3. Requirement for oxygen during formation of the tyrosyl radical. B2/HU was incubated in the presence or absence of oxygen together with a crude extract from *E. coli*, dithiothreitol, and Mg^{2+} (7). The ordinate gives the increase in the amplitude of the EPR spectrum characteristic for the tyrosyl radical.

Fractionation of the Radical-Introducing System

To obtain an understanding of the functional significance of these results, we began with the fractionation and characterization of the system responsible for radical introduction. It soon turned out that we were dealing with a complex system in which several proteins and nonprotein factors participate in the overall reaction (13). This chapter is not the proper place to describe in detail our efforts in this direction, but a summary of our present knowledge concerning the various components required for the reaction is given in Fig. 22.4 and Tables 22.1 and 22.2.

As shown in Fig. 22.4, protein fractionation on DEAE resulted in the separation of two components (fractions A and B) that were required together for catalytic activity. Fraction B was further subfractionated by gel exclusion

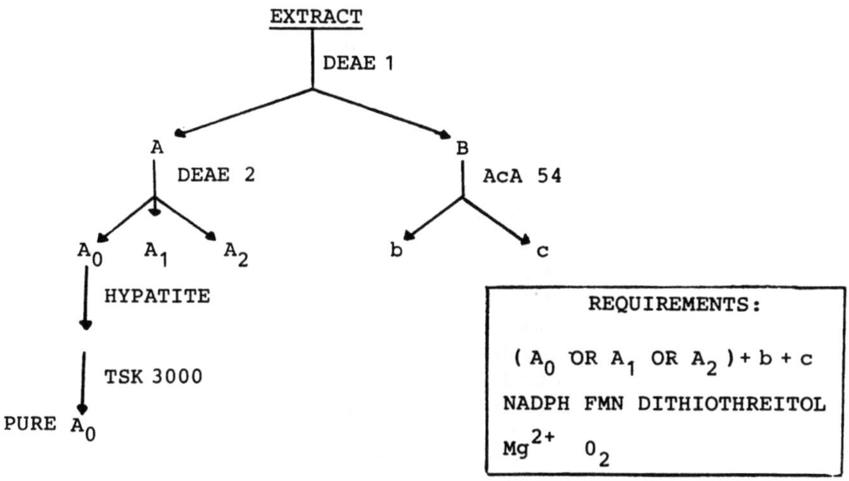

FIG. 22.4. Purification scheme for the various protein fractions required for radical introduction.

TABLE 22.1. Fractionation of the radical-introducing enzyme system and requirements for various protein fractions.

Experiment	Units[a]
Experiment 1	
Complete system (50 μg B + 7 μg A)	12.0
−A	2.0
−B	0.5
Experiment 2	
Complete system (9 μg b + 2 μg c + 0.4 μg A_0)	11.0
$-A_0$	1.5
$-A_0 + A_1$ (2 μg)	12.0
$-A_0 + A_2$ (2 μg)	12.0
$-A_0$ + superoxide dismutase[b] (0.1 μg)	13.0
Experiment 3	
Complete system (10 μg b + 4 μg c + 0.4 μg A_0)	12.0
−b	1.0
−c	3.0

[a] Units are defined as described by Eliasson et al. (13). The protein fractions were obtained as outlined in Fig. 22.3.
[b] Commercial material from beef erythrocytes.

chromatography into fractions b and c. All three fractions (A, b, and c) were now required together for radical formation. Rechromatography of fraction A on DEAE separated three active peaks (A_0, A_1, and A_2). Fraction A_0 was obtained in pure form after two further purification steps involving chromatography on hydroxylapatite and Ultropac TSK 3000 SW. From its amino acid sequence, A_0 was identified as manganese superoxide dismutase.

TABLE 22.2. Effect of pyridine and flavin nucleotides on enzymatic radical formation.

Experiment	Units[a]
Experiment 1[b]	
No addition	5
+NADPH (0.7 μM)	37
+NADH (0.8 μM)	8
+NADP (1.0 μM)	10
+NAD (0.9 μM)	8
Experiment 2	
+NADPH	8
+NADPH + FMN (5μM)	18
+NADPH + FAD (5 μM)	11

[a] Units are defined as described by Eliasson et al. (13). Incubations were made with 0.1 μg superoxide dismutase, 8 μg fraction b, and 0.8 μg fraction c.
[b] FMN 10 μM was present in all experiments.

In the presence of fractions b + c, each separated A fraction supported radical formation. The interchangeability of the three A fractions was explained when it was found (13) that they corresponded to the three forms of superoxide dismutase present in *E. coli*. Moreover, superoxide dismutase from beef erythrocytes fully substituted for *E. coli* fraction A_0. These interrelations are illustrated by the data in Table 22.1.

At this stage of protein purification the activity of the system depended on the addition of a boiled extract of *E. coli*, a requirement that could be replaced by the combined addition of NADPH and FMN. Although at early stages of protein purification NAD, NADH, and NADP were almost as active as NADPH, the latter nucleotide gave by far the best stimulation when tested with more highly purified protein fractions, and the inclusion of an NADPH regenerating system gave the best results. FMN (but not FAD) stimulated the reaction two- to threefold. Results from experiments demonstrating the requirements for NADPH and FMN are summarized in Table 22.2.

Neither protein fraction b or c have as yet been obtained in pure form. Although fraction b so far has turned out to be elusive and probably is not a single homogeneous protein, we obtained fraction c in a highly purified form by FPLC on a Mono Q column. We found that fraction c cochromatographed with ferric iron reductase activity (14), assayed by following the NADPH-dependent reduction of ferric citrate with ferrozine as a ferrous iron trap (15).

We then carried out parallel assays of fraction c activity and iron reductase activity during the purification of c. As demonstrated by the results of Table 22.3, the two activities co-purified over a 300-fold range. From these and other data we tentatively concluded that the two activities reside in the same protein.

TABLE 22.3. Co-purification of fraction c and ferric iron reductase activity.

Experiment	Specific Activity[a]		Fraction c/
	Fraction c	Iron Reductase	Iron Reductase
Experiment 1			
Bacterial extract	3.5	0.19	18
DEAE step	12.0	0.15	80
AcA 54 step	105.0	1.40	75
Experiment 2			
AcA 54 step	35.0	1.1	32
Mono Q step	1300.0	48.0	27

[a] Units were defined and determined as described in ref. 7 and 15. Specific activities were defined as units per milligram of protein. Values for fraction c activity varied among determinations, and experiments 1 and 2 are therefore not directly comparable.

Mechanism of Radical Introduction

At the present time a discussion of the mechanism of the reaction must remain speculative. Radical introduction can be represented in the following way:

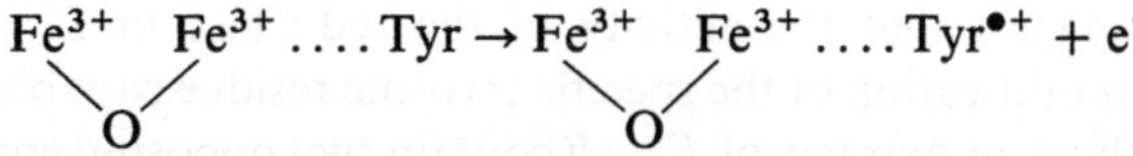

The reaction involves removal of an electron from the ring of a specific tyrosine-forming part of one of two identical polypeptide chains in protein B2. We should be able to explain both the chemistry involved in the generation of the radical and the remarkable specificity for one tyrosine-122.

Concerning the specificity, one may recall that radical introduction into apo B2 does not require enzymes and occurs with the same specificity. Therefore one might infer that specificity is provided by the structures of protein B2 and not by the radical-introducing enzyme system. The tyrosyl radical is located close to the iron center (16), and it is likely that the proximity of residue 122 to iron singles out this tyrosine for radical introduction.

With respect to the chemical mechanism, the overall reaction is oxidation. The requirement of the enzyme reaction for oxygen suggests that oxygen is the final electron acceptor. Fractionation of the system demonstrated that the reaction, in addition to oxygen, required NADPH and FMN, indicating a reductive step as an intermediate in the overall reaction.

The nature of this step is suggested from the finding that fraction c copurifies with a ferric iron reductase, requiring NADPH and FMN. Fraction c thus might be capable of reducing the iron center of protein B2. Returning to the spontaneous reactivation of apo B2, it is significant that this reaction occurs on addition of ferrous but not ferric iron (9), suggesting that in this case radical generation is coupled to the oxidation of the iron center. From these arguments the following reaction sequence may be inferred for the enzymatic radical generation:

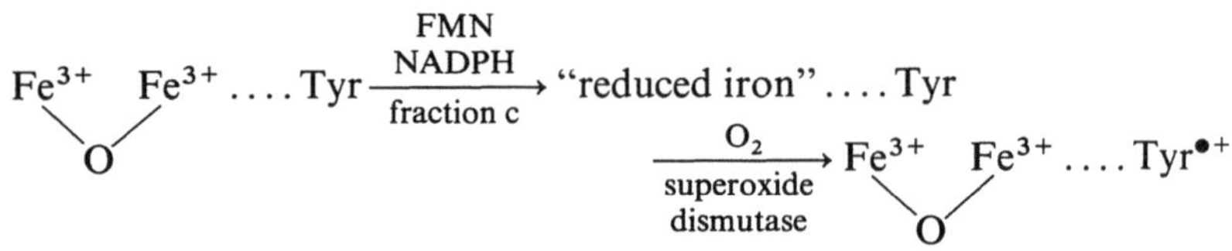

The Fe(III) center of B2 is reduced in the first step by fraction c in a reaction requiring NADPH and FMN. In the second step iron is reoxidized and the tyrosyl free radical is formed. The second step probably involves oxygen radicals, and superoxide dismutase may be required to protect a sensitive intermediate (13). Understanding the function of fraction b must await its purification.

General Conclusions

Active ribonucleotide reductase contains one tyrosyl free radical as part of its polypeptide structure. Molecules that lack the radical are devoid of enzyme activity. We suggest that the activity of the reductase in *E. coli* may be controlled by modification of the specific tyrosine residue capable of forming the tyrosyl radical, as extracts of *E. coli* contain two opposing enzyme activities that remove or introduce the tyrosyl radical. We have characterized the enzyme system that introduces the radical and thereby transforms an inactive form of the reductase into an active one. After fractionation we identified superoxide dismutase and ferric iron reductase as components of this system. Radical introduction involved both a reductive and an oxidative step and occurs probably by reduction of the iron center of the reductase followed by radicalization of a specific tyrosine residue coupled to the reoxidation of the iron.

Acknowledgments. This work was supported by a grant from the Swedish Medical Research Council to P. R. and by a Postdoctoral Research Fellowship from the Centre National de la Recherche Scientifique to M. F.

References

1. De Serres, F. J. (1985) *Genetic Consequences of Nucleotide Pool Imbalance.* Plenum Press, New York.
2. Thelander, L., and Reichard, P. (1979) *Annu. Rev. Biochem.* 48:133–158.
3. Brown, N. C., and Reichard, R. (1969) *J. Mol. Biol.* 46:39–55.
4. Ehrenberg, A., and Reichard, P. (1972) *J. Biol. Chem.* 247:3485–3488.

5. Sjöberg, B-M., Reichard, P., Gräslund, A., and Ehrenberg, A. (1978) *J. Biol. Chem.* 253:6863–6865.
6. Bianchi, V., Pontis, E., and Reichard, P. (1986) *Proc. Natl. Acad. Sci. USA* 83:986–990.
7. Barlow, T., Eliasson, R., Platz, A., Reichard, P., and Sjöberg, B-M. (1983) *Proc. Natl. Acad. Sci. USA* 80:1492–1495.
8. Sjöberg, B-M., Eklund, H., Fuchs, J. A., Carlson, J., Standart, N. M., Ruderman, J. V., Bray, S. J., and Hunt, T., (1985) *FEBS Lett.* 183:99–102.
9. Petersson, L., Gräslund, A., Ehrenberg, A., Sjöberg, B-M., and Reichard, P. (1980) *J. Biol. Chem.* 255:6706–6712.
10. Sjöberg, B-M., Loehr, T. M., and Sanders-Loehr, J. (1982) *Biochemistry* 21: 96–102.
11. Klotz, I. M., Klippenstein, G. L., and Hendrickson, W. A. (1976) *Science* 192:335–344.
12. Bishop, J. M. (1983) *Annu. Rev. Biochem.* 52:301–354.
13. Eliasson, R., Jörnvall, H., and Reichard, P. (1986) *Proc. Natl. Acad. Sci. USA.* 83:2373–2377.
14. Moody, M. D., and Dailey, H. A. (1985) *J. Bacteriol.* 163:1120–1125.
15. Moody, M. D., and Dailey, H. A. (1983) *Anal. Biochem.* 134:235–239.
16. Sahlin, M., Petersson, L., Gräslund, A., Ehrenberg, A., Sjöberg, B-M., and Thelander, L. (1987) *Biochemistry,* in press.

23
Differential Stabilization of Left-Handed Z-DNA and Z-RNA In Vitro and In Vivo

THOMAS M. JOVIN and DONNA J. ARNDT-JOVIN

The number of distinct conformations comprising the family of double helical nucleic acids continues to increase (1). The natural temptation is to consider the equilibration between these polymorphic states with the same formalisms and experimental approaches applied to classic helix-coil transitions. The latter, however, are thermodynamically much more dramatic because of the accompanying loss of secondary structure. Thus, although helix–helix interconversions can be highly cooperative and involve extensive structural rearrangements in the disposition of the sugar–phosphate backbone and ladder of basepairs, the equilibria are determined by the subtle interplay of intrinsic and extrinsic factors contributing rather small differences in free energy (2, 3). The prototypic reaction we consider here is the transition from right-handed helices (B-DNA and A-RNA) to the corresponding left-handed forms (Z-DNA and Z-RNA).

Perhaps the greatest difficulty faced by the experimentalist, be it the physical biochemist primarily interested in physical properties or the molecular biologist attempting to assess biological roles, relates to the fractional representation of the various conformations. For example, under conditions that may be regarded as physiological, left-handed DNA is generally not favored unless it is locally stabilized by topological stress (e.g., negative supercoiling) or by preferential ligands (e.g., Z-specific proteins). The presumably rare but distinct loci may be of great functional significance; their identification and characterization constitute a major challenge. These considerations motivate the systematic study of the energetics and kinetics associated with B-Z and A-Z transitions in vitro and (a much more difficult undertaking) in vivo. We report here on some efforts carried out in our laboratory in this area.

B-Z and A-Z Equilibria in Solution

The canonical B-Z transition is displayed by the poly[d(G-C)] sequence in salt solutions of increasing concentration (4). Thus the left-handed Z form (reviewed in 3, 5) is stabilized at high ionic strength, the transition midpoint in

NaCl being on the order of 2.3 *M* (4). Recently a formalism has been reported for predicting the corresponding phase diagram. It is based on consideration of the explicit distribution of the formal negative charges in the B and Z helices and their electrostatic interactions with environmental salt, for which a statistical mechanical representation incorporating hard sphere interactions is feasible (6). This approach has now been extended to other helix–helix interconversions (7) and integrated into programs for molecular mechanics calculations with the result that structural refinements can be performed for the macromolecule immersed in a realistic ionic milieu (R. Klement, E. von Kitzing, D. M. Soumpasis, and T. M. Jovin, submitted). The B-Z transition of poly[d(G-C)] can be quantitatively treated and potential surfaces as well as counterion distributions calculated for the two conformations at different ionic strengths (R. Klement, et al., in preparation).

Although it can be anticipated that the above results and their extensions should be quite useful in interpreting the binding of drugs and proteins with DNA and RNA, specific interactions of small ligands are also crucial. For example, the tervalent cation $Co(NH_3)_6{}^{3+}$ is an effective stabilizer of the Z conformation (8), possibly due to specific interactions with N7 and O6 of G and a phosphate group of DNA (9). An obvious strategy is to devise chiral compounds that would discriminate between right- and left-handed helices. Certain Co^{3+} and Ru^{2+} *tris o*-phenanthroline compounds exhibit some of the desired discrimination (10, 11). We have investigated a series of heteronuclear Co complexes with emphasis on the DNA binding properties of chemically resolved optically active disastereomers (J. Dabrowiak and T. M. Jovin, submitted). In the case of the hexavalent tetranuclear action Co_4en, *tris* [di-μ-hydroxo-*bis*(ethylenediamine)cobalt(III)]cobalt(III), a very effective stabilization of the Z form of poly[d(G-C)] is achieved in a two-step reaction involving an initial binding to the B conformation (at low ionic strength) followed by the induced transition to the Z form (J. Dabrowiak and T. M., Jovin, submitted). These compounds are of particular interest because they exhibit large circular dichroism signals in regions of the visible spectrum such that the state of the ligands and of the DNA can be assessed independently.

In addition to such external factors, intrinsic structural features of helical nucleic acids determine their tendency to undergo right-to-left (R-L) transitions. We have noted interesting contrasts in the effects of alterations at three loci of the macromolecule: (a) substitutions at the C5 position of the pyrimidines with methyl or halogen groups; (b) stereospecific substitution of nonbonding oxygen atoms in the phosphodiester bond; and (c) substitution of ribose (RNA) for deoxyribose (DNA). Table 23.1 summarizes the dramatic contrasts in the effects of these structural manipulations on the R-L equilibria. (see also 3). Thus, although the bulky C5 substituents greatly favor the Z form of DNA, in the case of RNA it is the right-handed A conformation that is more stabilized. Likewise, phosphorothioate substitution in the dGpC step (and in the stereospecific *R*p configuration) greatly facilitates the transition to the Z form, whereas the same substitution in the dCpG step blocks it entirely

TABLE 23.1. Effects of structural modifications on the B-Z and A-Z transitions of $(G\text{-}C)_n$ polymers.

Substitution	B-Z Transition (DNA)	A-Z Transition (RNA)[a]
C5 of C		
Methyl	↑[b]	↓
Bromine	↑Constitutive[c]	↓
Phosphorothioate		
Gp(S)C	↑[d]	↓
Cp(S)G	↓Blocked[d]	↓

[a] Unpublished data of T. Jovin. The temperature-dependent transition of the unmodified RNA, poly[r(G-C)], in concentrated $NaClO_4$ is documented in Hall et al. (14).
[b] Data from Behe and Felsenfeld (8). Arrows here and elsewhere indicate whether the modification leads to an increased stabilization (↑) or destabilization (↓) of the left-handed Z form.
[c] Data from Malfoy et al. (13). Constitutive: the Z form is stable at all ionic strengths.
[d] Data from Jovin et al. (12). The term "blocked" indicates failure to induce the Z form.

(12). In the case of the RNA A-Z interconversion (14), this pattern of specificity is almost exactly reversed. These results point to fundamental differences between DNA and RNA such that the respective right-handed forms (and probably to a lesser degree the common left-handed Z forms) are differentially affected by the chemical manipulations described above. One would surmise that subtle differences in ionic interactions, intramolecular hydrogen bonding, and hydration states are responsible for these phenomena. Their experimental elucidation requires high-resolution solution studies (e.g., using two-dimensional nuclear magnetic resonance (NMR) as well as additional crystallographic data.

Z-DNA in Nature

The fact that constituitive Z-polynucleotides are good immunogens has resulted in a battery of monoclonal and polyclonal antibodies, some having sequence or base specificity in addition to their global requirement for the Z conformation (15–18). Using these antibodies to Z-DNA we and others have been able to demonstrate that potential Z-DNA sequences occur in all organisms so far tested. The biological significance of this finding is unclear, as there has been no direct in vivo demonstration of Z-DNA. It appears that antibodies do not access the chromatin without prior fixation, a process involving either removal of proteins and/or local reversible denaturation (19, 20). Nonetheless, using a standard set of mild fixation conditions, dramatic differences in immunofluorescence intensity can be observed, e.g., between metaphase and interphase cells (Fig. 23.1) or upon short exposure of unfixed polytene chromosomes to a range of acidic pH values (20).

With an intensified television camera system attached to a conventional

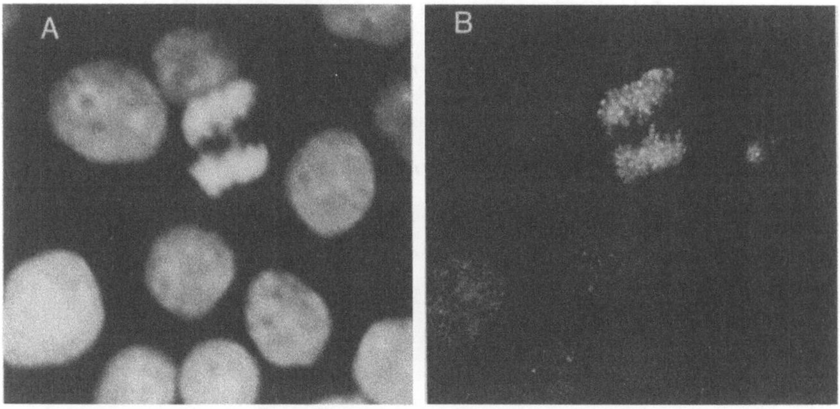

FIG. 23.1. Z-DNA antibody binding to mammalian metaphase and interphase cells. Human adenocarcinoma cells were fixed in methanol:acetic acid (3:1) for 10 min, washed in buffered saline, and stained with a polyclonal antibody against Z-DNA. *A*: Staining with DNA fluorescence dye H-33342. *B*: Same field observed for indirect immunofluorescence. T4 polyclonal anti-Z DNA IgG (20 μg/ml) fluoresceinated goat anti-rabbit IgG (GARIG).

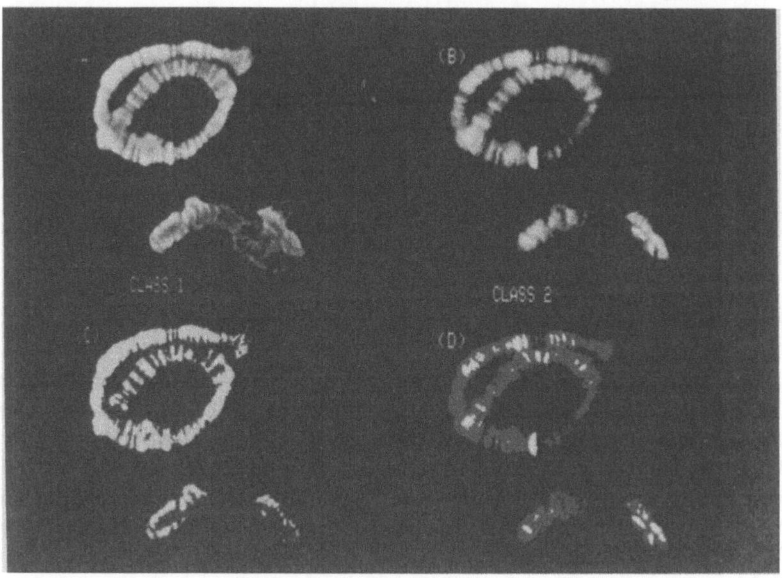

FIG. 23.2. GC-rich Z DNA sequences in polytene chromosome quantitation by digital image analysis. Double-stained polytene chromosomes 2 and 4 from *Chironomus thummi*. *A*: Polyclonal antibody (recognizing all Z sequences) at high concentration (95 μg/ml); indirect immunofluorescence of fluorescein GARIG. *B*: Monoclonal antibody (recognizing only Z-form d(G-C) sequences) at 15 μg/ml; indirect immunofluorescence of Texas red goat anti-mouse $F(ab)_2$. *C*, *D*: Bright regions in *A* and *B*, respectively, highlighted. Because the polyclonal antibody is in excess, the bands in *D* correspond to regions of the chromosomes rich in Z-(G-C) sequences.

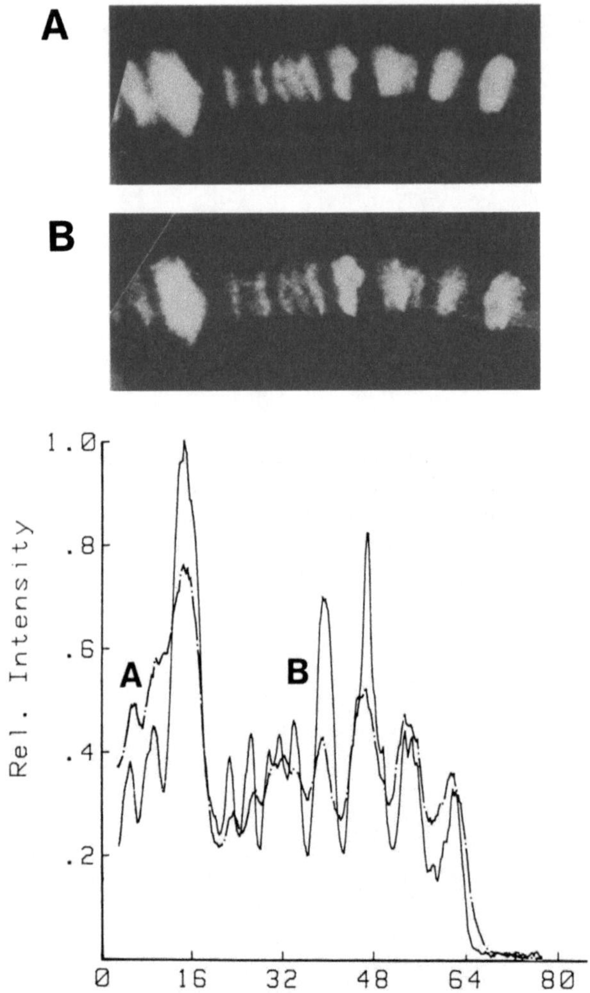

FIG. 23.3. Quantitation of Z-DNA and B-DNA in polytene chromosomes. A focused 2-μm spot was scanned mechanically at 25 μm/sec along the axis of a *Chironomus thummi* chromosome 3 stained with a Z-DNA specific polyclonal antibody and with H-33342 for total DNA. *A*: H-33342 fluorescence 420 to 460 nm, excitation 365 nm, and corresponding scan from region 3b4-3c4. *B*: Indirect immunofluorescence emission 520 to 580 nm, excitation 460 to 490 nm, and corresponding scan. Data acquired with a Zeiss UMSP 80 photometer.

epifluorescence microscope and integrated with our digital imaging system (GIPS), we have quantitated differences along polytene chromosomes that are particularly striking in the heterochromatic regions (21, 22). The latter contain up to 20 times higher frequencies of apparent Z-DNA than neighboring bands. The sequence-specific anti-Z-DNA antibodies have helped to determine that the telomeric regions of *Chironomus* contain predominantly

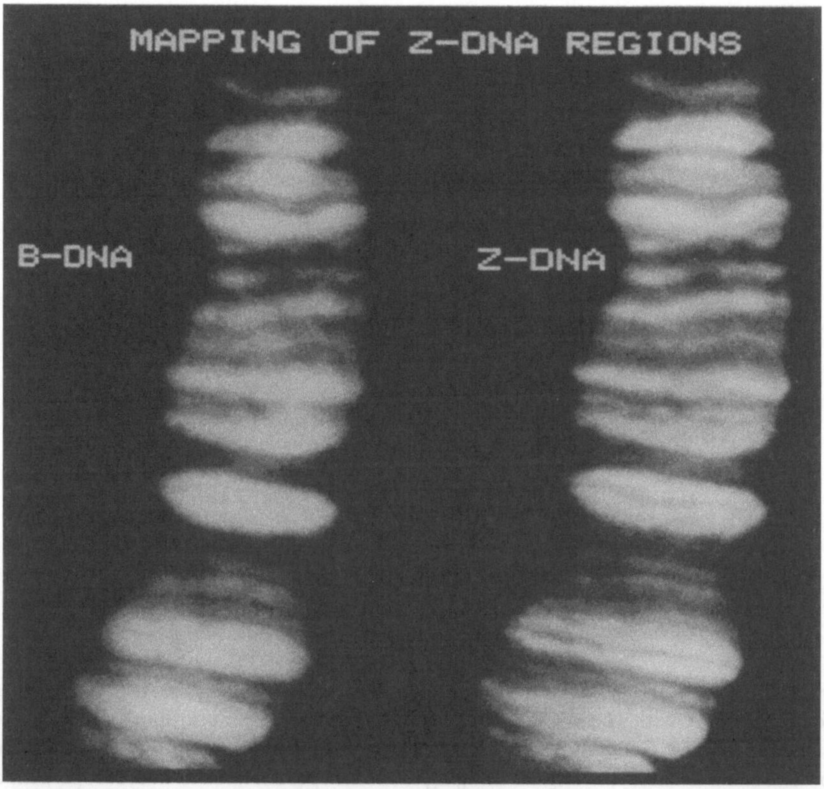

FIG. 23.4. High-resolution laser scanning microscopy of Z-DNA and B-DNA. Argon ion raster laser scanning microscope with digital image analysis mapping of Z-DNA regions in *Chironomus* chromosomes. *Left*: B-DNA staining with mithramycin, excitation 456 nm. *Right*: Z-DNA staining with anti-Z-DNA antibody, indirect rhodamine immunofluorescence, excitation 514 nm. Resolution 0.5 μm. Data acquired with a prototype Zeiss Laser Scan Microscope and Goettingen Image Processing System (GIPS).

clusters of alternating d(A-C), whereas other heterochromatic bands in *Chironomus* are enriched in d(G-C) segments (such as those seen in the No. 2 and 4 chromosomes in Fig. 23.2). Quantitative immunofluorescence using a laser spot scanner (Fig. 23.3) or a laser scanning microscope (Fig. 23.4) can resolve single bands (or genes). In nonheterochromatic regions (Fig. 23.4) the immunofluorescence pattern reflects the general DNA content but only at a measured frequency estimated to be 0.02 to 0.1% (23). We conclude from these measurements that (a) the heterochromatic regions may play a structural role in chromosome condensation, and (b) almost all genes in the insect contain one or a few potential Z-DNA sequences.

This fact that most, if not all, genes contain some potential Z-DNA sequences prompted us to examine developing embryos for clues about the role

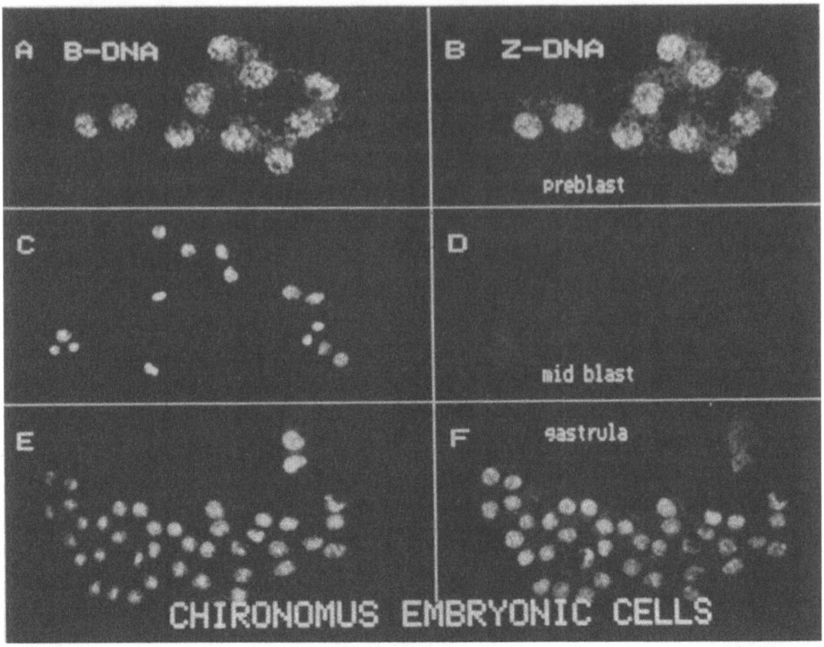

FIG. 23.5. Accessibility of Z-DNA sequences during embryogenesis. *A, C, E*: Staining for total DNA with H-33342. *B, D, F*: Z-DNA indirect immunofluorescence of the same fields. *A, B*: Cells from syncytial stage. C, *D*: Blastula stage. *E, F*: Gastrula stage of *Chironomus* embryos. Digital imaging from an epifluorescence microscope with a double-intensified Cohu TV camera using the Goettingen Image Processor System (GIPS).

of Z-DNA. In early embryogenesis massive changes in gene activity and DNA synthesis occur. We have found stage-specific changes in the relative capacity for binding anti-Z-DNA antibodies in the sea urchin, frog (*Xenopus laevis*), and the dipteran *Chironomus thummi* and *Drosophila melanogaster*. In all four systems and using a standard set of fixation conditions, antibodies to Z-DNA did not stain embryos in the midblastula transition period without exposure to high salt concentrations or acid extraction (as exemplified by *Chironomus* embryo cells in Fig. 23.5). We have shown that the apparent inability of the antibody to bind may be due to increased expression in the embryos of proteins masking the Z DNA sequences (S. Ritter, et al., in preparation).

The DNA-binding proteins from individual *Drosophila* embryos can be blotted with radioactive Z- or B-polynucleotides and stage-specific differences detected according to the following protocol. First we select stage-specific embryos after dechorionization under the dissecting microscope. The embryos are placed on a nylon mesh filter (100 μm), washed, recorded by digital image acquisition, and squashed onto a polycarbonate filter by exertion of pressure in the presence of 25% methanol. Further washing of these filters in buffered saline solution removes the cytoplasmic proteins, traps the vitelline membranes on the nylon, and retains the nuclei and chromosomes compactly on the polycarbonate disk. Nuclei of all stages of the cell cycle are efficiently trapped on the filters. The DNA-binding proteins are then eluted from the nuclei at various salt concentrations and attached to diazobenzyloxymethylcellulose (DBM) paper, which is subsequently blocked. The immobilized proteins are then blotted with ^{32}P-poly[d(G-m^{5}C)] in the Z conformation and in the presence of excess B-DNA. Binding is seen at the locations of the squashed embryos on the filter. We observe apparent changes during development in the expression of the Z-binding proteins, which are in the process of being purified from tissue culture cells and from discrete stage embryos.

What is the biological significance of the above observations? Clearly, the answer is related to the functional role(s) of Z-DNA (5, 24). Kmiec and Holloman (25) have shown that the Z conformation may be an intermediate in the recombination of DNA catalyzed by the *Ustilago* enzyme Rec-1, a finding in accordance with an old suggestion by Pohl (26) that left-handed DNA could facilitate recombination. Another possibility is that control regions along the genes influence transcriptional activity and that the latter is correlated with the expression of the Z conformation, a circumstance that might account for the changes in the immunofluorescence of the embryonic cells. Indeed, van de Sande and Jovin (27) and Butzow et al. (28) showed that *Escherichia coli* RNA polymerase transcribes polynucleotide templates in the Z conformation at a reduced (but finite: 10–40%) rate compared to the same DNAs in the B form. Durand et al. (29) extended these studies to the eukaryotic RNA polymerase from wheat germ and obtained similar results. Negative supercoiling also produces coordinate changes in structure and transcription. Using an alternating d(G-C)$_{16}$ insert downstream of a *lac* promoter in a supercoiled plasmid, Peck and Wang (30) demonstrated that transcription was hindered or blocked at superhelix densities exceeding the threshold required to drive the sequence into the Z conformation. However, a d[(A-C)$\cdot$(T-G)]$_{21}$ insert did not have this effect (30). The interpretation of these data is complicated because of the difficulty of assessing local structural features of the polymerase–DNA complexes, a problem that is of course general to all members of the polymorphic DNA family.

The various helical forms of DNA differ in tertiary but not primary or secondary structure. Thus, the *potential* for adopting the left-handed Z conformation exists everywhere, although in a hierarchy of probability determined by numerous intrinsic and extrinsic factors, many of which are probably as yet unknown. The primary experimental challenge is to devise strategies that avoid the use of potentially perturbing agents such as antibodies. It can be anticipated that success will derive from the combined use of spectroscopic and molecular biological methods.

Acknowledgments. We wish to thank Zeiss Instruments and in particular Drs. Bauerschmidt and Wilke for loan of the instrumentation and their assistance in the measurements with the UMSP 80 and the Laser Scan Microscope, respectively.

References

1. Zimmerman, S. B. (1982) *Annu. Rev. Biochem.* 51:395–427.
2. Soumpasis, D. M., and Jovin, T. M. (1987) In *Nucleic Acids and Molecular Biology*, F. Eckstein and D. M. Lilley, (eds.), pp. 85–111, Springer-Verlag, Berlin.
3. Jovin, T. M., Soumpasis, D. M., and McIntosh, L. P. (1987) *Annu. Rev. Phys. Chem.* 38:521–560.
4. Pohl, F. M., and Jovin, T. M. (1972) *J. Mol. Biol.* 67:375–396.
5. Rich, A., Nordheim, A., and Wang, A. H. J. (1984) *Annu. Rev. Biochem.* 53:791–846.
6. Soumpasis, D. M. (1984) *Proc. Natl. Acad. Sci. USA* 81:5116–5120.
7. Soumpasis, D. M., Wiechen, J., and Jovin, T. M. (1987) *J. Biomol. Struct. Dyn.* 4:535–552.
8. Behe, M., and Felsenfeld, G. (1981) *Proc. Natl. Acad. Sci. USA* 78:1619–1623.
9. Gessner, R. V., Quigley, G. J., Wang, A. H-J., van der Marel, A., van Boom, J. H., and Rich, A. (1985) *Biochemistry* 24:237–240.
10. Barton, J. K., Basile, L. A., Danishefsky, A., and Alexandrescu, A. (1984) *Proc. Natl. Acad. Sci. USA* 81:1961–1965.
11. Barton, J. K., and Raphael, A. L. (1985) *Proc. Natl. Acad. Sci. USA* 82:6460–6464.
12. Jovin, T. M., et al. (1982) *Cold Spring Harbor Symp. Quant. Biol.* 47:143–154.
13. Malfoy, B., Rousseau, N., and Leng, M. (1982) *Biochemistry* 21:5463–5467.
14. Hall, K., Cruz, P., Tinoco, I., Jr., Jovin, T. M., and van de Sande, J. H. (1984) *Nature* 311:584–586.
15. Lafer, E. M., Möller, A., Nordheim, A., Stollar, B. D., and Rich, A. (1981) *Proc. Natl. Acad. Sci. USA* 78:3546–3550.
16. Malfoy, B., and Leng, M. (1981) *FEBS Lett.* 132:45–48.
17. Stollar, B. D. (1986) *CRC Crit. Rev. Bichem.* 20:1–36.
18. Zarling, D. A., Arndt-Jovin, D. J., Robert-Nicoud, M., McIntosh, L. P., Thomae, R., and Jovin, T. M. (1984) *J. Mol. Biol.* 176:369–415.
19. Hill, R. L., and Stollar, B. D. (1983) *Nature* 305:338–340.
20. Robert-Nicoud, M., Arndt-Jovin, D. J., Zarling, D. A., and Jovin, T. M. (1984) *EMBO J.* 3:721–731.
21. Arndt-Jovin, D. J., Robert-Nicoud, M., Baurschmidt, P., and Jovin, T. M. (1985) *J. Cell Biol.* 101:1422–1433.
22. Arndt-Jovin, D. J., Robert-Nicoud, M., Kaufman, S. J., and Jovin, T. M. (1985) *Science* 230:247–256.
23. Arndt-Jovin, D. J., Robert-Nicoud, M., Zarling, D. A., Greider, C., Weimer, E., and Jovin, T. M. (1983) *Proc. Natl. Acad. Sci. USA* 80:4344–4348.
24. Jovin, T. M., McIntosh, L. P., Arndt-Jovin, D. J., Zarling, D. A., Robert-Nicoud, M., van de Sande, J. H., Jorgenson, K. F., and Eckstein, F. (1983) *J. Biomol. Struct. Dynam.* 1:21–57.
25. Kmiec, E. B., and Holloman, W. K. (1984) *Cell* 36:593–598.

26. Pohl, F. M. (1967) *Naturwissenschaften* 54:616.
27. Van de Sande, J. H., and Jovin, T. M. (1982) *EMBO J.* 1:115–120.
28. Butzow, J. J., Shin, Y. A., and Eichhorn, G. L. (1984) *Biochemistry* 23:4837–4843.
29. Durand, R., Job, D., Zarling, D. A., Tissière, M., Jovin, T. M., and Job, D. (1983): *EMBO J.* 2:1707–1714.
30. Peck, L. J., and Wang, J. C. (1985) *Cell* 40:129–137.

24 Interaction of Restriction Endonucleases with Phosphorothioate DNA

Fritz Eckstein

The exchange of one of the nonbridging oxygens of a phosphate group in nucleotides by sulfur yields the phosphorothioate analogues of nucleotides (1). These analogues differ in several aspects from the natural compounds: (a) When the phosphorothioate group is linked to two nonidentical groups, e.g., in the internucleotidic linkage or in the α or β positions of the triphosphate moiety of a nucleoside 5′-triphosphate, the phosphorus is chiral, and thus a pair of diastereomers is to be expected (Fig. 24.1). (b) For reasons that are not understood, the enzyme-catalyzed hydrolysis of phosphorothioate mono- or diesters is often extremely slow in comparison to the natural compounds. At the same time, however, transfer of the phosphorothioate group to groups other than water proceeds quite frequently with rates only marginally slower than with the phosphate group. (c) The difference in affinity between oxygen and sulfur for hard and soft metal ions offers the opportunity of selectively reacting phosphorothioate groups with some such ions. All of these properties have been exploited to gain some insight particularly into several stereochemical aspects of enzymatic phosphoryl and nucleotidyl transfer (1).

It was observed quite early, first with DNA-dependent RNA polymerase (2, 3) and later with various DNA-dependent DNA polymerases (4–8), that these enzymes efficiently polymerize the α-phosphorothioate analogues of the normal substrates, nucleoside (NTPαS) and deoxynucleoside (dNTPαS) 5′-*O*-1-thiotriphosphate, respectively. These enzyme-catalyzed reactions thus introduce phosphorothioate groups base-specifically into RNA or DNA, respectively. All of these enzymes accept only the Sp-diastereomer of the substrate and carry out the polymerization with inversion of configuration at phosphorus producing the Rp configuration at the internucleotidic phosphorothioate linkage (Fig. 24.1). This internucleotidic linkage is hydrolyzed considerably more slowly than the unmodified groups by quite a few nucleases, e.g., RNase A and RNase T_1 in the case of polyribonucleotides (2, 3, 9, 10) and 5′ → 3′ as well as 3′ → 5′ exonuclease of *E. coli* DNA polymerase I, exonuclease III, nuclease S1, and DNase I in the case of polydeoxynucleotides and DNA (4–6, 11–14). These results stimulated an investigation into the interaction of restriction endonucleases with phosphorothioate-containing

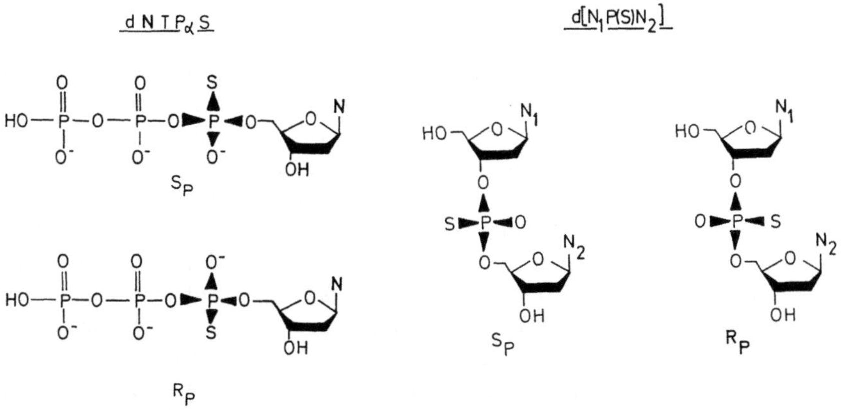

FIG. 24.1. Representative structures for the diastereomers of deoxynucleoside 5′-*O*-(1-thiotriphosphates) (left) and dinucleoside 3′-5′-phosphorothioates (right).

oligonucleotides and DNA as it was thought that there was the potential of blocking restriction sites in DNA.

Two oligonucleotides representing the recognition sequences for two restriction enzymes containing phosphorothioate groups at the point of cleavage—d(GGsAATTCC)[1] and d(GCsCCGGGC)—were synthesized (15, 16). Two strategies can be employed for the synthesis of such compounds, both of which make use of the polymer-supported phosphite triester approach. With the first strategy, sulfur is introduced at the desired position by replacing iodine oxidation by the addition of sulfur. As the stereochemistry of this reaction is not controlled, both diastereomers of the phosphorothioate are produced. Sometimes such a mixture of diastereomers is separable by preparative high-performance liquid chromatography (HPLC) on a reverse-phase column. With the second approach the required dinucleoside phosphorothioate is synthesized separately and the diastereomers separated by chromatography on SiO_2. After phosphitylation at the 3′-OH group, these diastereomers are introduced in the polymer-linked oligonucleotide at the desired position, and the synthesis is completed by further addition of monomeric nucleoside 3′-phosphites. This method produces the pure diastereomers of the phosphorothioate-containing oligomer.

We have synthesized the diastereomers of d(GGsAATTCC) by both methods and those of d(GCsCCGGCC) by the latter. Incubation of the two diastereomers of d(GGsAATTCC) with Eco RI showed that only the Rp isomer was hydrolyzed by this enzyme, and that the rate of hydrolysis was about 15 times slower than for the all-phosphate-containing oligomer (15). It

[1] d(GGsAATTCC): d(GGAATTCC) having a phosphorothioate group between dG and dA; the position of phosphorothioate in other oligonucleotides is indicated in the same way.

was also established that this hydrolysis proceeds with inversion of configuration at phosphorus, indicating that no covalent enzyme intermediate is involved in this reaction (17). Neither isomer of d(GCsCCGGG) was hydrolyzed by the enzyme Ava I under conditions where the all-phosphate-containing oligomer was cleaved to $\geqq 90\%$ (16). These admittedly limited results indicated that phosphorothioate groups exerted different effects on various restriction enzymes in terms of stabilization against hydrolysis.

These studies were extended to phosphorothioate-containing DNA (16, 18). The molecules constructed for these studies were prepared by using a viral (+) strand as template to which a primer was annealed and which was extended to form the (−) strand by enzymatic polymerization of a mixture of the three normal deoxynucleoside triphosphates, the fourth being a deoxynucleoside 5′-*O*-(1-thiotriphosphate). In the presence of T4 DNA ligase, such a reaction leads to the formation of circular covalently closed double-stranded DNA, which contains only phosphate groups in the (+) strand and a mixture of phosphate and phosphorothioate groups in the (−) strand, the latter being incorporated base-specifically by the choice of dNTPαS. Thus by employing, for example, dATPαS in the polymerization reaction, the phosphorothioate group is incorporated in the 5′ position to each deoxyadenosine in the (−) strand. As a consequence, two phosphorothioate groups are present, e.g., in the recognition sequence for Eco RI, one of them being located at the point of cleavage. Obviously, phosphorothioate groups can be incorporated into the cleavage site of any restriction enzyme by the proper choice of dNTPαS.

Such hybrid DNA has been investigated with a large number of restriction enzymes. The picture emerging from these studies (16, 18) is summarized in Table 24.1, where it can be seen that three classes of enzymes can be distinguished: those that only nick such DNA (class I), those where nicking is observed in the first phase of the reaction followed by linearization (class II), and those where nicking and linearization occur essentially simultaneously (class III). Interestingly, the two enzymes Eco RI and Ava I exhibit the behavior predicted from the studies with the oligonucleotides, i.e., slow linearization with Eco RI and nicking with Ava I. This finding makes it likely

TABLE 24.1. Summary of restriction enzyme reactions with M13 DNA containing phosphorothioate-modified cleavage sites in the (−) strand.

Reaction Type	Enzymes
Nicking only (class I)	Ava I, Ava II, Ban II, Fsp I, Hind II, Nci I, Pst I, Xho II
Linearization follows completion of nicking (class II)	Bam HI, Bgl I, Eco RI, Hind III, Hpa II, Taq I
Linearization starts before nicking is complete (class III)	Alu I, FnuD II, Hae II, Hae III, Hgi AI, Hha I, Hinf I, Mbo I, Msp I, Mst II, Pvu II, Rsa I, Sac I, Sal I, Sma I, Tha I

that nicking is the result of cleavage of a phosphodiester bond in the unmodified (+) strand. It is worth mentioning that apparently the conditions under which the enzymes are tested can determine in which class they fall (Table 24.1). Thus we have observed that Sma I under normal conditions (no salt) falls into class III, but when tested in the presence of 200 m*M* NaCl linearization was slowed considerably although the rate of nicking was unchanged, moving this enzyme into class I. Apparently nicking is not salt-dependent but linearization is.

The enzymes of class I are of particular interest for further study as they offer the opportunity of introducing a nick exclusively in the (+) strand of such hybrid DNA. This situation would represent a considerable advance over the existing methods of introducing nicks into double-stranded DNA by incubation with a restriction enzyme or DNase I in the presence of ethidium bromide as with this method the nicking occurs randomly in one of the two strands (19).

We have tried to take advantage of this exclusive nicking in the (+) strand for developing a method for the rapid generation of oligonucleotide-directed mutations at high frequency (20). There is considerable interest in such a method because oligonucleotide-directed mutagenesis normally yields mutants only at rather low frequency (1–10%) (21) for two reason: (a) The mutated sequence introduced by a mismatched oligonucleotide is present only in the (−) strand. As both strands are replicated roughly to a similar extent, only 50% of the new DNA can have the new sequence. (b) Efficient mismatch repair systems in the host cell eliminate the mismatch before replication has occurred. Several systems have been developed to eliminate these two interfering factors (22). All of these systems are based in one way or another on the use of particular viral or bacterial strains. We though that the phosphorothioate method could be developed for application with any wild-type strain.

The strategy followed is such that the mismatch oligonucleotide primer is annealed to the (+) strand and is extended in the presence of a DNA polymerase large fragment, T4 DNA ligase, three dNTP, and one dNTPαS, which is chosen so as to place a phosphorothioate at the cleavage site of a restriction enzyme of class I (Fig. 24.2A). We have so far investigated the enzymes Pvu I and Nci I, which require dCTPαS in the polymerization reaction (20, 23). The resulting RF IV DNA (covalently closed double-stranded circular DNA) is nicked by the restriction enzyme to produce RF II DNA (RF IV DNA with a nick in one of the strands). The nick is taken as the starting point for a gapping reaction with exonuclease III that destroys the (+) strand past the mismatch. Under the conditions we employ, a gap of 3000 nucleotides is obtained by incubation at 37°C in about 30 min. The gap is filled in by incubation with DNA polymerase I, T4 DNA ligase, and the four normal dNTP, yielding again RF IV DNA. This double-stranded DNA contains the new sequence in both strands and therefore is no longer a substrate for the mismatch repair system. Ideally a 100% mutation frequency is to be expected upon transformation. What we observe is 40 to 65% with the Pvu I

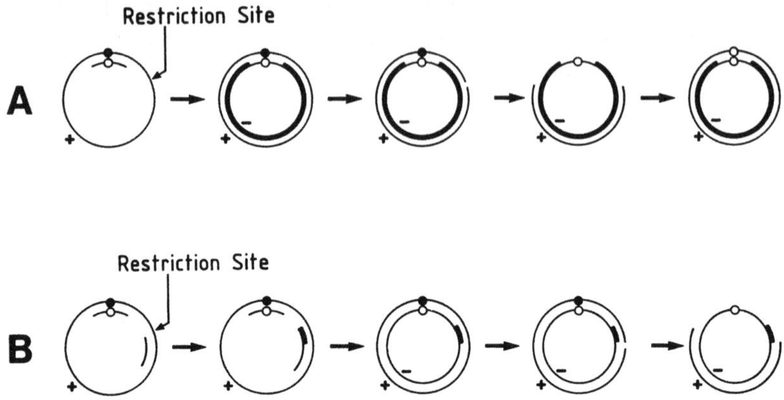

FIG. 24.2. Oligonucleotide-directed mutagenesis employing phosphorothioates. *A*: Method using a mismatch oligonucleotide primer only. *B*: Method using a mismatch oligonucleotide primer and a primer partially covering the restriction site to be protected. (•, o) mismatch: (o) wildtype, (•) new sequence. Heavy lines = stretches containing phosphorothioate groups.

(20) and 80 to 90% with the Nci I enzyme (23). We attribute the difference between the two results to the lower efficiency of the nicking reaction employing Pvu I, which is an enzyme with a low turnover number and which is easily inactivated under the incubation conditions.

For reasons we do not understand at present, we have observed that the transforming activity of the RF IV DNA resulting from Pvu I nicking (but not Nci I nicking) was about 50- to 100-fold lower than when no phosphorothioate was present in the (−) strand. This finding led us to develop a somewhat different strategy, which was to minimize the content of phosphorothioate groups but retain the resistance against restriction enzyme cleavage. This situation was achieved by employing a second primer that covers the recognition sequence of Pvu I up to the point of cleavage, so that upon short incubation with dGTP, dTTP, and dCTPαS only a small piece of (−) strand was synthesized, up to the position where dATP was required, placing, however, phosphorothioate at the cleavage site (Fig. 24.2B). (−) Strand synthesis was completed by adding a large concentration of all four dNTPs, diluting out dCTPαS. In this way only seven phosphorothioate groups are incorporated into the (−) strand. Indeed, the transforming activity of this DNA after nicking with Pvu I and gapping with Exo III is as high as with totally unmodified DNA, but the mutation frequency is again 50 to 70%. Applying the same strategy for the protection of the Nci I site, we observed that at least for this enzyme a phosphorothioate group exclusively at the cleavage site is not sufficient for protection against hydrolysis and thus for preventing linearization, but that at least one more phosphorothioate group next to it is required (23).

Obviously this method for the production of mutants at high frequency is not limited to the use of the restriction enzymes Pvu I and Nci I. Potentially, at least all of the enzymes of class I should be suitable for this purpose. As more restriction enzymes are tested for their ability to cleave phosphorothioate bonds, this list will probably increase and make this method rather versatile.

References

1. Eckstein, F. (1985) *Annu. Rev. Biochem.* 54:367–402.
2. Matzura, H., and Eckstein, F. (1968) *Eur. J. Biochem.* 3:448–552.
3. Eckstein, F., and Gindl, H. (1970) *Eur. J. Biochem.* 13:558–564.
4. Burgers, P. M. J., and Eckstein, F. (1979) *J. Biol. Chem.* 254:6889–6893.
5. Brody, R. S., and Frey, P. A. (1981) *Biochemistry* 20:1245–1252.
6. Romaniuk, P. J., and Eckstein, F. (1982) *J. Biol. Chem.* 257:7684–7688.
7. Brody, R. S., Adler, S., Modrich, P., Stec, W. J., Leznikowski, Z. J., and Frey, P. A. (1982) *Biochemistry* 21:2570–2572.
8. Bartlett, P. A., and Eckstein, F. (1982) *J. Biol. Chem.* 257:8879–8884.
9. DeClercq, E., Eckstein, F., Sternbach, H., and Merigan, T. C. (1970) *Virology* 42:421–428.
10. DeClercq, E., Eckstein, F., Sternbach, H., and Merigan, T. C. (1969) *Antimicrob. Agents Chemother.* 187–191.
11. Kunkel, T. A., Eckstein, F., Mildvan, A. S., Koplitz, R. M., and Loeb, L. (1981) *Proc. Natl. Acad. Sci USA* 78:6743–6738.
12. Gupta, A. P., Benkovic, P. A., and Benkovic, S. J. (1984) *Nucleic Acids Res.* 14:5897–5911.
13. Putney, S., Benkovic, S. J., and Schimmel, P. (1981) *Proc. Natl. Acad. Sci. USA* 78:7350–7354.
14. Schreiber, G., Koch, E. M., and Neubert, W. J. (1985) *Nucleic Acids Res.* 13:7663–7672.
15. Connolly, B. A., Potter, B. V. L., Eckstein, F., Pingoud, A., and Grotjahn, L. (1984) *Biochemistry* 23:3443–3453.
16. Taylor, J. W., Schmidt, W., Cosstick, R., Okruszek, A., and Eckstein, F. (1985) *Nucleic Acids Res.* 13:8749–8764.
17. Connoly, B. A., Eckstein, F., and Pingoud, A. (1984) *J. Biol. Chem.* 259:10760–10763.
18. Potter, B. V. L., and Eckstein, F. (1984) *J. Biol. Chem.* 25:14243–14248.
19. Österlund, M., Luthman, H., Nilsson, S. V., and Magnusson, G. (1982) *Gene* 20:121–125.
20. Taylor, J. W., Ott, J., and Eckstein, F. (1985) *Nucleic Acids Res.* 13:8765–8785.
21. Zoller, M. J., and Smith, M. (1983) *Methods Enzymol.* 100:469–500.
22. Smith, M. (1985) *Annu. Rev. Genet.* 19:423–462.
23. Nakamaye, K. and Eckstein, F. (1986) *Nucleic Acids Res.* 14:9679–9698.

25
Role of Conformational Dynamics of Myosin in Muscle Contraction

TIAN YOW TSONG, A. BERTAZZON, and W. F. HARRINGTON

During muscle contraction the energy released by cleavage of the γ-phosphodiester bond of ATP is effectively converted to mechanical work by the coupling of the ATP hydrolysis reaction to the conformational dynamics of muscle proteins. Much effort has been devoted to the study of various modes of motion in the intact contractile apparatus of muscle fibers and myofibrils and in purified muscle proteins, as well as the relation between these motions and the force-generating event.

Many models have been proposed. In the models of Huxley (1) and Huxley and Simmons (2), power stroke is postulated to occur by some type of structure change that results in rotation of the myosin head (S1) while it is transiently attached to the actin filament in an active cross-bridge cycle. This head rotation is assumed to be closely coupled to the ATP cleavage reaction (3, 4). A detailed search for head rotation in actively contracting muscle, however, has failed to detect such a process. For example, Cooke and coworkers (5) were unable to detect significant head rotation in activated glycerinated muscle fibers using EPR analysis of a spin label attached to the S1 subunit. The fluorescence depolarization studies of Yanagida (6, 7) showed that in isometrically contracting muscle the fluorescent label in the active site remained at the fixed orientation observed in rigor fibers. Huxley and co-workers (8) have investigated time-resolved changes in the x-ray diffraction patterns of contracting muscle following abrupt step-change in length. They observed little or no change in the intensity of reflections in the equatorial pattern that would signal head rotation. Here we consider an alternative mechanism for force generation in muscle, i.e., the helix-coil transition model, and provide some kinetic and thermodynamic evidence in support of the basic assumptions of the model.

Helix-Coil Transition Model

The important features of the helix-coil transition model are based on known physical chemical properties of fibrous, helical proteins (9, 10). Muscles are rich in this type of protein (11). The main structural components involved in the power stroke are myosin of the thick filament and actin of the thin filament. At a low temperature the tail segment of the myosin molecule, the myosin rod, is nearly 100% helical and is composed of two helical chains that wrap about each other to form the well-known α-helical coiled-coil structure (11). An order-to-disorder transition of a helical fibrous polymer is known to be accompanied by a substantial reduction in its end-to-end length according to the theory of Flory (9, 10). The elastic force produced by this melting process would be sufficient to account for force generation in muscle (9, 10). Based on structural studies of the thick filament, thin filament, and the cross bridge, as well as transient mechanical studies of activated muscle, Harrington proposed that the force-generating element of muscle resides in the S2 segment of myosin (11, 12). In the cross-bridge cycle, S2 is proposed to alternate between two states (Fig. 25.1A). In one state the S2 segment lies on the surface

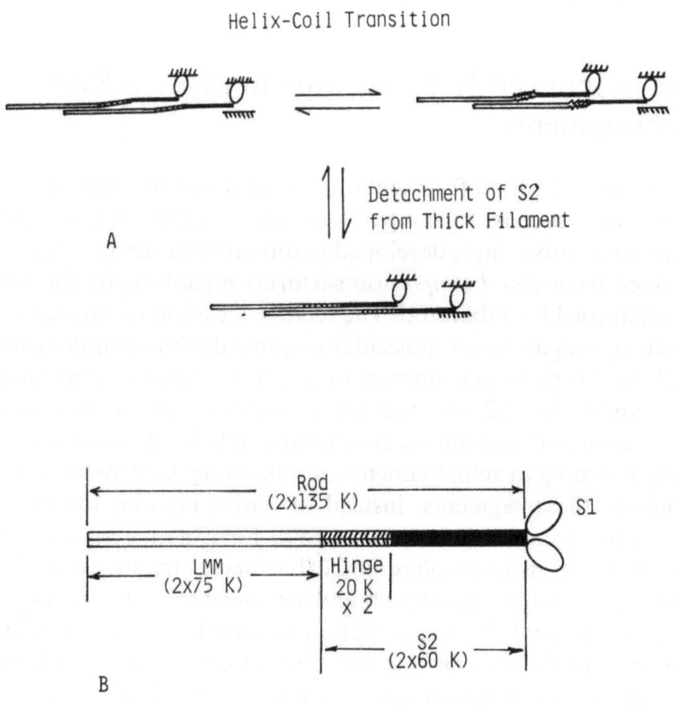

FIG. 25.1. Myosin and its arrangement in thick filaments. *A*: The rod of myosin in rigor state is in the coiled-coil α-helical structure. S1 is attached to actin in the thin filaments, and S2 (darkened + shaded blocks) lies on the surface of the thick filament. When muscle is activated, S2 detaches from the thick filaments. The loss of surface interaction destabilizes part of the S2 structure within the hinge domain (shaded block). This process immediately triggers a helix-to-coil transition of the hinge and generates contractile force. *B*: Definition of myosin fragments used in this work is shown. The rod, S2, and part of the hinge were prepared by proteolytic digestion followed by chromatographic separations. See text for details.

of the thick filament, and in the other S2 is released from its contact with the thick filament. In the latter position the loss of contact energy destabilizes the segment, resulting in partial melting of the helical structure. The net result is a shortening of the overall length of S2 during isotonic contraction and generation of elastic force during that part of the cycle when S1 is attached to the thin filament. There is now substantial biochemical evidence showing that this event indeed occurs. The rate of proteolytic cleavage at several widely spaced sites located in the hinge region of myosin increases about 100-fold when glycerinated muscle fibers are activated, compared to proteolysis in the rigor or relaxed state (13–15). This striking difference between activated and resting muscle is observed over the entire temperature range 5° to 40°C. For purified myosin and its subfragments, temperature- or pH-induced helix-coil transitions have been reported from several laboratories (16–20). Kinetics of some of these transitions have also been measured by the temperature-jump relaxation method (21, 22).

Thermally Induced Helix Melting in Myosin Rod and Subfragments

Rabbit myosin and its subfragments—rod, light meromyosin, S2, short S2, and hinge—were prepared from psoas muscle according either to published procedures or to procedures developed in our laboratories (21, 22). Frog rod was obtained from *Rand temporaria* sartorius myofibrils by the same procedure as that used for rabbit rod. The relative locations of various structural domains in myosin and their molecular weights (double-stranded) are shown in Fig. 25.1B. Of particular interest to us are the regions implicated in the helix-coil model, i.e., S2 and the hinge segments. In the organized thick filament structure these segments are located immediately next to S1, which is the moiety attaching to actin filament and exhibiting ATP hydrolysis activity. The choice of these segments, instead of native myosin, for the physical measurements (discussed below) is twofold. First, native myosin molecules denature and aggregate at physiological temperature to form dimer and higher weight assemblies in solution and are therefore not practical for these measurements. Second, by fragmentation we wish to assign and pinpoint the local region(s) in the myosin rod that gives rise to specific conformational changes. All of these fragments, with the exception of the hinge fragment, were highly ordered at temperatures below 20°C. Optical activity measurements confirm that they are in the α-helical form. When the temperature of solutions was raised, the helical structures of these fragments melted, with T_m ranging between 25° and 50°C, depending on the location of the fragment in the intact myosin rod as well as on the pH and the salt concentration.

Long S2 of rabbit showed two discernible transitions when optical rotation was monitored, one at 43°C and the other at 55°C. Rabbit rod, which comprises S2 and LMM (the segment of myosin that self-associates to form the interior core of the thick filament) showed similar low temperature and high temperature transitions. However, the high temperature transition was larger in amplitude suggesting that the coiled-coil segment (LMM) adjoining S2 has generally greater thermal stability and melts at a higher temperature. The low temperature transitions in these two fragments can be tentatively assigned to melting in the hinge domain based on the similar T_m values obtained from optical rotation measurements and enzyme-probe studies reported earlier by Ueno and Harrington (13–15). In frog rod the corresponding transitions occurred, respectively, at 36°, 43°, and 55°C. Here the lower-lying transition is split into two peaks, at 36° and 43°C, the 36°C peak dominating. Hence the lower-lying transition is about 7°C below that in rabbit rod, but the higher (55°C) transition is unchanged. This finding is of particular interest in view of the lower temperature at which the frog usually functions.

We have also carried out temperature jump studies on the various isolated subfragments of myosin. Temperature jumps of 5°C were made at temperatures over the range 25° to 60°C. At least two relaxation times were resolved, one at 0.5 ms and the other at 5.0 ms, for rabbit S2 and rod. Only a single fast process was detected in the hinge fragment with relaxation time of 0.15 ms. In addition to the signals for the two relaxations, an additional signal amounting to about 20% of the total amplitude at 235 nm occurred at much slower time and was not resolved. The relaxation time of the slow process was estimated to be several tenths of a second. Relaxations were monitored with two signal sources using the intrinsic absorption of the myosin fragments at 235 nm and the absorption of an indicator dye, phenol red, at 560 nm. The melting of the subfragments is accompanied by a release of protons: two for rod (16) and an unspecified number for the other fragments. In an unbuffered solution the temperature jump employed was sufficient to give a measurable absorption change in the pH-sensitive dye. The amplitude of the dye signal was substantially greater than that of the protein intrinsic absorption changes at 235 nm, with a better signal/noise ratio. Nevertheless, two methods for detecting melting gave closely similar kinetic rate constants and amplitudes. When the signal per degree T-jump was plotted against mean temperature, temperatures at which maximum signals were obtained were in good agreement with the T_m of the optical rotation melting curves. The results of these studies are compared in Table 25.1. It is interesting to note that the relaxations detected here are in the same time range observed for the recovery of tension (the "T_2 response") when a single activated muscle fiber is subject to an abrupt release or stretch (23).

Differential Scanning Microcalorimetric Study

The improved resolving power of the sensitive calorimetric method has been used to determine if the two transitions mentioned above can be cleanly separated. In addition, the cooperativity of each helix-to-coil transition can be

TABLE 25.1. Some thermodynamic and kinetic quantities of helix-to-coil transitions of myosin fragments.

A. Equilibrium measurements by optical rotation at 233 nm.

Sample	MW (kD)	Buffer[a]	% Helix at 5°C	T_m (°C)
Rabbit rod	2 × 135	1	90	47, 55
Long S2	2 × 60	2	87	43, 55
Short S2[b]	2 × 44	2	89	49
Hinge[c]	2 × 16	2	49	35.5

B. Temperature jump relaxations measured by optical absorption at 235 nm or by the color change of phenol red at 560 nm.

Sample	Buffer	τ_f (ms)	$T_{f,m}$[d] (°C)	τ_s (ms)	$T_{s,m}$[d] (°C)
Rabbit rod	3	0.8, 0.6	48, 55	10, 5	45, 54
Long S2	2	0.5, 0.6	30, 46	15	44
Short S2[b]	2	0.4	38	17	48
Hinge[c]	2	0.15	33	—	—

[a] 1 = 0.6 *M* NaCl, 20 m*M* phosphate, at pH 7.0. 2 = 0.1 *M* NaCl, 20 m*M* phosphate, at pH 7.0. 3 = 0.5 *M* NaCl, 20 m*M* phosphate, at pH 7.0.
[b] CNBr cleavage product of S2, containing 3 to 4 kD of the hinge.
[c] CNBr cleavage product, 3 to 4 kD shorter than the hinge.
[d] Temperatures at which the relaxation signals are maximal: τ_f fast relaxation time; τ_s = slow relaxation time.
Data adapted from Tsong et al. (22).

determined so that it is generally possible to decide if each transition detects melting of an independent domain. Assignment of the melting of different regions of myosin in a transition curve provides energetic information concerning the transition process. Figure 25.2 gives several typical thermal transition curves obtained with rabbit rods. From these thermograms it is clear that more than two transitions are discernible at all pH values studied. Consider first the thermogram obtained at pH 7.9. Here there are at least three distinguishable transitions. The major one at 43.5°C (A) has a heat uptake of 3.7 cal/g protein, or a ΔH of 1010 kcal/mol of 270 kD. ΔH for the slight hump at 48°C cannot be extracted from the thermogram because of its small magnitude. The higher temperature peak (B) occurring around 54°C has a heat uptake of 1.88 cal/g, or a ΔH of 510 kcal/mol. The total heat uptake in the temperature range 30° to 60°C is thus the sum of these transitions and is approximately 1600 kcal/mol. For the intact myosin rod with MW 270 kD, the total number of helical residues is around 2350. Assuming a mean residue weight of 115, it yields an enthalpy of 0.70 kcal per peptide residue. The transition A, which appears symmetrical, in fact, deviates considerably from a "two-state" transition.

There are several ways one can extract information from a calorimetric transition curve for each transition. First, if we assume that the melting of

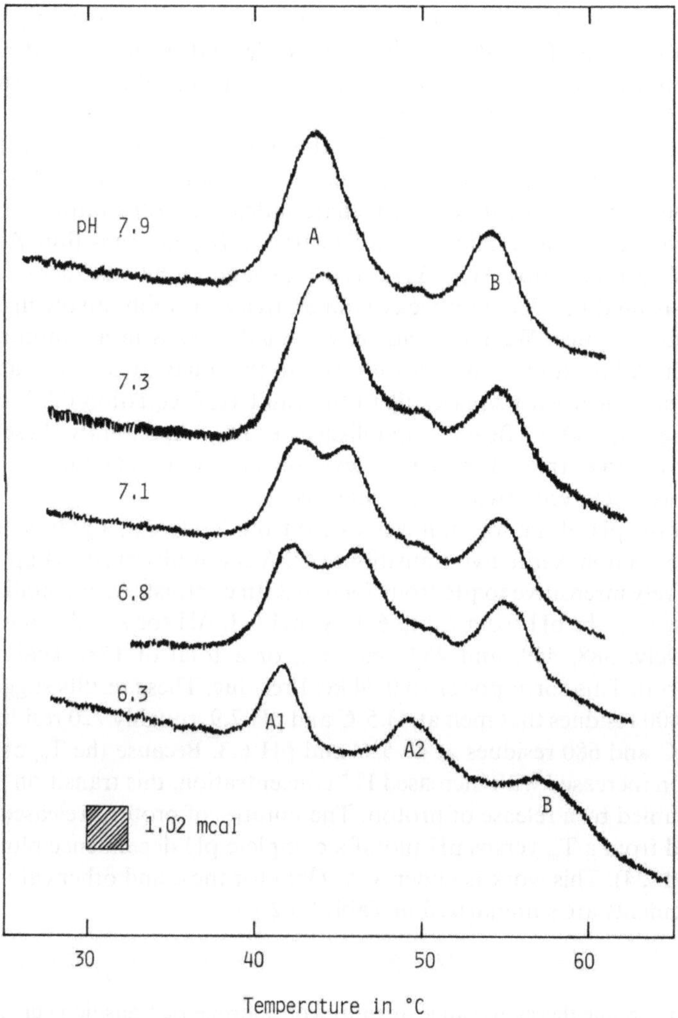

FIG. 25.2. Differential scanning microcalorimetric study of helix-to-coil transitions in rabbit myosin rod. Measurements were done with an MCI Microcalorimeter (Microcal, Amherst, MA). Rabbit rod, 0.9 ml (2.8 mg/ml), was sealed in the calorimeter, and the temperature was raised at a heating rate of 29.7°C per hour. Excess heat capacity was recorded against temperature. The buffer used was 0.5 *M* KCl, 20 m*M* potassium phosphate, 10 m*M* EDTA, adjusted to desired pH by HCl and NaOH. At pH 7.9, two major transitions (A and B) can be seen. When the pH was reduced, transition A splits into two transitions, A1 and A2. A2 was relatively sensitive to pH changes compared to A1 and B. Analysis and some useful quantities of these transitions are given in Table 25.2.

each residue contributes uniformly to the enthalpy change, transition A would reflect a melting of roughly 1400 residues. Second, from the width of the transition, the van't Hoff ΔH (ΔH_{VH}) can be calculated (24–26). The ratio of this ΔH to calorimetrically measured ΔH (ΔH_{cal}) is indicative of the cooperativity of the transition. When this ratio is unity, the transitions is completely cooperative and corresponds to a two-state transition. If the ratio is <1 the transition involves intermediate states, and if the ratio is >1 there is intermolecular interaction (26). Because ΔH_{VH} of transition A is 150 kcal/mol and the ratio $\Delta H_{VH}/\Delta H_{cal}$ is 0.15, roughly 1400 residues of the rod melt at around 43.5°C, with weak cooperativity, possibly involving several intermediate states. We have had only limited success in attempting to fit transition A into three and transition B into two quasi-two-state transitions using a nonlinear least squares fit to the van't Hoff equation (T. Y. Tsong, T. Bertazzon, and N. Beaty, unpublished results). Apparently these transitions are not to be treated as composites of only a few discrete transitions, each being a well behaved "two-state" transition.

When the pH of the solution was varied from 7.9 to 6.3, a gradual separation of transition A into two transitions (Al, A2) was observed. A1 appears to be relatively insensitive to pH from 7.9 to 6.3. In contrast, A2 is stabilized 6°C on decreasing the pH from 7.9 to 6.3. At pH 6.3, ΔH for A1, A2, and B are, respectively, 388, 459, and 432 kcal/mol, or a total of 1280 kcal/mol for myosin rod. This corresponds to 0.54 kcal/residue. These results suggest that of the 1400 residues that melt at 43.5°C and pH 7.9, roughly 720 residues melt at 41.5°C and 680 residues at 49.5°C and pH 6.3. Because the T_m of the A2 transition increased with increased H^+ concentration, this transition must be accompanied by a release of proton. The number of protons released can be obtained from a T_m versus pH plot if a complete pH dependence plot can be obtained (24). This work is under way. Data for these and other calorimetric measurements are summarized in Table 25.2.

TABLE 25.2. Some thermodynamic quantities of calorimetric transition curves of rabbit myosin rod.

Transition	T_m (°C)	$\Delta T_{1/2}$[a] (°C)	ΔH_{cal} (kcal/mol)	ΔH_{vH}[b] (kcal/mol)	R[c]
pH 7.9					
A	43.5	4.7	1010	150	0.15
B	54.0	3.8	508	197	0.39
pH 6.3					
A1	41.5	3.0	388	230	0.59
A2	49.5	5.2	459	140	0.31
B	57.0	5.9	432	130	0.30

[a] The width of a transition curve at the half-height of the peak.
[b] The van't Hoff $\Delta H_{vH} = 7.0\ T_m^2/\Delta T_{1/2}$. See ref. 26.
[c] The ratio $\Delta H_{vH}/\Delta H_{cal}$, indicative of the cooperativity of a thermal transition. See text for details.

Reduced Viscosity Detects a Low Temperature Transition in Rabbit Rod

In the calorimetric study we occasionally detected yet another low temperature transition for the intact rods of rabbit and frog. This transition is buried in the noise level of the calorimetric transition curve in most of the experiments and has a small ΔH. Although one can increase the concentration of the protein, it leads to aggregation (27) and consequent difficulty in the interpretation of the results. For this reason, we sought another physical method to demonstrate melting at low temperature in order to obtain information on cooperativities. Earlier work of Burke et al. (18) showed that melting of rabbit myosin rod at low temperature can be detected by viscosity measurement. Although there is no appreciable loss of helical content below 20°C, consistent with the calorimetric results, the reduced viscosity is quite temperature-sensitive. It shows a substantial drop on increasing the temperature between 5° and 20°C, suggesting that the rod undergoes a local perturbation in structure leading to hinging. Enzyme probe studies (13–15) of rabbit rod confirm that local melting occurs in the hinge region near the light meromyosin/heavy meromyosin junction at these temperatures. In the present study we reinvestigated the temperature dependence of reduced viscosity in order to make a quantitative comparison with our calorimetric results. The reduced viscosity of rabbit rod was determined at a low protein concentration (1 mg/ml) in order to minimize aggregation effects. Results at various temperatures are shown in Fig. 25.3.

The reduced viscosity of rabbit rod was 260 ml/g at 10°C and pH 7.3 (o). This value was less sensitive to pH changes near 7 and stayed constant from 7.3 to 6.3 (□). When the temperature was raised, there was a clear transition around 24°C, and the reduced viscosity decreased from 260 to 195 ml/g. The T_m (24.0°C) was unaffected by pH in the range studied, i.e., pH 7.3 to 6.3. The higher temperature transitions resembled optical rotation transitions and occurred in the range of calorimetric transitions. No attempt was made to resolve these transitions between 40° to 60°C.

The unusually high reduced viscosity for the rabbit rod is due to the shape of molecule. The reduced viscosity, $[\eta]_r$, equaled $\nu\phi/c$, where ν, ϕ, and c are, respectively, the viscosity increment, the fraction of volume occupied by the solute, and the concentration. If we use 0.74 ml/g for the partial specific volume of the protein, ν can be calculated to be 370 ml/g for an anhydrous rod. An elongated rod-shaped molecule may be approximated by a prolate ellipsoid. By using the Simha relation, one can calculate the axial ratio of the equivalent ellipsoid (28). An axial ratio of 75 is obtained for the rabbit rod. After the 24°C transition, the viscosity increment (ν) decreases to 280 ml/g. The apparent axial ratio drops to 64 following the low temperature transition.

One can also estimate the enthalpy of the transition from the reduced

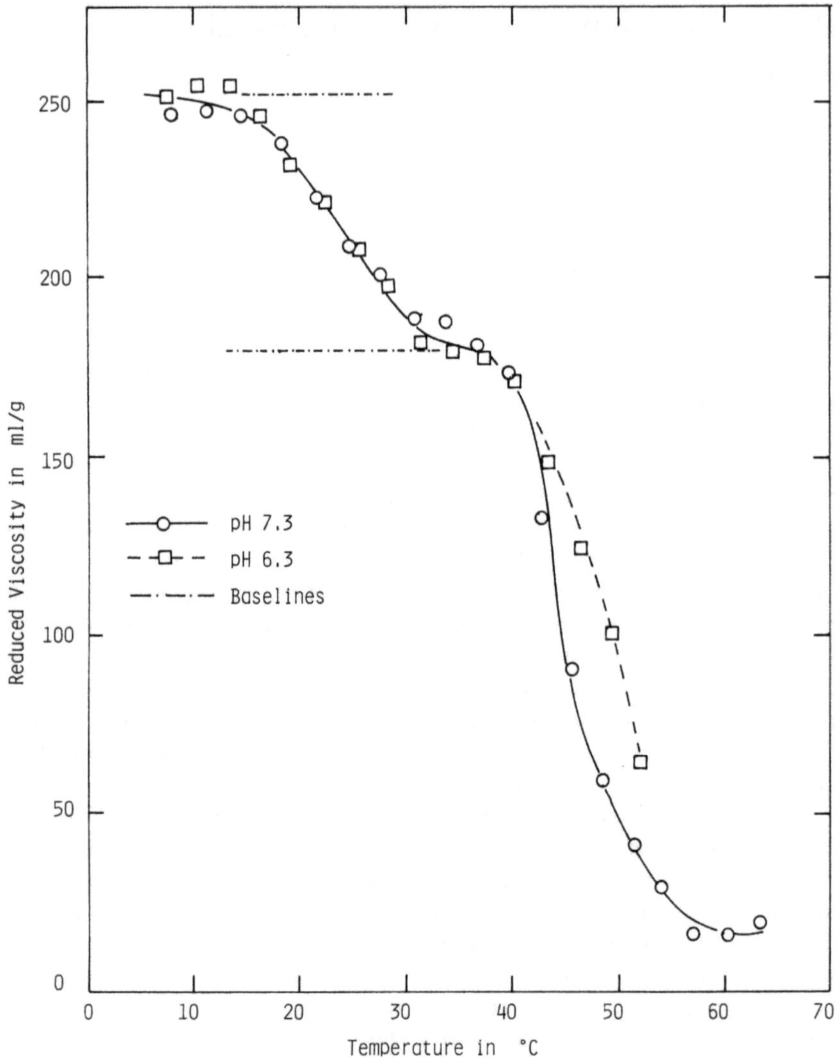

FIG. 25.3. Temperature-dependence of the reduced viscosity of rabbit rod. The reduced viscosity of rabbit rod at 1.05 mg/ml was measured with a Wescan-Ubbelohde semi-microviscometer with automatic timer (Wescan Instruments, Santa Clara, CA). The outflow time of the viscometer for the buffer was 94.5 sec at 10°C, and for the rod solution it was about 120 sec. The time resolution was 0.02 sec. The buffer composition was 0.5 *M* KCl, 20 m*M* potassium phosphate, 10 m*M* EDTA, adjusted to desired pH with HCl or NaOH. A low temperature transition at 24.0°C is seen for experiments done at pH 7.3 and 6.3. The van't Hoff ΔH of the transition is 25 kcal/mol and was calculated from the slope of the transition curve using the relation, $\Delta H_{VH}(T_m) = 4RT_m^2 \times (\text{slope at } T_m)$. See ref. 24 for details.

viscosity transition curve by using the van't Hoff equation if suitable baselines are assigned for the pre- and the posttransition states (24). As shown in Fig. 25.3, this step was undertaken. The enthalpy of the transition is estimated to be around 25 kcal/mol. Assuming a 0.6 kcal/residue change, this transition would amount to the melting of approximately 40 residues. This estimate is based on full cooperativity of the transition and is likely an underestimation. Because the cooperativity of the calorimetric transitions listed in Table 25.2 ranges between 0.15 and 0.59, it is safe to assume that roughly 100 residues are involved in this melting process. A nearly 30% reduction in the reduced viscosity with only a few percent of helix melting suggests that melting occurs near the center of the rod. Taken together with the proteolytic digestion experiments reported by Ueno and Harrington (13), we conclude that a part of the hinge domain melts at a temperature near 24°C when the segment is in direct contact with solvent, i.e., when it is released from the surface of the thick filament.

Conclusion

The low temperature transition detected by the reduced viscosity versus temperature profile (Fig. 25.3), which we attribute to hinging of the myosin rod, appears to be a common feature of muscle species that have markedly different working temperatures. Light scattering and optical rotation melting studies of rabbit, frog, scallop, and ice fish myosin rods (29) revealed that the radius of gyration versus the temperature profile is shifted to low temperature in each species by about 5° to 8°C with respect to the optical rotation melting curve, strongly suggesting hinging of the rod near its center. The melting temperatures of the rod fragments of these species parallel the working temperatures of their respective muscles.

The present results provide additional evidence from calorimetric and viscosity studies for an extended region of low thermal stability, the hinge domain, which is located near the center of the myosin rod. The results lend further support to the idea that a helix-coil transition in this region of a cycling cross-bridge underlies the tension-generating process in working muscle. A major question about the helix-coil mechanism is how the energy-producing process (ATP cleavage) is coupled with helix melting, which occurs some 40 to 50 nm distant from the ATPase-active site. Earlier enzyme-probe studies have shown that the myosin head in its activated ($M^* \cdot ADP \cdot P_i$) state must attach to actin before bridge release from the thick filament surface can occur (15). It seems clear that simple addition of ADP (with or without P_i) to rigor muscle fibers does not affect the disposition of the S2 segment of the bridges along the thick filament backbone. It appears from these and other studies (15) that attachment of the myosin head in its high energy state to actin results in changes in environment around the S1 subunit–thick filament interface, possibly by altering the local electrostatic charge near the S1–rod junction. In future studies it will be essential to establish the pathway of coupling the energy transducing reaction, ATP cleavage, with this process.

Acknowledgments. This work supported by NIH grants GM-37304 to T.Y.T. and AM-04349 to W.F.H.

References

1. Huxley, H. E. (1969) *Science*, 164:1356–1366.
2. Huxley, A. F., and Simmons, R. M. (1971) *Nature* 233:533–538.
3. Lymn, R. W., and Taylor, E. W. (1970) *Biochemistry* 9:2975–3983.
4. Eisenberg, E., and Hill, T. L. (1985) *Science* 227:999–1006.
5. Cooke, R., Crowder, M. S., and Thomas, D. D. (1982) *Nature* 300:776–778.
6. Yanagida, T. (1981) *J. Mol. Biol.* 146:539–560.
7. Yanagida, T. (1985) *J. Muscle Res. Cell Motil.* 6:43–52.
8. Huxley, H. E., Faruqi, A. R., Kress, M., Boards, J., and Koch, M. H. J. (1982) *J. Mol. Biol.* 158:637–684.
9. Flory, P. J. (1969) *Statistical Mechanics of Chain Molecules*, pp. 30–47. Interscience-Wiley, New York.
10. Flory, P. J. (1956) *Science* 124:53–60.
11. Harrington, W. F. (1979) *The Proteins*, Vol. 4, pp. 245–409. Academic Press, New York.
12. Harrington, W. F. (1971) *Proc. Natl. Acad. Sci. USA* 68:685–689.
13. Ueno, H., and Harrington, W. F. (1981) *J. Mol. Biol.* 149:619–640.
14. Ueno, H., and Harrington, W. F. (1984) *J. Mol. Biol.* 180:667–701.
15. Ueno, H., and Harrington, W. F. (1986) *J. Mol. Biol.* 190:59–68; 190:69–82.
16. Goodno, C. C., and Swenson, C. A. (1975) *Biochemistry* 14:873–877.
17. Swenson, C. A., and Ritchie, P. A. (1980) *Biochemistry* 19:2219–2224.
18. Burke, M., Himmelfarb, S., and Harrington, W. F. (1973) *Biochemistry* 12:701–708.
19. Potekhin, S. A., and Privalov, P. L. (1978) *Biofizika* 23:219–223.
20. Sutoh, K., Chen-Chiao, Y-C., and Harrington, W. F. (1978) *Biochemistry* 17:1234–1239.
21. Tsong, T. Y., Karr, T., and Harrington, W. F. (1979) *Proc. Natl. Acad. Sci. USA* 76:1109–1113.
22. Tsong, T. Y., Himmelfarb, S., and Harrington, W. F. (1983) *J. Mol. Biol.* 164:431–450.
23. Ford, L. E., Huxley, A. F., and Simmons, R. M. (1977) *J. Physiol.* (*Lond.*) 269:441–515.
24. Tsong, T. Y., Hearn, R. P., Wrathall, D. P., and Sturtevant, J. M. (1970) *Biochemistry* 9:2666–2677.
25. Privalov, P. L. (1982) *Adv. Protein Chem.* 35:1–104.
26. Sturtevant, J. M. (1974) *Annu. Rev. Biophys. Bioeng.* 3:35–51.
27. Sutoh, K., Sutoh, K., Karr, T., and Harrington, W. F. (1979) *J. Mol. Biol.* 126:1–22.
28. Tanford, C. (1961) *Physical Chemistry of Macromolecules.* Wiley, New York.
29. Rogers, M. E., and Harrington, W. F. (1986) *Biophys. J.* 49:220a.

26
Enzymatic Modulation of Cytoskeletal Self-Assembly: ADP Ribosylation of Microtubule Protein Components

Daniel L. Purich and Robin M. Scaife

The cytoskeleton is a lace-like protein fiber network spanning the eukaryotic cytoplasm and connecting the outer membrane with other interior organelles. These cytoskeletal fibers consist mainly of three types: actin filaments, intermediate filaments, and microtubules. They attach to various cellular components through such cross-bridging protein elements as spectrin, vinculin, fodrin, microtubule-associated proteins, and fiber-binding sites on centrioles and chromosomes. The microtubule cytoskeleton exerts considerable control over cell shape and movement, mitosis and meiosis, endocytosis, and membrane recycling, as well as cell–cycle and cell–cell interactions. For this reason, microtubules have become a focus for extensive characterization and mechanistic inquiry (1, 2). All tubule-linked processes are highly dynamic, and the need for their temporal and structural specification is essential for living cells to achieve excitatory and adaptive behavior. The interphase microtubule cytoskeleton, for example, must be disassembled and completely remodeled into the mitotic apparatus for chromosome disjunction and cytokinesis. Not surprisingly, then, there must be considerable control of all cytoskeletal behavior, and enzyme-catalyzed covalent modification may represent an important strategy for such regulation.

In this chapter we describe a series of experiments indicating that brain microtubule proteins can be efficiently ADP-ribosylated by an avian erythrocyte enzyme. Such modification leads to the inhibition of microtubule assembly through the synergistic action of ADP ribosyl (ADPR)–tubulin and ADPR–microtubule-associated protein-2 (MAP-2). It is the first documented case of a change in tubulin polymerization behavior arising from enzyme-catalyzed covalent modification of tubulin and a microtubule-associated protein (MAP).

Experimental Procedures

Materials

Sources for reagents used in microtubule preparation and assembly are described elsewhere (3, 4). Avian erythrocyte ADPR transferase was kindly provided by Dr. Joel Moss at the National Institutes of Health, and [^{32}P]NAD$^+$ and [^{3}H]GTP were products of ICN Radiochemicals and Amersham, respectively.

Methods

Bovine brain microtubule protein was prepared by the protocols of Karr et al. (4). SDS gel electrophoresis was carried out by the method of Laemmli (5), stained with Coomassie brilliant blue dye, and probed for radiolabeling using Kodak X-OMAT film with a Dupont Cronex enhancing screen at −70°C. Radioactivity incorporation in excised protein bands was also estimated by counting in a Beckman LS 8000 instrument. Assembly was estimated by turbidity (350 nm) or by uptake of [^{3}H]GTP radiotracer using the filtration assay of Wilson et al. (6).

Results

That the *Cholera* enterotoxin catalyzes mono(ADP ribosyl)ation of several GTP-binding proteins is well documented (7). Among these proteins are the GTP-regulatory proteins in the adenylate cyclase system (8–10), the GTP-dependent photo-signaling protein designated transducin (11), and several microtubule proteins (12–14). Nonetheless, the brain microtubule cytoskeletal proteins exhibit an extremely low stoichiometry of ADPR incorporation (i.e., $>$ 0.005 mol/mol tubulin). All published attempts to increase this stoichiometry have met with failure, and there was no apparent inhibition or activation of tubule assembly. Likewise, there was no evidence for selective uptake or exclusion of ADPR proteins during microtubule elongation (12–14). Other enzymes, including the arginine-specific avian erythrocyte mono-(ADPR)-transferase (15) modify several GTP-binding proteins. We reasoned that such vertebrate enzymes may display a different substrate specificity beyond the action of *Cholera* toxin. Accordingly, we used purified avian erythrocyte ADPR-transferase to explore the potential covalent modification of microtubule proteins.

Shown in Fig. 26.1 is the time course of NAD$^+$-dependent ADPR group incorporation into microtubule protein precipitated by trichloracetic acid. We have duplicated this result with brain tubulin in the presence and absence of MAPs. Furthermore, this modification reaction was observed with protein derived from who microtubule protein purification methods having different profiles of MAPs (4). The amplitude of the ADP ribosylation time course is directly proportional to the microtubule protein concentration, and we observed no modification in the absence of the avian erythrocyte enzyme. The reaction is also inhibited by thymidine, a common characteristic of vertebrate mono- and poly(ADP ribosyl)ation enzymes (7).

To identify the protein(s) modified by the turkey erythrocyte enzyme, we

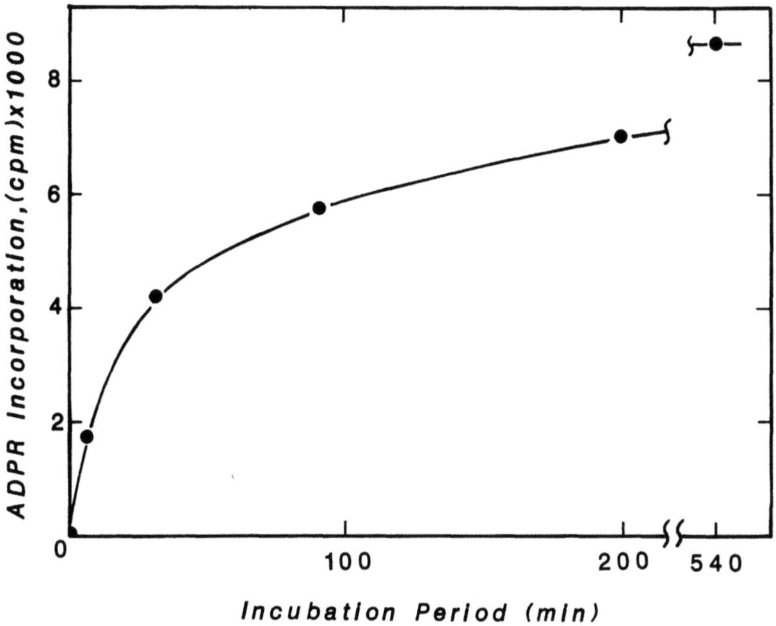

FIG. 26.1. Time course of microtubule protein modification by the avian erythrocyte ADPR transferase. Microtubule protein (2.4 mg/ml) was incubated with avian enzyme 1.9 μg/ml and 75 μM radiolabeled NAD^+ for the designated intervals at 30°C. Aliquots were quenched by the addition of trichloracetic acid (10% v/v final), and precipated fractions were counted.

subjected the covalently modified samples to SDS gel electrophoresis. The findings shown in Fig. 26.2. indicate that both tubulin and MAP-2 are modified. The autoradiogram in lane B also shows other radiolabeled protein species, and we have determined that most are proteolytic cleavage products of the high-molecular-weight MAPs. From knowledge of the concentration of both of these cytoskeletal proteins and the extent of radiolabel uptake, we estimate stoichiometries of 0.4 and 3 to 4 mol ADPR/mol of tubulin and MAP-2, respectively. This level of tubulin modification is about 200 times greater than the ADPR content previously achieved with the *Cholera* toxin (12–14). Notably, the conditions presented in the legend to Fig. 26.1 were chosen to be compatible with requirements for microtubule protein stability and polymerization. We did not find any higher stoichiometry for either tubulin or MAP-2, even upon subsequent additions of more enzyme or NAD^+ to the reaction mixtures. Likewise, we also observed that assembly of the microtubule protein prior to NAD^+ and enzyme addition served only to reduce the extent of ADP ribosylation (data not shown).

We next determined if ADP ribosylated microtubule protein displayed any altered assembly/disassembly behavior. To do so, we conducted assembly

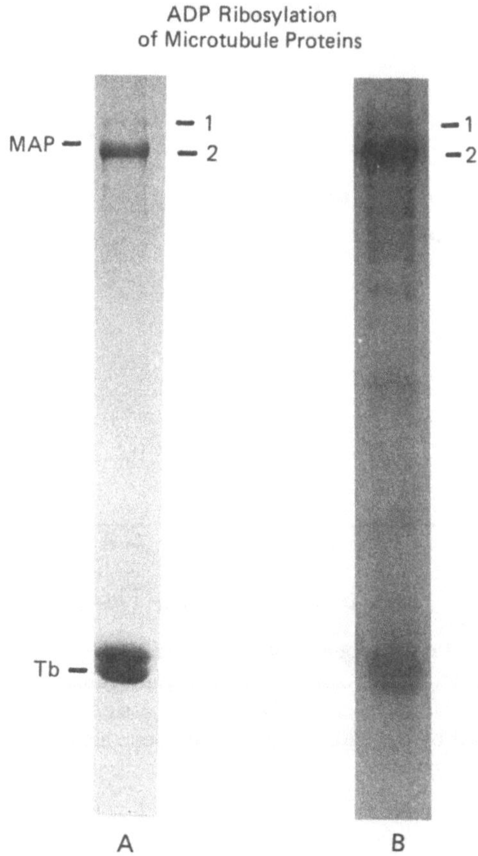

FIG. 26.2. Electrophoretic analysis of ADPR incorporation into brain microtubule proteins. Erythrocyte enzyme 2.1 μg/ml and [^{32}P]NAD$^+$ 76 μM were incubated with microtubule protein 1.15 mg/ml for 60 min at 30°C in tubule assembly buffer. After SDS polyacrylamide gel electrophoresis, gels were stained with Coomassie dye (A) and subjected to autoradiography (B). The high molecular weight microtubule-associated proteins (MAP-1 and MAP-2) and tubulin polypeptide chains are designated 1, 2, and Tb, respectively.

experiments at several levels of microtubule, protein ADP ribosylation. A significant finding was that at label incorporation corresponding to only 10 to 15% modification the assembly rate was inhibited by 65 to 70%. The data in Fig. 26.3 were obtained with microtubule protein modified to about 0.4 mol ADPR/mol tubulin and 2 to 3 mol ADPR/mol MAP-2. This potent inhibition was not observed with similar samples in the absence of NAD$^+$ and/or the avian erythrocyte transferase. We also found that recombination of ADPR-modified tubulin and unmodified tubulin always resulted in impaired assembly compared to control experiments with unmodified tubulin. Likewise,

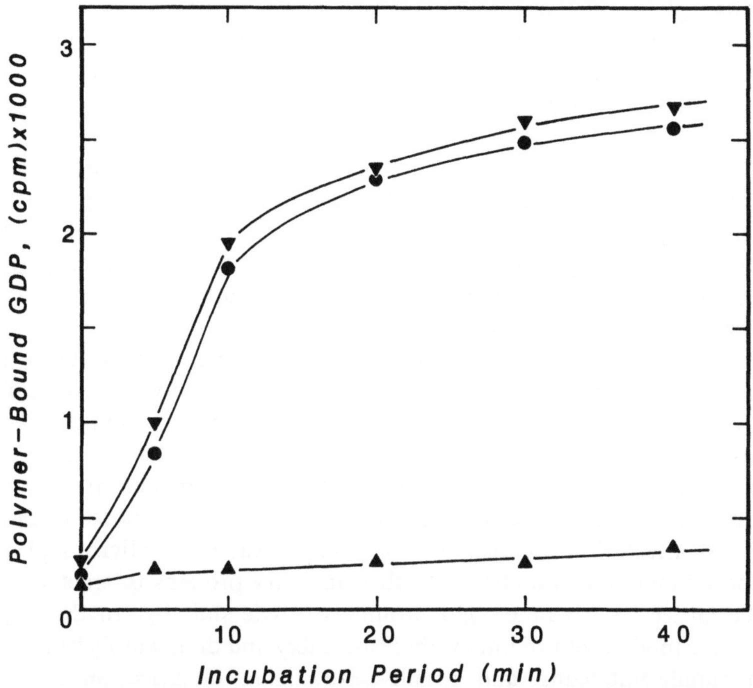

FIG. 26.3. Effect of ADP ribosylation on microtubule self-assembly. Microtubule protein 2.4 mg/ml was incubated for 60 min at 30°C in the absence or presence of either 80 μM NAD^+ or avian enzyme 5 μg/ml. Protein samples were then chilled (4°C) for 30 min before addition of 240 μM [^{3}H]GTP. Assembly was induced by warming to 30°C, and aliquots were filtered at designated intervals to partition polymerized and unpolymerized protein. (▾) Assembly in the absence of NAD^+ and the avian enzyme. (●) Assembly in the presence of NAD^+ only. (▴) Assembly in the presence of both NAD^+ and ADPR-transferase.

modified MAP-2 is about three to four times less efficient in promoting microtubule assembly with unmodified tubulin.

Discussion

The self-assembly and turnover of fibrous elements of the cytoskeleton offer many steps that may be altered to achieve effective control through enzyme-catalyzed covalent interconversion. In terms of microtubule assembly and disassembly, such discrete phases include nucleation, elongation, opposite-end assembly/disassembly, dynamic subunit exchange events, and the attachment of associated proteins to repetitive superhelical sites on the tubule lattice (1). For example, nucleation of microtubule assembly involves reactions of sufficiently high kinetic order to be disfavored, and covalent interconversion may alter the strength of interprotomeric bonding and influence the nucleation process so that more or fewer microtubules are formed. Elongation requires sequential repetitive tubulin protomer addition to microtubule ends, and both the kinetics and thermodynamics of protomer binding/release steps can be expected to be likewise affected by covalent modification reactions. There are other likely interactions at microtubule ends with tubulin, and unpolymerized tubulin also serves as an important target. Because tubulin gene expression is linked to the free unpolymerized tubulin level in cells (16), such covalent interconversion may even control the de novo synthesis of tubulin at appropriate times in the cell cycle.

We have described the enzyme-catalyzed modifications of tubule proteins (17, 18), which chiefly include tyrosination and detyrosination, phosphorylation and dephosphorylation, acetylation, and the limited *Cholera* toxin-mediated ADP ribosylation. In one case tubulin detyrosination and retyrosination occur by action of specific enzymes directed at the tubule polymer and tubulin protomer, respectively. Yet this modification reaction does not affect the assembly or disassembly reactions. Microtubules are composed mainly of tubulin, but other auxiliary proteins interact with the tubule lattice in a manner that promotes tubule stability. In this regard, enzymatic modulation of microtubule assembly and disassembly can involve both tubule-stabilizing and tubule-destabilizing modification reactions. Despite the variety of steps subject to likely control, characterization of the molecular mechanisms underlying such regulation is still largely circumstantial. One of the difficulties has been the absence of good model systems for analyzing such behavior. We believe that the avian erythrocyte enzyme may now provide such an opportunity.

The selective action of the avian erythrocyte ADPR transferase on several functionally unrelated GTP-binding proteins suggests the existence of a structurally related motif or binding site. Future work on the amino acid sequence surrounding the site(s) of ADPR modification may shed light on this issue. Likewise, we must also address the mechanistic basis of the inhibitory action in terms of effects on the tubulin GTPase activity. Loss of GTP-hydrolyzing capacity on ADP ribosylation is a recurring theme in the cases of adenylate cyclase and transducin (7). Finally, although we do not ascribe any physiological role for this ribosylation reaction, the high evolutionary relatedness of the tubulins suggest that ADP ribosylation may have a regulatory function.

References

1. Purich, D. L., and Kristofferson, D. (1984) *Adv. Protein Chem.* 36:133–212.
2. Hill, T. L., and Kirschner, M. W. (1983) *Int. Rev. Cytol.* 78:1–125.
3. Karr, T. L., Podrasky, A. E., and Purich, D. L. (1979) *Proc. Natl. Acad. Sci. USA* 76:5475–5479.

4. Karr, T. L., White, H. D., and Purich, D. L. (1979) *J. Biol. Chem.* 254:6107–6111.
5. Laemmli, U. K. (1970) *Nature* 227:680–685.
6. Wilson, L., Snyder, K. B., Thompson, W. C., and Margolis, R. L. (1982) *Methods Cell Biol.* 24:159–169.
7. Ueda, K., and Hayaishi, O. (1985) *Annu. Rev. Biochem.* 54:73–100.
8. Moss, J., and Vaughan, M. (1979) *Annu. Rev. Biochem.* 48:581–600.
9. Moss, J., and Richardson, S. H. (1978) *J. Clin. Invest.* 62:281–285.
10. Gill, D. M., and Meren, R. (1978) *Proc. Natl. Acad. Sci. USA* 75:3050–3054.
11. Abood, M. E., Hurley, J. B., Pappone, M-C., Bourne, H. R., and Stryer, L. (1982) *J. Biol. Chem.* 257:10540–10543.
12. Kaslow, H. R., Groppi, V. E., Abood, M. E., and Bourne, H. R. (1981) *J. Cell Biol.* 91:410–413.
13. Amir-Zaltsman, Y., Ezra, E., Scherson, T., Zutra, A., Littauer, U. Z., and Salomon, Y. (1982) *EMBO J.* 1:181–186.
14. Hawkins, D. J., and Browning, E. T. (1982) *Biochemistry* 21:4474–4479.
15. Moss, J., Stanley, S. J., and Watkins, P. A. (1980) *J. Biol. Chem.* 255:5838–5840.
16. Lau, J. T. Y., Pittenger, Havercroft, J. C., and Cleveland, D. W. (1986) Ann. N. Y. Acad. Sci. 466:75–88.
17. Terry, B. J., and Purich, D. L. (1982) *Adv. Enzymol.* 53:113–161.
18. Purich, D. L., and Scaife, R. M. (1985) *Curr. Top. Cell. Regul.* 27:107–116.

27
Human Angiogenin: An Organogenic Protein

BERT L. VALLEE

The first human tumor-derived protein with in vivo angiogenic activity to be obtained in pure form has been isolated from serum-free supernatants of an established human adenocarcinoma cell line (HT-29); it was named angiogenin (1-5). It was purified by cation-exchange and reverse-phase high-performance liquid chromatography. The yield was ~0.5 μg/l of medium. Biological activity of angiogenin was monitored throughout purification by using the chick embryo chorioallantoic membrane assay (1). Statistical evaluation demonstrates that it displays activity in this system with as little as 35 femtomoles per egg. Moreover, only 3.5 picomoles is required to induce extensive blood vessel growth in the rabbit cornea.

The amino acid sequence and disulfide bond pairing of human tumor-derived angiogenin have been determined by conventional sequencing techniques adapted and applied to nanomole and subnanomole levels of material (4). Angiogenin, obtained from such conditioned media, is a single-chain protein consisting of 123 amino acids (pI 9.5) and MW ~14,400. The amino-terminus is blocked by a pyroglutamyl residue, and the carboxyl-terminal residue is proline. The sequence is as follows: ⟨Glu-Asp-Asn-Ser-Arg-Tyr-Thr-His-Phe-Leu-Thr-Gln-His-Tyr-Asp-Ala-Lys-Pro-Gln-Gly-Arg-Asp-Asp-Arg-Tyr-Cys-Glu-Ser-Ile-Met-Arg-Arg-Arg-Gly-Leu-Thr-Ser-Pro-Cys-Lys-Asp-Ile-Asn-Thr-Phe-Ile-His-Gly-Asn-Lys-Arg-Ser-Ile-Lys-Ala-Ile-Cys-Glu-Asn-Lys-Asn-Gly-Asn-Pro-His-Arg-Glu-Asn-Leu-Arg-Ile-Ser-Lys-Ser-Ser-Phe-Gln-Val-Thr-Thr-Cys-Lys-Leu-His-Gly-Gly-Ser-Pro-Trp-Pro-Pro-Cys-Gln-Tyr-Arg-Ala-Thr-Ala-Gly-Phe-Arg-Asn-Val-Val-Val-Ala-Cys-Glu-Asn-Gly-Leu-Pro-Val-His-Leu-Asp-Gln-Ser-Ile-Phe-Arg-Arg-Pro-OH. Three disulfide bonds link the half-cystinyl residues 26–81, 39–92, and 57–107. The sequence is homologous to that of the pancreatic ribonucleases with 35% identity, and many additional residues are replaced conservatively. Similarities are especially apparent around the major active-site residues His-12, Lys-41, and His-119 of ribonuclease, which are conserved, as are three of the four disulfide bonds.

Human cDNAs coding for angiogenin were isolated from a cDNA library prepared from human liver poly(A) mRNA employing a synthetic oligonucleotide as a hybridization probe. The largest cDNA insert (697 basepairs) contained a short 5′-noncoding sequence followed by a sequence coding for a signal peptide of 24 (or 22) amino acids, 369 nucleotides coding for the mature protein of 123 amino acids, a stop codon, A 3′-noncoding sequence of 175 nucleotides, and a poly(A) tail. The gene coding for human angiogenin was then isolated from a genomic lambda charon 4A bacteriophage library employing the cDNA as a probe. The nucleotide sequence of the gene and the adjacent 5′- and 3′-flanking regions (4688 basepairs) was then determined. The coding and 3′-noncoding regions of the gene for human angiogenin were found to be free of introns, and the DNA sequence for the gene agreed well with that of the cDNA. The amino acid sequence of human angiogenin, as predictable from the gene sequence, was in complete agreement with that determined by amino acid sequence analysis (5).

The complete chemical characterization of a unique human organogenic messenger molecule, i.e., one that induces organ formation, accomplishes the first major objective of this long-term investigation of organogenesis in general and angiogenesis in particular. The unexpected homology to ribonuclease A suggests novel approaches to the investigation of the biological process of angiogenesis and may be relevant to the evolution of organogenic molecules (6). These provocative findings are thought to have important physiological implications.

References

1. Vallee, B. L., Riordan, J. F., Lobb, R. R., Higachi, N., Fett, J. W., Crossley, G., Buhler, R., Budzik, G., Breddam, K., Bethune, J. L., and Alderman, E. M. (1985) *Experientia* 41:1–15.
2. Alderman, E. M., Lobb, R. R., and Fett, J. W. (1985) *Proc. Natl. Acad. Sci. USA* 82:5771–5775.
3. Fett, J. W., Strydom, D. J., Lobb, R. R., Alderman, E. M., Bethune, J. L., Riordan, J. F., and Vallee, B. L. (1985) *Biochemistry* 24:5480–5486.
4. Strydom, D. J., Fett, J. W., Lobb, R. R. Alderman, E. M., Bethune, J. L., Riordan, J. F., and Vallee, B. L. (1985) *Biochemistry* 24:5486–5494.
5. Kurachi, K., Davie, E. W., Strydom, D. J., Riordan, J. F., and Vallee, B. L. (1985) *Biochemistry* 24:5494–5499.
6. Alderman, E. M., Lobb, R. R., Fett, J. W., Riordan, J. F., Bethune, J. L., and Vallee, B. L. (1985) *Biochemistry* 24:7866–7871.

28
Structure and Mechanism of Action of a Membrane-Bound Enzyme: Chloroplast Coupling Factor

GORDON G. HAMMES

In all organisms a transmembrane pH gradient is utilized as the source of free energy for the synthesis of ATP from ADP and P_i (1). The membrane-bound enzymes (coupling factors) that catalyze the synthesis and pump protons are remarkably similar from all sources: The basic structure consists of a ball on the surface of the membrane (F_1) attached to a stalk (F_0) that passes through the membrane. The minimal structure consists of five types of polypeptide chains in F_1, probably in the stoichiometry $\alpha_3\beta_3\gamma\delta\varepsilon$, and three types of polypeptide chains in F_0, with the stoichiometry still controversial. F_1 is an ATPase and contains all of the nucleotide binding sites but of course cannot make ATP catalytically when separated from the membrane. F_0 is a hydrophobic structure that probably serves as the proton channel.

This chapter considers only the chloroplast coupling factor from spinach. The plant enzymes appear to differ somewhat from the bacterial and mitochondrial enzymes in mechanistic detail, but the basic mechanistic concepts are probably the same. The nature and number of nucleotide binding sites is first considered. The structure of the enzyme, as elucidated by fluorescence resonance energy transfer measurements, is then described. Finally the mechanism of ATP hydrolysis and synthesis is discussed.

Nucleotide Binding Sites

If chloroplast coupling factor 1 (CF_1; M_r 400,000) is dialyzed extensively, a single ADP remains tightly bound (2). However, if radioactive ADP (or ATP) is added to the medium, radioactive ADP is found to be incorporated into the site containing tightly bound ADP. The explanation of this phenomenon is that the binding of nucleotide at an allosteric site promotes dissociation of tightly bound ADP. The rate of exchange depends on the concentration of nucleotide and becomes quite rapid during ATP hydrolysis. This site is referred to as site 1.

Two additional nucleotide binding sites are found on CF_1 (2). One of these binds MgATP almost irreversibly: The half-time for dissociation is days. This site, site 2, is quite specific for both Mg^{2+} and ATP. The remaining nucleotide site, site 3, binds di- and trinucleotide phosphates in the presence or absence of metals with an equilibrium dissociation constant in the micromolar range. Thus far no convincing evidence has been found for more than three nucleotide binding sites with dissociation constants <1 mM.

The locations of the nucleotide binding sites on CF_1 have been investigated with the photoaffinity labels arylazido-ADP/ATP (2) and 3′-*O*-(4-benzoyl) benzoyl ADP/ATP (3). The results obtained indicate that primarily the β polypeptide is labeled when sites 1 and 2 are occupied by the photoaffinity labels during photolysis. However, a small amount of the α polypeptide also appears to be labeled. With site 3, the α and β polypeptides are labeled in approximately equal amounts. If the three nucleotide binding sites are assumed to be approximately equivalent in a structural sense, the three nucleotide binding sites are primarily on the β polypeptide but near the interface of α and β polypeptides. The functional role of the three nucleotide sites is discussed later.

Fluorescence Resonance Energy Transfer

Ultimately, to understand the molecular mechanism of catalysis and regulation the structure of the coupling factor must be known. Although a high-resolution crystal structure is desirable, thus far high-quality crystals have not been obtained. [A low-resolution structure has been obtained for the rat liver mitochondrial enzyme (4).] Consequently, indirect methods must be used to obtain structural information. One of the most useful methods for complex biological systems is fluorescence resonance energy transfer (5). For this method to succeed, sites on the coupling factor must be labeled specifically with fluorescent or highly absorbing moieties. If the fluorescence emission of the energy donor overlaps the absorption spectrum of the energy acceptor, radiationless energy transfer can occur when the fluorescent donor is excited with light. The consequence of this energy transfer is that the fluorescence quantum yield and lifetime of the donor are reduced in the presence of an energy acceptor. Quantitation of this energy transfer allows the distance between the energy donor and the acceptor to be calculated. Typically distances of 10 to 100 Å can be measured with uncertainties of about 10 to 15%. The details of the theory have been frequently discussed (5–7). The most critical aspect of this method is the specific labeling of the protein; if nonspecific labeling occurs, even to a small extent, the distance measurements can be seriously compromised.

The specific sites that have been labeled on CF_1 are given in Table 28.1. The unique properties of the nucleotide binding sites allow them to be individually labeled with a combination of nucleotide analogues, ε-ADP/ATP, and trinitrophenyl-ADP/ATP (8). The β polypeptide chain has a unique tyrosine that can be specifically labeled; the enzyme is inactive after the labeling (9, 10).

The γ polypeptide has a single sulfhydryl group ("dark sulfhydryl") that can be labeled with a variety of maleimides (11). The enzyme remains active after the labeling. When isolated, CF_1 is a latent ATPase (12). If the single disulfide

TABLE 28.1. Sites on CF_1 used for energy transfer measurements.

Site	Polypeptide Location
Nucleotide site 1	$\beta(\alpha)$
Nucleotide site 2	$\beta(\alpha)$
Nucleotide site 3	$\beta - \alpha$
Tyrosine (3)	β
Disulfide	γ
Dark sulfhydryl (1)	γ
Light sulfhydryl (1)	γ
Sulfhydryl (1)	ε
Amino (1)	α

on the γ polypeptide is reduced by dithiothreitol or other reagents, it becomes an ATPase that prefers Ca^{2+} as an activating metal. The membrane-bound enzyme prefers Mg^{2+}. The disulfide can be specifically labeled after reduction if the "dark sulfhydryl" is first blocked with *N*-ethylmaleimide (10). Finally an additional sulfhydryl ("light sulfhydryl") on the γ polypeptide can be labeled if the maleimide label is present when thylakoids are flashed with light (11). This group cannot be modified in isolated CF_1, and modification inactivates the enzyme. The ε polypeptide has a single sulfhydryl that becomes labeled if the enzyme is reacted extensively with maleimides; unfortunately, this labeling is not specific (13). However, the ε polypeptide can be isolated, labeled, and reconstituted with ε-deficient CF_1 (7); the result is a specifically labeled, catalytically active molecule. Finally a single amino group on an α polypeptide has been labeled; the resulting enzyme is active (14).

With the above array of specific labels, virtually all of the possible combinations of fluorescence energy transfer measurements were made (7, 8, 14–17). The resulting collection of data was subjected to a least squares minimization to give a structural model of CF_1. The only sites not uniquely located by this procedure are the tyrosines on the β polypeptide (17); their locations have one degree of freedom that is not precisely defined. When carrying out this analysis, no individual measured distance had to be modified by more than 10% to be consistent with the final structure. In some cases the distance between labels on CF_1 and the surface of the membrane was determined in a reconstituted system (15, 16).

Some views of the structural model derived from the data are shown in Fig. 28.1. Several features of this structure merit attention. The nucleotide binding

FIG. 28.1. Spatial relations among the nucleotide binding sites (1, 2, 3), the γ-"light sulfhydryl" (L), the γ-"dark sulfhydryl" (D), the γ-disulfide (SS), the β-tyrosines (T), the ε-sulfhydryl (ε), an amino group on the α-polypeptide (LY), and the membrane surface (7, 14, 16, 17). The positions of the β-tyrosines are not uniquely determined. Distances are shown in angstroms. The distances to the membrane surface in the lowest structure could be shorter because the lines need not be perpendicular to the membrane. Views perpendicular and parallel to the membrane surface are shown. ▷

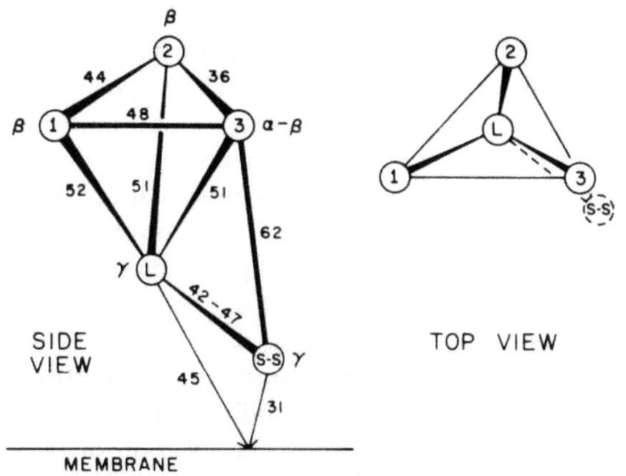
β
2
44
36
β
1
48
3
α-β
52
51
51
62
γ
L
42-47
SIDE VIEW
S-S
γ
45
31
MEMBRANE
TOP VIEW
2
L
1
3
S-S

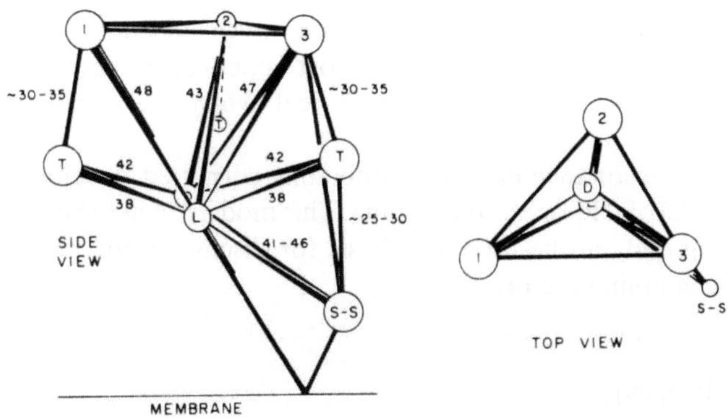
1
2
3
~30-35
48
43
47
~30-35
T
T
42
42
38
38
L
~25-30
SIDE VIEW
41-46
S-S
MEMBRANE
2
D
1
3
S-S
TOP VIEW

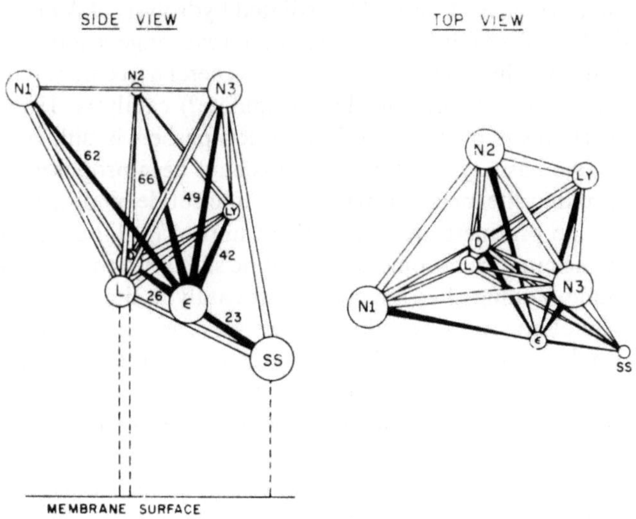
SIDE VIEW
TOP VIEW
N1
N2
N3
62
66
49
LY
42
L
26
ε
23
SS
MEMBRANE SURFACE
N2
LY
D
L
N3
N1
ε
SS

sites are far from the surface of the membrane (~90 Å), immediately suggesting that the coupling between catalysis and ion pumping is indirect. The nucleotide sites also are far from the tyrosines (β polypeptide) and disulfide (γ polypeptide) that modify the catalytic activity. Importantly, the structure is intrinsically asymmetrical; possible functional consequences of this asymmetry are considered below. The α, β, and γ polypeptides are essential for catalytic activity; the δ and ε polypeptides are not. The disulfide is directly below nucleotide site 3. The "light sulfhydryl" is approximately in the center of the molecule and may play a crucial role in the coupling of ion pumping and enzyme catalysis. The "light sulfhydryl" is close to, and probably protected by, the "dark sulfhydryl." The ε polypeptide is an inhibitor of the ATPase activity and is located close to the γ disulfide. This area can be regarded as a regulatory region of the molecule. The location of the δ polypeptide is not yet known, but this situation should soon be remedied. Dynamic fluorescence anisotropy measurements have given additional insight into the overall structure of CF_1: CF_1 is almost spherical if the δ polypeptide is removed but is significantly nonspherical when the δ polypeptide is present (18).

Fluorescence resonance energy transfer has permitted development of a structural model of CF_1 on the membrane. This model can now be subjected to further testing and can be used as a basis for discussing the mechanism of action of the coupling factor.

ATP Hydrolysis

The steady-state kinetics of the Ca^{2+}-mediated hydrolysis of ATP by CF_1 has been extensively studied (19, 20), and the oxygen exchange reactions catalyzed also have been investigated (21, 22). There is general agreement that interactions between nucleotide sites mediate (regulate?) catalysis. However, the molecular mechanism by which this is accomplished is still a matter of controversy. Two basic types of mechanisms have been proposed.

In the alternating site mechanism (22), one nucleotide binding site hydrolyzes ATP, but the products are not released until ATP binding occurs at a second site, which in turn becomes the catalytic site. With this mechanism the three nucleotide binding sites are equivalent catalytic sites, but rather than functioning simultaneously they alternatively serve as catalytic sites. A major difficulty with this mechanism is that the structure and the nucleotide binding sites are asymmetrical. It has been suggested that the sites become structurally equivalent during catalysis through rotation of the α-β polypeptides with respect to the other polypeptides (23). No convincing evidence exists for such a rotation, which appears to require a large energy input to move polypeptide chains extensively with respect to each other.

A second type of mechanism requires the binding of a nucleotide at one site to influence events (binding and/or catalysis) at another nucleotide binding site. For example, the assumption that ATP bound at site 1 promotes catalysis at site 3, whereas ADP binding at site 1 inhibits catalysis at site 3, adequately explains the steady state and exchange data (as does the alternating-site mechanism) (20, 24).

To elucidate the role of each nucleotide binding site in catalysis, the events at each site must be independently monitored. This has been partially done, although more extensive data are required. Site 2 clearly is not a catalytic site: If radioactive MgATP is bound to site 2, the rate of ATP hydrolysis is unaltered, and the MgATP is still bound after extensive steady-state turnover (19, 24). The possibility exists that ATP must be bound at this site for catalysis to occur, as it is not possible to carry out ATP hydrolysis with this site unoccupied. The situation with respect to sites 1 and 3 is not so clear-cut. If radioactive ADP is put into site 1, it dissociates with a rate constant that is smaller than the turnover number (20, 25). Moreover, the dependence of the rate of dissociation of ADP from site 1 on the ATP concentration appears to be hyperbolic, whereas the dependence of ATP hydrolysis on the ATP concentration is sigmoidal (25). Thus sites 1 and 3 interact, but an alternation of activity has not been shown. The dissociation of the first ADP from site 1 is abnormally slow; however, the rate of dissociation of nucleotide from site 1 during steady-state catalysis is similar in magnitude to the turnover number. A fair assessment of the current situation is that both site 1 and site 3 contribute significantly to the ATPase activity (25). In terms of the structure, site 3 is directly above the regulatory region of the molecule containing the γ disulfide and the ε polypeptide; site 1 is much farther away from this regulatory region.

ATP Synthesis

A possible criticism of the mechanistic studies of the Ca^{2+}-activated ATPase is that the biological function of coupling factors is to synthesize ATP. The mechanistic features of combined ion pumping and ATP synthesis may differ from those of ATP hydrolysis in the uncoupled system. Although this seems unlikely, detailed mechanistic studies of ATP synthesis obviously are of importance. When carrying out such studies, a major problem is maintaining a constant pH gradient during ATP synthesis. One method of accomplishing this goal is to co-reconstitute bacteriorhodopsin and CF_1-CF_0 into phospholipid vesicles (26). If light is shined on reconstituted bacteriorhodopsin, protons are pumped inside the vesicles. After a brief time (seconds), the pH gradient reaches a steady state. The magnitude of the pH gradient can be modulated by changes in the light intensity. Optimization of the co-reconstitution has given rates of ATP synthesis $>1\ \mu M$/min-mg $CF_1 \cdot CF_0$ (27). The rate of ATP synthesis is limited by the magnitude of the pH gradient that can be obtained: Bacteriorhodopsin can be saturated with light, and the amount of bacteriorhodopsin that can be reconstituted into a single vesicle is limited. The vesicle population also is heterogeneous with regard to size and protein composition.

The steady-state rate of ATP synthesis has been studied in the bacteriorhodopsin-$CF_1 \cdot CF_0$ reconstituted system over a range of substrate concentrations, external pH values, and pH gradients (28, 29). Some of the important conclusions reached are now summarized. By determination of the rate of ATP synthesis and hydrolysis at different pH gradients, the number of protons per ATP synthesized or hydrolyzed was determined to be 3.0. The dependence of the rate on external pH is independent of the pH gradient; conversely, the dependence of the rate on the pH gradient is independent of the external pH. If the gradient is assumed to be primarily associated with the ion pumping, and the external pH with catalysis, it suggests that the coupling of catalysis and ion pumping is probably indirect, i.e., through a conformational change. Further analysis of the data indicates that the ion pumping occurs after binding of the substrates and that the ion pumping is sequential, not concerted. If specific ionizable groups can be associated with ion pumping, the pK values must be >8 on the outside of the vesicle and <6 on the inside of the vesicle.

The data obtained with reconstituted systems require considerably more refinement: Individual steps in ion pumping and catalysis need to be studied, and new methods must be developed for this purpose. In this regard, phase-lifetime spectrophotometry should be noted as a possible tool for studying ion pumping (30). With this method the intensity of light is modulated, and the difference between the rate of modulation and ion pumping (or spectral changes reflecting ion pumping and/or protein conformational changes) is measured. This method has been used to study rate processes in the millisecond time range that accompany ion pumping in reconstituted bacteriorhodopsin (31).

Summary

The chloroplast coupling factor is a membrane-bound enzyme that synthesizes ATP from ADP and P_i while simultaneously pumping protons across the membrane. The structure of the enzyme can be approximated as a ball (CF_1) attached to a stalk (CF_0) that passes through the membrane. The structure of CF_1 has been probed with a variety of techniques, especially fluorescence resonance energy transfer. Several features are mechanistically significant: The structure is intrinsically asymmetrical; the nucleotide binding sites are far from the membrane ($\sim$90 Å); and the ε and γ polypeptide chains have regulatory domains that are far from the nucleotide sites. Each of the three nucleotide sites has distinct properties, which makes the sites distinguishable in terms of their binding properties. Steady-state kinetics of the Ca^{2+}-ATPase activity suggest complex cooperativity between the sites. A model system has been developed for the study of ATP synthesis. The coupling factor and bacteriorhodopsin have been co-reconstituted into phospholipid vesicles. The rate of synthesis of ATP by light in this system is ~ 1 μmol/mg-min. The rate is limited by the vesicle heterogeneity and the magnitude of the pH gradient that bacteriorhodopsin can produce. Studies with the reconstituted system indicate that the ATP synthesis and proton pumping are indirectly coupled and that three protons are pumped for each ATP synthesized or hydrolyzed.

The details of both the structure and mechanism of action of chloroplast coupling factor remain to be elucidated, although considerable progress has been made. Because of space limitations, the coupling factors of mitochondria and bacteria have not been discussed. These systems are quite similar; however the regulatory γ disulfide is absent, and six nucleotide binding sites are found, three of which are involved in catalysis/regulation. Many reviews of these systems are available (32–34). Finally, the precise nature of the coupling between catalysis and ion pumping is not yet known. However, all ion-pumping ATPases appear to use a similar mechanism in which the enzyme switches between two or more conformations during catalysis, with ion pumping accompanying the shifts in conformation (24, 35).

Acknowledgment. This work was supported by a grant from the National Institutes of Health (GM 13292).

References

1. Mitchell, P. D. (1961) *Nature* 191:144–148.
2. Bruist, M. F., and Hammes, G. G. (1981) *Biochemistry* 20:6298–6305.
3. Kambouris, N. G., and Hammes, G. G. (1985) *Proc. Natl. Acad. Sci. USA* 82:1950–1953.
4. Amzel, L. M., McKinney, M., Naryanan, P., and Pederson, P. L. (1982) *Proc. Natl. Acad. Sci. USA* 79:5852–5856.
5. Hammes, G. G. (1981) In *Protein–Protein Interactions*, edited by C. Frieden and L. W. Nichol, pp. 257–287. Wiley, New York.
6. Dale, R. E., Eisinger, J., and Blumberg, W. E. (1979) *Biophys. J.* 26:161–194.
7. Richter, M. L., Snyder, B., McCarty, R. E., and Hammes, G. G. (1985) *Biochemistry* 24:5755–5763.
8. Cerione, R. A., and Hammes, G. G. (1982) *Biochemistry* 21:745–752.
9. Deters, D. W., Nelson, N., Nelson, H., and Racker, E. (1975) *J. Biol. Chem.* 250:1041–1047.
10. Cantley, L. C., and Hammes, G. G. (1976) *Biochemistry* 15:9–14.
11. McCarty, R. E., and Fagan, J. (1973) *Biochemistry* 12:1503–1507.
12. Farron, F., and Racker, E. (1970) *Biochemistry* 9:3829–3836.
13. Holowka, D. A., and Hammes, G. G. (1977) *Biochemistry* 16:5538–5545.
14. Nalin, C. M., Snyder, B., and McCarty, R. E. (1985) *Biochemistry* 24:2318–2324.
15. Cerione, R. A., McCarty, R. E., and Hammes, G. G. (1983) *Biochemistry* 22:769–776.

16. Snyder, B., and Hammes, G. G. (1984) *Biochemistry* 23:5787–5795.
17. Snyder, B., and Hammes, G. G. (1985) *Biochemistry* 24:2324–2331.
18. Schinkel, J. E., and Hammes, G. G. (1986) *Biochemistry* 25:4066–4071.
19. Cantley, L. C., and Hammes, G. G. (1975) *Biochemistry* 14:1968–1975.
20. Bruist, M. F., and Hammes, G. G. (1982) *Biochemistry* 21:3370–3377.
21. Kohlbrenner, W. E., and Boyer, P. D. (1983) *J. Biol. Chem.* 258:10881–10886.
22. Boyer, P. D., and Kohlbrenner, W. E. (1981) In *Energy Coupling in Photosynthesis*, edited by B. R. Selman and S. Selman-Reimer, pp. 231–240. Elsevier/North Holland, New York.
23. Gresser, M. J., Myers, J. A., and Boyer, P. D. (1982) *J. Biol. Chem.* 257: 12030–12038.
24. Hammes, G. G. (1982) *Proc. Natl. Acad. Sci. USA* 79:6881–6884.
25. Leckband, D., and Hammes, G. G. (1986) *Biochemistry* 26:2306–2312.
26. Racker, E., and Stoeckenius, W. (1974) *J. Biol. Chem.* 249:662–663.
27. Krupinski, J., and Hammes, G. G. (1986) *Proc. Natl. Acad. Sci. USA* 83: 4233–4237.
28. Dewey, T. G., and Hammes, G. G. (1981) *Proc. Natl. Acad. Sci. USA* 78:7422–7425.
29. Dewey, T. G., and Hammes, G. G. (1981) *J. Biol. Chem.* 256:8941–8946.
30. Takabe, T., and Hammes, G. G. (1981) *Biochemistry* 20:6859–6864.
31. Krupinski, J., and Hammes, G. G. (1985) *Biochemistry* 24:6963–6972.
32. Senior, A. E., and Wise, J. G. (1983) *J. Membr. Biol.* 73:105–124.
33. Walker, J. E., Saraste, M., and Gay, N. J. (1984) *Biochim. Biophys. Acta* 768:164–200.
34. Pederson, P. L., and Amzel, L. M. (1985) In *Achievements and Perspectives of Mitochondrial Research*, Vol. 1, pp. 169–189. Elsevier, New York.
35. Jencks, W. P. (1980) *Adv. Enzymol.* 51:75–106.

29

Electrostatic Effects and the Dynamics of Multienzyme Reactions at the Surface of Plant Cells

JACQUES RICARD and GEORGES NOAT

Enzymes may behave differently depending on whether they are in free solution or bound to an insoluble polyelectrolyte. One of the main reasons for this difference of dynamic behavior is the existence of electrostatic interaction effects between the fixed charges of the matrix and the mobile charges of the solution. Owing to electrostatic attraction or repulsion of mobile ions by the fixed charges of the polyelectrolyte, the local concentration of these mobile ions inside this charged matrix may be quite different from their corresponding concentrations in the bulk phase. This difference necessarily modifies the enzyme activity, particularly if the substrate of the enzyme is itself an ion.

The primary plant cell wall behaves as a polyanion, and therefore one may expect the electrostatic effects to be involved in the dynamics of cell wall enzyme reactions. These enzymes are involved in two basic biological processes: the transport of organic solutes across the cell wall, and the extension and building up of this cell wall. Owing to the polyanionic nature of the cell wall, one may expect these effects to be controlled by the fixed charge density of the wall, as well as by the concentrations of the mobile ions present in the bulk phase.

The aim of this chapter is to discuss these problems at the theoretical and experimental levels. Most of the results presented have been published elsewhere in several research papers (1–6).

Electrostatic Effects and Systemic Cooperativity of Bound Cell Wall Enzymes

If an enzyme that follows Michaelis-Menten kinetics is buried in a polyanionic matrix, and if the substrate of this enzyme is a monoanion, the enzyme reaction rate is

$$v = \frac{V_{max}[S_i^-]}{Km + [S_i^-]} \quad [1]$$

where $[S_i^-]$ is the local concentration of substrate. This local concentration is related to the bulk concentration $[S_0^-]$ through the relation

$$\Pi = \frac{[S_0^-]}{[S_i^-]} = \exp(-F\Delta\psi/RT) \qquad [2]$$

where Π is the electrostatic partition coefficient, F the Faraday, $\Delta\psi$ the difference of electrostatic potential between the inside and the outside of the matrix, and R and T have their usual significance. Inserting Eq. 2 into Eq. 1 yields

$$v = \frac{V_{max}[S_0^-]}{Km\Pi + [S_0^-]} \qquad [3]$$

Obviously, as the bulk substrate concentration is varied, Π varies as well, in such a way that Eq. 3 is not the equation of a hyperbola but mimics a sort of cooperativity.

As the ionic strength of the bulk phase is increased, the difference of electrostatic potential between the inside and the outside of the matrix tends to decrease, and therefore the electrostatic partition coefficient Π tends to approach unity. The bound enzyme then tends to behave exactly as if it were free in solution.

A more realistic situation, at least with unbroken cell walls or with cell wall

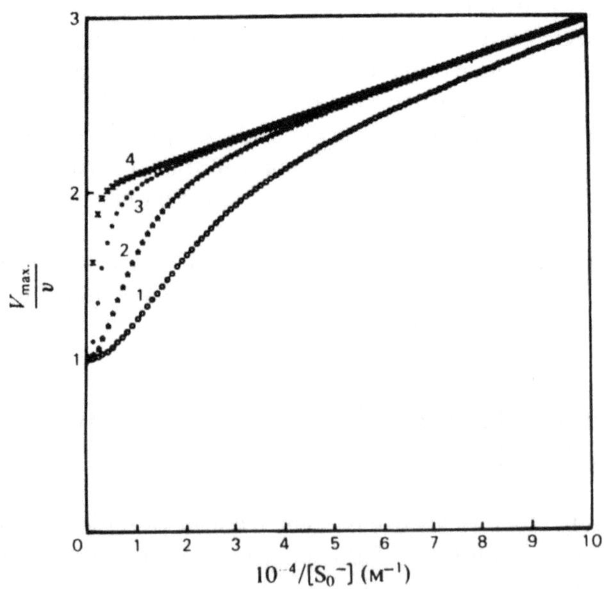

FIG. 29.1. Apparent cooperative kinetics of an enzyme heterogeneously distributed within a polyanionic matrix. Curves 1 to 4 correspond to increasing fixed charge densities. [See Ricard et al. (2).]

fragments, occurs when enzyme molecules are buried in the matrix and others are located "outside" the matrix, in close contact with the bulk phase. The overall reaction rate pertaining to this system is then

$$v = \frac{V_{max_1}[S_0^-]}{Km + [S_0^-]} + \frac{V_{max_2}[S_0^-]}{Km\Pi + [S_0^-]} \qquad [4]$$

where V_{max_1} and V_{max_2} are directly related to the "concentration," or to the number of the enzyme molecules, located at the surface of the matrix or buried in that matrix. This type of complex situation may give rise to either an apparent negative kinetic cooperativity or to a succession of positive and negative cooperativities (mixed cooperativity). This situation is illustrated in Fig. 29.1. These theoretical predictions may be submitted to experimental control with an acid phosphatase bound to sycamore, or soybean, cell walls.

Because of the presence of polygalacturonic acids in the cell walls, this organelle behaves as an insoluble polyanion. It may be shown easily by increasing the ionic strength of a suspension of sycamore or soybean cell wall fragments and titrating the efflux of protons in the bulk phase (Fig. 29.2). By removing calcium from cell wall fragments, one considerably enhances the

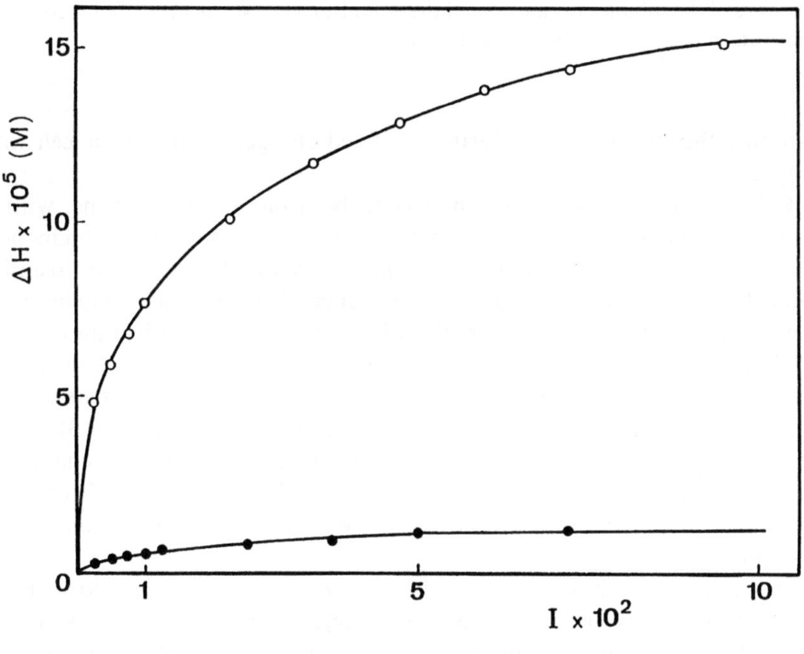

FIG. 29.2. Proton efflux from cell wall fragments induced by ionic strength jumps. The lower curve (●) shows the efflux from "native" cell walls loaded with calcium. The upper curve (○) shows the proton extrusion from "calcium-deprived" cell wall fragments. The pH is measured in the bulk phase. [See ref. Crasnier et al. (3).]

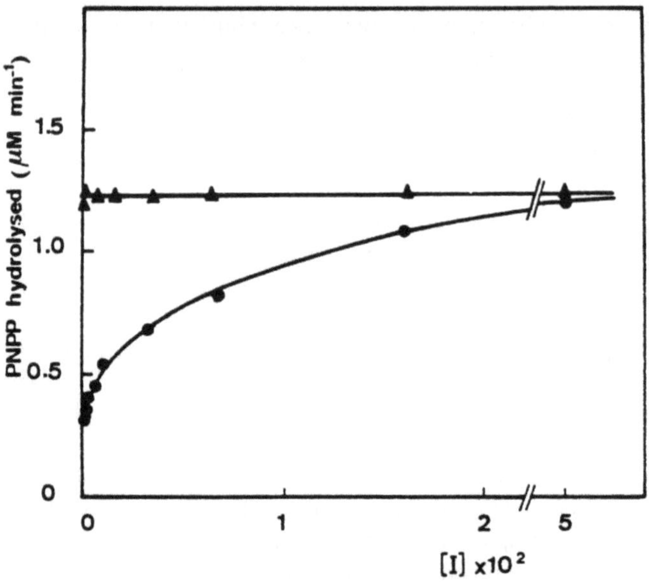

FIG. 29.3. Ionic strength dependence of cell wall acid phosphatase activity. The enzyme activities of the soluble enzyme (▲) and the bound enzyme (●) are measured at constant substrate concentration. [See Noat et al. (1).]

proton efflux by simply increasing the fixed charge density of the cell wall (Fig. 29.2).

Cell wall acid phosphatase appears to be a monomeric enzyme, with a molecular weight close to 100,000, which follows Michaelis-Menten kinetics. At "high" ionic strength, the enzyme bound to cell wall fragments or to entire unbroken cells displays the same kind of kinetic behavior. One may observe, however, that if the ionic strength is increased at fixed substrate concentration, the steady-state reaction velocity increases and reaches a plateau (Fig. 29.3). Conversely, in the same range of ionic strength the enzyme in free solution does not display any variation of reaction velocity (Fig. 29.3). These results are clearly in agreement with the view that the fixed negative charges of the cell wall matrix tend to repel the substrate of the enzyme (paranitrophenyl phosphate). This repulsion effect becomes weaker as the ionic strength in the bulk phase is increased.

If the kinetic behavior of the bound enzyme is studied at "low" ionic strength, one may observe an apparent negative cooperativity that is decreased, then suppressed, by "high" ionic strengths as the bound enzyme becomes more and more active. Indeed these effects are not intrinsic molecular properties of the enzyme but, rather, systemic properties. The good fit that occurs between Eq. 4 and the experimental results suggests that acid phosphatase molecules are heterogeneously spread in the cell wall, i.e., that some enzyme molecules are buried in the cell wall matrix, whereas others are directly in contact with the bulk phase. Because, according to this theoretical model, departure from Michaelis-Menten kinetics is explainable by the existence of electrostatic repulsion effects, one may estimate the extent of these effects by the extent of the deviation from Michaelis-Menten kinetics.

Because this apparent negative cooperativity of the bound acid phosphatase is dependent on the electrostatic repulsion effects of the fixed charges of the matrix to the substrate one may expect that changing this fixed charged density would modify the extent of negative cooperativity of the bound enzyme. This modification is what is actually observed. If calcium is removed from cell wall fragments, negative cooperativity is enhanced. Loading cell walls with calcium results in decreased cooperativity (Fig. 29.4). Departure from linearity of the reciprocal plots allows estimation of the electrostatic partition coefficient (Π), from either "native" or "calcium-deprived" cell wall fragments. These experimental results are clearly consistent with the predictions of the model discussed above.

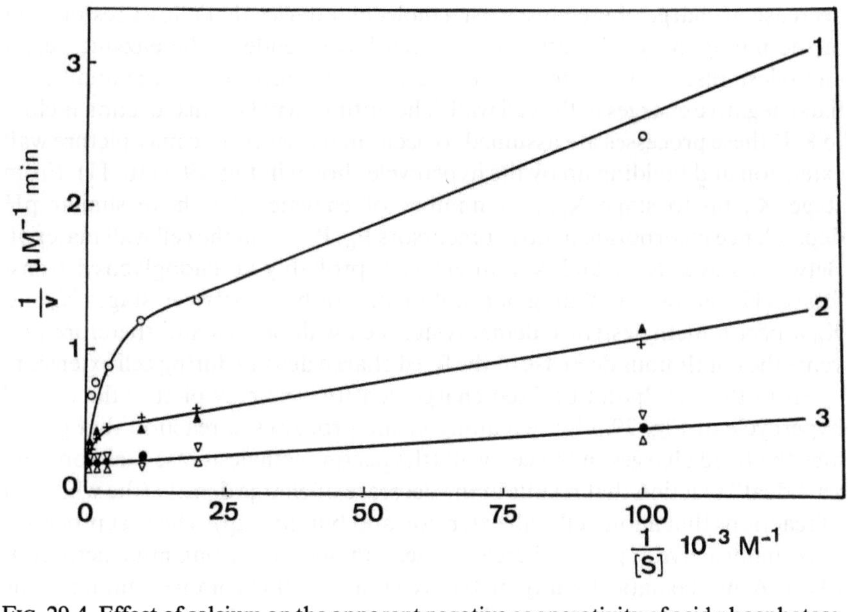

FIG. 29.4. Effect of calcium on the apparent negative cooperativity of acid phosphatase bound to plant cell walls. Curve 1 shows the apparent cooperativity of acid phosphatase bound to "calcium-deprived" cell walls. Curve 2 shows the apparent negative cooperativity of acid phosphatase bound to either "native" (+) or "calcium-deprived" (▲) wall fragments supplemented with $CaCl_2$. Curve 3 shows the suppression of the cooperativity of the bound enzyme at "high" ionic strength. [For experimental conditions and numerical values, see Crasnier et al. (3).]

Dynamics of Multienzyme Systems on the Cell Surface and the Extension of Plant Cell Walls

Electrostatic effects that have been discussed so far certainly play a major role in the mechanism of plant cell wall extension and building up. Isolated soybean cells in sterile culture represent a good system to study cell wall elongation and building up because when clumps of soybean cells are transferred on fresh culture medium the cells first extend and then divide in a more or less synchronous way. The detailed biochemical mechanisms of plant cell wall elongation are still not clear (7, 8), but there is little doubt that several biochemical events take place during the process of wall extension and building up. Neutral and methylated carbohydrate precursors are incorporated in the cell wall. Moreover, cellulose microfibrils must slide during cell growth, which must occur together with a loosening and autolysis of the wall. These processes are, at least in part, enzymatic in nature and are maximally active around pH 5. Because the incorporation of polysaccharide precursors occurs in neutral, methylated form, the fixed charge density of the wall should continuously decrease during elongation. In order to avoid this continuous decrease of charge, there must exist a molecular device that allows restoration of the initial charge density after the cell has extended. The enzyme pectin methylesterase, which demethylates pectins (9), generates a large number of fixed negative charges in the cell wall. The optimum pH of this reaction is close to 8. If these processes are assumed to occur in sequence, one may picture wall extension and building up by the hypercycle shown in Fig. 29.5 (10, 11). From stage X_0 up to stage X_{n-1}, a number of enzymes that have similar pH dependence incorporate neutral precursors $P_0, P_1, \ldots$ in the cell wall material. Between stages X_{n-1} and X_n, an enzyme, probably an endoglycosyl transferase (12), allows the sliding of cellulose microfibrils. Between stages X_n and X_0 a pectin methylesterase demethylates cell wall pectins and therefore prevents the continuous decrease of the fixed charge density during cell extension.

From the standpoint of fixed charge density, one may picture the overall hypercycle of Fig. 29.5 by two antagonistic processes: a reaction that generates the fixed charges in the cell wall (the pectin methylesterase reaction) and an "ideal" reaction that results in the decrease of charge density (the whole set of reactions that allow cell wall extension and building up). The first process is maximally active at pH 8, whereas the second one has its maximum activity at pH 5. A mathematical study of the dynamics of this complex multienzyme system shows the occurrence of a steep charge transition for a narrow pH range (Fig. 29.6). This situation typically corresponds to a phenomenon of zero-order ultrasensitivity or of high cooperativity (13–16) that takes place because in the same pH range, pectin methylesterase is inhibited by protons, whereas the wall-loosening enzymes are activated, or conversely.

This general model of the ionic control of plant cell wall extension and building up has three important implications that may be submitted to experimental control:

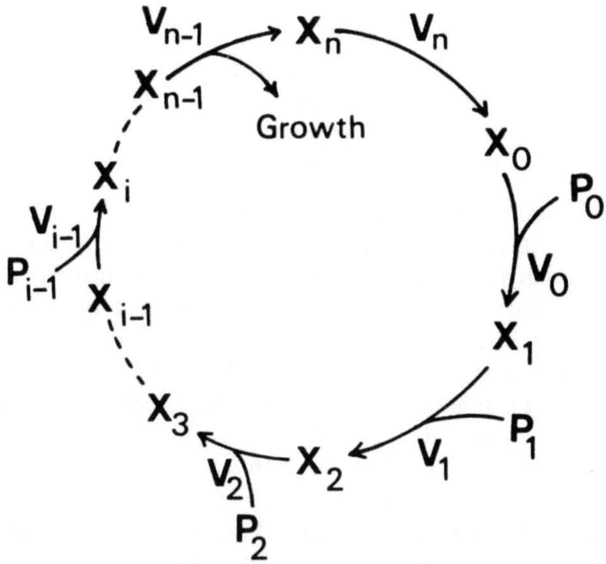

FIG. 29.5. Model hypercycle leading to quantitative expression of the ionic regulation of cell wall enzymes.

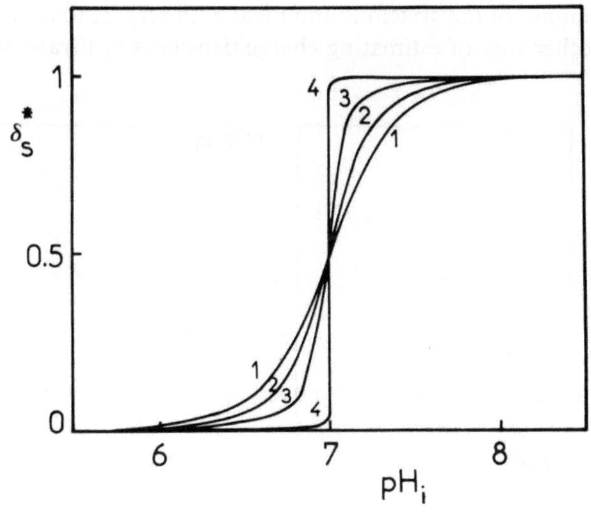

FIG. 29.6. Ultrasensitivity of the charge density to pH changes. Curves 1 through 4 correspond to decreasing values of the Michaelis constants of the various enzymes involved in cell wall extension and building up. [For numerical values, see Ricard and Noat (4).]

1. Pectin methylesterase in situ generates the fixed negative charges of the cell wall.
2. The building up of this cell wall charge density should display a strong cooperativity to slight pH changes.
3. There must exist an interplay between the Donnan potential of the cell wall and the activity of wall-loosening enzymes.

Upon raising the ionic strength in the bulk phase, the amplitude of proton efflux may be considered as an estimate of the fixed charge density of the cell wall. Pectin methylesterase has been isolated and purified from soybean cell walls. The enzyme is a monomeric protein, with a molecular weight of ~33,000, that is maximally active around pH 8. If this enzyme is really involved in the building up of the cell wall Donnan potential, one may expect that by stimulating its activity the cell wall charge density is increased—which is exactly what has been observed. If soybean cell wall fragments are prepared at pH 4.5, incubated at a pH higher than 4.5, and finally transferred to the initial pH 4.5 medium, the proton efflux observed when raising the ionic strength in the bulk phase is increased (Fig. 29.7). This increase shows that during incubation of cell wall fragments at a pH higher than 4.5 the increase in activity of pectin methylesterase in situ has resulted in a larger Donnan potential of the cell wall.

Ultrasensitivity in the appearance of a cell wall Donnan potential implies that the response of the system is amplified with respect to a slight variation of pH. Another way of estimating charge density is to titrate unmethylated

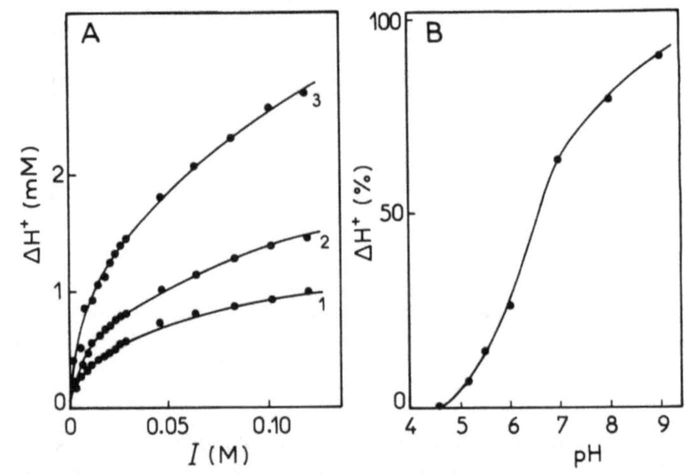

FIG. 29.7. Proton efflux from the cell wall and the fixed charges generated by the pectin methylesterase reaction. *A*: Proton efflux from cell wall fragments. Curves 1 through 3 correspond to cell wall fragments that have been preincubated at pH 4.5, 7.0, and 8.5, respectively. *B*: Variation of the amplitude of the proton efflux as a function of the pH of preincubation. [See Moustacas et al. (5).]

uronic acids in the cell wall. When the cell wall volume increases, the fixed charge density should decrease and therefore the local pH should increase, resulting in activation of pectin methylesterase and generating an increased charge density. If this process were not cooperative, one would have expected the fixed charge density to remain constant during cell elongation. Alternatively, if there is a cooperativity of the response, one would expect an increase of charge density together with an increase of the cell wall volume. This situation is exactly what may be observed. If the cell wall volume as well as the percent of unmethylated uronic acids in the wall are estimated, it is seen that during 5 days of cell culture, the percent of unmethylated uronic acids (i.e., the fixed negative charge density) increases together with the cell wall volume. After the fifth day of culture, the cell wall volume still increases, but the fixed charge density decreases (Fig. 29.8). As noted above, the simultaneous increase of cell wall volume and charge density necessarily implies the response of the system to slight pH changes to be cooperative. One may explain that, after the fifth day of culture, the charge density decreases as the cell wall extends by assuming that the local pH is then outside (below or above) the narrow pH range where the system sharply responds to slight pH changes.

Cell wall extension and building up involves wall loosening and autolysis. The extent of cell wall autolysis may then be considered a rough estimate of the activity of the "growth enzymes." When raising the ionic strength of the

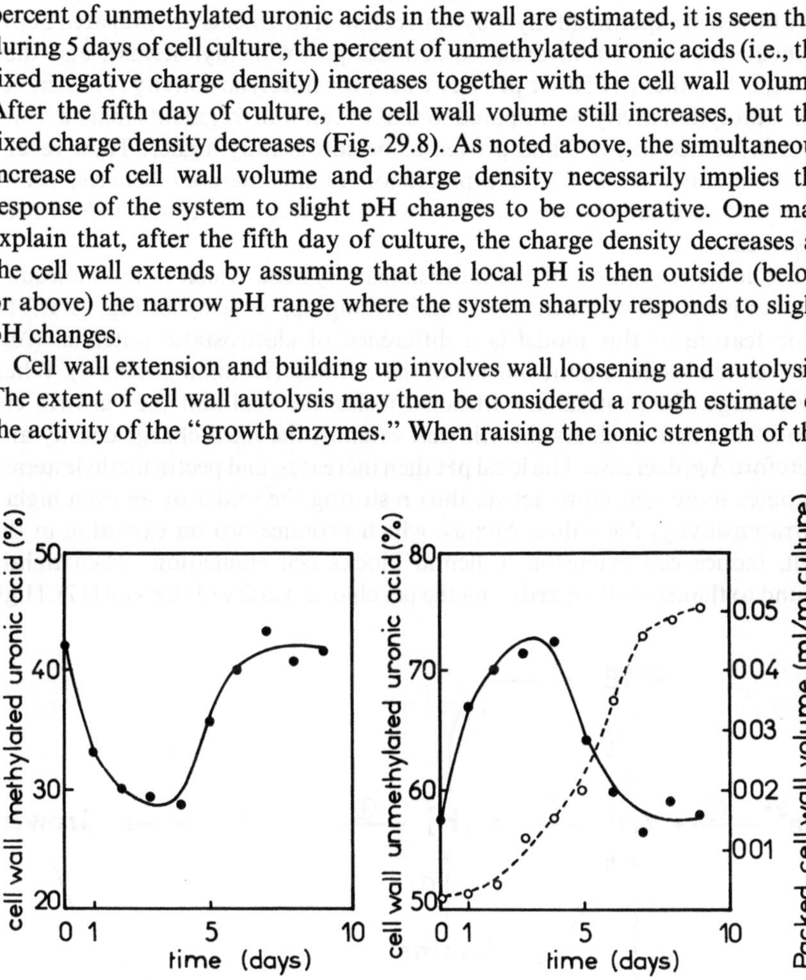

FIG. 29.8. Variation of methylated and unmethylated uronic acids during cell wall growth. *Left*: Variation of methylated uronic acids (as a percentage of total uronic acids). *Right*: Variation of unmethylated uronic acids (as a percentage of total uronic acids) (●) and of cell wall volume (○). The amount of unmethylated uronic acids represents an image of the fixed charge density. [For experimental conditions, see Moustacas et al. (5).]

bulk phase, the optimum pH of cell wall autolysis is lowered by about one pH unit. This result simply implies that the local proton concentration in the wall is larger at low than at high ionic strength. Therefore one would expect that after activating pectin methylesterase in situ the cell wall charge density should be increased, thus attracting protons and other cations in the cell wall. The local pH should then be lowered, which may in turn activate wall-loosening enzymes. The existence of this interplay between the activity of pectin methylesterase and that of wall-loosening enzymes is predicted by the model and may be tested experimentally. Soybean cell wall fragments are prepared at pH 5, incubated at pH 8 in order to activate pectin methylesterase, and then brought to either pH 5.0 or pH 5.5, where the autolytic activity is measured. This autolytic activity is compared to that of cell wall fragments that have not been incubated at pH 8 and proved to be consistently higher. These results unambiguously show that the prediction of an interplay between pectin methylesterase activity and wall-loosening enzyme activity has indeed been observed.

The ionic regulation of the multienzyme system, which is the molecular basis of plant cell wall extension and building up, is shown in Fig. 29.9. The basic feature of this model is a difference of electrostatic potential ($\Delta\psi$) between the inside and the outside of the cell wall (Donnan potential). When $\Delta\psi$ is large, the protons are attracted inside the wall and the "growth enzymes" are activated. As the cell wall extends, the fixed charge density and therefore $\Delta\psi$, decrease. The local pH then increases, and pectin methylesterase becomes more and more active, thus restoring the initial or an even higher (ultrasensitivity) $\Delta\psi$ value. Auxins, which produce proton extrusion in the wall, induce cell extension. Calcium blocks cell elongation, when tightly bound to the cell wall, by reducing the net charge density of the wall (17). High

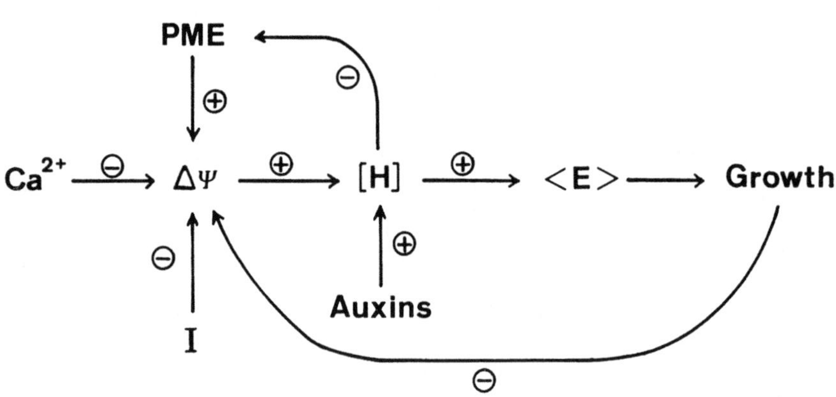

FIG. 29.9. Tentative model of the electrostatic and ionic control of the cell wall enzyme system. PME = pectin methylesterase; (+) activation or increase; (−) inhibition or decrease; I = ionic strength.

ionic strengths inhibit growth by suppressing the difference of electrostatic potentials between the inside and the outside of the wall.

Conclusions

An enzyme that acts on a charged substrate and that follows Michaelis-Menten kinetics in free solution may display cooperative kinetics when bound to an insoluble polyelectrolyte. These effects are due to electrostatic attraction or repulsion of substrate by the fixed charges of the matrix and are controlled by the ionic strength of the bulk phase.

Acid phosphatase from plant cell walls displays Michaelis-Menten behavior when free in solution but exhibits apparent negative cooperativity when bound to sycamore cell walls. This cooperativity disappears upon raising the ionic strength. This behavior is apparently due to repulsion of the substrate by the fixed negative charges of the cell wall. Calcium, which may be tightly bound to plant cell walls, modulates the fixed charge of the walls and therefore the kinetic cooperativity of acid phosphatase.

Plant cell wall extension is controlled by the Donnan potential of the wall, which regulates the activity of both the wall-loosening enzymes and pectin methylesterase. Activation of wall-loosening enzyme results in wall extension. When the cell wall extends, its fixed charge density decreases and pectin methylesterase becomes activated. The properties of these enzymes in the wall are controlled by their microenvironment.

References

1. Noat, G., Crasnier, M., and Ricard, J. (1980) *Plant Cell Environ.* 3:225–229.
2. Ricard, J., Noat, G., Crasnier, M., and Job, D. (1981) *Biochem. J.* 195: 357–367.
3. Crasnier, M., Moustacas, A. M., and Ricard, J. (1985) *Eur. J. Biochem.* 151: 187–190.
4. Ricard, J., and Noat, G. (1986) *Eur. J. Biochem.* 155:183–190.
5. Moustacas, A. M., Nari, J., Diamantidis, G., Noat, G., Crasnier, M., Borel, M., and Ricard, J. (1986) *Eur. J. Biochem.* 155:191–197.
6. Nari, J., Noat, G., Diamantidis, G., Woudstra, M., and Ricard, J. (1986) *Eur. J. Biochem.* 155:199–202.
7. Cleland, R. E. (1981) In *Encyclopedia of Plant Physiology, New Series: Plant Carbohydrates II*, Vol. 138, edited by W. Tanner and F. A. Loewus, pp. 255–273. Springer, Berlin.
8. Taiz, L. (1984) *Annu. Rev. Plant Physiol.* 35:585–657.
9. Goldberg, R. (1984) *Physiol. Plant* 61:58–63.
10. Eigen, M. (1971) *Naturwissenschaften* 33:465–523.
11. Eigen, M., and Schuster, P. (1979) *The Hypercycle*. Springer, Berlin.
12. Albersheim, P. (1976) In *Plant Biochemistry*, edited by J. Bonner and J. E. Varner, pp. 225–274. Academic Press, New York.

13. Chock, P. B., Rhee, S. G., and Stadtman, E. R. (1980) *Annu. Rev. Biochem.* 49:813–843.
14. Chock, P. B., and Stadtman, E. R. (1980) *Methods Enzymol.* 64B:297–325.
15. Goldbeter, A., and Koshland, D. E. (1981) *Proc. Natl. Acad. Sci. USA* 78:6840–6844.
16. Goldbeter, A., and Koshland, D. E. (1984) *J. Biol. Chem.* 259:14441–14447.
17. Demarty, M., Borvan, C., and Thellier, M. (1984) *Plant Cell Environ.* 7:441–448.

30
Energy Transduction by Electroconformational Coupling

R. DEAN ASTUMIAN, P. BOON CHOCK, HANS V. WESTERHOFF, and T. Y. TSONG

Free energy transduction is a fundamental biological process. Virtually everything an organism does involves converting energy from one form to another, and a function of many enzymes is the channeling of free energy through metabolic pathways in which it can be used to do work. The mechanism by which diverse chemical reactions are coupled into driving energy-requiring processes remains the topic of active and lively discussion. Bearing on this problem is the observation that application of an oscillating electric field (≈ 1 kHz) to a suspension of erythrocytes induced the Na^+,K^+-ATPase to mediate active transport of Rb^+ without detectable hydrolysis of ATP (1, 2), although the time average of the field was zero. This phenomenon was explained theoretically by Tsong and Astumian (3), who showed that an enzyme catalytic cycle that includes an electrically sensitive step can transduce energy from an oscillating electric field even if the reaction catalyzed is not directly coupled to the field. Many transmembrane proteins do in fact have large dipole moments, or net charge, that make them susceptible to influence by variation in the transmembrane potential.

Subsequently, we have refined and extended these calculations, introducing a physically explicit four-state model for translocation (4, 5). It was found that the primary requirement necessary for this behavior is an asymmetry in the stability of the four enzyme states, and that many transmembrane proteins can be expected to have characteristics allowing them to transduce energy from an oscillating (4) or stochastically fluctuating (5) electric field. It was not necessary that the process being catalyzed depend on the electric field directly. These considerations led us to propose a mechanism for free energy transduction based solely on coulombic interactions between two enzymes catalyzing a priori uncoupled reactions (5, 6).

In order to understand how electrical interactions between enzymes can give rise to free energy coupling, we formulate and analyze an physically explicit model in this chapter. First, the enzyme model and its kinetic parameters on which subsequent considerations are based is described. Then it is shown that the qualitative behavior of this enzyme when exposed to stochastic electrical fluctuation can be predicted by use of the King-Altmann-Hill diagram method. A quantitative numerical analysis demonstrates that our simple scheme, based solely on enzyme–enzyme coulombic interactions, is competent to transduce energy with efficiency comparable to that observed experimentally for physiological systems. Finally, we discuss the implications of our analysis in terms of other models for free energy transductions and in light of experimental results.

The theory described provides a general mechanism for free energy transduction via macromolecules. One characteristic differentiating this mechanism from many others is that the oscillation or nonequilibrium fluctuation of a thermodynamic parameter is sufficient to mediate free energy transduction, and that there need be no direct chemical or energetic coupling between the driving and driver reactions. Also, here we stress the perturbation of conformational equilibria *within* the normal enzyme catalytic cycle.

Enzyme Model

We consider a model (kinetic diagram) for an enzyme-catalyzed reaction that is sufficiently general to allow broad application and yet simple and specific enough to facilitate numerical calculation. We have chosen the catalyzed transport of an uncharged substance (S) across a membrane as the particular reaction to be emphasized (3). We show that in many cases this system "catalyzes" the transport of S against its gradient, thereby doing work, when exposed to *nonequilibrium* electric fluctuations. In order to analyze how this phenomenon occurs, we must use a model that is both simple and realistic. Such a mechanism requires at least four states: two in which the binding site is unoccupied but exposed to inside or outside, and two similar conformations with the binding site occupied by ligand (S). For a pump mechanism, as opposed to diffusion through a passive channel, this situation represents the minimum number of states physically required, and there could be additional states within the catalytic cycle. Also, it should be noted that in some cases it is allowable to consider only two or three states in arriving at a reasonable approximate mathematical description of the kinetic behavior of the model. In the description here, we have retained four states, as the transitions that involve translocation of the binding site from inside to outside exposure may depend on the membrane potential but not the ligand concentration, and those involving binding or dissociation depend on ligand concentration but are taken to be independent of the membrane potential. Under a simple assumption discussed below, all of the rate constants can be written in terms of three parameters. This number represents a considerable reduction over the a priori number of coefficients necessary to describe the system and results in a much clearer understanding of how the various parameters interact to allow energy transduction from a fluctuating electric field. The generality of the conclusions drawn from this work is undisturbed by the assumption.

A specific physical conceptualization of the catalytic mechanism discussed (4, 6, 7) is shown in Fig. 30.1A, where we have also given the rate constants for

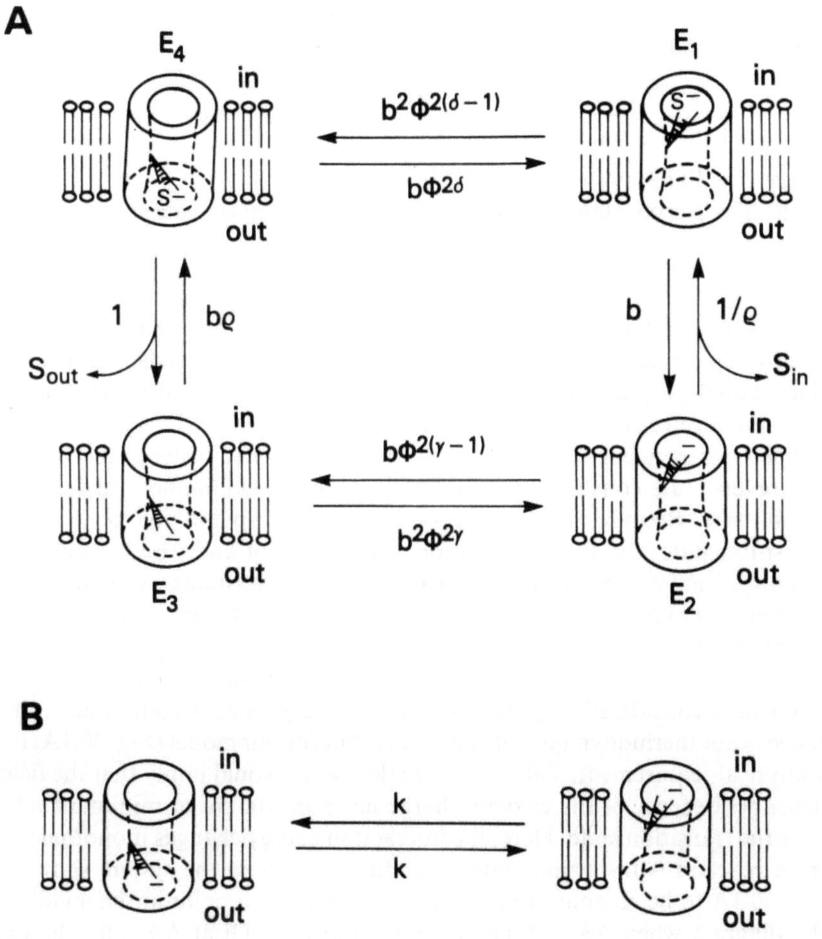

FIG. 30.1. Diagrams for the "translocator" and "generator." *A*: Physical conceptualization of an enzyme cycle for the translocation of "S" across a membrane. The only conformational difference between enzyme states 4 and 1 (and 3 and 2) is the location of a "point charge" on the protein. Hence the transitions $4 \rightleftharpoons 1$ and $3 \rightleftharpoons 2$ are influenced by the electrical field. We consider that the charge is displaced by exactly the membrane thickness (d) between these pairs of states. The rate constants were determined as described in the text and in ref. 4. *B*: "Generator" charge that can move back and forth within the membrane, the electrical field of which influences the transitions in *A*.

the transitions. The dependence of the equilibrium coefficients on the membrane potential ($\Delta\psi$) may be explicitly written (4):

$$\frac{K_{14}(\Delta\psi)}{K_{14}(O)} = \frac{K_{23}(\Delta\psi)}{K_{23}(O)} = \exp\left[\frac{F\Delta\psi}{(RT)}\right] \equiv \phi^2 \qquad [1]$$

where F = Faraday's constant
R, T = gas constant and Kelvin temperature, respectively
$\phi \equiv \exp\left(\frac{-F\Delta\psi}{2RT}\right)$

We assume that the transitions $1 \rightleftharpoons 2$ and $3 \rightleftharpoons 4$ do not involve a change in the macroscopic polarizability and hence do not depend on $\Delta\psi$. Also, the equilibrium ratio S_{in}/S_{out} does not depend on $\Delta\psi$, and thus there is no direct coupling between the electric field and the translocation of S.

The $\Delta\psi$ dependence expressed in Eq. 1 is apportioned between the forward and reverse rate constants depending on the electrical moment of the transition state. Because the binding of S may perturb the character of the transition state, the two distributions for the $1 \rightleftharpoons 4$ and $2 \rightleftharpoons 3$ transitions, given by δ and γ, need not be identical; δ and γ are constants we assume to be independent of $\Delta\psi$. This assumption implies that the effective dipole moments of "reactant," "transition," and "product" states do not change upon shifts in $\Delta\psi$, and only the relative energies between the three states vary.

We have considered only the case where δ and γ vary between 0 and 1. This range is not thermodynamically necessary, but for our model (Fig. 30.1A) it is a physical requirement. Values outside this range would imply that the field-dependent energy of the enzyme charge has a maximum or minimum value near the transition state. Here, the interaction energy changes monotonically from reactant → transition state → product. We chose the system shown in Fig. 30.1A to be, at equilibrium, symmetrical about a point in the middle of the diagram when $\Delta\psi = 0$, i.e., [4] = [2] and [1] = [3] at $\Delta\psi = 0$. The two parameters b (the "bias" factor) and ρ (the "output" factor) (4) are given by:

$$b \equiv \left(\frac{\overline{[4]}}{\overline{[3]}}\right) = \left(\frac{\overline{[2]}}{\overline{[1]}}\right) = \left(\frac{\alpha_{34}}{\alpha_{43}}\right)_{\rho=1} = \left(\frac{\alpha_{12}}{\alpha_{21}}\right)_{\rho=1} \qquad [2]$$

and

$$\rho = \exp\left[\frac{\Delta\mu_s}{(2RT)}\right] \qquad [3]$$

where α_{ij} = general transition coefficients for the transition $i \rightleftharpoons j$
b = constant for a particular enzyme
$\overline{[i]}$ = probability of the enzyme to be in state i if the system were at equilibrium in the absence of an electrical field

Arbitrarily normalizing by setting $\alpha_{43} = 1$ and using the condition of microscopic reversibility (8), the rate coefficients were determined. The term b is a characteristic of the enzyme and represents the fact that S may bind with different affinities on the two sides of the membrane. $RT \cdot \ln(b)$ is essentially the same concept as that which Jencks has referred to as interaction energy (9). ρ is a measure of how far the catalyzed reaction is from equilibrium. The parameter ϕ deals with the differences in interactions between the various enzyme states and an external parameter, or with interactions between two enzymes or subunits. In the present work, the external parameter was taken as the transmembrane potential ($\Delta\psi$) or as intermolecular coulombic forces, but ϕ may be interpreted much more generally.

Using this model we can now discuss how an enzyme can act as a mediator for free energy conversion between two reactions that are neither chemically nor energetically coupled.

Diagram Method Analysis of Energy Coupling Mediated by the Interactions Between Two Proteins

Two reactions may be coupled by interactions between the enzymes catalyzing them, even if the reactions themselves are a priori neither chemically nor energetically coupled (5, 6). From this point of view, we might expect that free-energy transduction can be understood in terms of the "basic" enzyme parameters of two proteins (b in our model) and interaction energies that depend on the reciprocal forces exerted on one enzyme by the other, or on an enzyme by an "external" parameter. We have denoted the normalized exponent of this interaction energy (E_{int}) by $\phi^2 = \exp(E_{int}/RT)$. If the catalytic cycle for enzyme A consists of the four states 1, 2, 3, and 4 and enzyme B of the four states 1′, 2′, 3′, and 4′, the complex AB would be described by a total of 16 states: 11′, 12′, 13′, etc., as shown in Fig. 30.2A. The equilibrium constants between any two states of enzyme A (or B) would be modified from those of the isolated enzyme by an interaction energy with enzyme A (or B). For example, if the physical presence of enzyme A in state 1 stabilizes state 2′ of enzyme B more than 1′, the equilibrium $11' \rightleftharpoons 12'$ would be shifted further to the right than the isolated B equilibrium $1' \rightleftharpoons 2'$, and we could write $K_{11',12'} = K_{1',2'} \cdot \phi^2{}_i$. One can assign ϕ values for each transition and apportion the influence to the rate constants.

The steady-state behavior (cycle fluxes, state probabilities, etc.) of a system described by such a diagram may be determined analytically by the so-called diagram method (8, 10). Unfortunately, for a 16-state model the algebra involved would be extremely tedious. Such a model can, however, be solved in a mathematically equivalent fashion by the matrix-inversion method (11); and numerical results obtained using this method are presented later. Here let us focus on a simpler model for which analytical equations describing the steady-state behavior can be obtained. We consider the model arising from the combination of a four-state diagram (Fig. 30.1A) with a two-state diagram (Fig. 30.1B), yielding an eight-state model, shown with appropriate and

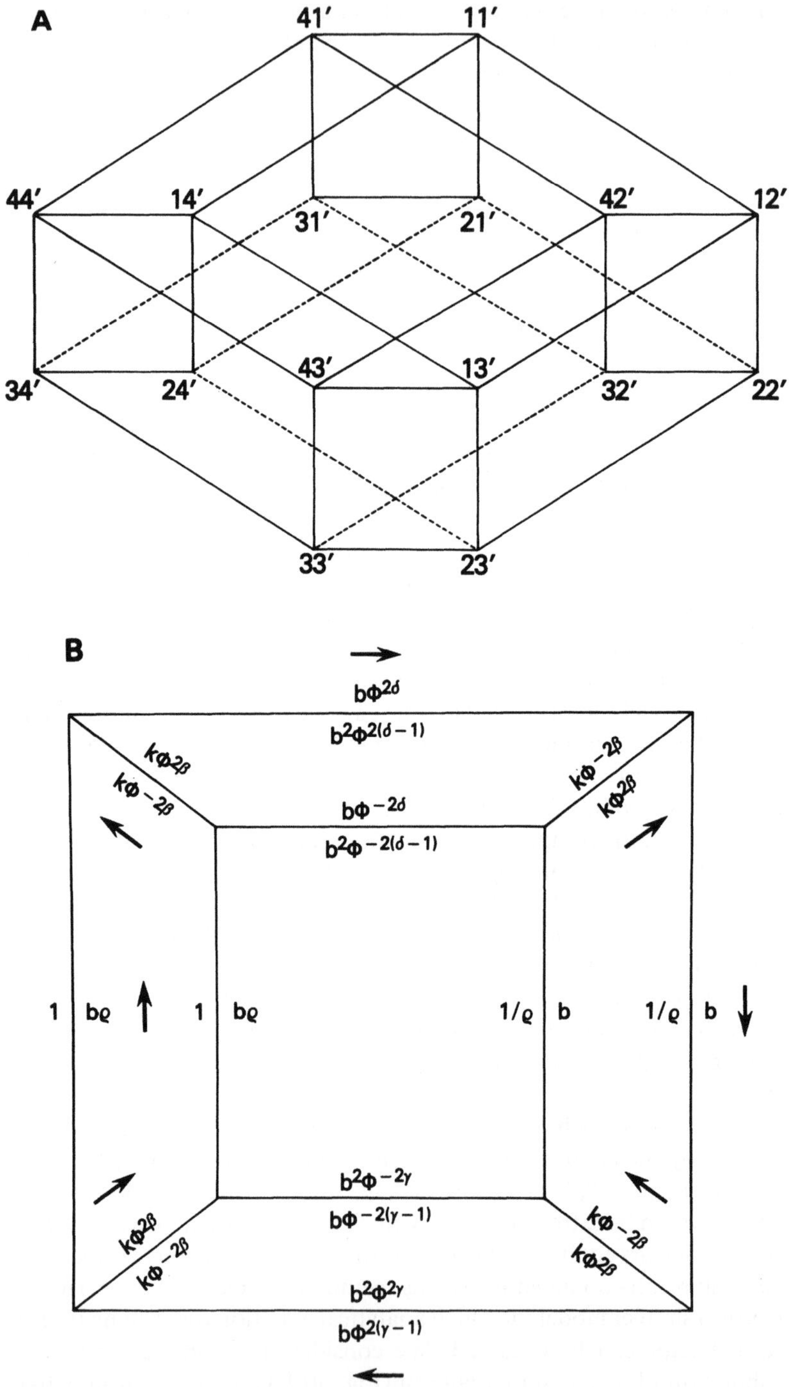
A
41′
11′
44′
14′
31′
21′
42′
12′
34′
24′
43′
13′
32′
22′
33′
23′
B
bΦ^{2δ}
b²Φ^{2(δ−1)}
kΦ^{2β}
kΦ^{−2β}
bΦ^{−2δ}
b²Φ^{−2(δ−1)}
kΦ^{−2β}
kΦ^{2β}
1
bϱ
1
bϱ
1/ϱ
b
1/ϱ
b
b²Φ^{−2γ}
bΦ^{−2(γ−1)}
kΦ^{2β}
kΦ^{−2β}
kΦ^{−2β}
kΦ^{2β}
b²Φ^{2γ}
bΦ^{2(γ−1)}

thermodynamically consistent rate coefficients in Fig. 30.2B. Such a model could result from complexing a four-state enzyme, or translocator (Fig. 30.1A), with any "field generator" protein, such as that shown in Fig. 30.1B (5), electron transport protein [e^-_{in} or e^-_{out} being the two states (4)], or a channel (3). In order to be able to calculate all the ϕ values, we use the combination of Fig. 30.1A with Fig. 30.1B in our derivation. A numerical value for ϕ may be obtained from knowledge of the distances between the two point charges on the protein, their magnitudes, and the effective dielectric constant, by use of Coulomb's law. For simplicity we assumed both $\alpha_{1'2'}$ and $\alpha_{2'1'}$ to be equal to k, as shown in Fig. 30.1B for the isolated protein, and assigned the interaction parameter, ϕ^{β}. At equilibrium $\beta = 1$, and the interactions would be totally reciprocal [we have termed it "endogenous noise" (5, 6)]. If the generator protein (Fig. 30.1B) were being driven by an external energy source, we might have $\beta = 0$ such that the transition from + to − (in to out) would be independent of the translocator state, a situation we have termed "autonomous noise" (5, 6), which is similar to the case of chemical oscillations in which one oscillator is unaffected by the other (12). Here, we have considered only the case of fluctuations symmetrical about zero membrane potential. Modification to the case of arbitrary background potential can be treated simply by inserting a constant $\Delta\psi$ into the equations (6, 7, 13).

King-Altmann-Hill Diagram Method

Having specified all of the rate coefficients for the diagram in Fig. 30.1A (see also Fig. 30.2B), we may use the King-Altmann-Hill diagram method to solve for the steady-state cyclic cluxes of the system. When the inner or outer box are considered separately, $\rho_s > 1$ (< 1) would cause net clockwise (counter-clockwise) cycling. If we look at the diagram as a whole and find cycling in a direction opposite to that expected based on the value of ρ_s, it would be

◁ FIG. 30.2. Combined translocator and generator diagrams. *A*: The result of combining two four-state translocators to allow for the coupling of one process to the other. In the calculations for Fig. 30.4B, the downhill translocation of a substance P was shown to drive the thermodynamically uphill translocation of S, even though the two processes, in the absence of considerations about the enzyme, are unrelated. This coupling was due solely to electrostatic interactions between the two translocators. *B*: Model describing the influence of noise on an enzyme that would result from combining Figs. 30.1A and B. If the system were at equilibrium, β would be equal to 1. In this case, it was shown that none of the individual cycles (Fig. 30.3) would have non-zero driving forces, and we termed the generator fluctuations "endogenous" noise. In the case of "autonomous" noise (a situation in which the generator is uninfluenced by the transitions of the translocator), β would be 0, and transport of S against its gradient was shown to occur, demonstrating free energy transduction from the nonequilibrium electrical fluctuations.

indicative of free energy transduction from the fluctuations mediated by the field generator (e.g, the protein shown in Fig. 30.1B). Because the individual cyclic fluxes and the net transport flux are continuous functions of all the parameters, an alternative way of proving energy transduction is to show that at $\rho = 1$("equilibrium") there still exists net cyclic flux (in either direction).

In order to solve for the steady-state fluxes, we first decompose the diagram into its individual single cycles. We find that there are 28 cycles (shown in Fig. 30.3). The thermodynamic driving force (X_κ) for any cycle (κ) is given by (8):

$$X_\kappa = RT \cdot \ln\left(\frac{\Pi_{\kappa,+}}{\Pi_{\kappa,-}}\right) \qquad [4]$$

$\Pi_{\kappa,+}$ + ($\Pi_{\kappa,-}$) is the product of the (pseudo) unimolecular rate constants around the cycle in a clockwise (counterclockwise) direction. The net flux around the cycle is:

$$J_\kappa = N \cdot (\Pi_{\kappa,+} - \Pi_{\kappa,-}) \cdot \frac{\sum_\kappa}{\sum} = N \cdot \Pi_{\kappa,-} \cdot \left[\exp\left(\frac{X_\kappa}{RT}\right) - 1\right] \frac{\sum_\kappa}{\sum} \qquad [5]$$

where N = number of molecules of enzyme
$\sum$ = sum of all the state probabilities

X_κ and $\Pi_{\kappa,-}$ for our system are given in Table 30.1. We have also tabulated the product $\Pi_{\kappa,-}[\exp(X_\kappa/RT) - 1]$, which contains the contribution from the kinetic parameters of the transitions *within* cycle κ, and the first eight cycle probabilities ($\sum_\kappa$), which contain information about the flux into that cycle.

Of the 28 cycles in Fig. 30.3, only 16 (Nos. 1–16) involve net transport, and 8 (Nos. 1–8) of these cycles have driving forces dependent on ϕ and ρ. Owing to microscopic reversibility, the driving force for each cycle must individually be zero at equilibrium. This fact immediately tells us that when $\beta = 0$ the system must be away from equilibrium. Input of free energy is necessary to maintain $\beta = 0$. We do not concern ourselves with the source of this input free energy for the time being. For now, let us simply consider whether the net flux, $J_{trans} = \sum_{J_\kappa}$, is not zero when $\beta = 0$, thereby demonstrating free energy transduction. Parameters for calculating J_κ have been given in Table 30.1, as well as the equation for the unnormalized transport flux. From the special cases we see that so long as $b \neq 1$, energy-driven fluctuations induce net cyclic flux at $\rho = 1$. Even if $b = 1$, having $\gamma \neq \delta$ is sufficient to allow for cycling (13). Interestingly, under this condition the relative magnitude of γ versus δ determines the direction of cycling. What we have proved is that our model is competent to transduce free energy near $\rho = 1$, and that the asymmetry necessary is not very restrictive. Chen (13) has extended our earlier work (5), discussing a few additional symmetry considerations in the context of a more general model. He also showed that, as expected, analysis in terms of a stochastic formulation confirms our findings. His considerations did not address the question of whether such an enzyme system could (at $\rho \ll 1$; $\rho = 0.13$ is the equivalent of a 10-mV electrochemical gradient) mediate efficient energy transduction.

TABLE 30.1. Parameters for calculating cyclic fluxes ($k = 1, \beta = 0$).

κ	X_κ	$\pi_{\kappa,-}$	$\pi_{\kappa,-}[\exp(X_\kappa/RT) - 1]$	$\sum_\kappa$ for Cycles 1–8
1–4	$2RT\ln(\rho \cdot \phi^2)$	$b^4\phi^{2(\delta-\gamma-1)}\rho^{-1}$	$b^4\phi^{2(\delta-\gamma+1)}(\rho - \phi^{-4}\rho^{-1})$	$\sum_1 = [(b^2\phi^{-2(\delta-1)} + k + b)(b^2\phi^{2\gamma} + k + b\rho)]$
5–8	$2RT\ln(\rho/\phi^2)$	$b^4\phi^{-2(\delta-\gamma-1)}\rho^{-1}$	$b^4\phi^{2(\gamma-\delta+1)}(\rho\phi^{-4} - \rho^{-1})$	$\sum_2 = [(b\phi^{-2\delta} + k + 1)(b\phi^{2(\gamma-1)} + k + 1/\rho)]$
9, 13, 15, 16	$2RT\ln(\rho)$	$b^4\phi^{-2(\delta+\gamma-1)}\rho^{-1}$	$b^4\phi^{-2(\delta+\gamma-1)}(\rho - \rho^{-1})$	$\sum_3 = [(b\phi^{-2\delta} + k + 1)(b^2\phi^{-2(\delta-1)} + k + b)] - b^3\phi^{-2(2\delta-1)}$
10–12, 14	$2RT\ln(\rho)$	$b^4\phi^{2(\delta+\gamma-1)}\rho^{-1}$	$b^4\phi^{2(\delta+\gamma-1)}(\rho - \rho^{-1})$	$\sum_4 = [(b^2\phi^{2\gamma} + k + b\rho)(b\phi^{2(\gamma-1)} + k + 1/\rho)] - b^3\phi^{2(2\gamma-1)}$
17–20, 25, 27	0	—	0	$\sum_5 = [(b\phi^{2\delta} + k + 1)(b\phi^{-2(\gamma-1)} + k + 1/\rho)$
21	$4RT\ln(\phi)$	$b^4\phi^{-2}\rho$	$b^4(\rho\phi^2 - \rho\phi^{-2})$	$\sum_6 = [(b^2\phi^{2(\delta-1)} + k + b\rho)(b^2\phi^{-2\gamma} + k + b)$
22	$-4RT\ln(\phi)$	$b^4\phi^2\rho$	$b^4(\rho\phi^{-2} - \rho\phi^2)$	$\sum_7 = [(b^2\phi^{-2\gamma} + k + b\rho)(b\phi^{-2(\gamma-1)} + k + 1/\rho)] - b^3\phi^{-2(2\gamma-1)}$
23	$-4RT\ln(\phi)$	$b^4\phi^2\rho^{-1}$	$b^4(\rho^{-1}\phi^{-2} - \rho^{-1}\phi^2)$	$\sum_8 = [(b\phi^{2\delta} + k + 1)(b^2\phi^{2(\delta-1)} + k + b)] - b^3\phi^{2(2\delta-1)}$
24	$4RT\ln(\phi)$	$b^4\phi^{-2}\rho^{-1}$	$b^4(\rho^{-1}\phi^2 - \rho^{-1}\phi^{-2})$	
26	$4RT\ln(\phi)$	$b^4\phi^{-2}$	$b^4(\phi^2 - \phi^{-2})$	
28	$-4RT\ln(\phi)$	$b^4\phi^2$	$b^4(\phi^{-2} - \phi^2)$	

$$[J_{\text{Trans}} \cdot \textstyle\sum/(b^5k^2)]_{\rho=1} = Z = (\phi^2 - \phi^{-2}) \cdot \{b^3(\phi^2 - \phi^{-2}) + [b^2 + (2k+1)b] \cdot (\phi^{2(1-\gamma)} + \phi^{2\delta} - \phi^{-2\delta} - \phi^{-2(1-\gamma)}) - [b + (2k+1)](\phi^{2(1-\delta)} + \phi^{2\gamma} - \phi^{-2\gamma} - \phi^{-2(1-\delta)}) - b(\phi^2 - \phi^{-2})\}$$

If $\gamma = \delta = 0.5$: $Z = (b-1)(\phi^2 - \phi^{-2})^2 \cdot [b(b+1) + 2(b+2k+1)/(\phi + \phi^{-1})]$

If $\gamma = \delta = 0$, or $\gamma = \delta = 1$: $Z = (\phi^2 - \phi^{-2})^2[b^3 + b^2 + (2k-1)b - (2k+1)]$

If $\gamma = 1, \delta = 0$: $Z = (\phi^2 - \phi^{-2})^2[b^3 - 3b - 4k - 2]$

If $\gamma = 0, \delta = 1$: $Z = (\phi^2 - \phi^{-2})^2[b^3 + 2b^2 + (4k+1)b]$

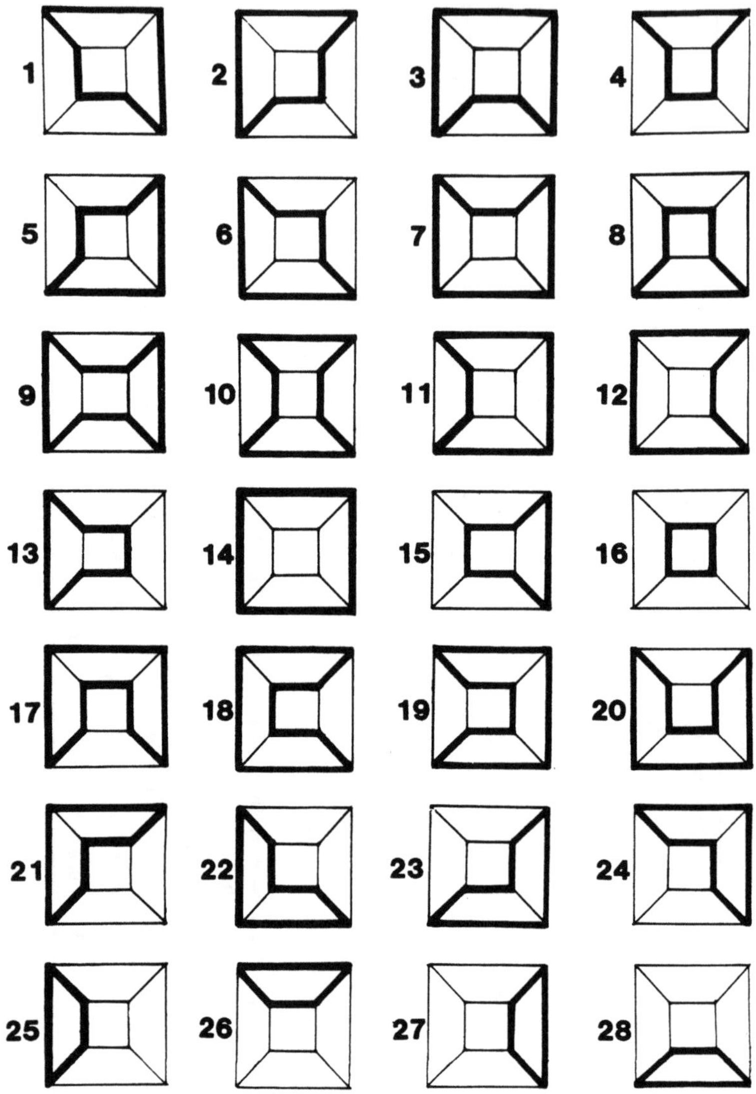

FIG. 30.3. Individual cycles resulting from Fig. 30.2B. The driving forces and unnormalized fluxes for these cycles are given in Table 30.1, along with the net flux for some special cases.

The steady-state properties for dynamic systems can also be determined analytically by inversion of the matrix resulting from the set of pseudo-first-order differential equations describing the kinetics of the system. This matrix inversion can be facilely accomplished by computer, and thus we can readily explore the properties of both the 8- and 16-state models at arbitrary input and output free energies.

If our 8-state model were to describe the work done by an enzyme due to the stochastic opening and closing of an ion channel, the input free energy would be $F(\Delta\psi_{closed} = \Delta\psi_{open}) = RT\ln(\phi^2)$. If this fluctuation were symmetrical about zero potential, we could calculate from our model the output flux at arbitrary ρ, with the output free energy, $\Delta G_{out} = -RT\ln(\rho^2)$. Results of a sample calculation in which the transport flux (J_{out}/ν) and an effectiveness (ξ) are plotted versus $RT \cdot \ln(\phi^2)$ are shown in Fig. 30.4A. The opening frequency (ν) is equal to the steady-state total flux from inner to outer box (or vice versa), which is 0.5 k. ξ is calculated according to the equation

$$\xi = -J_{out} \cdot \frac{\ln(\rho^2)}{\ln(\phi)} \qquad [6]$$

The parameter values used in the calculation are given in the legend to Fig. 30.3A. The standard measure of thermodynamic efficiency, $\eta = J_{out} \cdot \Delta G_{out} / (J_{in} \cdot \Delta G_{in})$, cannot be calculated for "autonomous noise," as the assumption of $\beta = 0$ implies an essentially infinite input free energy. Nevertheless, there are many cases (e.g., perhaps redox driven electron transport) in which the input free energy-induced fluctuations may reasonably be approximated as "autonomous noise."

We have also performed calculations for the 16-state model, described elsewhere (5, 6), for which η can be calculated, and the results are shown in Fig. 30.4B. Here, two translocator enzymes, one for "S" and one for "P," are considered to be separated by a distance of ≈ 15 Å, with each enzyme carrying a -2 charge. Using Coulomb's law, ϕ is calculated to be approximately 1600. The calculation demonstrates that a gradient of S or P can drive the formation of a gradient of the other, where the only interaction between the systems is the mutual coulombic forces between the two enzymes. The model can equally well describe a proton pump coupled to an ATPase, where the output reaction would be ATP synthesis (3, 6).

The point we emphasize here is that the efficiency of energy conversion can be as high as values seen experimentally, even though we have considered only the coulombic interactions between two neighboring enzymes. Also, we see that essentially complete coupling is achieved with physiologically reasonable values for the various parameters. The model on which these calculations are based has been discussed in greater detail elsewhere (5, 6), and a more complete thermodynamic analysis will be published in a forthcoming paper (R. D. Astumian and H. V. Westerhoff, in preparation).

Conclusion and Summary

Mechanisms for free energy transducion can be distinguished based on the means by which the "input" and "output" reactions interact. In both chemical (14) and direct chemiosmotic mechanism (15), the interactions are primarily between the substrates and products of the coupled reactions, and the main function of the enzymes involved is to facilitate those reactions. Most other

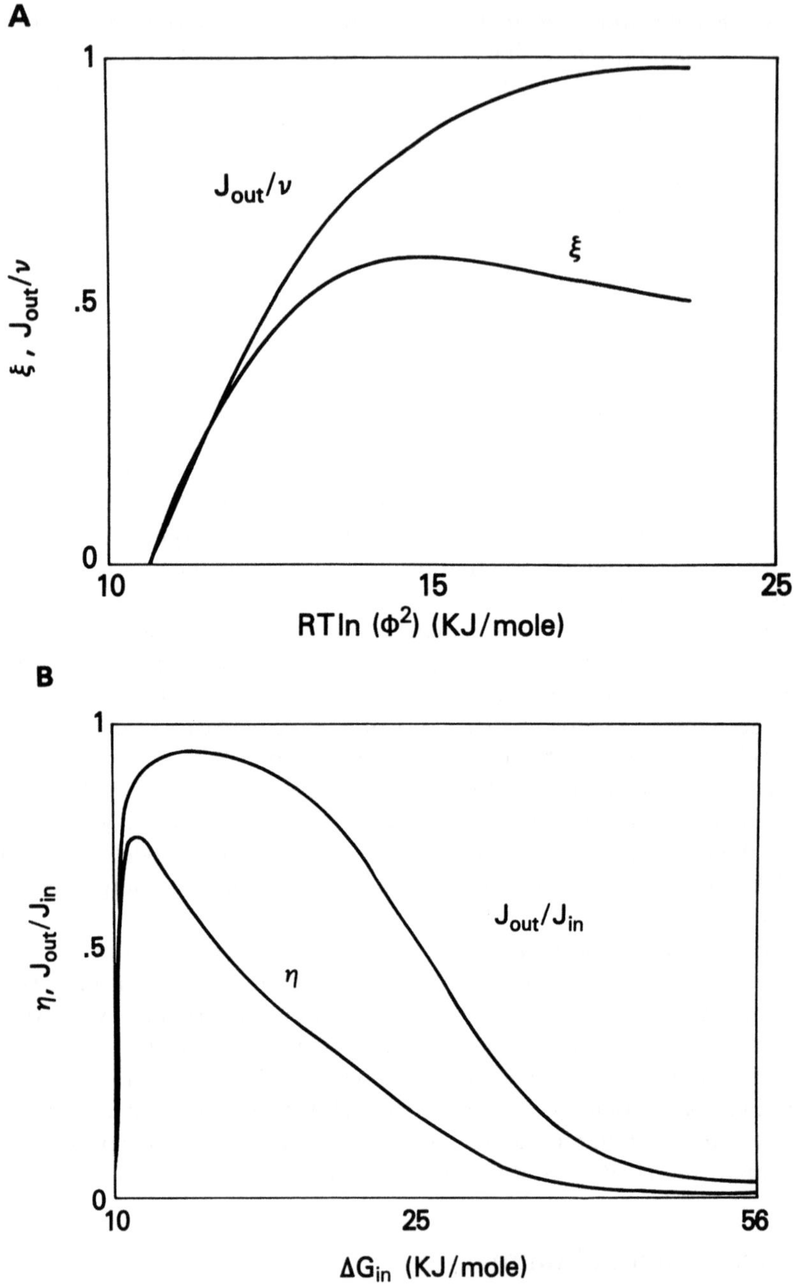

FIG. 30.4. Results of calculations for Figs. 30.2A and B. *A*: Plot of effectiveness (ξ) and the normalized flow (J_{out}/ν) versus $R \cdot T \cdot \ln(\phi^2)$ for the diagram in Fig. 30.2B. The parameter values were $b = 500$, $\rho = 0.13$, $k = 1$, and $\gamma = \delta = 0.5$. *B*: Plot of efficiency (η) and flow ratio (J_{out}/J_{in}) versus input free energy (ΔG_{in}) for the diagram in Fig. 30.2A. The parameter values were $b = 5000$, $\phi = 1600$, and $\rho_{out} = 0.13$; all the field-dependence apportionment factors (i.e., γ and δ) were 0.5.

mechanisms involve obligatory conformational coupling steps. For example, in the Ca^{2+}-ATPase, it is proposed that phosphorylation of the enzyme by ATP of necessity involves a conformational change in which the binding site for Ca^{2+} goes from being exposed to the "outside" to being exposed on the "inside" (16). Such a mechanism may well hold for many energy-transducing enzymes (17), where conformational transitions at one site go "hand in hand" with transitions at another site, as proposed by Boyer for the mechanism of ATP synthesis by F_0F_1-ATPase (18).

In the mechanisms we have discussed, the input and output reactions are neither chemically nor energetically coupled. The two active sites catalyzing these reactions are mechanically independent, and all possible states arising from their permutations are explicitly considered. For the sake of simplicity, the only *thermodynamic* interactions considered were coulombic, and yet calculations revealed that such a mechanism can give rise to efficient free energy transduction, with flow ratios approaching unity. The ideas exposed here allow for an understanding of how two fundamentally unrelated processes can be enzymatically coupled. The specific physical force (coulombic) responsible for the coupling and properties sufficient for an enzyme to have in order for it to mediate efficient energy transduction were discussed. These basic characteristics of our model can be compared to general schemes for energy coupling (9, 16, 19, 20). For example, the kinetically "blocked" transitions discussed by Hammes (17) can be seen to arise in our case from coulombic repulsions experienced in transitions leading to states with both charged groups on the same side of the membrane. Also, the intrinsic asymmetry in our model, denoted by "b," is equivalent to the interaction energy stressed by Jencks (9). Thus although the scheme described here is not general by any means, it might be of some use in analyzing some of the physical interactions that might play a role in determining energy coupling mechanisms.

A nonequilibrium fluctuation in any thermodynamic parameter is in principle sufficient to drive many enzymes to do work (5). Let us look at a specific case by which such nonequilibrium fluctuations would occur and contrast it to a source of equilibrium fluctuations. Phosphorylation-dephosphorylation of an enzyme, where the net chemical reaction is ATP $\rightarrow$ ADP + P_i, would produce nonequilibrium electrical fluctuations (the ligands are charged). The reciprocal forces preventing equilibrium fluctuations from doing work would be overcompensated by the energy of hydrolysis of ATP. On the other hand, the simple binding and release of an allosteric ligand, e.g., Mg^{2+}, would result in equilibrium noise (the binding of Mg^{2+} would be influenced by the state of the enzyme as much as the enzyme conformational equilibrium would be influenced by the binding of ligand), which cannot do work. This statement would be true for nonmembrane and membrane proteins alike. It represents one more possible advantage of regulation by covalent modification over regulation by small molecule allosteric ligands (R. D. Astumian and P. B. Chock, unpublished data).

We hope that the physical conceptualization of a possible energy-transducing cycle will stimulate subsequent research efforts into understanding the physical interactions behind energy transduction.

Acknowledgments. The authors thank Drs. Yi-der Chen, Richard Hendler, and Frits Kamp of the NIH and Dr. Zoltan Schelly of the University of Texas, Arlington for many stimulating and fruitful discussions. This work was partially supported by NIH grant GM28795 to T. Y. T.

References

1. Serpersu, E. H., and Tsong, T. Y. (1983) *J. Membr. Biol.* 74:191–201.
2. Serpersu, E. H., and Tsong, T. Y. (1984) *J. Biol. Chem.* 259:7155–7162.
3. Tsong, T. Y., and Astumian, R. D. (1986) *Biolect. Bioenerg.* 15:457–476.
4. Westerhoff, H. V., Tsong, T. Y., Chock, P. B., Chen, Y., and Astumian, R. D. (1986) *Proc. Natl. Acad. Sci. USA* 83:4734–4738.
5. Astumian, R. D., Chock, P. B., Tsong, T. Y., Chen, Y., and Westerhoff, H. V. (1987) *Proc. Natl. Acad. Sci. USA* 84:434–438.
6. Westerhoff, H. V., and Astumian, R. D. (1987) In *Towards a Cellular Enzymology*, edited by G. R. Welch et al. (in press).
7. Tsong, T. Y., and Astumian, R. D. (1987) *Prog. Biophys. Mol. Biol.* (in press).
8. Hill, T. L. (1977) *Free Energy Transduction in Biology*. Academic Press, New York.
9. Jencks, W. P. (1980) *Adv. Enzymol.* 51:75–106.
10. King, E. L., and Altman, C. (1956) *J. Phys. Chem.* 60:1375–1380.
11. Noble, B. (1969) *Applied Linear Algebra*. Prentice-Hall, Englewood Cliffs, NJ.
12. Hudson, J. L., Lamba, P., and Mankin, J. C. (1986) *J. Phys. Chem.* 90:3430–3434.
13. Chen, Y. (1986) *Proc. Natl. Acad. Sci. USA* 84:729–733.
14. Slater, E. C. (1953) *Nature* 172:975–978.
15. Mitchell, P. (1961) *Nature* 191:144–148.
16. Pickart, C. M., and Jencks, W. P. (1984) *J. Biol. Chem.* 259:1629–1643.
17. Hammes, G. G. (1982) *Proc. Natl. Acad. Sci. USA* 79:6881–6884.
18. Boyer, P. D. (1984) In *H^+-ATPsynthase: Structure, Function, Biogenisis*, edited by S. Papa, K. Altendorf, L. Ernster, and L. Packer, pp. 329–338. Adriatica Editrice, Bari, Italy.
19. Hill, T. L., and Eisenberg, E. (1981) *Q. Rev. Biophys.* 14:463–511.
20. Westerhoff, H. V., and Van Dam, K. (1987) *Thermodynamics and Control of Biological Free Energy Transduction*. Elsevier, Amsterdam.

31
Topics in Petroleum Biotechnology

CHING T. HOU

Biotechnology research in the petroleum industry has been actively pursued for more than two decades. In the past attention was directed toward single cell protein production and dewaxing of petroleum fractions. After the first energy crisis in 1973, more attention was focused on biopolymers for enhanced oil recovery. Studies of possible production of chemicals, especially fine chemicals, by biological means were initiated during the 1970s. The production of epoxides, alcohols, ketones, diacids, acrylamide, and metabolites of aromatic and cyclic compounds has been studied intensively. Some of this biotechnology has been commercialized in certain countries but not in all because of the social and economic conditions involved. Studies on the environmental impact of petroleum fractions are also an important area of biotechnological studies.

In this chapter three topics in petroleum biotechnology are addressed. Two are concerned with the production of chemicals, and the third is in the area of enhanced oil recovery.

Epoxides

Epoxides are extremely valuable industrial chemicals because of their ability to polymerize, forming epoxy homopolymers and copolymers. Ethylene oxide and propylene oxide constitute the two most important commercial epoxides.

Microbial production of epoxides was first demonstrated by Van der Linden in 1963 (1). Since then, four microbial systems have been studied for epoxide production.

Production of Epoxides by *n*-Octane-Grown Bacteria

Production of epoxides by *n*-octane-grown bacteria was studied by Cardini and Jurtshuk using *Corynebacterium* sp. 7ElC and by Exxon Research Center using *Pseudomonas oleovorans* (Table 31.1).

Large-scale production of epoxides was examined by Schwartz and McCoy

TABLE 31.1. Microbial epoxidation of alkenes.

Parameter	*Pseudomonas oleovorans*[a]	*Methylococcus* sp. CRL M1[b]	*Brevibacterium* sp. CRL 61[c]	*Nocardia corallina* B-276[d]
Growth substrate	*n*-Octane	Methane	Propane	Glucose
Best reaction substance	1-Decene	Propylene	Propylene	1-Tetradecene
Product	1,2-Epoxydecane	Propylene oxide	Propylene oxide	1,2-Epoxytetradecane
Reaction rate (μmol/h/mg cells)	0.05	2.5	2.8	2.6 g/l/5 days
Stereospecificity	Pref. (+)[e]	(±)	—	Pref. (+)

[a] Abbott and Hou (25).
[b] Hou et al. (7).
[c] Hou et al. (14).
[d] Furuhashi (15).
[e] May and Schwartz (5).

(3) using a two-phase fermentor. In the presence of cyclohexane, a 90 mol% conversion of substrate to product was accomplished within 72 h, compared with an 18 mol% conversion in the absence of cyclohexane. The products were simultaneously separated and concentrated in the organic phase.

The enzyme involved in the hydroxidation/epoxidation of alkanes/olefins was purified by Peterson and Coon in 1968 (4). It consists of three protein components plus a cofactor, NADH.

Stereospecificity of the product 7,8-epoxy-l-octene was examined by May and Schwartz (5) using the *n*-octane-grown *Pseudomonas* system. A >80% stereoselectivity for the (R)-(+) isomer was reported.

Production of Epoxides by Methanotrophs

Whole cell suspension of methanotrophs converting alkenes, alkanes, and aromatic compounds to their corresponding oxygenated products was discovered independently by us and other groups (6–8). Although the conversion of hydrocarbons by a cell-free system of methanotrophs had been reported earlier (9), the whole cell system has been considered to be important for industrial application. The product epoxide accumulated extracellularly. Substrate specificity studies showed that ethylene and propylene were better substrates for epoxidation. This situation is different from the *n*-octane-grown *Pseudomonas* system, which preferred 1-decene as the substrate (Table 31.1).

Stereospecificity of the epoxide products was examined by Subramanian (10) using 1-butene as the substrate. Equal amounts of both isomers were found.

A gas-solid bioreactor has been constructed to produce propylene oxide from propylene using immobilized whole cells of methanotrophs (11). The catalytic capacity can be regenerated in situ by methanol vapor. The production of propylene oxide was found to proceed at a constant rate of 18 μmol/h. The conversion for propylene was 2.7% for a single pass. The biocatalyst required regeneration after 10 h of operation. Although the bioreactor in its present form is not economically feasible for the production of propylene oxide, it demonstrates the potential of using this new type of bioreactor for the production of chemicals from gaseous substrates.

The enzyme system involved in this epoxidation reaction has been purified (12, 13). It consists of three protein components.

Production of Epoxides by Propane-Grown Bacteria

Cell suspensions of ethane- or propane-grown bacteria also epoxidized alkenes to their corresponding epoxides (Table 31.1) (14). Substrate specificity studies showed that C2 to C4 gaseous alkenes are the better epoxidation substrates. The enzyme involved in this epoxidation system has not been studied.

Production of Epoxides by Glucose-Grown Bacteria

Furuhashi of Bio Research Center (a subsidiary of Nippon Mining K. K.) developed a method to convert C3 to C20 olefins to their corresponding epoxides by using glucose-grown *Nocardia corallina* B-276 (15). It is a fermentation operation in which epoxide is produced from alkene in 5 days in the presence of a growth substrate (glucose). The best epoxidation substrates for this biosystem is tridecene-1 and tetradecene-1 with yields of 2.0 and 2.6 g/l, respectively.

The product epoxides are optically active. The optical purities are 75% for propylene oxide and 86% for 1,2-epoxytetradecene.

With regard to propylene oxide production, Furuhashi used a bubbling column and cell suspension of glucose-grown *Nocardia* cells. The production rate reached its maximum level after 10 h but decreased rapidly thereafter. A continuous supply of glucose to the reaction mixture together with gaseous substrate maintained the maximum reaction rate (around 400 mmol/l/day) for 100 h. No study was conducted on this system at the enzyme level. Small quantities of optically active epoxides (prepared by this bioepoxidation process) are available commercially from Nippon Mining K. K.

Production of Methyl Ketones

Methyl ketones are important industrial solvents. Several years ago we discovered that cell suspensions of methylotrophs converted both alkanes (16) and secondary alcohols (17) to their corresponding methyl ketones (Table 31.2). Among secondary alcohols, 2-propanol and 2-butanol were oxidized at a high rate. The reaction is catalyzed by a novel enzyme, secondary alcohol

TABLE 31.2. Microbial ketone formation.

Parameter	*Methylosinus* sp. CRL 15[a]	*Hansenula polymerpha*[b]	*Methylococcus* sp. CRLM1[b]	*Psueudomonas fluorescens*[c]
Growth substrate	Methane	Methanol	Methanol	Propane
Best reaction substance	Propane	2-Propanol	2-Butanol	2-Butanol
Product	Acetone	Acetone	1-Butanone	2-Butanone
Reaction rate (μmol/hr/mg cells)	0.3	5.9	5.0	15.0
Stereospecificity	—	Pref. (−)[d]	Pref. (−)	(±)

[a] Patel et al. (16).
[b] Hou et al. (17).
[c] Hou et al. (21).
[d] Hou et al. (19).

dehydrogenase (18). Secondary alcohol dehydrogenases were purified from both methylotrophic bacteria (19) and yeast (20). The (−) enantiomer of 2-butanol was oxidized preferentially over its (+) isomer.

A thermostable secondary alcohol dehydrogenase was found in propane-grown *Pseudomonas fluorescens* NRRL B-1244 (21). The optimum temperature for this enzyme activity was 60°to 70°C. Unfortunately, this enzyme was equally active on both stereo isomers of substrate secondary alcohols.

The system for the production of methyl ketones from secondary alcohols has the potential for further industrial exploitation. The production of stereospecific secondary alcohol from either *n*-alkane or a racemic mixture of secondary alcohol may be attractive industrially.

Biopolymers for Enhanced Oil Recovery

Biopolymers or synthetic polymers are required as fluid viscosity enhancers in enhanced oil recovery (EOR) operations. Screening of new biopolymers and synthesis of new polymers are active research programs in industries involved in EOR. Xanthan biopolymer and hydrolyzed polyacrylamide are currently the only two commercially available water-soluble polymers to have been used in actual field tests for EOR. In Exxon's 1980–1982 Loudon EOR pilot test, process conditions were carefully controlled in an attempt to prevent microbial activity. Despite these controls, xanthan-containing drive water suffered a loss of viscosity at 0.65 PV injection. Studies showed that microorganisms were responsible for the loss of polymer solution viscosity (22). Thus the prevention of xanthan biodegradation during EOR operations is important.

Although it was reported that xanthan-degrading microbes are rare, a salt-tolerant bacterial culture, HD 1, that degraded xanthan was isolated. Culture HD 1 is a mixed culture consisting in a bacillus and a short, fat, nonmotile, rod-shaped bacterium. Xanthan-degrading enzymes are produced extra-

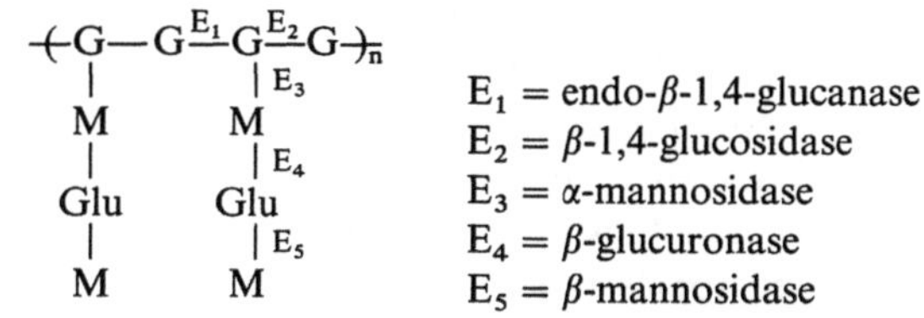

FIG. 31.1. Possible enzymes involved in xanthan biodegradation. (Reprinted from Hou, C. T., Barnabe, N., and Greaney, K. [1986] *J. Soc. Industr. Microbiol.* 1:31; with permission.)

TABLE 31.3. Properties of purified xanthan depolymerase.

Property	Measurement
Molecular weight	60,000 daltons
Optimum pH	5.5
Optimum temperature	30°–40°C
Reaction rate	650 cps/min/mg protein
Cleavage site	Endo-β-1.4-glucosidic linkage
Products	Oligosaccharides and polysaccharides

Hou et al. (24).

cellularly. Growth on xanthan is a requirement for producing xanthan depolymerase, indicating that the enzyme is inducible (23). Products of xanthan degradation by this crude enzyme preparation were identified as glucose, mannose, pyruvated mannose, acetylated mannose, and unidentified oligo- and polysaccharides. The activities of xanthan-degrading enzymes were not affected by the presence or absence of air or by the presence of $Na_2S_2O_4$ or common biocides used in the oil fields, e.g., formaldehyde and 2, 2-dibromo-3-nitrilopropionamide.

It is important to understand the initial attack on the xanthan molecule by these xanthan-degrading enzymes, and especially to identify the one that reduces the viscosity of the xanthan solution. A novel xanthan depolymerase was isolated and purified from culture HD 1 grown on xanthan (24). The molecular weight of this enzyme is 60,000 daltons. Optimum pH was 5, and the optimum temperatures were 30° to 35°C for xanthan depolymerase activity. Xanthan depolymerase cleaves the endo-β-1,4-glucosidic linkage of the xanthan molecule, freeing reducing groups of some sugars and decreasing viscosity of the polymer solution. Only the backbone of β-1,4-linked glucans with sidechains or other substituents were cleaved. No monosaccharide was produced by the action of this enzyme. The oligosaccharide(s) in the low-molecular-weight fraction of the reaction products consisted of 15 to 58 monosaccharide units. The enzymic reaction resulted in a decrease in weight-average molecular weight of xanthan from 6.5×10^6 to 8.0×10^5 daltons in

0.5 h. This enzyme alone could not degrade xanthan to single or multiple pentasaccharide unit(s). Although it was believed that xanthan is a homogeneous polymer composed of pentasaccharide units, results obtained from our studies suggested that there may be regions inside the xanthan molecule that are susceptible to the attack of this xanthan depolymerase.

The above discussed topics reflect only a part of the broad petroleum biotechnology. In general, studies of microbial metabolism/bioconversion of petroleum fractions could lead to the production of useful fine chemicals and/or environmental pollution control. Studies of enzyme catalysis could provide vital information to chemists for synthesizing third-generation catalysts and lead to the use of a new catalyst that mimics the biological system. Recombinant DNA techniques are tools for stimulating the progress of such biotechnology research.

References

1. Van der Linden, A. C. (1963) *Biochim. Biophys. Acta* 77:157–159.
2. Cardini, G., and Jurtshuk, P. (1970) *J. Biol. Chem.* 245:2789.
3. Schwartz, R. D., and McCoy, C. J. (1977) *Appl. Environ. Microbiol.* 34:47–49.
4. Peterson, J. A., and Coon, M. J. (1968) *J. Biol. Chem.* 243:329–334.
5. May, S. W., and Schwartz, R. D. (1974) *J. Am. Chem. Soc.* 96:4031.
6. Dalton, H. (1980) *Adv. Appl. Microbiol.* 26:71–87.
7. Hou, C. T., Patel, R. N., Laskin, A. I., and Barnabe, N. (1979) *Appl. Environ. Microbiol.* 38:127–134.
8. Higgins, I. J., Hammond R. C., Sariaslani, F. S., Best, D., Davies, M. M., Tryhorn, S. E., and Taylor, F. (1979) *Biochem. Biophys. Res. Commun.* 89: 671–676.
9. Colby, J., Stirling, D. I., and Dalton, H. (1977) *Biochem. J.* 165:395–402.
10. Subramanian, M. (1985) *Society of Industrial Microbiology Proceedings*, p. 28.
11. Hou, C. T. (1984) *Appl. Microbiol. Biotechnol.* 19:1–4.
12. Green, J., and Dalton, H. (1985) *J. Biol. Chem.* 260:15795–15801.
13. Woodland, M. P., and Dalton, H. (1985) *J. Biol. Chem.* 259:53–59.
14. Hou, C. T., Patel, R. N., Laskin, A. I., Barnabe, N., and Barist, I. (1983) *Appl. Environ. Microbiol.* 46:171–177.
15. Furuhashi, K. (1981) *Hakko Kogyo* 39:1029–1036.
16. Patel, R. N., Hou, C. T., Laskin, A. I., Felix, A., and Derelanko, P. (1980) *Appl. Environ. Microbiol.* 39:727–733.
17. Hou, C. T., Patel, R. N., Laskin, A. I., Barnabe, N., and Marczak, I. (1979) *Appl. Environ. Microbiol.* 38:135–142.
18. Hou, C. T., Patel, R. N., Lakin, A. I., Barnabe, N., and Marczak, I. (1979) *FEBS Lett.* 101:179–183.
19. Hou, C. T., Patel, R. N., Barnabe, N., and Marczak, I. (1981) *Eur. J. Biochem.* 119:359–364.
20. Patel, R. N., Hou, C. T., Laskin, A. I., and Derelanko, P. (1981) *J. Appl. Biochem.* 3:218–232.
21. Hou, C. T., Patel, R. N., Laskin, A. I., Barist, I., and Barnabe, N. (1983) *Appl. Environ. Microbiol.* 46:98–105.

22. Bragg, J. R., Maruca, S. D., Gale, W. W., Gall, L. S., Wernau, W. C., Beck, D., Goldman, I. M., Laskin, A. I., and Nasland L. A. (1983) Paper SPE 11989, Society of Petroleum Engineering of AICHE.
23. Hou, C. T., Barnabe, N., and Greaney, K. (1986) *J. Soc. Industr. Microbiol.* 1:31–37.
24. Hou, C. T., Barnabe, N., and Greaney, K. (1986) *Appl. Environ. Microbiol.* 52:37–44.
25. Abbott, B. J., and Hou, C. T. (1973) *Appl. Microbiol.* 26:86–91.

32
Experimental Determination of Rate Constants in Enzymatic Reactions

CARL FRIEDEN and MICHAEL H. PENNER

Perhaps the best way to understand the kinetic mechanism of an enzymatic reaction is to be able to determine the rate constants that are associated with each step along the catalytic pathway. Such an understanding is important in defining not only the kinetic pathway but also which steps reflect the rate-limiting process and the catalytic strategy of the enzyme. We have been developing, over the past few years, techniques that allow the determination of individual rate constants by kinetic methods. Primarily, the method involves analysis of the full time course of the enzymatic reaction by computer simulation. As is shown in this chapter, this type of analysis has proved useful: We have been able to explore questions related to the size of bimolecular rate constants in the formation of binary or ternary complexes, the rate of the chemical transformation, the evaluation of isotope effects, the rates of conformational changes, the rates of the association or dissociation of an enzyme with itself or another protein to give a species with different kinectic parameters, the rate and extent of formation of abortive complexes, the nature of product inhibition, as well as many other kinetic characteristics. This chapter deals with some of these issues.

Materials and Methods

Computer Simulation

Kinetic analyses have been carried out using a computer program called KINSIM developed in our laboratory (1). Although kinetic simulation systems have been developed in the past (1, and references therein), the program we have developed is quite flexible and largely machine-independent (although it does not yet run on the smaller personal computer). Essentially we set up a proposed mechanism of the type E + S goes to ES, etc. directly on the terminal. The computer compiles the mechanism into a set of differential equations assuming that each of the reaction steps is governed by mass action. The computer simulation system then solves these equations by numerical integration. The system is set up in such a way that it utilizes an adjustable integration time step, so it becomes reasonable to solve the equations in real time and to interact with the program rapidly. In practice we display real data and the simulated curve together and adjust parameters until a good visual fit is obtained to all real data curves over a wide range of conditions.

Dihydrofolate Reductase

Wild-type *Escherichia coli* dihydrofolate reductase was a gift of Drs. J. Kraut and E. Howell (University of California, San Diego). Dihydrofolate was prepared by the method of Blakley (2) and was stored at −70°C in 5 mm HCl, 50 m*M* β-mercaptoethanol. Actitivy measurements were performed at 340 nm in 0.05 *M* Bis-Tris buffer, pH 7.2, at 20°C, containing 1 m*M* dithioerythritol and 40 μ*M* EDTA using a Durrum model D-110 stopped flow spectrophotometer. The millimolar extinction coefficient difference between substrates and products is 11.8 cm^{-1} (3). Stopped flow quenching of protein fluorescence was performed on the same equipment in the fluorescence mode using an excitation wavelength of 280 nm and a Corning 0–54 filter on the emission side.

All other chemicals were obtained from Sigma, with the NADPH used made by enzymatic methods. Deuterated NADPH (labeled in the A position) was made enzymatically. Tetrahydrofolate was made by the dihydrofolate reductase reaction and purified by DEAE chromatography.

Results and Discussion

Kinetic studies of dihydrofolate reductase in which measurements of initial velocities are attempted are complicated by several factors. First, the Michaelis constant for either substrate is quite low (<1 μ*M*) (3). Second, the enzyme shows at least two types of substrate-induced rate enhancement (4, 5); and third, abortive complexes are easily formed. Using computer simulation of the full time course of stopped flow measurements, we have overcome most of these problems and have been able to determine many of the individual rate constants in the mechanism. Our work supplements (a) previous studies with *E. coli* dihydrofolate reductase that have shown that either substrate, NADPH or DHF, can bind to the enzyme in the absence of the other (6) and (b) kinetic studies (3) which concluded that the enzyme mechanism approached that of a rapid equilibrium random Bi Bi system.

Biomolecular Steps

Binary Complex Formation

The rate of binding of either NADPH or DHF can be measured by following the quenching of enzyme fluorescence (6). Previous studies along these lines using NADPH or DHF (6) showed that quenching occurs in two phases. Cayley et al. (6) found that only the fast phase was apparent at NADPH concentrations <0.5 times the enzyme concentration. The authors interpreted these results as showing that the apoenzyme existed in two forms, only one of which could bind NADPH. Our experiments do not confirm this result. Two phases exist even when the NADPH concentration is 0.3 times the enzyme concentration, although the size of the slow phase was only $\sim 7\%$ of the fast phase under these conditions. Such results are more consistent with a mechanism in which the ligand (either NADPH or DHF) binds the apoenzyme and induces a conformational change. The rate of the conformational change (regardless or the mechanism) appears to be related to a lag in catalytic activity when both substrates are added simultaneously to the enzyme (5), i.e., a hysteretic effect.

Ternary Complex Formation

The bimolecular rate constants for the formation of ternary complexes can be measured using single turnover experiments. Experimentally, NADPH or DHF is preincubated with the apoenzyme, and a less than stoichiometric amount of either DHF or NADPH is added to initiate the enzymatic reaction. The preincubation is necessary to convert the enzyme into a single form (as described above). The single turnover experiment would be expected to represent the bimolecular rate constant associated with ternary complex formation, provided the rate of the chemical transformation (and associated steps prior to the chemical transformation) is relatively rapid compared to ternary complex formation. This condition is satisfied for the DHFR reaction. We found the initial rate of the single turnover reaction to be described by k[E] [S] as we varied the enzyme and substrate concentrations separately and simultaneously. To further verify that the rate constants obtained represent bimolecular steps, we performed the same experiments at different viscosities. It was found that buffers containing sucrose or glycerol (giving twice the relative viscosity of the control buffer) produced a decrease in the observed rate by about 50%, whereas a buffer containing ficoll at a relative viscosity of 2 did not affect the observed rate. Sucrose and glycerol are small molecules that affect the microviscosity, and ficoll is a polymer that affects the macroviscosity of the solution. It would be expected that sucrose and glycerol would affect diffusion rates (i.e., the bimolecular step), but ficoll would not. These results are consistent with the notion that the step examined in the single turnover reaction is the bimolecular rate constant for ternary complex formation.

Chemical Transformation

Single turnover experiments can yield, as indicated above, bimolecular rate constants for the formation of ternary complexes, as the substrate concentrations used are below the enzyme concentration. Similar experiments can be performed at much higher substrate levels as well. If the rate-limiting step occurs at a point after the chemical transformation, at high substrate concentrations there will be a burst of substrate disappearance representing the chemical transformation. Indeed, this result is observed for dihydrofolate reductase. The rate of the burst at high substrate levels ($> \sim 10\ \mu M$) is independent of the substrate concentration. In addition, it shows a fairly large deuterium isotope effect in contrast to that observed for the steady-state deuterium isotope rate effect of about 1.2-fold. These results suggest that the rate-limiting step for the reaction occurs at a step after the chemical transformation. Different mechanisms have been suggested (7, 8). Stone and Morrison (7) have proposed, based on pH studies, that the rate-limiting step of the overall reaction is a conformational change associated with proton transfer to DHF. Similarly, Howell et al. (8) interpreted their data from studies of site-directed mutant enzymes to propose the following relative magnitudes of individual rate constants: proton transfer < hydride transfer < product release. Our data show that product release or a step prior to product release but after hydride transfer is the slow step.

Beyond the Chemical Transformation

Because the values for the Michaelis constants for either substrate are below 1 μM, the steady-state rate at high substrate concentration should be dominated by the value of the rate-limiting step in the enzymatic process. This value is 16 sec^{-1} and is about 12-fold lower than the rate constant for the chemical transformation. At this time it is not clear if the rate-limiting step is a conformational change that precedes product release or if it is product release itself. Equilibrium binding studies indicate a strong affinity of DHFR for THF ($K_D \sim 0.2\ \mu M$), which may reflect a slow dissociation rate of THF from the enzyme. A rate-limiting THF off-rate for this enzyme has been proposed by Benkovic (personal communication). However, the patterns of the full time course of the reaction compared at various substrate levels well above the Michaelis constants (15–80 μM substrates) indicate that isomerization prior to product release, rather than produce release per se, is the rate-limiting step. Thus we have assumed that a conformational change prior to product release is rate-limiting.

Abortive Complex Formation

Figure 32.1 shows the full time course of the reaction for two real data curves (A and C) compared to simulated full time courses expected if there were no abortive complex formation (B and D). In curve A the DHF concentration is 10 μM and the NADPH concentration 20 μM. In curve C the NADPH concentration is 10 μM and the DHF concentration 20 μM. In both cases these concentrations are well above the Michaelis constants. Comparing the simulated data to the real data shows that, although the initial portions of the curves are identical, there is eventual inhibition of substrate disappearance in both cases and that the two real data curves are more different from each other

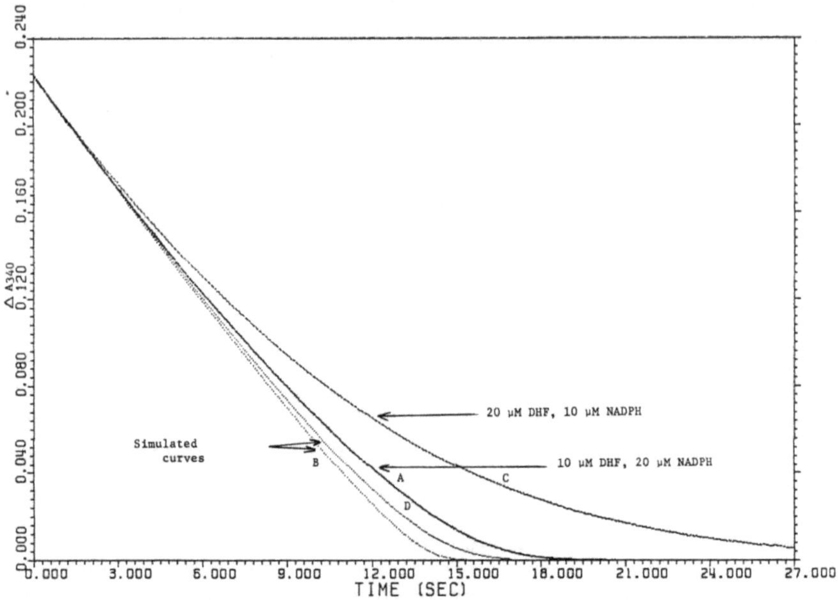

FIG. 32.1. Comparison of the full time courses of the reaction. Curves A and C are real data curves, whereas curves B and D represent computer simulations of the reaction that do not include the abortive complex formation. Experiments were performed in 50 mM bis-tris, pH 7.2, 20°C, containing 1 mM dithioerythritol, 40 μM EDTA, and an enzyme concentration of 0.065 μM. Curves A and B are 10 μM DHF and 20 μM NADPH, and curves C and D are 20 μM DHF and μM NADPH.

than would be expected from simple substrate depletion. This difference could be explained either on the basis of product inhibition or abortive complex formation. Progress curve analysis by computer simulation indicates that the pattern of inhibition seen in the real data is best described by inhibition due to abortive complex formation. As indicated by the rate constants in the legend to Fig. 32.2, the difference between the two real data curves is probably due to the relative differences in the rates of association and dissociation of substrates and products regarding abortive complex formation. When fitting the data it was found that two abortive complexes could describe the data (E-DHF-NADP and E-THF-NADPH) and that both of the complexes were formed in an ordered manner with product adding to the enzyme–substrate binary complex. The most effective abortive complex is that formed by the reaction E-DHF + NADP.

KINETIC MECHANISM

The information above describes a mechanism for the DHFR reaction that predicts kinetic data over a wide range of conditions. This mechanism is

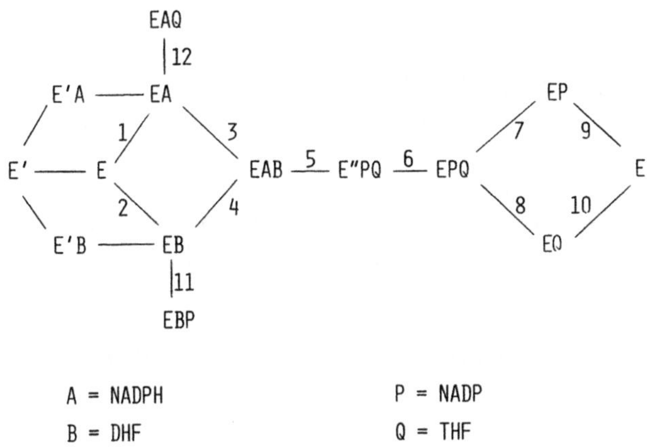

FIG. 32.2. Kinetic mechanism for dihydrofolate reductase as determined by the data presented here. The numbers represent steps described by the rate constants given below. Values for the unnumbered steps or the right side of the mechanism have not yet been accurately determined. Values given below may change slightly with better fitting of the data.

Step	Forward	Reverse
1	65 μM^{-1} sec^{-1}	6 sec^{-1}
2	90 μM^{-1} sec^{-1}	2 sec^{-1}
3	60 μM^{-1} sec^{-1}	25 sec^{-1}
4	30 μM^{-1} sec^{-1}	55 sec^{-1}
5	200 sec^{-1}	7 sec^{-1}
6	16 sec^{-1}	1 sec^{-1}
7	300 sec^{-1}	50 μM^{-1} sec^{-1}
8	300 sec^{-1}	50 μM^{-1} sec^{-1}
9	300 sec^{-1}	60 μM^{-1} sec^{-1}
10	300 sec^{-1}	60 μM^{-1} sec^{-1}
11	34 μM^{-1} sec^{-1}	45 sec^{-1}
12	8 μM^{-1} sec^{-1}	30 sec^{-1}

shown in Fig. 32.2, which includes a list of the rate constants used for fitting all the kinetic data so far described. The mechanism is a random order of substrate addition where these steps do not equilibrate prior to the chemical transformation step. The rate-limiting step for the reaction is either a conformational change after the chemical transformation or release of one of the products. The full time course of the reaction is markedly affected by the formation of two abortive complexes that form in a preferred order.

We believe that delineation of the individual rate constants as described here should provide the necessary background for explaining the changes in kinetic parameters observed with enzymes that differ in a single amino acid residue as created by the technique of site-directed mutagenesis.

Addendum

More recent data obtained by Fierke (personal communication) under our conditions show that THF dissociation from the enzyme is slow (5 sec^{-1}). To fit the full time course data, it is found that the dissociation of THF is enhanced in the E · NADPH · THF abortive complex (17 sec^{-1}). Thus, a rate-limiting conformational change of the ternary complex is not necessary and the mechanism is similar to that proposed by Fierke et al. (9).

Acknowledgment. This work was supported by USPHS grant AM 13332.

References

1. Barshop, B. A., Wrenn, F. R., and Frieden, C. (1983) *Anal. Biochem.* 130:134–145.
2. Blakley, R. L. (1960) *Nature* 188:231–232.
3. Stone, S. R., and Morrison, J. F. (1982) *Biochemistry* 21:3757–3765.
4. Baccanari, D. P., and Joyner, S. S. (1981) *Biochemistry* 20:1710–1716.
5. Penner, M. H., and Frieden, C. (1985) *J. Biol. Chem.* 260:5377–5369.
6. Cayley, P. J., Dunn, S. M. J., and King, R. W. (1981) *Biochemistry* 20:874–879.
7. Stone, S. R., and Morrison, J. F. (1984) *Biochemistry* 23:2753–2758.
8. Howell, E. E., Villafranca, J. E., Warren, M. S., Oatley, S. J., and Kraut, J. (1986) *Science* 231:1123–1128.
9. Fierke, C. A., Johnson, K. A. and Benkovic, S. J. (1987) *Biochemistry* 26:4085–4092.

33 Applications of Alternative Substrate Kinetics: In Vivo and In Vitro Data on Aminoglycoside Antibiotic Inactivating Enzymes

DEXTER B. NORTHROP

The study of mechanisms of enzymatic reactions is composed of two lines of inquiry. The first addresses the *kinetic mechanism*, which describes the order of the addition of substrates and the release of products. It also defines the central complexes between substrates and the enzymes that are possible; hence it is often the first to be undertaken. The second addresses the *chemical mechanism*, which describes processes of bond making and breaking that occur within central complexes and which define the roles of the amino acid side chains within the active site of the enzyme. This chapter will address the former and describe a new method for their determination. The usual kinetic mechanisms for enzymes catalyzing reactions between two substrates are shown in Fig. 33.1 (1). The first, a ping-pong kinetic mechanism, consists in the formation of only binary enzyme–substrate complexes, within two independent half-reactions, separated by a chemically modified form of free enzyme. The other three mechanisms are all sequential, and the kinetic mechanism consists simply in defining whether the binding sequence leading to and from the ternary enzyme–substrate complex is ordered or random. The Theorell-Chance mechanism is a special variant of ordered, in which the reaction segment controlling the release of the final product is rate-limiting.[1]

Kinetic mechanisms of enzymatic catalysis have been determined traditionally by initial velocity experiments, where one substrate is varied against different *concentrations* of a co-substrate. The data are plotted as reciprocal

---

[1] The concept of a rate-limiting step has been much abused in enzymology. Following Johnston (2) and Boyd (3), the concept as used here is restricted to reaction processes that are first order from their point of initiation during a single turnover and end in an irreversible step. Because these processes may consist of several separate and reversible chemical steps, they are referred to as "segments" of a reaction. For example, a ping-pong mechanism normally has only two segments, each encompassing one half reaction. An ordered mechanism also has two segments; one includes all steps following the binding of the second substrate up to and including the release of the first product, and the other is simply the release of the second product (plus perhaps a conformational change of the enzyme occurring between the release of the two products).

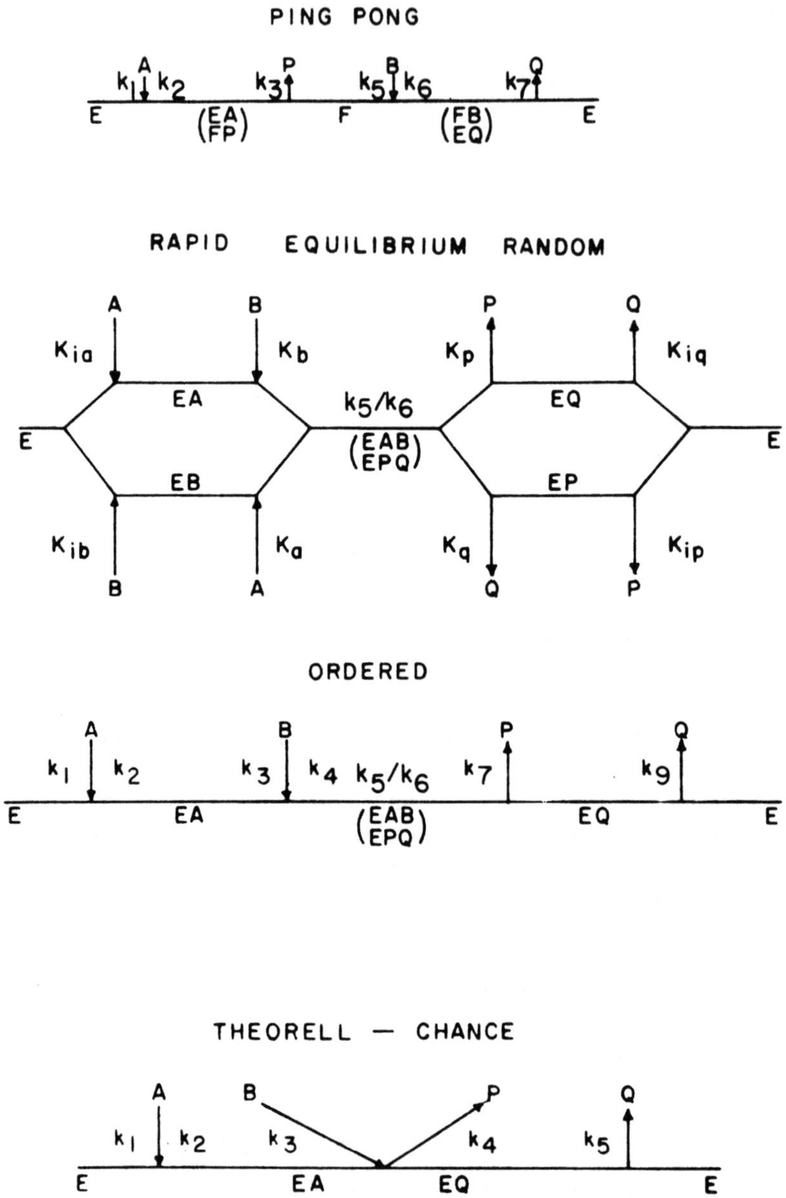

FIG. 33.1. Kinetic mechanisms for enzymes catalyzing reactions involving two substrates and two products. Horizontal lines represent enzyme, arrows represent the comings and goings of reactants, and movement along the line from left to right represents reaction progress. This nomenclature follows that of Cleland, W. W. (1964) *Biochim. Biophys. Acta* 67 : 104–137.

initial velocities versus reciprocals of the concentration of the varied substrate at different fixed levels of the co-substrate, forming a set of Lineweaver-Burk plots. Enzymes obeying ping-pong kinetic mechanisms yield a pattern of parallel lines, whereas the sequential mechanisms produce patterns of intersecting lines. In order to distinguish between sequential mechanisms, it is necessary to conduct additional kinetic experiments, such as product inhibition or isotopic exchange studies. Often it is difficult to assay enzymes at much below K_m levels, so the change in velocity as a function of one of the substrates within a pattern is only a factor of 2 or so; with both substrates at or below K_m levels, the reaction rate slows quickly after just a few turnovers because of substrate depletion, so assays with marginal sensitivity tend to greatly underestimate initial velocities, which in turn distort the pattern in its most critical area. As a result, it is sometimes difficult to tell if an initial velocity pattern is parallel or intersecting.

For example, the data in Fig. 33.2 show one of several initial velocity

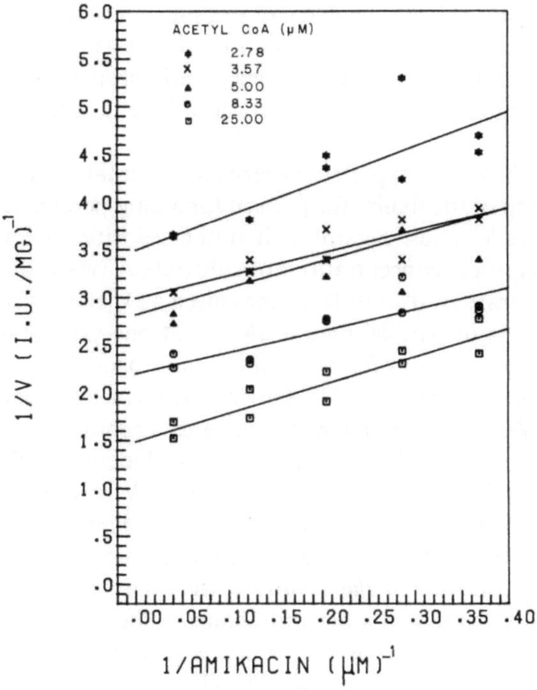

FIG. 33.2. Initial velocity pattern obtained with kanamycin acetyltransferase. Double reciprocal plots of velocities versus varied concentrations of amikacin at various fixed concentrations of acetyl-CoA. A fit of the data to the rate equation for a sequential mechanism by nonlinear regression analysis gave: $V = 0.82 \pm 0.04$ IU/mg, $K_a = 2.40 \pm 0.36$ μM, $K_b = 5.58 \pm 0.35$ μM, and $K_{ia} = -0.34 \pm 0.34$ μM. The simplest interpretation would be that the K_{ia} term is insignificant, and therefore the pattern is parallel, supportive of a ping-pong kinetic mechanism; but, in fact, the correct mechanism is rapid equilibrium random sequential.

patterns from an investigation of the kinetic mechanism of kanamycin acetyltransferase. Despite repeated attempts to increase the precision of the initial velocity measurements, we found it impossible—given the limits of the usual spectrophotometric assay—to determine whether the pattern was parallel or intersecting.

Alternative Substrate Kinetic Theory

If an enzyme accepts more than one substrate and co-substrate, it may be possible to deduce a kinetic mechanism from alternative substrate data instead of relying solely on initial velocity patterns (4). Instead of varying one substrate against different concentrations of the co-substrate, as above, different *identities* of co-substrates may be used (5). This approach has some surprises and can have considerable advantage over the more traditional method. For example, because the alternative substrate is held at saturating levels, one does not have the problem of running out of both substrates at the same time, which adds to the precision of the measurement; and because alternative substrates may differ in their kinetic properties by considerably more than a factor of 2, the interpretation of kinetic data can be much less ambiguous.

Figure 33.3 shows the expected patterns for alternative substrate kinetics. Shown at the top of the figure, the pattern for a ping-pong mechanism is seen as a set of parallel lines, regardless if initial velocities are measured as a function of reciprocal concentrations of substrate A versus different alternatives of co-substrate B, B′, and B″ or measured as 1/[B] versus A, A′, and A″ (6). A pair of sets of parallel lines is also what one expects to see in initial velocity experiments—and for much the same reason.[2] The slopes of these lines are apparent reciprocals of V/K for the substrate whose concentrations were varied. V/K includes all rate constants encompassed by the initial binding of the varied substrate up to and including the first subsequent irreversible step. With ping-pong mechanisms the V/K for one substrate ends with the irreversible release of a product *before* the binding of the other; hence the V/K (and subsequent slope of a line) for each substrate is independent of changes in concentration *or the identity* of the other.[3]

In the random sequential kinetic mechanism, shown in the middle of Fig. 33.3, neither set of lines is parallel, which is reminiscent of the intersecting initial velocity patterns of *all* sequential mechanisms. Here, however, only the

---

[2] It should be noted that the pair of initial velocity patterns are simply different plots of the same data, but that two different data sets are necessary to generate a pair of alternative substrate plots.

[3] An exception should be noted: If an alternative to substrate A results in a change in the chemical group being transferred—and in the other form of free enzyme, from E to F—this change *may* result in a change in the V/K for the second half reaction.

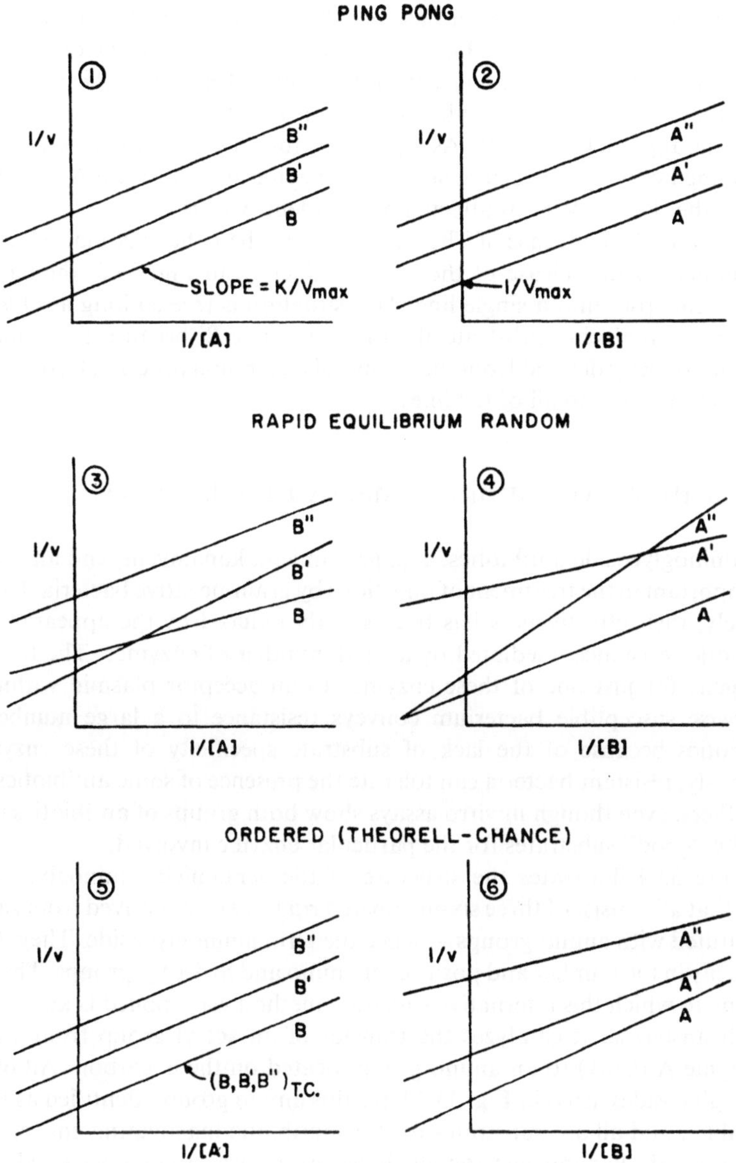

FIG. 33.3. Kinetic patterns for alternative substrate kinetics. Double reciprocal plots of velocities versus concentrations of one substrate at saturating levels of alternatives of the other substrate, represented by primes. Subscript T. C. in No. 5 refers to Theorell-Chance.

random sequential mechanism has this characteristic. In the ordered case, shown at the bottom of Fig. 33.3, one pattern is parallel and the other is not. The reason is that when the concentration of the second substrate approaches saturation, i.e., alternative substrates B or B′, etc., the V/K of the first substrate approaches k_1 (7). As k_1 is a diffusion-controlled rate constant determined only by characteristics of the enzyme and substrate A, it follows that neither k_1 nor V/K_a is affected by the identity of B.

Finally, in the limit case of Theorell-Chance, where the maximal velocity is determined by the release of the final product Q, in Fig. 33.2, the parallel pattern collapses into a single line. This situation is true so long as Q is not changed by replacing substrate B with alternative B′ because the common maximal velocity derived from the release of a common product, Q, dictates a common intercept to all of the lines.

Applications to Antibiotic-Inactivating Enzymes

The aminoglycoside antibiotics, e.g., gentamicin, kanamycin, and amikacin, are important in the treatment of infections by gram-negative bacteria. Unfortunately, their effectiveness has been greatly reduced by the appearance of antibiotic resistance, mediated by a small number of enzymes. The transfer of a gene for just one of these enzymes to an acceptor plasmid within an otherwise susceptible bacterium conveys resistance to a large number of antibiotics because of the lack of substrate specificity of these enzymes. Curiously, resistant bacteria can tolerate the presence of some antibiotics but not others, even though in vitro assays show both groups of antibiotics to be equally "good" substrates for the particular enzyme involved.

Figure 33.4 illustrates the structure of the gentamicin aminoglycosides. Note that all consist of three six-membered rings; two are derived from sugars substituted with amino groups—hence the term aminoglycoside. They differ primarily in the number and position of amino and hydroxyl groups. The first enzyme to which this alternative substrate method was applied is kanamycin acetyltransferase. It catalyzes the transfer of an acetyl group from acetyl-coenzyme A (CoA) to an amino group located on the 6′ carbon. All of the aminoglycosides listed in Fig. 33.4 have this amino group (identified as R_5 in the table), and all are substrates for this antibiotic-inactivating enzyme—in addition to the kanamycins, for which the enzyme was originally named.[4] The second enzyme subjected to this kinetic method, gentamicin adenylyltransferase, catalyzes the transfer of an adenine monophosphate group from ATP to the 2″ hydroxyl. Again, all of the aminoglycosides listed in the table of Fig.

[4] Kanamycin acetyltransferase and gentamicin adenylyltransferase are trivial names for aminoglycoside acetyltransferase (6′)-4 and aminoglycoside nucleotidyltransferase (2″)-1, abbreviated AAC (6′)-4 and ANT (2″)-1, respectively, by the plasmid group nomenclature (5).

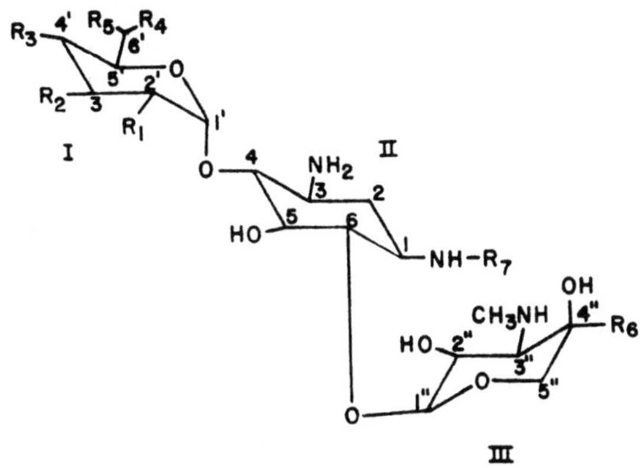

COMPOUND	R_1	R_2	R_3	R_4	R_5	R_6	R_7
GENTAMICIN B	OH	OH	OH	H	NH_2	CH_3	H
GENTAMICIN B_1	OH	OH	OH	CH_3	NH_2	CH_3	H
GENTAMICIN C_{1a}	NH_2	H	H	H	NH_2	CH_3	H
SISOMICIN	NH_2	H	$\Delta_{4'}$	H	NH_2	CH_3	H
N-ETHYL SISOMICIN	NH_2	H	$\Delta_{4'}$	H	NH_2	CH_3	CH_2CH_3

FIG. 33.4. Structures of aminoglycoside antibiotics related to gentamicin and sisomicin.

33.4 have the hydroxyl at carbon 2″, and all are substrates for the enzyme. Both antibiotic-inactivating enzymes have been purified (9, 10), and both have been found to accept alternative nucleotides as well as a variety of antibiotics as substrates and co-substrates.

Table 33.1 lists kinetic data for 7 of 12 aminoglycoside antibiotics shown to be modified by acetyltransferase (11). Note that the maximal velocities differ by a factor of more than 3, and V/K values by nearly 20. Using fixed and saturating levels of gentamicin C_{1a}, amikacin, and neomycin B, the first alternative substrate pattern was determined and is shown in Fig. 33.5. Each of the three lines represent the changes in reaction velocity as a function of the concentration of the same substrate; that is, each is a Lineweaver-Burk plot for butyryl-coA. However, the lines differ by the identity of antibiotic used. The pattern shown is clearly an intersecting one, consistent with a sequential kinetic mechanism, ruling out the possibility of ping-pong and resolving the ambiguity caused by Fig. 33.2.

TABLE 33.1. Aminoglycoside and nucleotide kinetic constants for kanamycin acetyltransferase and gentamicin adenylyltransferase.

Substrate	V_{max}	V/K
Kanamycin Acetyltransferase		
Aminoglycosides		
Sisomicin	0.59	0.23
→ Amikacin	0.59	0.35
Kanamycin A	0.35	0.54
Tobramycin	0.70	1.00
Neomycin C	0.33	3.35
→ Gentamicin C_{1a}	0.64	1.39
→ Neomycin B	0.22	4.35
Nucleotides		
→ Acetyl-CoA	0.55	0.112
→ Propionyl-CoA	0.28	0.034
→ Butyryl-CoA	0.04	0.003
Gentamicin Adenylyltransferase		
Aminoglycosides		
Netilmicin	3.2	0.002
Kanamycin A	5.3	0.040
→ Gentamicin A	5.2	0.040
Kanamycin B	7.4	0.230
Gentamicin B_1	12.9	0.230
Tobramycin	10.2	0.240
→ Gentamicin C_1	8.9	0.360
Sisomicin	10.1	0.540
Gentamicin B	4.8	0.640
Gentamicin C_2	3.9	3.700
→ Gentamicin C_{1a}	3.5	6.000
Nucleotides		
→ Mg:ATP	4.7	3.6
Mg:UTP	3.9	1.4
Mg:CTP	3.2	0.9
Mg:GTP	3.0	6.4
→ Mg:dATP	2.6	9.6
Mg:dTTP	2.1	3.6
Mg:dCTP	1.9	1.6
→ Mg:dGTP	0.9	6.4

The next question is which of the three possible sequential mechanisms applies to the acetyltransferase. The answer can be found by examining the pattern of alternative co-substrates. Table 33.1 lists kinetic data for nucleotides that serve as acyl donors for the enzyme (11). Again, the differences in the kinetic constants for the nucleotides are quite large (> 10-fold in both parameters), and these differences are reflected in the pattern of alternative co-

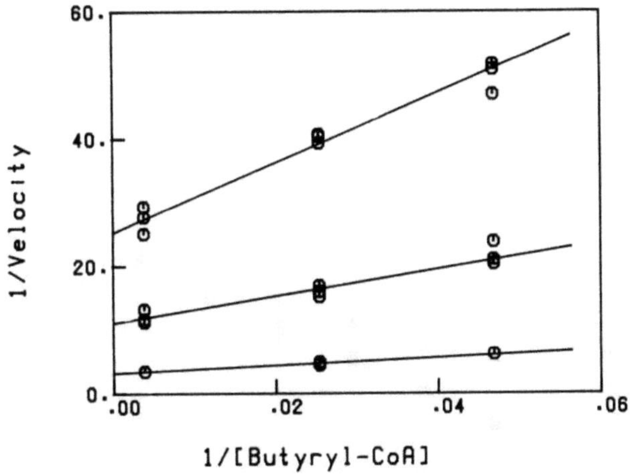

FIG. 33.5. Alternative substrate pattern obtained with kanamycin acetyltransferase. Double reciprocal plots of velocities versus nucleotide concentrations at saturating levels of various antibiotics. Velocity units are micromoles per minute per milligram; concentration units are micromoles. The antibiotics were (top to bottom) gentamicin C_{1a}, amikacin, and neomycin B, respectively.

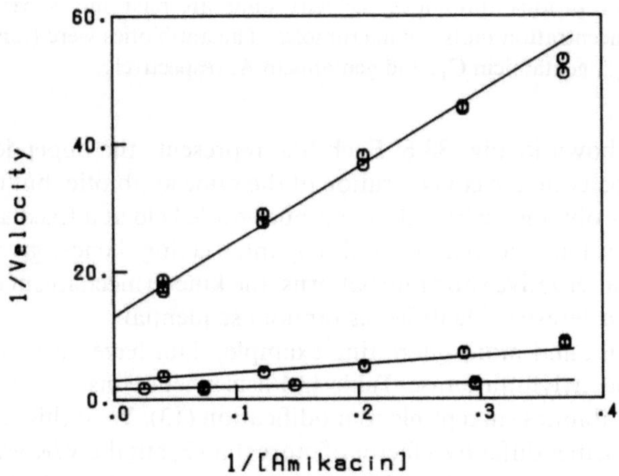

FIG. 33.6. Alternative co-substrate pattern obtained with kanamycin acetyltransferase. Double reciprocal plots of velocities versus aminoglycoside concentrations at saturating levels of various nucleotides. Velocity units are micromoles per minute per milligram; concentration units are micromoles. The co-substrates were (top to bottom) *n*-butyryl-CoA, *n*-propionyl-CoA, and acetyl-CoA, respectively.

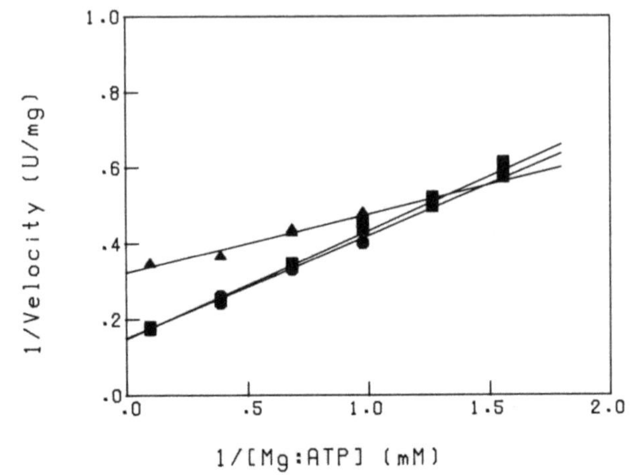

FIG. 33.7. Alternative substrate pattern obtained with gentamicin nucleotidyltransferase. Double reciprocal plots of velocities versus nucleotide concentrations at saturating levels of various antibiotics. Velocity units are micromoles per minute per milligram; concentration units are micromoles. The antibiotics were (top to bottom) gentamicin C_{1a}, gentamicin C_1, and gentamicin A, respectively.

substrates, shown in Fig. 33.6. Each line represents the dependence of the reaction velocity on the concentration of the same antibiotic, but the data for each line was obtained with a different nucleotide held at a fixed and saturating concentration. The pattern is clearly intersecting; hence, given a *pair* of intersecting alternative substrate patterns, the kinetic mechanism of kanamycin acetyltransferase is identified as random sequential.[5]

As a second and more interesting example, data have been collected for gentamicin adenylyltransferase. Table 33.1 lists kinetic constants for 11 aminoglycoside antibiotics susceptible to modification (13). With this enzyme also, maximal velocities differ by a factor of more than 3, but the V/K values extend over a wide range (more than three orders of magnitude), which is curious. Selecting saturating levels of gentamicins A, C_1, and C_{1a}, the first of the second pair of alternative substrate patterns is shown in Fig. 33.7. These patterns are again Lineweaver-Burk plots for the same varied substrate, ATP, but using different antibiotics. At first glance, the pattern may appear to be intersecting, but one must consider the magnitude of differences between the lines compared to the differences in kinetic constants for each antibiotic. Two

[5] Using other kinetic data, the mechanism was further established to be a *rapid equilibrium*, random sequential mechanism (12).

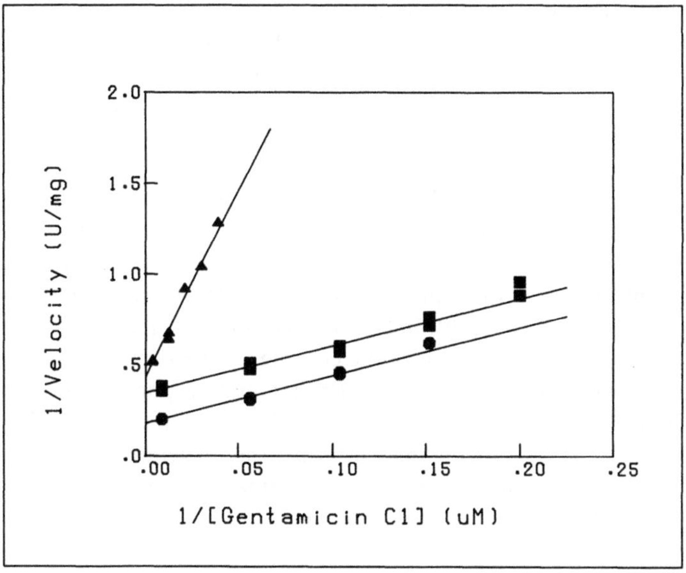

FIG. 33.8. Alternative co-substrate pattern obtained with gentamicin nucleotidyltransferase. Double reciprocal plots of velocities versus aminoglycoside concentrations at saturating levels of various nucleotides. Velocity units are micromoles per minute per milligram; concentration units are micromoles. The co-substrates were (top to bottom) dGTP, dATP, and ATP, respectively.

of the lines are coincidental, and the third differs in slope and intercept by less than a factor of 2, but this finding contrasts with a change in V/K for the antibiotic of 150-fold. On that scale, the kinetic differences between the antibiotics are expressed at only a 1% level in the lines; hence all three are considered to be coincident lines.

For the companion alternative co-substrate plot, nucleotide specificity was investigated and the results listed in Table 33.1 (13). Using saturating levels of ATP, dATP, and dGTP varied against gentamicin C_1, an intersecting pattern was again obtained, shown in Fig. 33.8. The three nucleotides had kinetic differences in a range from three- to five-fold, and differences in slopes and intercepts of similar magnitude are expressed. This second pair of patterns—one intersecting and one a set of coincidental lines—argues strongly for a Theorell-Chance kinetic mechanism.[6]

The Theorell-Chance kinetic mechanism (14) is illustrated in Fig. 33.9; nucleotide binds to the enzyme first, followed by antibiotic; pyrophosphate is the first product released, followed by AMP–gentamicin (in this example).

[6] Supplementary data from substrate and dead-end inhibition and from the kinetic effects of changes in pH and viscosity also support a Theorell-Chance kinetic mechanism (13).

What makes the mechanism Theorell-Chance is the slow release of the final product relative to catalysis. That this situation should be so is understandable, considering the number of binding ligands present in this rather large product. V_{max} therefore is determined solely by this final step of the reaction, whereas the other kinetic parameters are determined by other, isolated portions of the mechanism: V/K for the substrate is determined solely by the binding of the nucleotide, and V/K for the co-substrate includes binding of the antibiotic, catalysis, and release of pyrophosphate. This fact now explains a number of curiosities, e.g., why the V/K values should vary so widely for different antibiotics whose maximal velocities are so similar.[7] Given the orderedness of substrate addition (i.e., nucleotide first and then antibiotic), it follows that the nucleotide–antibiotic complex, should be released in a two-step fashion of opposite order because it contains portions of each. The antibiotic portion of the product swings out first, probably followed by a slow conformational change of the enzyme, with the nucleotide portion released last.[8] Hence the rate-limiting release of the final product is only marginally dependent on the identity of the antibiotic.

In Vivo Alternative Substrate Kinetics

The mechanism and kinetic results may also be used to explain the curious selective tolerance of resistant bacteria for some of the antibiotics that are "good" substrates. During early studies into antibiotic susceptibility of resistant bacteria and specificity of antibiotic-inactivating enzymes, an "excess" of antibiotic was used in the latter. Thus without doing so intentionally, antibiotics designated "good" substrates were those with the large maximal velocities.

To ascertain specifically which kinetic parameter of the inactivating enzymes determines the level of antibiotic resistance, precise measurements of minimal inhibitory concentrations (MICs) were obtained for sets of five antibiotics acting against separate strains of resistant *Escherichia coli*, which produced either the adenylyl- or the acetyltransferase enzyme. Table 33.2 shows the results of the former compared to the kinetic parameters V and V/K. (Because the Michaelis constant is meaningless in the Theorell-Chance mechanism, only V and V/K were considered.)

The MIC values cover nearly two orders of magnitude and are matched by more than that in V/K, whereas maximal velocities vary hardly at all in

[7] It also explains why K_m values were not included in the tables: Michaelis constants are meaningless in this extreme kinetic mechanism because the two kinetic parameters from which they are derived, V and V/K, are completely unrelated to each other.

[8] Partial uncompetitive substrate inhibition by antibiotic versus nucleotide, and slope-linear, intercept-hyperbolic, partial noncompetitive dead-end inhibition by antibiotic support the hypothesis of a three-step, ordered release of the nucleotidylated antibiotic (13).

TABLE 33.2. In vivo versus in vitro kinetics for gentamicin adenylyltransferase and kanamycin acetyltransferase.

Antibiotic	MIC[a] (μg/ml)	V_{max}	V/K
Gentamicin Adenylyltransferase			
Gentamicin C_2	2630	3.9	3.700
Gentamicin C_{1a}	1980	3.5	6.000
Tobramycin	1650	10.2	0.240
Sisomicin	1310	10.1	0.540
Netilmicin	40	3.2	0.002
Kanamycin Acetyltransferase			
Kanamycin	800	0.81	6.2
Tobramycin	640	0.93	2.9
Amikacin	480	0.46	2.8
Gentamicin C_{1a}	100	1.35	2.6
Sisomicin	80	3.33	0.6

[a] Minimum inhibitory concentration.

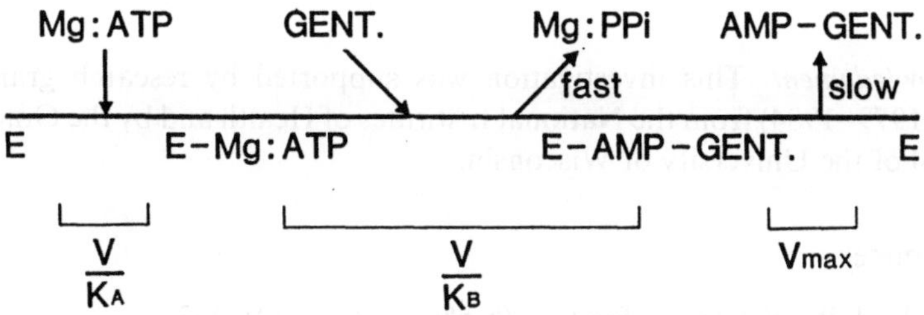

FIG. 33.9. Theorell-Chance kinetic mechanism of gentamicin nucleotidyltransferase.

comparison. More dramatic, perhaps, are similar results obtained with the acylating enzyme, shown in Table 33.2. Again, MIC and V/K values vary together over a 10-fold range, whereas the changes in maximal velocities are actually in the wrong order: They decrease as resistance increases. (Michaelis constants followed neither trend.) Because this enzyme obeys the more common, random kinetic mechanism, its pattern of resistance shows that the dependence of antibiotic resistance on V/K values is a general phenomenon, not dependent on a specific kinetic mechanism.

With hindsight, these results make good sense. Looking again at the kinetic mechanism (Fig. 33.9), it is reasonable at low concentrations of unmodified antibiotic in the cell (a condition that must be maintained if the enzyme is to provide protection) that the central portion of the reaction mechanism (i.e., getting from free antibiotic to release of pyrophosphate, the part encompassed by V/K_b) would be the important part of turnover—not the actual time for each complete turnover. In other words, what we typically would identify as the rate-limiting step for this enzyme is of no significance to its physiological function within the bacterium.

Conclusion

The history of enzymology is dominated by two highly remarkable features of enzymes: a rapid rate of catalysis and a high degree of specificity. Remarkable as these features are, however, we must not be misled by awe, tradition, or a history based on too few examples. We should not close our minds to the fact that many enzymes act on more than just one or a few substrates, and much can be learned by exploiting differences between alternative substrates. Nor should we assume that the common measure of catalysis known as the maximal velocity (V_{max} or k_{cat}) is necessarily an important property in vivo (or in immobilized systems). Moreover, the results reported above suggest that other enzymatic processes within the bacterial cell may be wholly dependent on another kinetic parameter; hence targeting of enzymes in, for example, drug design might be more successful if aim were taken on V/K rather than on catalytic turnover.

Acknowledgment. This investigation was supported by research grant AI 1108 (1979–1984) from the National Institutes of Health and by the Graduate School of the University of Wisconsin.

References

1. Cleland, W. W. (1964) *Biochim. Biophys. Acta* 67:104–137.
2. Johnston, H. S. (1966) *Gas Phase Reaction Theory*, Ch. 16. Roland Press, New York.
3. Boyd, R. K. (1978) *J. Chem. Educ.* 55:84–89.
4. Huang, C. Y. (1979) *Methods Enzymol.* 63:486–500.
5. Radika, K., and Northrop, D. B. (1984) *Anal. Biochem.* 141:413–417.
6. Webb, B. C., Todhunter, J. A., and Purich, D. L. (1976) *Arch. Biochem. Biophys.* 173:282–292.
7. Frieden, C. (1957) *J. Am. Chem. Soc.* 79:1894–1896.
8. Mitsuhashi, S. (1975) *Drug Action and Drug Resistance*, p. 269. University of Tokyo Press, Tokyo.
9. Radika, K., and Northrop, D. B. (1984) *Arch. Biochem. Biophys.* 233:272–285.
10. Van Pelt, J. E., and Northrop, D. B. (1984) *Arch. Biochem. Biophys.* 230:250–263.
11. Radika, K., and Northrop, D. B. (1984) *Biochemistry* 23:5115–5122.
12. Radika, K., and Northrop, D. B. (1984) *J. Biol. Chem.* 259:12543–12546.
13. Gates, C. A. (1986) *Kinetic Mechanism of Aminoglycoside Nucleotidyltransferase 2″-I.* University of Wisconsin School of Pharmacy, Madison.
14. Theorell, H., and Chance, B. (1951) *Acta Chem. Scand.* 5:1127–1144.

34
Kinetics of Irreversible Modification of Enzyme Activity

WEI LIU and C. L. TSOU

Enzyme inhibition has always been an important field of study owing not only to its usefulness in providing information on such fundamental aspects of biochemical problems as enzymatic catalysis and metabolic pathways but also to its implications in pharmacology and toxicology. In most textbooks on enzyme kinetics, chapters on enzyme inhibition have been devoted almost entirely to reversible inhibition kinetics with barely a passing mention on irreversible inhibition. However, developments have shown that irreversible inhibition of enzyme activity is equally important. Chemical modification studies with irreversible inhibitors can provide definitive information on the nature of the functional groups involved in enzymatic catalysis, which is not possible to obtain with reversible inhibitors. On the other hand, many chemotherapeutic agents and pesticides are also irreversible or quasi-irreversible enzyme inhibitors.

Some years ago a systematic study on the kinetics of irreversible modification of enzyme activity was presented. From the equations derived for the substrate reaction in the presence of the modifier (1, 2), it has been shown that the apparent rate constant for the irreversible modification of enzyme activity can often be obtained in a single experiment. It now appears that the importance of the kinetics of irreversible enzyme modification reactions is beginning to be appreciated, and experimental studies based on this approach have produced useful results not only for inhibition kinetics (3–5), but also for activation kinetics in this and other laboratories (6).

General Considerations

The following scheme applies to both reversible and irreversible modifications of enzyme activity by the modifier Y, including activation and inhibition.

$$\begin{array}{ccccc} & \mathrm{Y} & & \mathrm{Y} & \\ & + & & + & \\ \mathrm{S} + & \mathrm{E} & \underset{k_{-1}}{\overset{k_{+1}}{\rightleftharpoons}} & \mathrm{ES} & \xrightarrow{k_{+2}} \mathrm{E} + \mathbf{P} \\ & k_{+0} \Big\downarrow\Big\uparrow k_{-0} & & k_{+0'} \Big\downarrow\Big\uparrow k_{-0'} & \\ \mathrm{S} + & \mathrm{EY} & \underset{k_{-1'}}{\overset{k_{+1'}}{\rightleftharpoons}} & \mathrm{EYS} & \xrightarrow{k_{+2'}} \mathrm{EY} + \mathbf{P} \end{array} \quad [1]$$

If Km and Km′ are the Michaelis constants and the modification reactions are relatively slow compared with the setup of the respective steady states of the enzymatic reactions, the rate of modification is

$$-\frac{d[E]_T}{dt} = \left[\frac{[Y](k_{+0}Km + k_{+0'}[S])}{Km + [S]} + \frac{k_{-0}Km' + k_{-0'}[S]}{Km' + [S]}\right][E]_T - \frac{[E]_0(k_{-0}Km' + k_{-0'}[S])}{Km' + [S]} \quad [2]$$

where $[E]_T$ and $[E]_T'$ = total concentrations of the unmodified and modified enzymes, respectively
$[E]_0$ = total enzyme concentration

The result can be written as:

$$-\frac{d[E]_T}{dt} = (A[Y] + B)[E]_T - B[E]_0 \quad [3]$$

where A and B = apparent rate constants for the binding and the dissociation of the modifier to the enzyme, respectively

For inhibitors leading to complete suppression of enzyme activity, i.e., EYS is inactive, the expression for the substrate reaction in the presence of the modifier is:

$$\frac{d[P]}{dt} = \frac{k_{+2}[E]_0[S]}{(Km + [S])(A[Y] + B)}[B + A[Y]e^{-(A[Y]+B)t}] \quad [4]$$

For irreversible modifications where both k_{-0} and $k_{-0'}$ equal zero, (hence B = 0) and when boundary conditions are so chosen that the change in [S] does not affect significantly the [E]/[ES] and [EY]/[EYS] ratios, integration of Eq. 4 gives:

$$[P] = \frac{v}{A[Y]}(1 - e^{-A[Y]t}) \quad [5]$$

The amount of product formed when t approaches infinity, $[P]_\infty$, is given by v/A[Y]. Figure 34.1 shows a plot of [P] versus t for the inhibition of acetylcholinesterase by paraoxon. The asymptote gives the value of $[P]_\infty$. As [Y] is known and v can be determined in experiments without the inhibitor, the

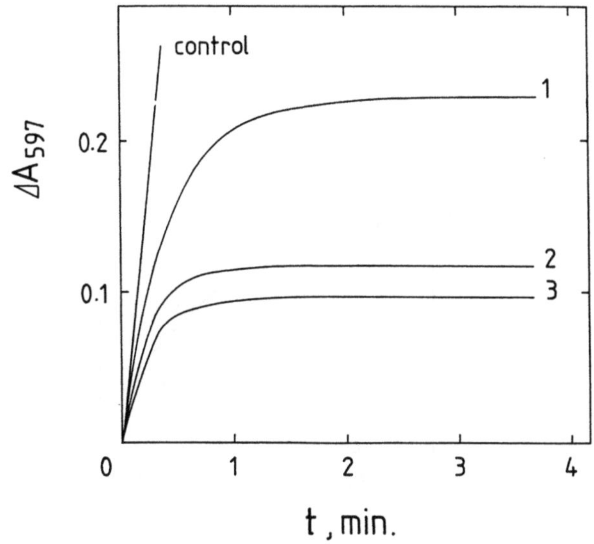

FIG. 34.1. Course of product formation during the action of acetylcholinesterase in the presence of paraoxon. The hydrolysis of acetylcholine at pH 8.5 and 25°C was followed by a pH change indicator (thymol blue) method in the presence of the inhibitor of different concentrations (0.51, 1.01, and 1.26 m*M* for curves 1, 2, and 3, respectively).

apparent rate constant A can be determined in a single experiment. Alternatively, it can be easily shown that a plot of $\log([P]_\infty - [P])$ versus t gives a straight line, as shown in Fig. 34.2, and A can be obtained from the slope that is 0.43A[Y]. From the above it can be easily shown that the microscopic rate constants k_{+0} and k_{+0}' can be determined from suitable plots of $1/[P]_\infty$ versus 1/[S].

The general equations obtained above apply to both inhibition and activation, and the substrate reaction during enzyme activation can be treated in a similar way. If the enzyme is completely inactive before activation, as in most cases of zymogen activation, i.e., only [EYS] is active, the course of product formation is

$$[P] = \frac{k_{+2}'[E]_0[S]A[Y]}{(Km' + [S])(A[Y] + B)}[t - \frac{1}{A[Y] + B}(1 - e^{-(A[Y]+B)t})] \qquad [6]$$

When t is sufficiently large, the exponential term can be neglected the curve approaches a straight line that cuts the abscissa at 1/(A[Y] + B). Both A and B can be obtained from a plot of the reciprocal of the intercepts at different values of [Y]. The slope of the straight line gives the rate of the activated enzyme.

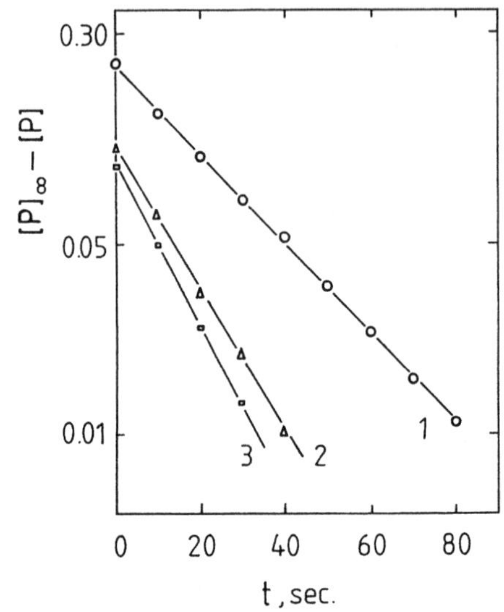

FIG. 34.2. Semilogarithmic plots for determination of the apparent rate constants of inhibition. Data were taken from Fig. 34.1 for the plots. The numbering of the lines is also as for Fig. 34.1.

Inhibitors of the Complexing Type

Affinity labeling has been used extensively to study the active sites of enzymes with modifiers that complex reversibly to enzymes to form EY prior to an irreversible modification leading to EY′. As the complexing inhibitors are usually sufficiently similar in structure to the substrate or the transition state intermediate, they probably occupy the substrate binding site so that the ternary complex EYS does not form. For irreversible inhibitions, it can be shown that the apparent rate constant (A) and the corresponding equation for [P] are given by

$$A = \frac{k_{+3}Km}{KmK_0 + K_0[S] + Km[Y]} \qquad [7]$$

$$[P] = \frac{VK_0[S]}{k_{+3}Km[Y]}(1 - e^{-A[Y]t}) \qquad [8]$$

where k_{+3} = rate constant for EY → EY′ conversion
K_0 = dissociation constant for Y

From the above it is clear that the [P] versus t plot also approaches a finite value, $[P]_\infty$, as in the case of noncomplexing inhibitions, but the expression for $[P]_\infty$ contains the rate constant k_{+3}, instead of A. The apparent rate

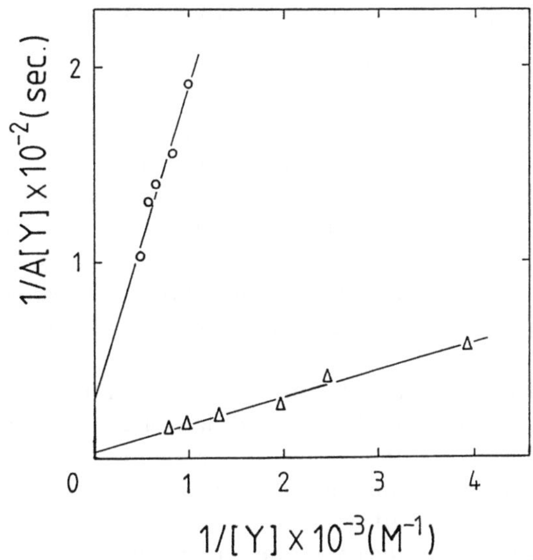

FIG. 34.3. The 1/(A[Y]) versus 1/[Y] plot showing complexing of the inhibitors with the enzyme. (o) DEP; (Δ) paraoxon.

constants are different for complexing and noncomplexing inhibitions in that [A] is independent of [Y] for noncomplexing inhibitions, but a plot of 1/A versus [Y] gives a straight line cutting at the ordinate for complexing inhibitions. Figure 34.3 shows such a plot from the results obtained with acetylcholinesterase.

Enzyme Reactions Involving Two Substrates

Most enzyme reactions involve more than one substrate, and previous kinetic treatments have dealt almost entirely with the irreversible modification of enzymes involving a single substrate. The corresponding kinetics of enzyme reactions involving two substrates has now been considered. The equations derived depend on the mechanism of the substrate binding sequences and are considerably more complex than those for single substrate reactions. However, the methods proposed above for the evaluation of the apparent rate constant also apply to reactions involving two substrates. The microscopic rate constants can be obtained by suitable secondary plots. For complexing inhibitors in all cases of sequential substrate binding mechanisms, the plots of the reciprocal of the apparent rate constant A versus [Y] give straight lines that cut at the ordinate and can thus be distinguished from the noncomplexing inhibitors.

Concept of Substrate Competition for Reversible Inhibition: Application to Irreversible Inhibition

For reversible inhibition it has been common practice to distinguish three kinds of inhibition—competitive, noncompetitive, and uncompetitive—depending on the effect of the inhibitor on the apparent Km. As Scheme 1 applies to both reversible and irreversible inhibitions, it is evident that the concept of substrate competition also applies to both cases. Detailed kinetic criteria have now been given to irreversible inhibitors. Table 34.1 summarizes the kinetic criteria for substrate competition of both reversible and irreversible inhibitions.

The same set of definitions has been employed for both reversible and irreversible inhibitions, and similar conditions are used to distinguish the three types of substrate competition. In competitive, noncompetitive, and uncompetitive irreversible inhibitions, as for reversible inhibitions, substrate binding prevents, does not affect, or promotes inhibitor binding, respectively. For reversible inhibitions, the conditions are based on the effect of substrate on the dissociation constants between E and Y, whereas in irreversible inhibitions these conditions are based on the effect of substrate on the rate constants for the formation of EY. It should be pointed out that for reversible inhibition it is required that $K_0 = K_0'$. Although strictly speaking $k_{+0} = k_{+0}'$ is not necessarily required for the two dissociation constants to be equal, as an operational definition it would be more convenient to use this criterion for irreversible noncompetitive inhibition. Figure 34.4 shows that the isopropyl fluorophosphate (DFP) and paraoxon inhibitions of acetylcholinesterase are competitive.

As the site of action of irreversible inhibitors can usually be determined, the demonstration of a strict competitive irreversible inhibition is useful in locating the site for substrate binding. Competitive irreversible inhibition has been hitherto referred to as substrate protection without a strict kinetic criterion.

From the last column of Table 34.1 it can be seen that the three types of substrate competition in irreversible inhibitions can be distinguished by suitable plots of [A] and [S]. Alternatively, as $[P]_\infty$ is directly obtainable from inhibition experiments, the substrate competition types can also be ascertained by suitable plots of $[P]_\infty$ and substrate concentrations.

Experimental Studies

The above approach of irreversible inhibition kinetics by following the course of substrate reaction in the presence of the inhibitor has been applied to systems that have been thoroughly studied by conventional methods, the kinetics of inhibition of chymotrypsin (2) and the inhibition of acetylcholinesterase by DFP and paraoxon (5). The rate constants obtained are in general

TABLE 34.1. Kinetic criteria for various types of inhibition.

		Reversible		Irreversible	
Inhibition Type	Definition	Condition	Fractional Activity at Steady-State	Condition	A
Competitive	Y does not bind to ES	$1/K_0' = 0$	$\frac{1 + [S]/Km}{1 + [S]/Km + [Y]/K_0}$	$k_{+0}' = 0$	$\frac{k_{+0}}{1 + [S]/Km}$
Noncompetitive	Y does not affect ES binding	$K_0 = K_0'$	$\frac{1}{1 + [Y]/K_0}$	$k_{+0} = k_{+0}'$	k_{+0}
Uncompetitive	Y binds to ES only	$1/K_0 = 0$	$\frac{1 + [S]/Km}{1 + [S]/Km + [Y]/K_0'Km}$	$k_{+0} = 0$	$\frac{k_{+0}'[S]}{Km + [S]}$

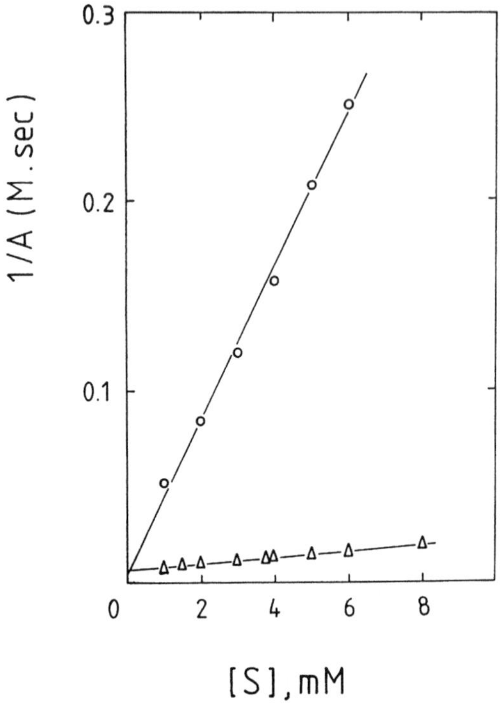

FIG. 34.4. The 1/A versus [S] plot showing competition of DFP and paraoxon with the substrate. Final concentrations for the inhibitors were 2 mM for DFP (○) and 0.51 mM for paraoxon (△).

accord with those obtained by conventional methods. As the apparent rate constant for the irreversible inhibition of an enzyme can be obtained in a single experiment, the above approach should be particularly useful in a comparison of the inhibitory power of a series of similar inhibitors, as shown in the studies by Harper et al. (7, 8) on the comparison of the inhibition rate constants of isocoumarins on the serine proteases.

Reactivation Kinetics

Acetylcholinesterase is an important enzyme involved in nerve conduction and consequently has been studied extensively. It is inactivated as a result of phosphorylation at the active site serine residue, and the inhibited enzyme can be readily reactivated by nucleophilic reagents such as pyridine-2-aldoxime methochloride (PAM). Although the kinetics of the reactivation have been extensively studied, the studies have usually been carried out in such a way that a fast initial phase of the reaction can be easily missed.

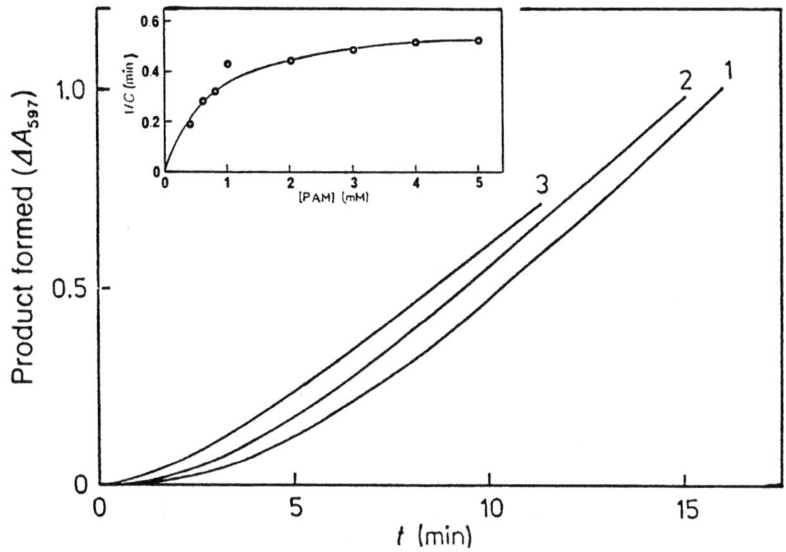

FIG. 34.5. Course of reactivation of phosphoryl enzyme by PAM. The substrate reaction was followed in the presence of the reactivator in final concentrations of 0.4, 0.6, and 5.0 m*M* for curves 1, 2, and 3, respectively. *Inset*: 1/C against [Y] plot.

In the present approach, the reactivation has been followed by the substrate reaction in the presence of PAM (6). The plots of [P] versus t with different [PAM] give a series of curves in agreement with Eq. 6 for product formation during enzyme activation in the presence of both the substrate and the activator. When t is sufficiently large, the curves approach straight lines with slopes giving the rate of the reactivated enzyme. The straight line portions of the curves should cut the *x*-axis at 1/(A[Y]), and a plot of the reciprocal of the intercepts against [Y] should yield a straight line with a slope giving the apparent reactivation rate constant. However, in the present case such a plot gave a curve similar in shape to a substrate saturation curve, suggesting complex formation and saturation at high modifier concentrations (Fig. 34.5).

From the known mechanism of catalysis of this enzyme and the fact that the reactivator, PAM, is also a reversible competitive inhibitor that may well complex with the enzyme prior to reactivation, the scheme for the reactivation reactions in the presence of the substrate can be described as:

$$\begin{array}{ccccccccccc}
EI & \xrightleftharpoons[k_{-4}]{k_{+4}[Y]} & EIY & \xrightarrow[IY]{k_5} & E & \xrightleftharpoons[k_{-1}]{k_{+1}[S]} & ES & \xrightarrow[P_1]{k_2} & ES' & \xrightarrow[H_2O]{k_3} & E + P_2 \\
k_{-7} \Big\Uparrow \Big\Downarrow k_{+7[S]} & & & & k_{-6} \Big\Uparrow \Big\Downarrow k_{+6[Y]} & & & & & & \\
EIS & & & & EY & & & & & &
\end{array} \qquad [9]$$

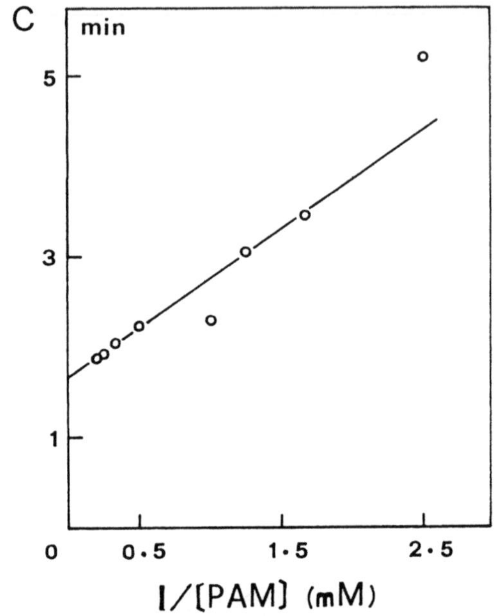

FIG. 34.6. C versus 1/[Y] plot. Experimental results obtained for Fig. 34.5 are replotted here to show the linear relation in agreement with Eq. 10.

The rate equation, for product formation during reactivation contains two exponential terms:

$$[P] = \frac{K[S][EI]_0}{Km(1 + [Y]/K_6) + [S]}\left[t + \frac{A_2}{A_1 G}e^{-A_1 t} - \frac{A_1}{A_2 G}e^{-A_2 t} - C\right], \qquad [10]$$

$$C = \frac{(K'/D) + [Y]}{k_5[Y]}$$

When t is sufficiently large, both exponential terms can be neglected and the curve approaches a straight line. It can be easily seen that a plot of the *x*-intercept (C) versus 1/[Y] should give a straight line, as has indeed been found to be the case (Fig. 34.6).

Equation 10 can be written in the following form.

$$\frac{[P]}{v} - t + C = B_1 e^{-A_1 t} - B_2 e^{-A_2 t} \qquad [11]$$

A semilogarithmic plot of the left-hand side terms against the right hand side indeed gives a curve that can be resolved into two straight lines as is found to be the case (Fig. 34.7). Using the classic approach of taking aliquots at intervals and measuring the activity recovery, the initial phase of the reactivation process can be easily missed. Using the present approach, the apparent

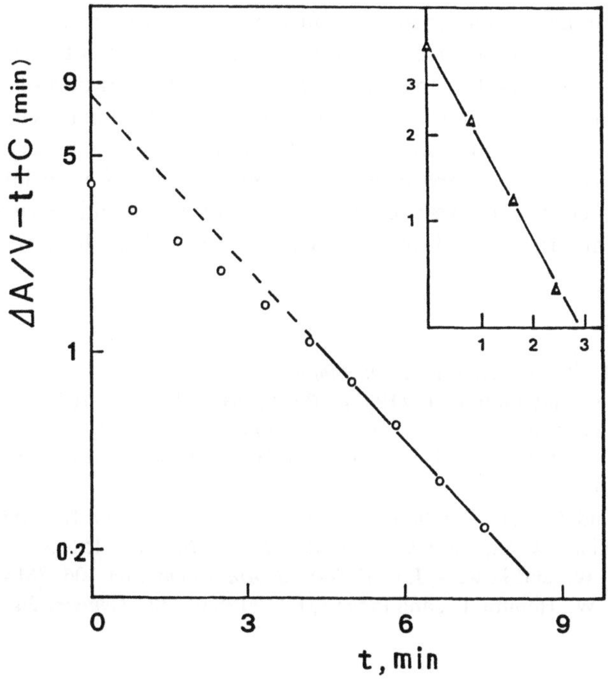

FIG. 34.7. Semilogarithmic plot of the course of reactivation. *Inset*: Semilogarithmic plot of the difference of the extrapolated straight line with the actually observed points in the initial portion of the curve.

rate constant can be obtained in a single experiment, and with a stopped-flow apparatus first-order rate constant of the order of 10 s^{-1} can be easily obtained.

The simplicity of the present method would be a great advantage when comparing the inhibitory power of a large number of compounds on a key enzyme as could be the case in the screening of possible chemotherapeutic agents. In this respect, the presence of substrate during determination of the rate constant for the modification reaction is desirable, as during the in vivo action of these chemotherapeutic agents the presence of substrate is inevitable. The rate constant for the binding of the modifier with the enzyme in the absence of the substrate can be obtained after ascertaining the inhibition type with experiments that employ different substrate concentrations and extrapolating the substrate concentration to zero in suitable plots.

The expression I_{50} (the inhibitor concentration that produces 50% inhibition) has been indiscriminately used for the quantitative assessment of inhibitory power. For a comparison of the results obtained by different authors under unavoidably different conditions, employing a term as little perturbed by experimental conditions as possible is desirable. In this respect, the use of

I_{50} is unfortunate for its dependence on substrate concentration of reversible inhibitions during its determination, which may or may not be the same in different laboratories. For irreversible inhibitors, the duration of incubation of the inhibitor with the enzyme should also be added, which, as is frequently the case, may vary considerably with different authors. Consequently it is almost impossible to compare the results obtained in different laboratories. As the apparent rate constant is easily determined, it could serve as a useful basis for comparing the inhibitory power of irreversible inhibitors.

References

1. Tsou, C. L. (1965) *Acta Biochim. Biophys. Sin.* 5:398–417.
2. Tian, W. X., and Tsou, C. L. (1982) *Biochemistry* 21:1028–1032.
3. Forsberg, A., and Puu, G. (1984) *Eur. J. Biochem.* 140:153–156.
4. Leytus, S. P., Toledo, D. L., and Mangel, W. F. (1984) *Biochim. Biophys. Acta* 788:74–86.
5. Liu, W., and Tsou, C. L. (1986) *Biochim. Biophys. Acta* 870:185–190.
6. Liu, W., Zhao, K. Y., Tsou, C. L. (1985) *Eur. J. Biochem.* 151:525–529.
7. Harper, J. W., and Powers, J. C. (1984) *J. Am. Chem. Soc.* 106:7618–7619.
8. Harper, J. W., Hemmi, K., and Powers, J. C. (1985) *Biochemistry* 24:1831–1843.

35
Applications of Stable Isotopes in Biochemistry

MARION H. O'LEARY

The use of radioactive isotopes has been very important in the development of modern biochemistry and is currently standard practice in almost all biochemistry laboratories. The use of stable isotopes, on the other hand, has developed more slowly, perhaps because analysis of materials labeled with stable isotopes is generally more difficult than analysis of materials labeled with radioactive isotopes.

However, in the last few years, biochemical applications of stable isotopes have increased. Tracer studies using nonradioactive isotopes sometimes have significant advantages over studies using radioactive isotopes. Kinetic isotope effects using stable isotopes are increasingly used in studies of enzyme reaction mechanisms. Part of this increase in interest results from a desire to find alternatives to classic radiochemical procedures, and part results from advances in the technology of measuring stable isotopes of carbon, oxygen, nitrogen, and other elements (principally by nuclear magnetic resonance or by mass spectrometry). In addition, highly-enriched isotopic compounds (>90% enrichment) are now available at reasonable prices. The purpose of this chapter is to survey applications of stable isotopes in biochemistry. Examples are drawn principally from our own work. We describe both tracer studies and isotope effect studies.

Isotope-Labeling Studies

Nuclear magnetic resonance (NMR) is often the method of choice for studying carbon-13-enriched compounds. The position of the isotope label within a metabolite can often be determined from the chemical shift in the NMR spectrum, unlike the case with radioactive labels, whose positions must generally be determined by degradation. The concentration of the label can sometimes be determined from the intensity of the NMR signal (although a number of problems make this a difficult procedure). Studies of labeled compounds attached to proteins, or even of the proteins themselves, are now possible.

Ribulose bisphosphate carboxylase, the first enzyme in the Calvin cycle of

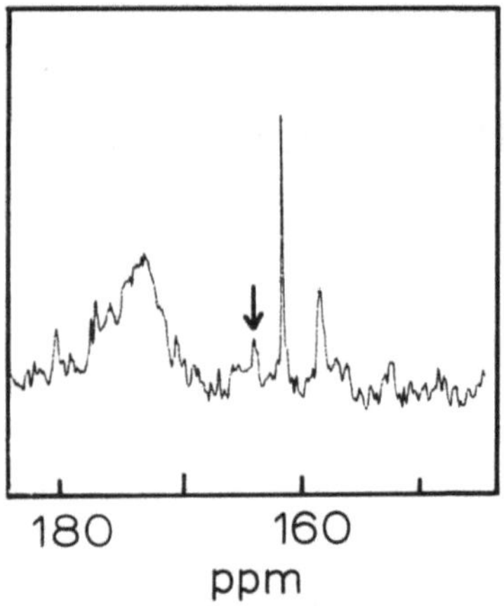

FIG. 35.1. ^{13}C NMR spectrum of the 175 ppm region of ribulose bisphosphate carboxylase from *Rhodospirillum rubrum* following activation with $^{13}CO_2$ and Mg^{2+}. Protein concentration = 50 mg/ml. The peak indicated by the arrow is the carbamate. The tallest peak is free HCO_3^-.

CO_2 fixation in plants (1), must be activated by CO_2 and Mg^{2+} before it is catalytically active. If the enzyme is activated with $^{13}CO_2$, then ^{13}C NMR can be used to demonstrate that CO_2 forms a carbamate by reacting with the amino group of a lysine residue on binding to the enzyme (Fig. 35.1) (2):

$$E{-}NH_2 + CO_2 + Mg^{2+} \rightleftharpoons E{-}NH{-}CO_2^-Mg^{2+}$$

This structure has subsequently been confirmed by esterification and isolation of the carbamate (3).

Isotope tracer studies using oxygen are more difficult than those using carbon because oxygen does not have a useful radioactive isotope and the common isotopes (^{16}O and ^{18}O) are NMR-silent. However, NMR can be used to study oxygen-18 provided a high-field NMR spectrometer is available. The presence of attached oxygen-18 causes a very small change (~1 Hz) in the chemical shift of an attached carbon, phosphorus, or other magnetic nucleus (4). For example, by use of high-field ^{13}C NMR, separate signals can be seen for a ^{13}C-labeled carboxyl group with zero, one, or two attached oxygen-18 atoms (Fig. 35.2) (5).

This fact has permitted development of a new method for determining whether CO_2 or HCO_3^- is the substrate for a carboxylase. If a carboxylase is supplied with $^{13}C^{18}O_2$ under conditions where CO_2 hydration is slow, then the oxygen-18 content of the product reflects the nature of the carboxylation step. If the carboxylation requires CO_2, then the carboxylic acid product is formed with full retention of the ^{18}O label. However, if the carboxylation requires HCO_3^-, then one-third of the label is lost following the conversion of $^{13}C^{18}O_2$ to $H^{13}C^{18}O_2{}^{16}O^-$ (Fig. 35.3). Analysis of the ^{18}O content can be accomplished by ^{13}C NMR of the product carboxylic acid.

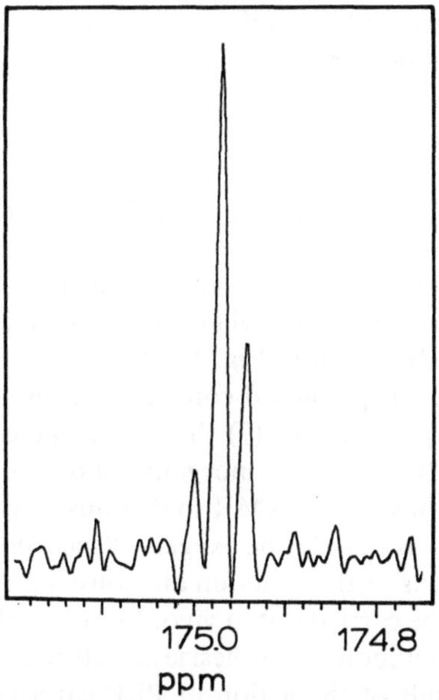

FIG. 35.2. ^{13}C NMR spectrum of the carbon-4 carboxyl region of the spectrum of malic acid formed by carboxylation of PEP with $^{13}C^{18}O_2$. The three peaks represent malic acid containing (from left to right) zero, one, and two atoms of ^{18}O.

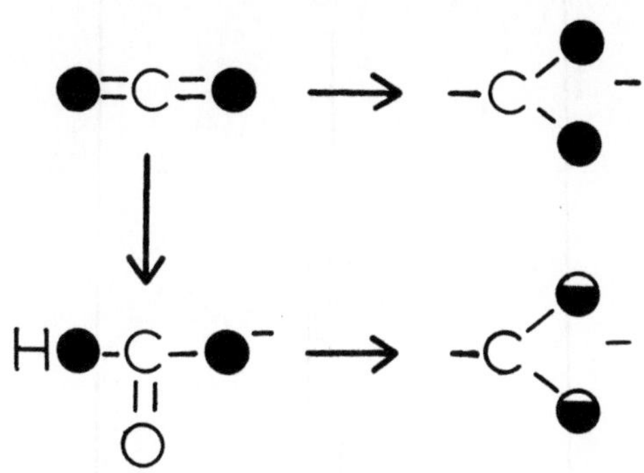

FIG. 35.3. Predicted labeling pattern for a carboxyl group formed by carboxylation with $C^{18}O_2$. If CO_2 is directly incorporated into the carboxylic acid, the extent of labeling in the acid is the same as that in the CO_2 (top line). However, if CO_2 must be converted to HCO_3^- prior to carboxylation, the extent of labeling of the carboxylic acid is two-thirds that of the CO_2 (bottom line).

This method has been used to confirm the previously reported (6) substrate requirement for phosphoenolpyruvate (PEP) carboxylase. This enzyme catalyzes the reaction (6)

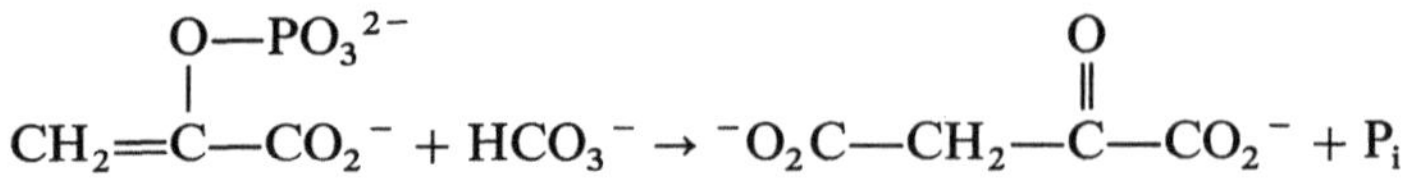

Carboxylation of PEP using $^{13}C^{18}O_2$ containing 88% ^{18}O gave oxalacetate containing 59% ^{18}O in the γ-carboxyl group, as expected if CO_2 is converted to HCO_3^- prior to carboxylation (Fig. 35.2).

A similar technique has permitted confirmation of an old observation. It was shown by mass spectrometry (7) that when the carboxylation of PEP is carried out using $HC^{18}O_3^-$ the inorganic phosphate produced contains oxygen-18. We used phosphorus NMR to demonstrate that this observation is correct and that there is exactly one oxygen-18 incorporated (Fig. 35.4). The mechanistic implications of this reaction are considered below.

Similar studies can be carried out in vivo. For example, plants that engage in Crassulacean acid metabolism synthesize and store large quantities of malic acid at night as a result of the action of PEP carboxylase and malate dehydrogenase. If the plants are exposed to $^{13}CO_2$ at night and then the malic acid is extracted and examined by NMR, it is found that carbon-1 and carbon-

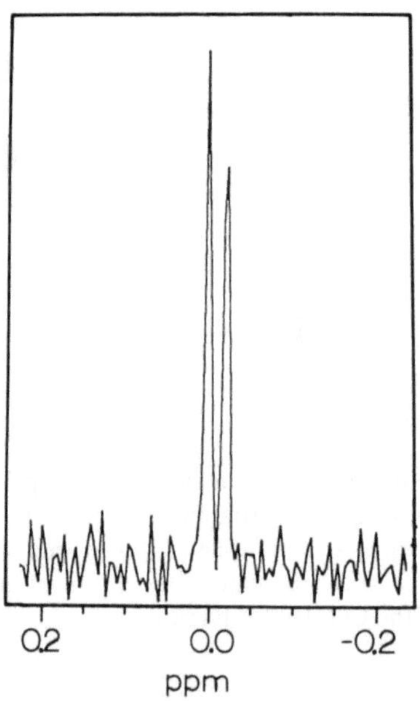

FIG. 35.4. ^{31}P NMR spectrum of inorganic phosphate formed by carboxylation of phosphoenolpyruvate with $HC^{18}O_3^-$. The left peak is for P_i containing no ^{18}O; the right peak is for one ^{18}O.

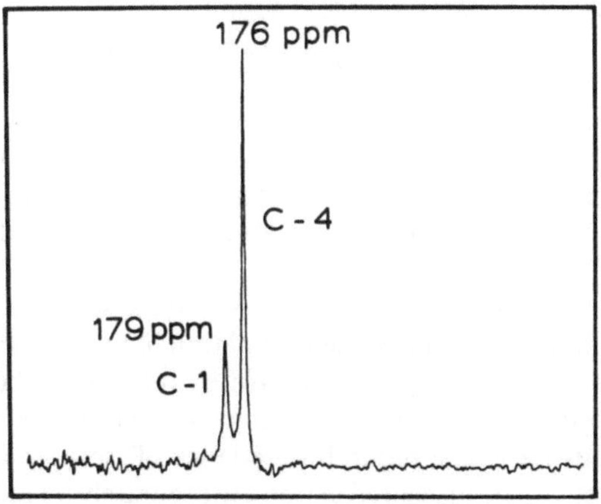

FIG. 35.5. ^{13}C NMR spectrum of the carboxyl region from an extract of the CAM plant *Kalanchoe daigremontiana* following overnight exposure to $^{13}CO_2$. The peaks at 176 and 179 ppm are, respectively, carbon-4 and carbon-1 of malic acid.

4 are labeled in a ratio of about 1:2 (Fig. 35.5), presumably as a result of the action of fumarase on 4-^{13}C-malate (5):

$$^{-}O_2{}^{13}C\text{—}CH_2\text{—}\underset{}{\overset{\large OH}{\overset{|}{C}H}}\text{—}CO_2{}^{-} \rightleftharpoons {}^{-}O_2{}^{13}C\text{—}CH\text{=}CH\text{—}CO_2{}^{-}$$

$$\rightleftharpoons {}^{-}O_2C\text{—}CH_2\text{—}\overset{\large OH}{\overset{|}{C}H}\text{—}{}^{13}CO_2{}^{-}$$

Analysis of the trimethylsilyl derivative of malate by mass spectrometry confirms this observation and further demonstrates that no malic acid is labeled in both carbon-4 and carbon-1, thereby eliminating the possibility that malate is formed by action of RuBP carboxylase followed by PEP carboxylase (8). Because ^{14}C is always used as a trace label, this last experiment cannot be performed with ^{14}C; instead, ^{13}C, which can be used in stoichiometric quantities, is required. If the same experiment is carried out using $^{13}C^{18}O_2$, either NMR or mass spectrometry can be used to demonstrate that the malic acid contains oxygen-18. Such data permit calculation of the level of carbonic anhydrase in the plant (5).

Kinetic Isotope Effects

Kinetic isotope effects are rate differences resulting from isotopic substitution. When deuterium is substituted for hydrogen, rate reductions of up to tenfold (and occasionally larger) may result. For example, the enzyme formate dehydrogenase catalyzes the reaction

$$HCO_2^- + NAD^+ \rightarrow CO_2 + NADH$$

Deuteration of formate leads to a hydrogen isotope effect of $k^H/k^D = 2.1$ on V_{max} and 2.3 on V_{max}/K_m (9). For isotopes of elements other than hydrogen, the differences are always small, never exceeding a few percent. For example, substitution of carbon-13 into formate leads to an isotope effect $k^{12}/k^{13} = 1.042$ (9).

Because heavy-atom kinetic isotope effects are always small, such effects are usually not measured by direct determination of rate constants for the two isotopic species. Instead, they are obtained by measuring the change in the isotopic composition of either the starting material or the product as the reaction proceeds (10). The natural abundance of isotopes is usually used. Such abundances can be measured extremely accurately by means of isotope-ratio mass spectrometry. We draw examples from studies of our own on enzymes involved in photosynthesis.

The dehydration of HCO_3^- and its reverse, the hydration of CO_2, have been studied for many years:

$$H^+ + HCO_3^- \rightleftharpoons CO_2 + H_2O$$

The reaction occurs spontaneously with a half-time of about 30 s near room temperature. The reaction catalyzed by carbonic anhydrase is many orders of magnitude faster than this.

Carbon isotope effects on the spontaneous dehydration of HCO_3^- were measured by conducting the reaction under vacuum so that the product CO_2 was removed from solution as rapidly as it was formed (11, 12). The isotope effect was calculated by comparison of the isotopic composition of the product CO_2 with that of the substrate HCO_3^-. The validity of the method was checked by showing that when the reaction was conducted in water containing ^{18}O the CO_2 produced contained no ^{18}O. In addition, in separate experiments the isotope effect on the hydration of CO_2 was measured by coupling the HCO_3^- formed to PEP carboxylase and malate dehydrogenase. Decarboxylation of the malate thus formed with malic enzyme permitted calculation of the isotope effect for CO_2 hydration. The ratio of this value to that for the dehydration of HCO_3^- gives the equilibrium isotope effect, as expected (11).

Table 35.1 shows some experimental data from our laboratory for the spontaneous dehydration of HCO_3^-. These data were obtained by two independent investigators measuring the isotope effect under the same conditions. The data illustrate the level of precision that can be achieved in isotope effects with properly controlled systems.

Carbon isotope effects on the spontaneous dehydration of HCO_3^- under a variety of conditions are shown in Table 35.2 (11, 12; P. Paneth, unpublished data). These isotope effects are smaller than those observed in other reactions

TABLE 35.1. Carbon isotope effects on the dehydration of HCO_3^- at pH 8.2, 24°C, in aqueous solution.

Isotope Ratio $^{13}CO_2/^{12}CO_2(\times 10^6)$		
CO_2 Product	HCO_3^- Substrate	k^{12}/k^{13}
	Data of Marlier and O'Leary (11)	
11256	11407	1.0140
11239	11410	1.0156
11243	11410	1.0152
11254	11405	1.0139
11245	11404	1.0146
		1.0147 ± 0.0007 (mean ± SD)
	Data of P. Paneth (unpublished)	
11239	11393	1.0142
11228	11393	1.0152
11233	11391	1.0143
11235	11395	1.0147
11230	11392	1.0149
		1.0147 ± 0.0004 (mean ± SD)

TABLE 35.2. Carbon isotope effects on the spontaneous and enzymatic dehydration of HCO_3^- at 24°C under various conditions.

	k^{12}/k^{13}	
Conditions	Spontaneous Reaction	Enzymatic Reaction
In H_2O	1.0199 ± 0.0004	1.0101 ± 0.0004
In D_2O	1.0224 ± 0.0004	1.0107 ± 0.0007
In H_2O at high pH	1.0103 ± 0.0003	

where CO_2 is formed (such isotope effects are usually in the range $k^{12}/k^{13} = 1.04 - 1.07$) and indicate that the reaction occurs by a two-step mechanism:

$$H^+ + HO{-}CO_2^- \underset{k_2}{\overset{k_1}{\rightleftharpoons}} H_2O^+{-}CO_2^- \overset{k_3}{\rightarrow} H_2O + CO_2$$

For such a mechanism, the observed isotope effect is smaller than the isotope effect on the carbon-oxygen bond breaking step. The observed isotope effect is given by

$$\text{Observed } k^{12}/k^{13} = (k_3{}^{12}/k_3{}^{13} + k_3/k_2)/(1 + k_3/k_2)$$

We have assumed that there is no carbon isotope effect on k_1 or k_2. The size of the isotope effect suggests that k_3/k_2 is 1 to 2. The change in isotope effect in D_2O occurs because D_2O decreases k_3/k_2. At higher pH the mechanism changes to direct decomposition of HCO_3^-, and the isotope effect decreases.

For the enzymatic reaction, the isotope effect (13) is similar in size to that observed in the high pH mechanism and is nearly independent of reaction conditions (Table 35.2). The isotope effects are consistent with other evidence (14) which indicates that the reaction follows ping-pong kinetics, with HCO_3^- dehydration and proton transfer occurring in separate half-reactions. For the dehydration half-reaction, the mechanism seems to involve kinetically significant binding and dissociation steps:

$$E + HCO_3^- \rightleftharpoons E\text{—}HCO_3^- \rightleftharpoons E\text{—}CO_2 \rightleftharpoons E + CO_2$$

In this case, the fact that the isotope effect is always the same indicates that the relative rates of the three steps are independent of D_2O, viscosity, and metal ion.

Phosphoenolpyruvate carboxylase has been studied in detail by means of isotope-effect techniques (15). As noted above, the substrate for the enzyme is HCO_3^-, rather than CO_2, and an atom of ^{18}O is transferred from $HC^{18}O_3^-$ to P_i in the course of the reaction. For the carbon atom of HCO_3^- the carbon isotope effect is $k^{12}/k^{13} = 1.0029$ at pH 7.5, 25°C, and this value is independent of HCO_3^- concentration. The small size of this isotope effect indicates that the carbon–carbon bond-forming step is not rate-determining. Instead, the reaction appears to occur by a two-step mechanism:

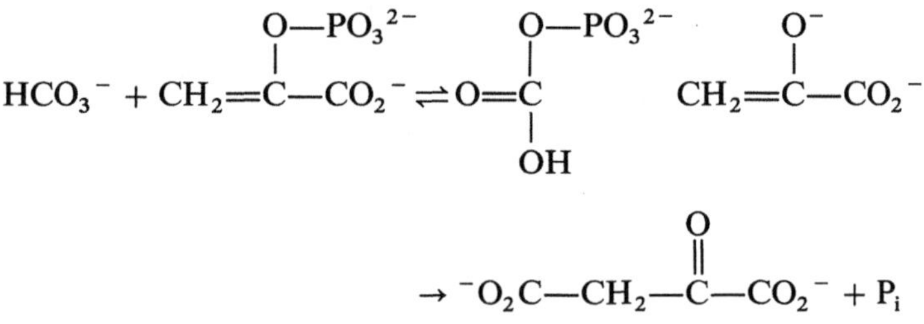

The first step is transfer of phosphate from PEP to HCO_3^-, forming the anhydride carboxyphosphate and the enolate of pyruvate. This step is rate-determining. In the second step, carboxylation occurs, forming oxalacetate and P_i. Consistent with this view is the fact that the oxygen isotope effect for the bridging oxygen of the phosphate ester is $k^{16}/k^{18} = 1.0047$ (P. Paneth, unpublished data). The isotope effect gets smaller as the concentration of HCO_3^- increases, indicating that substrate addition to the enzyme is ordered, with PEP binding first. This last isotope effect was measured by the "remote label" technique (16), which is often used for studying isotopic sites that are not amenable to direct isotope-ratio mass spectrometry.

Ribulose bisphosphate carboxylase, the first enzyme of the Calvin cycle, catalyzes the following reaction (1):

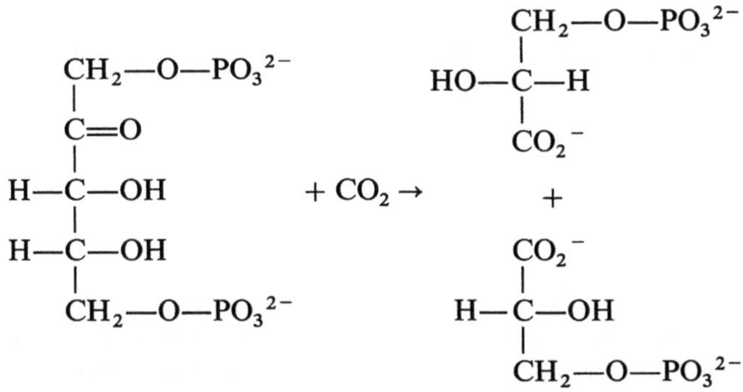

Isotope effects on this reaction are of interest for two reasons. First, these isotope effects are partially responsible for the large isotope fractionation seen during photosynthetic CO_2 fixation in C_3 plants (17). Second, these isotope effects can help us understand the mechanism of this carboxylation.

Measurement of the carbon isotope effect (for the carbon arising from CO_2) is made more difficult by the fact that the reaction produces two molecules of 3-phosphoglyceric acid, of which the carboxyl carbon of one, but not the other, arises from CO_2. It was necessary to find a method for distinguishing between these two carbons. This task was accomplished by using the fact that in the presence of Co^{2+} this enzyme catalyzes the oxygenation of ribulose bisphosphate, forming 3-phosphoglyceric acid from carbons 3 to 5 of ribulose bisphosphate, and this material can be used to obtain the needed isotopic composition (18).

Carbon isotope effects on the carboxylation of ribulose bisphosphate (RuBP) (Table 35.3) indicate that the reaction probably proceeds via the enediol intermediate first suggested by Calvin (19). Significant hydrogen isotope effects are observed for the hydrogen attached to carbon-3 (J. V. Schloss, personal communication). These isotope effects indicate that car-

TABLE 35.3. Carbon isotope effects for the carbon arising from CO_2 on the carboxylation of ribulose bisphosphate at 25°C.

Conditions	k^{12}/k^{13}
Enzyme from spinach	
pH 8.0	1.0288 ± 0.0013
pH 8.0, RuBP-3-d	1.0214 ± 0.0029
pH 9.0	1.0264 ± 0.0006
pH 7.0	1.0297 ± 0.0008
Enzyme from *Rhodospirillum rubrum*	
pH 8.0	1.0178 ± 0.0008

boxylation is partially, but not entirely, rate-determining. It appears that both carboxylation and hydrogen abstraction are partially rate-determining. It is interesting that the isotope effect is smaller with the enzyme from *Rhodospirillum rubrum* than with the enzyme from spinach (20).

Conclusions

Stable isotopes are finding increasing use in both fundamental and applied areas of biochemistry. NMR studies of carbon-13-labeled compounds have been used to demonstrate, for example, the formation of a carbamate at the active site of ribulose bisphosphate carboxylase. Study of oxygen-18-induced shifts in carbon-13 NMR spectra has allowed development of a new method for determining whether carbon dioxide or bicarbonate is the substrate for a variety of carboxylases. Studies of oxygen-18-induced isotope shifts in phosphorus NMR spectra has permitted confirmation of the early observation that label is transferred from oxygen-18-labeled bicarbonate ion into inorganic phosphate during the carboxylation of phosphoenolpyruvate. Kinetic isotope effects are now being used in a variety of studies of enzyme reaction mechanisms. Both the spontaneous and the enzyme-catalyzed dehydration of bicarbonate ion have been studied by this technique, and mechanisms for these processes have been elucidated. Carbon and oxygen isotope effects on the carboxylation of phosphoenolpyruvate have been interpreted in terms of a two-step mechanism. Carbon isotope effects on the carboxylation of ribulose bisphosphate by the ribulose bisphosphate carboxylase from spinach indicate that the reaction probably proceeds via the accepted enediol intermediate. The carboxylation step is partially, but not entirely, rate-determining.

Acknowledgments. I acknowledge the contributions of my many graduate students and postdoctoral associates whose work is discussed here, including Jeff Hermes, Joe Holtum, Robert Jaworski, Piotr Paneth, James Rife, and Chrissl Roeske. Financial support was provided by the National Science Foundation, the U.S. Department of Agriculture, the U.S. Department of Energy, and the University of Wisconsin Graduate School.

References

1. Miziorko, H. M., and Lorimer, G. H. (1983) *Annu. Rev. Biochem.* 52:507–535.
2. O'Leary, M. H., Jaworski, R. J., and Hartman, F. C. (1979) *Proc. Natl. Acad Sci. USA* 76:673–675.
3. Lorimer, G. H., and Miziorko, H. M. (1980) *Biochemistry* 19:5321–5328.
4. Risley, J. M., and Van Etten, R. L. (1979) *J. Am. Chem. Soc.* 101:252–253.
5. Holtum, J. A. M., Summons, R., Roeske, C. A., Comins, H. N., and O'Leary, M. H. (1984) *J. Biol. Chem.* 259:6870–6881.

6. O'Leary, M. H. (1982) *Annu. Rev. Plant Physiol.* 33:297–315.
7. Maruyama, H., Easterday, R. L., Chang, H. C., and Lane, M. D. (1966) *J. Biol. Chem.* 241:2405–2412.
8. Cockburn, W., and McAulay, A. (1975) *Plant Physiol.* 55:87–89.
9. Hermes, J. D., Morrical, S. W., O'Leary, M. H., and Cleland, W. W. (1984) *Biochemistry* 23:5479–5488.
10. O'Leary, M. H. (1980) *Methods Enzymol.* 64:83–104.
11. Marlier, J. F., and O'Leary, M. H. (1984) *J. Am. Chem. Soc.* 106:5054–5057.
12. Paneth, P., and O'Leary, M. H. (1985) *J. Am. Chem. Soc.* 107:7381–7384.
13. Paneth, P., and O'Leary, M. H. (1985) *Biochemistry* 24:5143–5147.
14. Silverman, D. N., and Vincent, S. H. (1984) *CRC Crit. Rev. Biochem.* 14:207–255
15. O'Leary, M. H., Rife, J. E., and Slater, J. D. (1981) *Biochemistry* 20:7308–7314.
16. O'Leary, M. H., and Marlier, J. F. (1979) *J. Am. Chem. Soc.* 101:3300–3306.
17. O'Leary, M. H. (1981) *Phytochemistry* 20:553–567.
18. Roeske, C. A., and O'Leary, M. H. (1984) *Biochemistry* 23:6275–6284.
19. Calvin, M. (1956) *J. Chem. Soc.* 1895–1915.
20. Roeske, C. A., and O'Leary, M. H. (1985) *Biochemistry* 24:1603–1607.

36
Enzyme Dynamics in Nonaqueous Media at Subzero Temperatures

Pierre Douzou

Despite the continuous development of new data acquisition techniques that reduce the time and improve the precision of recording, it is not possible to directly observe the nature and structure of most intermediates or to determine rates of conformational changes that accompany each activation process, stepping the reaction pathway either in solution or in the crystalline state. In order to circumvent this difficulty, we set up a procedure to investigate enzyme-catalyzed reactions at subzero temperatures in which the reaction rates are dramatically slowed according to the classic Arrhenius relation. In particular, reactions that normally occurred in a few seconds or minutes could be sufficiently slowed to permit enzyme–substrate intermediates to be stabilized and eventually isolated by column chromatography, and then studied by available spectroscopic techniques, including x-ray diffraction by enzyme crystals. The physicochemical basis of the procedure as well as its preliminary results have been reviewed elsewhere (1–3).

Experimental Procedures

A number of organic solvents (primary alcohols, polyols, etc.) can be used to depress markedly the freezing point of water; the higher their concentration, the lower is the freezing point of the mixture (1). In addition to the property of preventing water from freezing over a broad range of subzero temperatures, these solvents must preserve the dissolving power of the solutes in water, i.e., in the present case, enzymes and their substrates, neutral salts, and buffers. When each of the co-solvents listed in Table 36.1 (and many others) is present (up to 50% v/v), enzymes remain soluble at concentrations that allow their investigation (4), and suitable neutral salts and buffers can be used to recreate a suitable ionic strength.

Knowledge of changes in the pK values of buffers induced by organic solvents and selected subzero temperatures permits one to adjust the concentration of the buffer to recreate the actual proton activity (pH*). Thus one can investigate under the pH* conditions optimal for enzyme catalysis (5, 6).

TABLE 36.1. Binary solvent: volume ratios and freezing points.

Solvent	Freezing Point (°C) at Various Volume Ratios (v/v)								
	10/90	20/80	30/70	40/60	50/50	60/40	70/30	80/20	90/10
Ethylene glycol–water	−4	−10	−17	−26	−44	−69	−100	−83	−50
Methanol–water	—	—	—	−40	−49	−67	−85	−100	—
Polyethylene glycol–water	−35	−8	−17	−38	SC	SC	SC	SC	SC
Glycerol–water	−3	−8	−145	−29	SC	SC	SC	SC	SC
Dimethylsulfoxide–water	−3	−12	−19	−41	SC	SC	SC	−38	−7

The higher the concentration in cosolvent, the lower the freezing point, allowing investigation in fluid (but viscous) media at subzero temperatures.
SC = "supercooling" of solutions.

Under these conditions, most enzyme systems investigated so far have retained catalytic activity and can be investigated by available kinetic techniques over a broad range of subzero temperatures. However, careful trials must be carried out in each case to check this activity compared to that observed in aqueous solution. It is generally observed that in the presence of high cosolvent concentration the enzymatic activity to altered, and in most cases it inhibits. Such changes can be due to rather "trivial" causes, e.g., changes in the pK values of the ionizing groups at the active site of the enzymes; such changes can be compensated for by adusting the pH. (7). Diffusion-controlled reactions can be affected by abnormal viscosities of mixed solvents (3). In any case it must be checked that the effect of the organic solvent on enzyme activity does not increase as a function of time and is fully reversible by infinite dilution or dialysis; otherwise progressive denaturation of the enzyme would occur.

Because the goal of the low-temperature procedure is to investigate enzyme mechanisms, it is necessary to check that the actual pathway is identical to that which occurs under normal conditions of medium and temperature by establishing (a) the number of intermediates in the catalytic cycle; (b) the pH activity profiles, which must be characterized by curves of strictly identical shape in water and in mixed solvents (3); and (c) that the Arrhenius plots are linear over a broad range of normal and subzero temperatures (3).

The effect of temperature on the velocities of enzyme-catalyzed reactions can be estimated according to their activation energy (E_A) using the classic Arrhenius relation:

$$k = A \exp(-E_A/RT)$$

where k = rate constant of the reaction
A = freezing factor

According to this expected reduced rate, reactions that normally occur in the microsecond range (which can be monitored only by relaxation techniques) can be reduced to a millisecond time range and studied by rapid mixing techniques; and reactions that occur in the millisecond range and normally require stopped-flow techniques can be reduced to the time scale of seconds and thus can be studied by spectrophotometry at an ordinary speed of response.

Results

Temperature as a Tool to Investigate Enzyme–Substrate Intermediates

All of these expectations have been verified and reviewed (2, 3, 8, 9). One of the main goals of the subzero temperature approach was to stabilize the "groūnd-state" enzyme–substrate (ES) intermediates normally metastable and to better resolve the intermediates in the reaction pathway. Suitable cooling ⇀ warming "cycles" might permit the temporal resolution of some reactions to be obtained and allow structural analysis of their stabilized intermediates to be carried out.

The first temporal resolution of an enzyme-catalyzed reaction was carried out in 1970 in this laboratory (10). Since this preliminary work, our group and others working in collaboration with us or independently have successfully applied the subzero temperature approach to a variety of solubilized systems, including most classes of enzymes; in a number of cases the normally metastable enzyme–substrate complexes have been stabilized, sometimes purified, and then used as starting reagents to investigate some elementary steps. Most results have been described in reviews (11–14). Thus a large number of normally metastable ES intermediates can be handled at subzero temperatures, in solution and sometimes after purification by chromatography (15, 16), or even as crystals where x-ray diffraction permits determination of their three-dimensional structure (9, 17). x-ray diffraction can provide atomic resolution information about the intermediates, but the pathway is much more difficult to unravel, and no dynamic information is obtained from the series of "stop-action" pictures of the most highly populated complexes along this reaction pathway.

We have now witnessed the development of many physical techniques for probing the dynamics of proteins "at work," but such work is still in its formative stages, is not easy,and involves different routes. In this laboratory, we decided to use "antifreeze" solvents as tools to undertake such dynamic investigations.

Cryosolvents as Tools to Undertake Single-Step Kinetic Analysis

Kinetic Studies

We exploited the fact that the organic solvents usually used as antifreeze may reversibly and selectively alter the equilibrium and rate constants of the elementary steps in such a way as to lead to the temporal resolution of kinetically significant steps. Several reactions have been successfully investigated in this way (17–20) and have allowed us to carry out "single-step" kinetic studies over a wide range of temperatures (normal and subzero) and pressures.

The use of selected high concentrations of a variety of cryosolvents (primary alcohols, polyols, dimethylformamide) permitted the induction of controlled and reversible changes in equilibrium and rate processes that were in most cases dramatically slowed. Under these conditions the initial binding of substrates such as ATP to myosin (17), D-amino acid to D-amino acid oxidase (18), peroxides to peroxidases (19), and camphor and linalool to bacterial monooxygenases (20) appeared to be a two-step process involving, first, the formation of a collision complex (E + S) and, second, a conformational change leading to a reactive complex (ES*), in agreement with the induced-fit theory of Koshland (20). Such results may be translated by the following equations:

$$E + S \overset{K}{\rightleftharpoons} ES \text{ (collision complex)}$$

$$ES \underset{k_{-2}}{\overset{k_2}{\rightleftharpoons}} ES^* \text{ ("induced-fit" complex through a conformational change)}$$

It is best to clearly differentiate the two steps of a binding process and to "resolve" the classic bell-shaped curve schematizing this binding into a more realistic diagram (Fig. 36.1). It is also important to record in isolation the values of K and k_2 (sometimes with the help of subzero temperatures and using an analytical technique specific for a given intermediate) and then to test their possible variations under the influence of various conditions of medium (pH, ionic strength, ligands) to obtain further information about some subsequent regulatory processes.

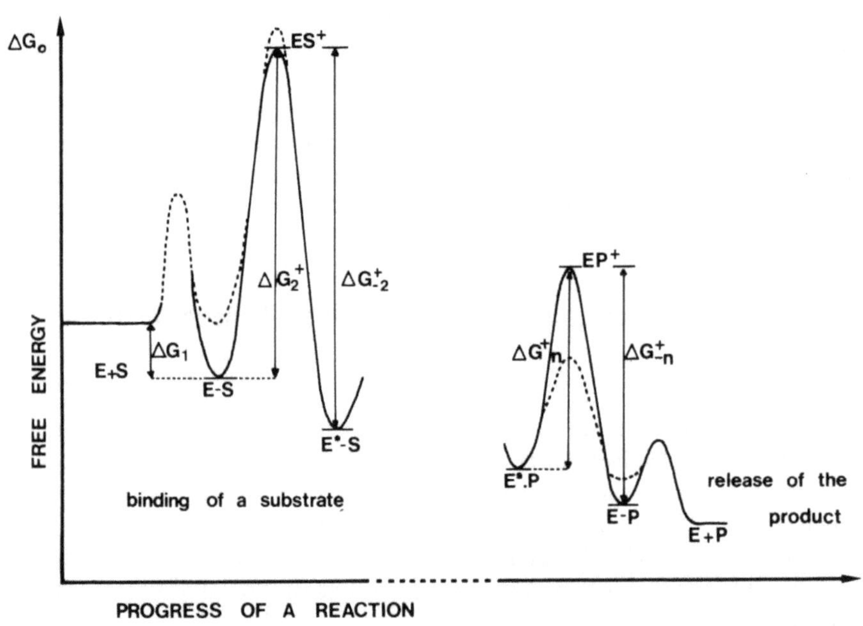

FIG. 36.1. Resolution of the binding steps ($E + S \rightleftharpoons ES \rightleftharpoons E^* \cdot S$) and of the steps involved in the release of product ($E^*P \rightarrow EP \rightarrow E + P$). The "activated complexes" $E\text{–}P^+$ and their free energy of activation (AG^+) are shown; their values can be deduced from experiments. The rate of progress (pace) of the reaction depends on how rapidly the protein can undergo the reversible conformational changes that result from enzyme–substrate (E–S) and enzyme–product (E–P) interactions and are termed $E\text{–}S \rightleftharpoons E^*\text{–}S$ and $E^* \cdot P \rightleftarrows E\text{–}P$, respectively. Dotted lines indicate that these conformational changes may be influenced by the environment and ligands, modifying the free energy of activation (AG^+).

Thus by merely changing the solvent composition, we were able to reveal the step $ES \rightleftharpoons ES^*$, which in water was hidden. The handling of rate constants K_1 and k_2 led us to the conclusion that a collision complex such as myosin-ATP is much more stable in the presence of 40% ethylene glycol (because of the increase of electrostatic interactions caused by decreasing the dielectric constant of the solvent), whereas the collision complex of horseradish peroxidase–hydrogen peroxide is insensitive. The formation of the induced-fit complex ES^* (k_2) is strongly influenced by the solvent and to a lesser extent by the type of substrate (H_2O_2, ROOH).

Thus co-solvents permit one to study these elementary steps in isolation and in fact to carry out "single-step" kinetic analysis in varying physical chemical conditions, e.g., pH and ionic strength. Table 36.2 shows how the kinetics of steps 1 and 2 (K_1 and k_2) of myosin ATPase are changed at different pH and ionic strengths in 40% ethylene glycol. This solvent, and indeed the pH and ionic strength, do not affect the turnover (i.e., k_{cat}) of the overall reaction. Consequently, these findings clearly show that modulation of a reaction pathway by the environment can affect its rate of progress ("pace") but not its overall rate through some of its elementary steps. Protons and cations act as "pace-setters" and are thus responsible for fine regulatory mechanisms

TABLE 36.2. Effect of solvent composition, pH, and ionic strength on rate constants of the SF1 ATPase reaction pathway.

A. Myosin (M) ATPase activity – subfragment 1 (SF1)

$$ATP \rightarrow ADP + P_i$$

B. Postulated reaction pathway consists in seven steps:

$$M + ATP \underset{}{\overset{k_1}{\rightleftharpoons}} M \cdot ATP \underset{k_{-2}}{\overset{k_2}{\rightleftharpoons}} M^* \cdot ATP \underset{k_{-3}}{\overset{k_3}{\rightleftharpoons}} M^{**} \cdot ADP \cdot P_i$$

binding (fast) isomerization chemical step

$$M^{**} \cdot ADP \cdot P_i \underset{k_{-4}}{\overset{k_4}{\rightleftharpoons}} M^*ADP \cdot P_i \underset{k_{-5}}{\overset{k_5}{\rightleftharpoons}} M^*ADP + P_i \nearrow \underset{k_{-6}}{\overset{k_6}{\rightleftharpoons}} M \cdot ADP \underset{k_{-7}}{\overset{k_7}{\rightleftharpoons}} M + ADP \nearrow$$

isomerization (fast) isomerization (fast)

C. Solvent composition kinetic constants of elementary steps (temperature 15°C)

KCl conc. (mM)	pH	k_1 ($10^3\ M^{-1}$)	k_2 (s^{-1})	k_3 (s^{-1})	k_{-3} (s^{-1})	k_4 (s^{-1})	k_{cat} (s^{-1})
Water							
5	8.0	5	160	13	6	0.12	0.08
5	8.0	125	16	11	11	0.12	0.06
Ethylene glycol 40% (v/v)							
150	8.0	9	45	14	7	0.09	0.06
150	6.4	37	8	—	—	0.06	0.04
150	9.0	—	—	—	—	0.12	0.07

closely associated with the conformational changes that result from enzyme–substrate interactions (ES → ES*) and, at least in some cases, enzyme–product interactions (EP* → EP).

An example of such a mechanism at the level of an enzyme–product interaction is provided by the myosin ATPase system, in which a key step is the transient interactions of actin with some enzyme–product intermediate, which through an increase in V_{max} by a factor of 200 leads to the rapid liberation of the products ADP and inorganic phosphate. It is the result of a triggered conformational change (EP* → EP), a step not well understood and currently under close examination in this laboratory by exploiting the "pace-breaking" effect of a cryosolvent and temperature on this reaction pathway.

In the myosin-ATPase system, actin plays the role of a "pace-maker" loosening enzyme–product complexes, which lead to the rapid release of products via the sequence

$$EP^* + A \rightarrow EP^* - A \rightarrow EP \cdot A \rightarrow E + P + A$$

where E = enzyme
P = product
A = actin

For the enzyme–product complex to decompose in the absence of the specific ligand, the system must "pump in" the free energy of thermal motion so as to overcome the relatively large potential energy barrier that controls the triggered conformational change needed for product release. It is by the binding of a specific ligand that the supply of this free energy in ensured.

Pace-making and pace-setting may be common occurrences in enzyme regulation. Two examples are given here:

1. Phosphorylase b is allosterically activated by AMP and surprisingly by certain solvents that have the effect of stabilizing an active conformation of the enzyme (21). These solvents can be looked on as "artificial" pace-makers of phosphorylase b.

2. In the case of the highly organized but complex systems involved in DNA, RNA, and protein synthesis, the forward steps proceed in the presence of certain specific cations and protein factors that regulate the pace and direction of the synthesis by subtle conformational changes of the ternary and quaternary structures of the systems. As with phosphorylase b, the effectors can be replaced by certain organic solvents at critical concentrations (22–25). Thus the cosolvent appears to "mimic" the physiological effectors by affecting (activating or inhibiting) certain reaction steps involved in the synthesis of these polymers. It is thought that the mechanisms involved by the physiological and artificial effectors are similar (26–28).

Further Thermodynamic Studies

In both of these examples it is striking that, despite their different natures, the artificial effectors behave similarly to the physiological effectors. Little is known of the structural changes involved (presumably protein conformational changes), but with the single-step kinetic techniques now available it might be possible to obtain structural and energetic information on these structural changes.

The thermodynamic constants ΔG^{+}, ΔH^{+}, ΔS^{+}, ΔV^{+} (obtained on steps in isolation by conducting kinetic recordings over a broad range of temperatures and pressures) should reflect the energetics as well as the structural features of the labile "activated" complexes ES^{+} or EP^{+} occurring between the successive complexes ES and ES*, or EP* and EP. These constants form an integral part of a full dynamic description of the conversion of ES to ES*, or of EP* to EP. Although their derivation may be a relatively easy matter in our experimental conditions their interpretation is notoriously difficult. Such an interpretation is of great importance to understanding enzyme dynamics. We are involved in a tentative approach toward this end.

Conclusion

Results so far obtained on several model enzyme systems through single-step kinetic studies in nonaqeous clearly confirm that the rate of progress (pace) of a reaction depends on how rapidly the protein can undergo reversible conformational changes that result from enzyme–substrate and enzyme–product interactions, which are influenced by ligands and physiological solutes acting as pace-setters or pace-makers. Such influence on the rate of conformational changes represents a fine regulatory mechanism involving little-understood protein structural dynamics now under examination using the single-step kinetic procedure over a wide range of temperatures and pressures.

Summary

Kinetic control is at the heart of enzyme catalysis, and kinetic data are therefore at the heart of enzymology. Despite continuous refinements in acquisition systems, however, these data often fail to give information about the most intimate regulatory mechanisms and their dynamics at the molecular level.

Using cosolvents as both antifreeze and perturbing agents of the rate of progress of a reaction, it was possible to obtain the temporal resolution of various systems and to undertake single-step kinetic analysis. With these conditions, it is possible to influence the "pace" (rate of progress) of some essential steps by conducting experiments in various physicochemical conditions to determine the "pace-setters" and "pace-makers" regulating the rate of progress of a reaction. Further single-step kinetic studies carried out over a wide range of temperatures and pressures will provide thermodynamic data about the activation process, as well as some structural (dynamic) and energetic features of the interconversion of kinetically significant intermediates.

Acknowledgments. I thank my colleagues C. Balny, T. E. Barman, G. Hui Bon Hoa, and F. Travers, who worked with me in this laboratory on the experiments reported herein. I owe them much for the large amount of data and the illuminating discussions to which they contributed. The work was supported by the Museum National d'Histoire Naturelle, the Institut National de la Recherche Agronomique (INRA), and the Institut National de la Santé et de la Recherche Médicale (INSERM). Part of the work was supported by a contract granted by the Ministère de la Recherche et de la Technologie (MRT contract 85-T, 0717-1985).

References

1. Douzou, P. (1977) *Cryobiochemistry: An Introduction.* Academic Press, New York.
2. Douzou, P. (1974) *Methods Biochem. Anal.* 22:401–512.
3. Douzou, P. (1977) *Adv. Enzymol.* 45:157–272.
4. Douzou, P., Hui Bon Hoa, G., Maurel, F., and Travers, F. (1976) In *Handbook of Biochemistry and Molecular Biology*, 3rd ed. Vol. I, (edited by G. D. Fasman) pp. 520–539. CRC Press,
5. Maurel, P., Hui Bon Hoa, G., and Douzou, P. (1975) *J. Biol. Chem.* 250:1376–1382.
6. Larroque, C., Maurel, P., Balny, C., and Douzou, P. (1976) *Anal. Biochem.* 73:9–19.
7. Maurel, P., and Douzou, P. (1975) *J. Biol. Chem.* 250:2678–2682.
8. Cox, R. P. (1977) *Biochem. Soc. Trans.* 6:689–697.
9. Douzou, P., and Petsko, G. A. (1984) *Adv. Protein Chem.* 36:245–361.
10. Douzou, P., Sireix, R., and Travers, F. (1970) *Proc. Natl. Acad. Sci. USA* 66:787–790.
11. Barman, T. E., and Travers, F. (1985) *Methods Biochem. Anal.* 31:1–59.
12. Douzou, P. (1982) *Proc. R. Soc. Lond.* 217:1–28.
13. Fink, A. L. (1976) *J. Theor. Biol.* 61:419–445.
14. Douzou, P. (1979) *Q. Rev. Biophysi.* 14:521–569.
15. Debey, P., Balny, C., and Douzou, P. (1976) *FEBS Lett.* 69:231–236.
16. Fink, A. L. (1973) *Biochemistry* 12:1736–1742.
17. Biosca, J. R., Travers, F., Hillaire, D., and Barman, T. E. (1984) *Biochemistry* 151:47–50.
18. Yagi, K., Lange, R., and Douzou, P. (1980) *Biochem. Biophys. Res. Commun.* 97:370–374.
19. Balny, C., Travers, F., and Douzou, P. (1987) Submitted.
20. Marden, M. C., and Hui Bon Hoa, G. (1982) *Eur. J. Biochem.* 129:111–117.
21. Dreyfus, M., Vandenbunder, B., and Buc. H. (1978) *FEBS Lett.* 95:185–189.
22. Brody, E. N., and Leautey, J. (1973) *Eur. J. Biochem.* 36:347–361.
23. Crepin, M., Cukier-Kahn, R., and Gros, F. (1973) *Proc. Natl. Acad. Sci. USA* 72:333–337.

24. Hamel, E., and Nakamoto, T. (1977) *Biochemistry* 11:223–238.
25. Nakanishi, S., Adhya, S., Gottesman, M., Pastan, L. (1974) *J. Biol. Chem.* 249:4050–4056.
26. Ballesta, J. P., and Vasquez, D. (1973) *Biochemistry* 12:5063–5068.
27. Voigt, J., Sander, G., Nagel, K., and Parmeggiani, A. (1974) *Biochem. Biophys. Res. Commun.* 57:1279–1286.
28. Hui Bon Hoa, G., Begard, E., Beaudry, P., Maurel, P., Grunberg-Manago, M., and Douzou, P. (1980) *Biochemistry* 19:3080–3088.

37
Kinetic–Structural Organization of Enzyme Systems

B. N. GOLDSTEIN

Many enzyme systems respond to chemical signals (substrates or effectors) by drastically changing their activities. Two phenomena are essential in this kinetic behavior: cooperativity of oligomeric enzymes and covalent modification of enzymes (1–3). Both phenomena can be easily understood from the structure of the kinetic schemes using graph-theoretical methods (4). This chapter is concerned with two applications of the graph-theoretical method: (a) cooperative equilibrium ligand–enzyme interactions; and (b) multiple steady-state transitions in phosphorylated enzyme systems.

Cooperativity in Oligomeric Enzymes

As an example, we consider a simple model suggested by Saifullin and Goldstein (5) to interpret the complex cooperativity in aspartate transcarbamoylase (6). Three conformational states have been postulated:

1. Fully symmetrical enzyme (R), binding in an ordered manner carbamoylphosphate (S_1) and succinate (S_2) at six sites,
2. Half-symmetrical enzyme (T), binding carbamoylphosphate only and at only half of the catalytic sites, and
3. Fully inactive enzyme (T′).

The construction in accordance with the rules presented by Volkenstein and Goldstein (4) for ligand binding is the following:

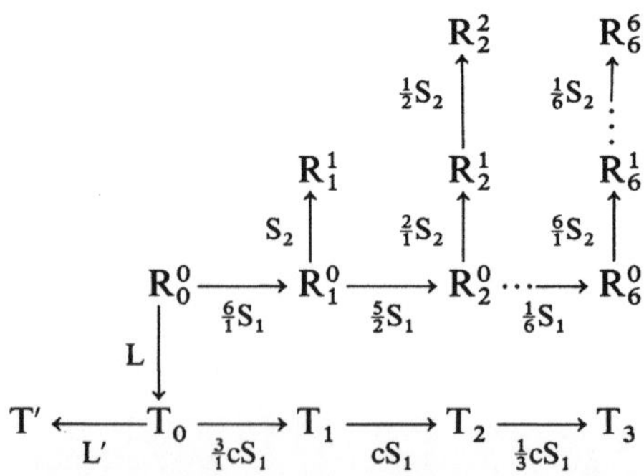

[1]

Here R_i^j and T_i are different enzyme species with S_1 bound at i sites and S_2 bound at j sites, which refer to different graph "nodes".

The following rules simplify the graph-theoretical analysis:

1. Enzyme-containing species are referred to as graph nodes 0, 1, ..., i, etc.
2. The graph for equilibrium ligand–enzyme interactions does not contain closed cycles; therefore the single pathway leads from the initial node (0) to each i-th node.
3. Each branch relating to ligand binding is assigned the dimensionless ligand concentration S with coefficient p/q, where p and q are the numbers of unbound and bound enzyme sites, respectively. Dimensionless concentrations are concentrations divided by corresponding microscopic dissociation constants (K_R, K_T, etc.). Branches relating to conformational transitions are assigned constants.
4. The path (path value) leading from the initial node (0) to each node (i) is equal to the product of its branches: $P_{0i} = P_{01}P_{12}\ldots P_{i-1,i}$.

The saturation function for ligand S is obtained according to the rule:

$$Y(S) = \frac{\sum_i P_{0i}^{(S)}}{1 + \sum_i P_{0i}}, \; i = 1, 2, \ldots \qquad [2]$$

where $P_{0i}^{(S)}$ is the path leading to each i-th node referred to the S-containing species.

Using the algorithm [2], we have obtained (5) the following saturation functions expressed in a compact form for the graph (1):

$$Y(S_1) = S_1 \frac{(1 + S_2)[1 + S_1(1 + S_2)]^{n-1} + \frac{1}{2}Lc(1 + cS_1)^{(n/2)-1}}{[1 + S_1(1 + S_2)]^n + L(1 + cS_1)^{n/2} + LL'} \qquad [3]$$

$$Y(S_2) = S_1 S_2 \frac{[1 + S_1(1 + S_2)]^{n-1}}{[1 + S_1(1 + S_2)]^n + L(1 + cS_1)^{n/2} + LL'} \qquad [4]$$

where $n = 6$

$c = K_R/K_T$

L, L' = allosteric constants

Functions 3 and 4, fitted to the experimental data presented by Rosenbuch and Griffin (6), show the complex cooperativity, which is initially positive, then negative, and then positive again with a increase in S_1 (5).

Multiple Steady-States in Phosphorylated Bifunctional Enzymes

The graph-theoretical method can also be applied to the time-dependent kinetics near the steady states for single-enzymic (7) as well as multienzymic (8) systems. This method is based on construction of the characteristic polynomial for the linearized kinetic equations.

A similar method has been proposed (9) and developed (10) for nonenzymic reactions. Ivanova and Goldstein (11) have used the results of Ivanova (10) to reveal the "critical fragments" responsible for critical phenomena (multiple steady states or self-oscillations) in graphs for enzyme systems. These critical fragments introduce negative terms into characteristic polynomial coefficients. Thus the critical phenomena can be interpreted in terms of interactions in enzyme systems.

Let us consider, for example, the bifunctional enzyme 6-phosphofructo-2-kinase/fructose 2,6-*bis*-phosphatase (12). The substrate of this system, fructose 2,6-bisphosphate, is a strong regulator of glycolysis/gluconeogenesis, and the cellular response to glucagon is mediated through the reversible phosphorylation of this enzyme.

The following reaction stages can be written (11) in accordance with available experimental data (12, 13):

$E_1 + S_1 \rightarrow E_1 + S_2$ (forward-reaction)

$E_2 + S_2 \rightarrow E_1 + S_1$ (reverse reaction combined with rapid enzyme dephosphorylation)

$E_3 + S_2 \rightleftarrows E_3S_2$ (inhibition of protein kinase E_3 by substrate S_2)

$E_1 + E_3 \rightarrow E_2 + E_3$ (enzyme phosphorylation)

External fluxes $\rightarrow S_1 \rightarrow$ and $\rightarrow S_2 \rightarrow$ can also be included, but they are not essential for our model. Here S_1 is fructose-6-phosphate, S_2 is fructose-2,6-bisphosphate, and E_1 and E_2 are dephosphorylated and phosphorylated enzyme species, respectively.

Mutually interconvertible enzymic and substrate species are in this case referred to as o-nodes, and all stages as □-nodes in the graph. All nodes are connected by arrows in accordance with the kinetic mechanism. Such graph contains "positive paths" with equally directed branches (o→□→o) and "negative paths" with oppositely directed branches (o→□←o) (10). For the system under discussion, we consider the following critical fragment of the full graph:

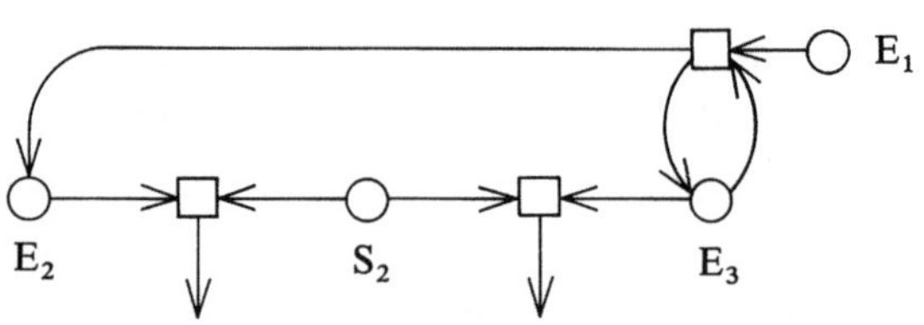

We have revealed here two "even cycles", each containing an even number of negative paths: E_2 →□← S_2 →□← E_3 →□→ E_2 (two negative paths) and E_3 →□→ E_3 (no negative paths) (11).

The critical phenomena in many systems may be produced by the typical combination of two even cycles, passing through common o- and □-nodes

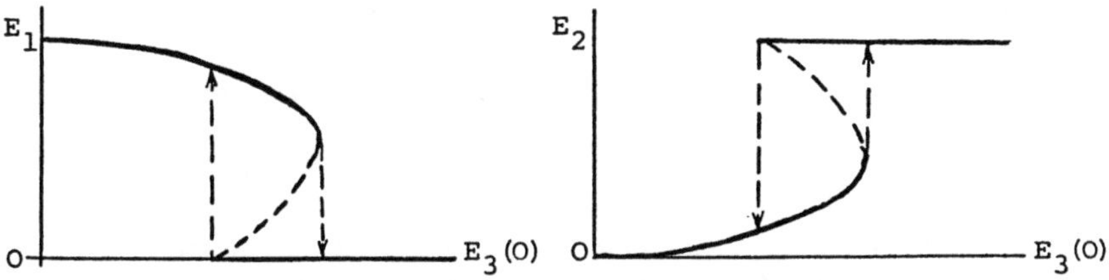

FIG. 37.1. Dependence of two stationary concentrations, E_1 and E_2, on protein kinase initial concentration, E_3(o).

(11), e.g., as it is shown in this scheme. All reactants of these cycles must refer to independent variables only.

The flux through the critical fragment must be great compared with other fluxes to produce critical phenomena. Therefore the modificatory reaction $E_1 \rightarrow E_2$ involved in the critical fragment must be rapid compared with the external, reaction $S_1 \rightarrow S_2$ in order to produce multiple steady states in the system discussed (14). Moreover, no irreversible effluxes (○→□) must originate from this fragment to others.

The curve shown in Fig. 37.1, computed for this system (14), reflect the possible hysteretic dependence of two reciprocal enzyme activities on glucagon added, as E_3(o) rises with glucagon addition (12).

The discontinuous (multiple steady-state) transitions shown in Fig. 37.1 by dotted arrows indicate the high "cooperativity" of the system. Hence small changes of glucagon concentrations can produce great changes in carbohydrate metabolism. Thus this kinetic model predicts amplification of hormonal signals directed through the bifunctional enzyme to glycolysis/gluconeogenesis. Moreover, this system may remain in the "stimulated" state even after the stimulus is removed (in the hysteresis domain) and may have a "memory" for hormonal signals.

Acknowledgments. The author extends his sincere thanks to Drs. S. R. Saifullin and A. N. Ivanova for very helpful assistance during this work.

References

1. Goldbeter, A., and Koshland, D. E., Jr. (1981) *Proc. Natl. Acad. Sci. USA* 78:6840–6844.
2. Stadtman, E. R., and Chock, P. B. (1977) *Proc. Natl. Acad. Sci. USA* 74:2761–2765.
3. Crabtree, B. (1985) *FEBS Lett.* 187:193–195.
4. Volkenstein, M. V., and Goldstein, B. N., (1966) *Biochim. Biophys. Acta* 115:471–485.

5. Saifullin, S. R., and Goldstein, B. N. (1986) *Molek. Biol.* 20:224–232 [in Russian].
6. Rosenbusch, J. P., and Griffin, G. H., (1973) *J. Biol. Chem.* 248:5063–5066.
7. Goldstein, B. N., (1983) *J. Theor. Biol.* 103:247–264.
8. Goldstein, B. N., and Shevelev, E. L. (1985) *J. Theor. Biol.* 112:493–503.
9. Clarke, B. L. (1974) *J. Chem. Phys.* 60:1481–1491.
10. Ivanova, A. N. (1979) *Kinet. Catalysis* 20:1019–1022 [in Russian].
11. Ivanova, A. N., and Goldstein, B. N. (1986) *Molek. Biol.* 20:1522–1529 [in Russian].
12. Stewart, H. B., El-Maghrabi, M. R., and Pilkis, S. J. (1985) *J. Biol. Chem.* 260:12935–12941.
13. El-Maghrabi, M. R., Fox, E., Pilkis, J., and Pilkis, S. J. (1982) *Biochem. Biophys. Res. Commun.* 105:794–802.
14. Goldstein, B. N., and Ivanova, A. N. (1987) *FEBS Lett.* (in press).

38
Relation Between Structure and Function in cAMP-Dependent Protein Kinases

SUSAN S. TAYLOR, JOSÉ BUBIS, JANUSZ SOWADSKI, JEAN A. TONER, and LAKSHMI D. SARASWAT

A primary mechanism for regulation in eukaryotic cells involves protein phosphorylation. Although much of our fundamental knowledge of protein phosphorylation and its role in regulation stemmed initially from the discovery of cAMP (1), the protein kinases now are known to represent a large and varied family of molecules that differ in size, substrate specificity, cellular location, and mechanism of activation (2, 3). Nevertheless, within this family of ATP-protein phosphotransferases, the best understood enzyme is still the protein kinase that is activated by cAMP (4). cAMP is a second messenger that is synthesized intracellularly in response to several mediators, including peptide hormones and steroids. The mechanisms for activating adenylate cyclase are specific to each cell type; however, in all instances the primary receptor for cAMP within the cell is cAMP-dependent protein kinase. In the absence of cAMP the enzyme exists as an inactive tetramer containing two regulatory (R) and two catalytic (C) subunits. When cAMP levels are elevated, the cyclic nucleotide binds with high affinity (K_D 10^{-8} M) to the regulatory subunit, thereby promoting dissociation of the complex into an R_2 dimer and two monomeric C subunits that are catalytically active. The proteins that are targeted for phosphorylation by the C subunit also vary with cell type; however, each has a common sequence that is recognized by the active site of the enzyme (5). This sequence includes two basic residues that are usually arginines, a small nonspecific intervening residue, and finally a serine or threonine that serves as the phosphate acceptor.

The potential roles that protein phosphorylation plays in the regulation of cellular processes were extended even further with the discovery that several oncogenes coded for enzymes that had intrinsic protein kinase activity (6). The subsequent finding that these oncogene kinases specifically transfer phosphate to tyrosine residues identified a new subclass of protein kinases (7, 8). These oncogene kinases also have protooncogene counterparts that are homologues; they are expressed in the normal untransformed cells and presumably play a role in normal growth regulation. Another class of molecules have also been added to this category of kinases that specifically phosphorylate tyrosine residues. They include the growth factor receptors, e.g., epidermal growth factor (9) and the insulin receptors (10). Although phosphorylations that occur on tyrosine residues are quantitatively less significant ($<0.5\%$) than those that occur on serine and threonine residues, these modifications nevertheless appear to play crucial roles in growth regulation, mitogenesis, and transformation (11, 12).

Our work has focused on better understanding the functional sites that are associated with cAMP-dependent protein kinases. The utilization of affinity labeling, group-specific labeling, and limited proteolysis has provided information about specific residues that contribute to function. The kinase subunits have been crystallized, and genes for the subunits have been cloned and expressed in bacteria. The combined tools of x-ray crystallography and directed mutagenesis provide new opportunities for understanding the structure of these proteins and provide a framework for better understanding the entire family of protein kinases.

Catalytic Subunit

There are at least three distinct functional sites associated with the catalytic subunit of cAMP-dependent protein kinase: the ATP binding site, the substrate or peptide recognition site, and the site where catalysis actually occurs. In the absence of a crystal structure for the enzyme, information about each of these sites must be deduced indirectly. A variety of approaches have been used, including binding of analogues (13), nuclear magnetic resonance (14), affinity labeling, and modification of the enzyme by group-specific reagents. Whereas the former methods provide general criteria for binding, only the latter methods have the potential to identify specific residues that contribute to binding or catalysis.

Several specific residues in the catalytic subunit have been shown to be localized at the active site of the enzyme. One of the first to be identified was lysine 72, which can be covalently labeled by the ATP analogue fluorosulfonyl benzoyl 5′-adenosine (FSBA), as summarized in Fig. 38.1 (15). Covalent modification of this residue leads to loss of activity, and both MgATP and R subunit can protect against inactivation. Although holoenzyme is not inactivated by FSBA, it is covalently modified on the R subunit, indicating the close proximity of the alkylating group to the subunit interaction site. Because the R subunit in some cases is also capable of being autophosphorylated, it is likely that the reactive fluorosulfonyl group is binding in a position that is analogous to the γ-phosphate of ATP even though FSBA does not retain the negative charges of ATP.

The role that sulfhydryl groups play in the catalytic subunit is still not clear. One cysteine residue, Cys 199, is located in close proximity to the active site, although it may not play a direct role in catalysis. This residue can be readily alkylated in the free C subunit, but both MgATP and R subunit are capable of protecting this residue against alkylation (16). A peptide analogue was syn-

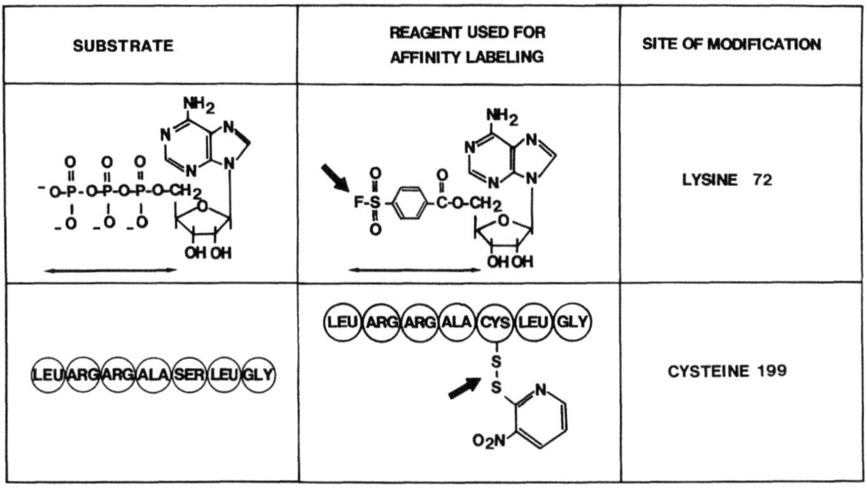

FIG. 38.1. Summary of affinity labeling of the catalytic subunit. Both ATP and peptide analogues have been used for affinity labeling. The thick arrows indicate the reactive group on each affinity analogue. The long, thin arrows below ATP and FSBA indicate the approximate distance between the ribose ring and the γ-phosphate of ATP or the fluorosulfonyl group of FSBA, which are roughly equivalent.

thesized that contains a sulfhydryl reactive group at a position that would be analogous to the phosphoserine (Fig. 38.1). This peptide inhibits the C subunit by forming a covalent bond with Cys 199, thus providing further support for localization of this residue at the active site (17). On the other hand, covalent modification of this residue by cyanylation did not lead to inactivation, implying that its role may not be essential for catalysis (18).

Other evidence has shown that treatment of the C subunit with a water-insoluble carbodiimide also leads to irreversible inhibition of catalytic activity. The holoenzyme is not inactivated by dicyclohexylcarbodiimide (DCCD), and MgATP also protects against inactivation. The specificity of this reaction has not yet been established; however, when tryptic peptides of the native enzyme are compared to those obtained from the DCCD-inhibited protein, a modified tryptic peptide can be identified. This peptide corresponds to residues 166 through 189 in the sequence contains four carboxylic acid residues, one of which is presumably the target for the action of DCCD.

Since the original sequencing of the bovine C subunit (19), several other cellular kinases have been sequenced including the cGMP-dependent protein kinase (20), phosphorylase kinase (21), and myosin light-chain kinase (22). In addition, protein sequences have been deduced from the DNA sequences of numerous oncogenes that have kinase activity, e.g., the gene product pp60$^{v\text{-}src}$ for the oncogene in the Rous sarcoma virus (23). Finally, protein sequences have been deduced for several growth factor receptors, e.g., insulin receptor

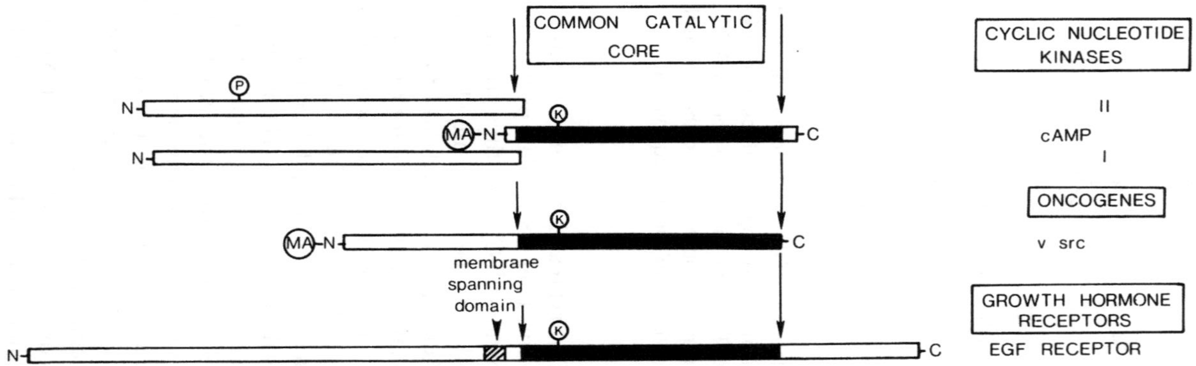

FIG. 38.2. Sequence homologies in various classes of protein kinases. Sequences are aligned on the basis of homology with the C subunit of cAMP-dependent protein kinases and are indicated from the NH_2-terminal end (left) to the COOH-terminal end (right) of the polypeptide chain. Homologies with $pp60^{v\text{-}src}$ were first recognized by Barker and Dayhoff (27). K = the conserved lysine residue; MA = myristic acid.

(24) and epidermal growth factor receptor (25), each of which has intrinsic protein kinase activity that is specific for transferring phosphate to tyrosine. These kinases represent a wide variety of proteins that differ in size, specificity, mechanism of ligand activation, and cellular localization. Nevertheless, each protein contains a common core that is homologous to the C subunit of cAMP-dependent protein kinase (Fig. 38.2). A closer analysis of these homologies provides an additional perspective for evaluating the functional sites associated with the C subunit.

In terms of the ATP binding site, lysine 72 is conserved in every protein that has been shown to be a protein kinase, thus confirming the likelihood that this residue plays an essential role. The homologous lysine residue in pp60$^{v\text{-}src}$ also was covalently modified by FSBA, indicating that ATP is binding in an analogous manner in both proteins (26). Modification of this lysine in pp60$^{v\text{-}src}$ by site-directed mutagenesis leads to loss of kinase activity, loss of transforming activity, and loss of tumorogenicity, which further supports the essential nature of this residue (27, 28). Another feature that is conserved is a series of three glycine residues corresponding to Gly_{50}, Gly_{52}, and Gly_{55} in the C subunit (Fig. 38.3). These glycine residues are conserved in all adenine nucleotide binding sites (30, 31). The importance of their role in maintaining secondary structure was first established with the comparison of the crystal structures of lactate dehydrogenase and glyceraldehyde-3-phosphate dehydrogenase, and led to the hypothesis that a nucleotide fold structure would be preserved in adenine nucleotide binding sites (32). This conjecture has proved to be the case and has been confirmed repeatedly as new crystal structures have been solved. It is the secondary structure that is conserved in the proteins rather than the amino acid sequence. The conserved glycine residues are an

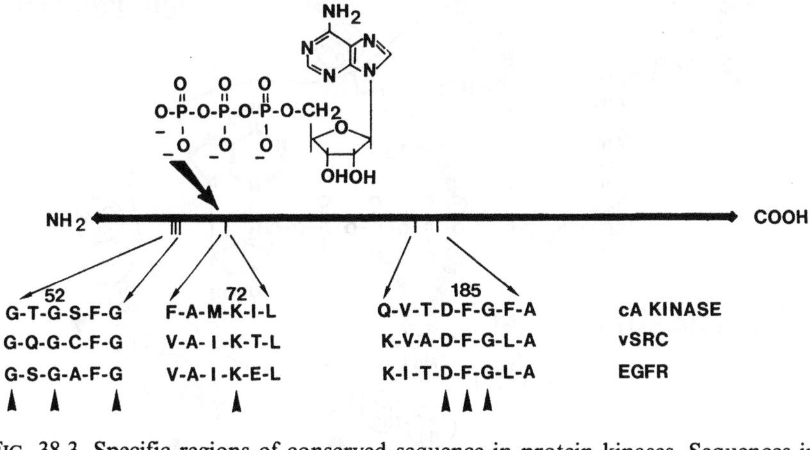

FIG. 38.3. Specific regions of conserved sequence in protein kinases. Sequences indicated are from the C subunit of cAMP-dependent protein kinase (19), the transforming gene product pp60$^{v\text{-}src}$ from the Rous sarcoma virus oncogene (23), and the epidermal growth factor receptor (25).

exception and represent a bend between a β sheet and an α-helix that comes close to the ribose ring of the adenosine moiety. The conserved glycines and the lysine support a structure for the C subunit that has the ATP binding site located in the amino terminal region of the protein.

Another sequence that is conserved in all protein kinases is Asp-Phe-Gly (residues 184–186). The role of the conserved Asp-Phe-Gly sequence is less clear, although this sequence is contained in the peptide that is modified by DCCD. It is also the only carboxylic acid that is conserved in every protein kinase. Although this sequence is conserved in all protein kinases, it is not found in other adenine nucleotide binding proteins, suggesting that it plays a role in kinase function but not necessarily in the binding of adenine nucleotides.

As indicated earlier, one of the sulfhydryl groups in the C subunit is in close proximity to the active site, although the role it plays is uncertain. Unlike lysine 72, this cysteine is not conserved in other kinases, which suggests that it is not essential for catalysis. It seems unlikely that the catalytic mechanism would differ so fundamentally within this family of related proteins. Instead, it is more likely that there is a nucleophilic group in close proximity to the serine hydroxyl group that tends to make the serine oxygen better able to attack the γ-phosphate group of ATP. The identity of this nucleophile is unknown; however, it undoubtedly is conserved in all of the kinases. One possibility is the aspartic acid 184 discussed above.

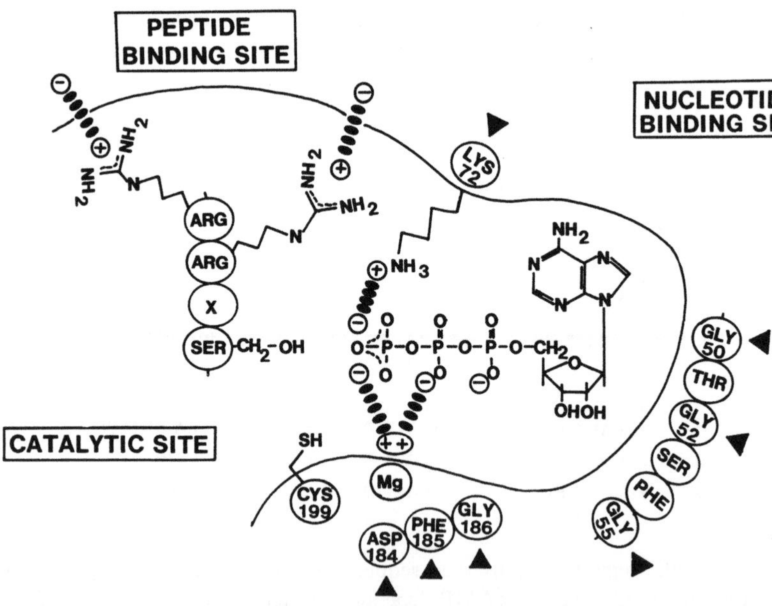

FIG. 38.4. Model for the active site of the C subunit of cAMP-dependent protein kinase. Amino acid residues known to be in close proximity to the active site are indicated. Those conserved in all protein kinases are indicated by ▲.

PROTEIN FORM	C-SUBUNIT ALONE	C-SUBUNIT plus AMP-PNP
CRYSTAL FORM	MONOCLINIC	CUBIC
SPACE GROUP	P21 a=64 Å; b =144 Å; c=48 Å	$P4_132$ a=b=c=170 Å
	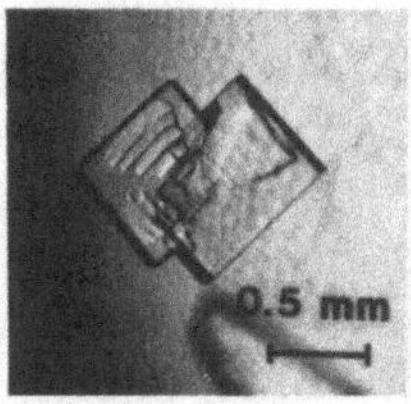	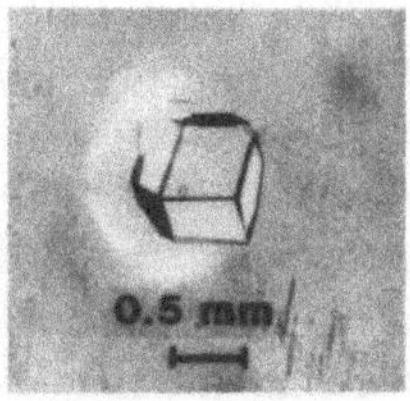

FIG. 38.5. Crystallization of the C subunit. General conditions for growing two distinct crystal forms are indicated as well as general properties of each crystal form.

The peptide recognition site also is not well defined. In contrast to the ATP binding site, this region is not anticipated to be highly conserved within the family of kinases, as the characteristic recognition sequence surrounding the phosphorylation site differs as does the phosphate acceptor itself (33). A model of the active site is summarized in Fig. 38.4, where the conserved features are indicated.

Because of the homologies that exist within this large and diverse family of enzymes, it is clear that knowing a structure for one kinase provides a valuable framework for all protein kinases. It is essential now to obtain a more detailed structure of these molecules. Crystals of the C subunit have been grown that are suitable for x-ray crystallographic analysis (34), and these results are summarized in Fig. 38.5. Initial efforts to crystallize the free C subunit yielded monoclinic crystals that diffracted to 3.5 Å. Because these crystals were somewhat fragile, alternative crystallization conditions were developed. Two distinctly different crystal forms have been observed subsequently. Cubic crystals that diffract to 2.8 Å have been grown in the presence and absence of AMP-PNP, a nonhydrolyzable analogue of ATP. Data are now being collected on these crystals using the multiwire area detector (35). A third crystal form has been grown in the presence of MgATP and a peptide inhibitor that was derived from the heat-stable protein kinase inhibitor (36). Because the conditions for growing these crystals are identical to the conditions used for growing the cubic crystals, it is likely that this distinct crystal form represents a different conformational state of the protein.

Regulatory Subunit

Although the regulatory subunit of cAMP-dependent protein kinase lacks catalytic activity, it does have a number of functional sites, including two cAMP binding sites, a C subunit recognition site, and in some cases an autophosphorylation site. In addition, there is the interaction site between the two protomers of the dimer. Finally, it is the R subunit that contains most of the antigenic determinants for the holoenzymes. Whereas the C subunit appears to be similar, if not identical, for all cAMP-dependent protein kinases isolated from various tissues (37), the R subunit clearly includes several unique gene products. The various holoenzymes have been classified as types I and II based on the order of elution from DEAE-cellulose, and the unique properties of each holoenzyme can be attributed to the R subunits (38, 39). The type I classification most likely includes one gene product (40), whereas at least two unique gene products are included in the type II classification (41, 42). One of the most sensitive means of distinguishing the R subunit forms is antigenically. For example, monoclonal antibodies to heart R^{II} were used initially to demonstrate that the major form of R^{II} in brain was not the same as the heart form (42).

Despite differences in the R subunits, each retains a common and well-defined domain structure. In general, each protomer can be divided into three equal segments: an amino terminal segment and two cAMP binding domains at the carboxy-terminus. It is the amino-terminus where all of the antigenic sites are localized, and these sites are equally exposed in the holoenzyme and in the dissociated R dimer (43, 44). When it was demonstrated that monoclonal antibodies to heart R^{II} did not cross-react with the major variant of R^{II} in brain, the brain R^{II} was sequenced in the region known to contain the antigenic site. This sequence established that these two proteins represent two distinct gene products (41). This sequence, however, also demonstrated the validity of the type I and type II classifications, as brain and heart R^{II} are far more homologous to one another than to R^{I} (Fig. 38.6). Other properties, e.g., autophosphorylation, appear to be uniquely associated with the type II R subunits. Isolation of a cDNA clone encoding for bovine R^{I} subunit also supports the finding that there is one major form of R^{I} (40); however, it is still not clear how many unique variants fall within the type II category. It also is not apparent at this point why such diversity exists.

The cAMP binding domain includes the carboxy-terminal two-thirds of the protein and contains two in-tandem, gene-duplicated sequences that represent two cAMP binding sites. Our understanding of the cAMP binding regions of the R subunit has been greatly facilitated by our knowledge of the catabolite gene activator protein (CAP), the major cAMP binding protein in *Escherichia coli* (45). In the presence of cAMP, CAP binds with a high affinity to the *lac* operon, thus providing a binding site for RNA polymerase and allowing transcription of the gene to proceed. CAP also has a well defined domain structure that includes a cAMP-binding domain at the amino-terminus, and

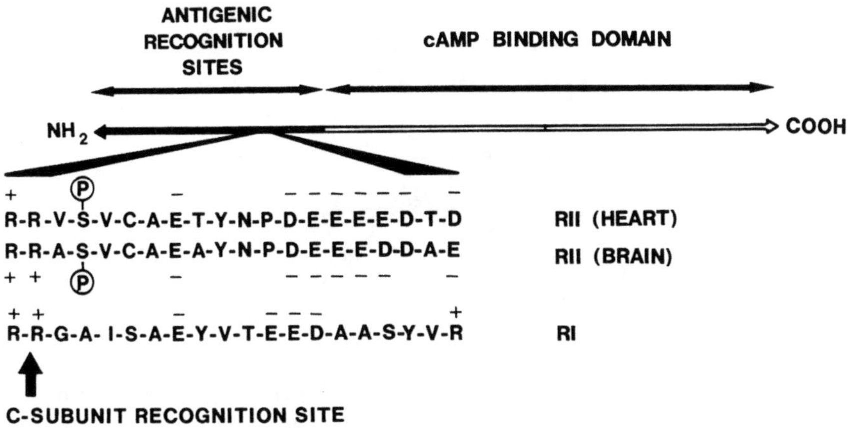

FIG. 38.6. Sequence homologies in the regulatory subunits of cAMP-dependent protein kinases. The linear sequence is indicated at the top, with the antigenic and cAMP-binding domains indicated by arrows. Sequences flanking the autophosphorylation site in R^{II} are indicated below as well as the sequence of the homologous region in R^{I}.

the crystal structure of CAP is known (46). It is the cAMP binding domain that is homologous in its amino acid sequence to the cAMP binding domains of the R subunit (47).

Affinity labeling has been used to identify specific residues that contribute to the cAMP binding sites in both R^{I} and R^{II}. A single residue, tyrosine 381, is labeled stoichiometrically in heart R^{II} with the photoaffinity analogue 8-N_3-cAMP (48). In the case of R^{I}, two residues are labeled: tyrosine 371, which is homologous to the R^{II} site, and tryptophan 260 (49). A model of the cAMP binding domains of R^{I} and R^{II} has been constructed by superimposing the R sequence onto the crystal structure of CAP (Fig. 38.7). This model has then been correlated with the photolabeling described above in order to establish whether the R subunit polypeptide chain is folding in a manner that is analogous to CAP. With this model, the tyrosine residue in both R subunits would lie in the region that corresponds to the long C helix of CAP. This C helix is indeed the major site of interaction of cAMP in CAP and thus is consistent with cAMP binding in an analogous manner in both proteins. Because tryptophan 260 is part of the second cAMP binding domain, it was possible that this residue was also in close proximity to tyrosine 371, so that both residues could potentially be labeled by the same 8-N_3-cAMP molecule. The model based on CAP, however, places these two residues at a distance of 20 Å. On the other hand, by placing two cAMP binding domains in tandem, as is found in the R subunit, the tryptophan actually can come into close proximity to the 8-N_3-cAMP that is bound to domain A (53).

To test this hypothesis, photolabeling was carried out under conditions where 8-N_3-cAMP was selectively displaced by analogues of cAMP that have

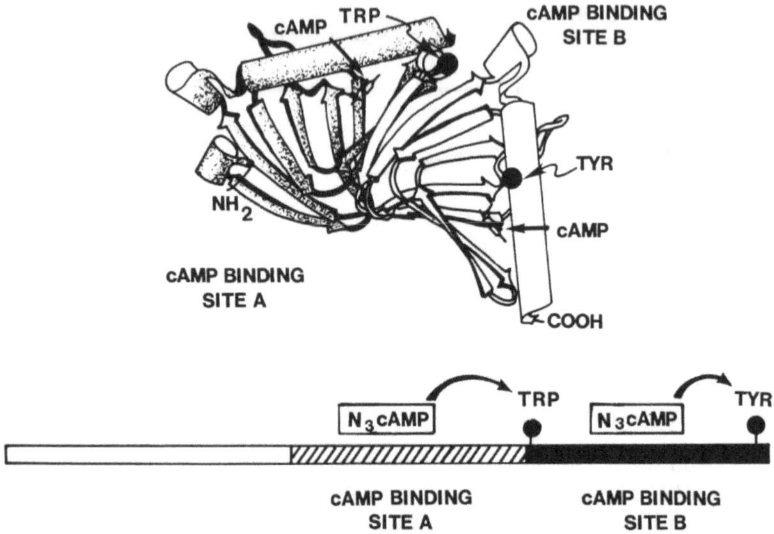

FIG. 38.7. Correlation of photolabeling with 8-N_3-cAMP and cAMP binding sites in the type I regulatory subunit of cAMP-dependent protein kinase. Locations of the photolabeled residues, Trp 260 and Tyr 371, are indicated below. Shown above is a model of the two in-tandem cAMP binding domains in R^I. The model is based on the crystal structure of CAP, a homologous cAMP binding protein. cAMP bound to domain B is aligned in good position to photolabel Tyr 371. Trp 260, on the other hand, is close to the interaction site between domain A and B and is most likely photolabeled by cyclic nucleotide bound to domain A.

a high preference for either site A or site B. Corbin and others have demonstrated that analogues of cAMP that are substituted at the C8 position have a high preference for one site, whereas N_6-substituted analogues frequently prefer the other site (50, 51). Both sites of the R^I subunit were initially saturated with 8-N_3-cAMP, and then 50% of the radiolabeled analogue was displaced with N_6-aminohexyl cAMP prior to photolysis. Under these conditions, photolabeling of tyrosine 371 was diminished only slightly, whereas photolabeling of tryptophan 260 was abolished. These results, summarized in Fig. 38.7, confirm the model that tryptophan 260 is modified by 8-N_3-cAMP bound to site A and that tyrosine 371 is modified by 8-N_3-cAMP bound to site B (Fig. 38.6). In addition, it provides a point of contact between the two domains (49).

Expression of Kinase Subunits in *E. coli*

In order to more fully understand the structures of these subunits it is essential to characterize them using x-ray crystallographic techniques. In conjunction with crystallographic studies, cDNA clones for the kinase subunits have been isolated and site-directed mutagenesis studies initiated. As a first step toward introducing specific mutations into the protein, it is essential to construct an expression vector so that suitable quantities of the protein can be isolated.

Using a cDNA clone that codes for the bovine R^{I} subunit (36), an expression vector has been constructed that yields 4 to 20 mg of R^{I} per liter of *E. coli* culture (52). The vector was constructed by inserting the full-length cDNA clone for R^{I} into the Hinc II site of pUC 7. This vector utilizes the *lac* promotor; however, the expression of R^{I} did not appear to follow the conventional mechanism for promotion that is seen for control by the *lac* promotor. Activation is normally mediated by the addition of isopropyl β-D-thiogalactopyranoside (IPTG), a lactose analogue that binds to the *lac* repressor. However, in the case of the R^{I} construct, the addition of IPTG actually inhibited production of R^{I}. Instead, if the cells were allowed to grow into the late log phase, R^{I} accumulated in the absence of IPTG. R^{I} was expressed initially in *E. coli* JM105, which is a constitutive overproducer of repressor. Expression in this strain yielded 2 to 4 mg/l of R subunit culture. Maximum expression was seen in *E. coli* 222, a strain that lacks adenylate cyclase (cya^{-}) and contains a mutation in CAP (crp*). This mutation allows

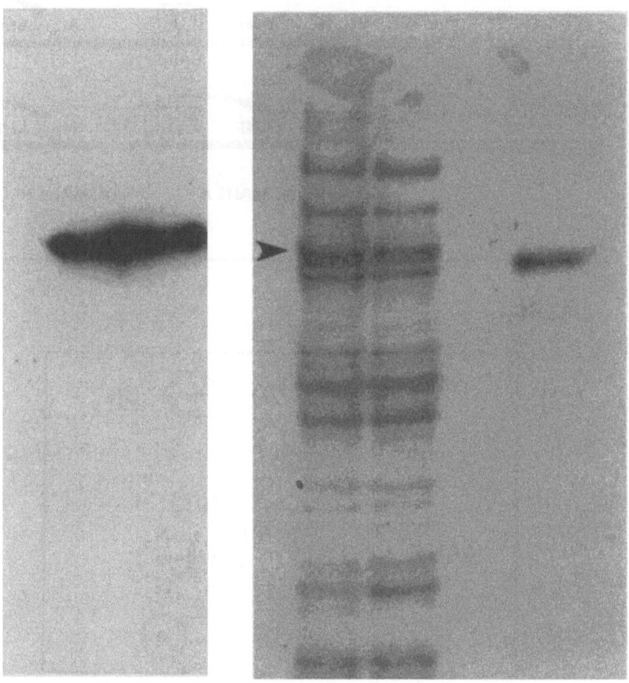

FIG. 38.8. Expression of R subunit in *E. coli* 222. Following gel electrophoresis of the total cell extract, a portion of the lanes were stained directly with Coomassie blue (right). The remaining lanes were electrotransferred to nitrocellulose, blocked with Tween-20, and photolabeled with 8-N_3-[^{32}P]-cAMP. Photolabeled protein was detected by autoradiography (left). The mobility of bovine R^{I} is also indicated.

CAP to bind to DNA with a high affinity in the absence of cAMP. With this strain as much as 20 mg of R subunit per liter accumulated (Fig. 38.8).

The expressed R subunit is a soluble dimer that behaves like the R^{I} that had been purified from bovine tissues even though it represents a fusion protein. The expressed R subunit contains an additional 10 amino acids at the amino-terminus, which was confirmed by amino acid sequencing of the purified protein. These additional residues do not appear to interfere with either function or stability, and the purified protein has subsequently been crystallized (52).

Having constructed an expression vector, the introduction of mutations into selected sites on the polypeptide chain can be initiated. The homology of the R subunit with CAP has provided a valuable framework for initiating these mutagenesis studies in the absence of a crystal structure of the R subunit. In particular, there are conserved features of each cAMP binding site that can be targeted, e.g., the arginine residue, which presumably interacts electrostatically with the negative charge on the cyclic phosphate ring (46). Another question that arises from our photolabeling studies in conjunction with binding of cAMP analogues is whether there exists a dipole–dipole interaction

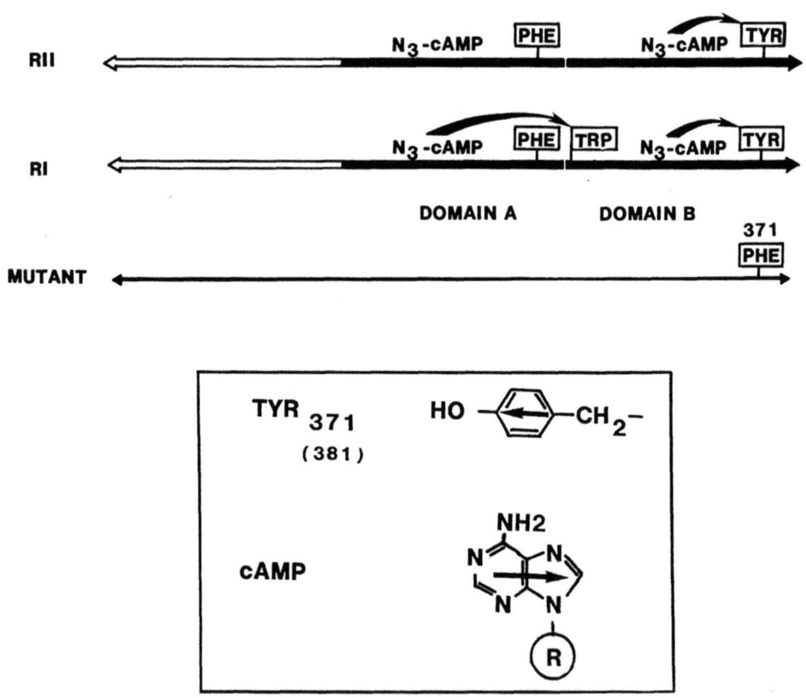

FIG. 38.9. Design for mutation of Tyr 371 to Phe 371. The positions of the amino acid residues that are covalently modified with 8-N_3-cAMP in R^{I} and R^{II} are indicated below. The potential dipole–dipole interaction between the tyrosine ring and the adenine ring of cAMP is also indicated below.

between the adenine ring of cAMP bound to domain B and the tyrosine ring that is photolabeled so efficiently. In order to address this question, a single base change has been introduced into the cDNA clone so that the mRNA now codes for a phenylalanine instead of a sine (Fig. 38.9). Introduction of this change yields a stable protein that binds cAMP, and these properties are now being characterized.

The eventual goal of these studies is not only to use directed mutagenesis to better correlate structure with function but also to eventually introduce altered genes into eukaryotic cells as a means of studying more selectively the physiological events that are regulated by cAMP.

Acknowledgment. Supported by U.S. Public Health Service grant GM 19301 (SST) and American Cancer Society grant NP-419 (SST). J. B. was supported by the Consejo Nacional de Investigaciones Científicas y Technológicas (CONCIT), Caracas, Venezuela; L. D. S. was supported in part by U.S. Public Health Training Grant AM-07233.

References

1. Sutherland, E. W., and Rall, T. W. (1958) *J. Biol. Chem.* 232:1077–1091.
2. Krebs, E. G., and Beavo, J. A. (1979) *Annu. Rev. Biochem.* 48:923–960.
3. Flockhart, D. A., and Corbin, J. D. (1982) *CRC Crit. Rev. Biochem.* 133–186.
4. Walsh, D. A., Perkins, J. P., and Krebs, E. G. (1968) *J. Biol. Chem.* 243:3763–3765.
5. Kemp, B. E., Benjamini, E., and Krebs, E. G. (1976) *Proc. Natl. Acad. Sci. USA* 73:1038–1042.
6. Collett, M. S., and Erikson, R. L. (1978) *Proc. Natl. Acad. Sci. USA* 75:2021–2024.
7. Hunter, T., and Sefton, B. M. (1980) *Proc. Natl. Acad. Sci. USA* 77:1311–1315.
8. Collett, M. S., Purchio, A. F., and Erikson, R. L. (1980) *Nature* 285:167–169.
9. Ushira, H., and Cohen, S. (1980) *J. Biol. Chem.* 255:8363–8365.
10. Kasuga, M., Zick, Y., Blithe, D., Crettaz, M., and Kahn, C. (1982) *Nature* 298:667–669.
11. Bishop, J. M. (1983) *Annu. Rev. Biochem.* 52:301–354.
12. Hunter, T. (1984) *Sci. Am.* 70–79.
13. Hoppe, J., Freist, W., Marutsky, R., and Wagner, K. G. (1977) *Eur. J. Biochem.* 80:369–373.
14. Granot, J., Mildvan, A. S., and Kaiser, E. T. (1980) *Arch. Biochem. Biophys.* 205:1–17.
15. Zoller, M. J., Kerlavage, A. R., and Taylor, S. S. (1979) *J. Biol. Chem.* 254:2408–2412.
16. Nelson, N. C., and Taylor, S. S. (1981) *J. Biol, Chem.* 256:3743–3750.
17. Bramson, H. N., Thomas, N., Matsuida, R., Nelson, N. C., Taylor, S. S., and Kaiser, E. T. (1982) *J. Biol. Chem.* 257:10575–10581.
18. Peters, K. A., Demaille, J. G., and Fischer, E. H. (1977) *Biochemistry,* 16:5691–5697.

19. Shoji, S., Ericsson, L. H., Walsh, K. A., Fischer, E. H., and Titani, K. (1983) *Biochemistry* 22:3702–3709.
20. Takio, K., Dale, R. D., Smith, S. B., Krebs, E. G., Walsh, K. A., and Titani, K. (1984) *Biochemistry* 23:4207–4218.
21. Reimann, E. M., Titani, K., Ericsson, L. H., Wade, R. D., Fischer, E. H., and Walsh, K. A. (1984) *Biochemistry* 23:4185–4192.
22. Takio, K., Blumenthal, D. K., Edelman, A. M., Walsh, K. A., Krebs, E. G., and Titani, K. (1985) *Biochemistry* 24:6028–6037.
23. Schwartz, D. E., Tizard, R., and Gilbert, W. (1983) *Cell* 32:853–869.
24. Ullrich, A., Bell, J. R., Chen, E. Y., Herrera, R., Petruzzelli, L. M., Dull, T. J., Gray, A., Coussens, L., Liao, Y. C., Tsubokawa, J., Mason, A., Seebrug, P. H., Grunfeld, C., Rosen, O. M., and Ramachandran, J. (1985) *Nature* 756–761.
25. Ullrich, A., Coussens, L., Hayflich, J. S., Dull, T. J., Gray, A., Tam, A. W., Lee, J., Yarden, Y., Libermann, T. A., Schlessinger, J., Downware, J., Mayes, E. L. V., Whittle, N., Waterfield, J. D., and Seeburg, P. G. (1984) *Nature* 309:418–425.
26. Kamps, M. P., Taylor, S. S., and Sefton, B. M. (1984) *Nature* 310:589–592.
27. Snyder, M. A., Bishop, M. J., McGreth, J. P., and Leuinson, A. D. (1985) *Mol. Cell. Biol.* 5:1772–1779.
28. Kamps, M. P., and Sefton, B. M. (1986) *Mol. Cell. Biol.* 6:751–757.
29. Barker, W. C., and Dayhoff, M. O. (1982) *Proc. Natl. Acad. Sci. USA* 73:2836–2839.
30. Hol, W. G. J., Van Duijhne, P. Th., and Berendsen, H. J. C. (1978) *Nature* 273:443–446.
31. Wierenga, R. K., Terpstra, P., and Hol, W. G. J. (1986) *J. Mol. Biol.* 187:101–107.
32. Rossmann, M. G., Moras, D., and Olsen, K. W. (1974) *Nature* 250:194–199.
33. Hunter, T. (1982) *J. Biol. Chem.* 257:4843–4848.
34. Sowadski, J. M., Xuong, N. H., Anderson, D., and Taylor, S. S. (1985) *J. Mol. Biol.* 182:617–620.
35. Xuong, N. H., Freer, S., Hamlin, R., Nielsen, C., and Vernon, W. (1978) *Acta Cryst.* 34:345–361.
36. Cheng, H. C., Van Patten, S. M., Smith, A. J., and Walsh, D. A. (1985) *Biochem. J.* 231:655–661.
37. Zoller, M. J., Kerlavage, A. R., and Taylor, S. S. (1979) *J. Biol. Chem.* 254: 2408–2412.
38. Hoffman, F., Beavo, J. A., Bechtel, P. T., and Krebs, E. A. (1975) *J. Biol. Chem.* 250:7795–7801.
39. Rosen, O. M., and Ehrlichman, J. (1975) *J. Biol. Chem.* 250:7788–7794.
40. Lee, D. C., Carmichael, D. F., Krebs, E. G., and McKnight, G. S. (1983) *Proc. Natl. Acad. Sci. USA* 80:3608–3612.
41. Weldon, S. L., Mumby, M. C., and Taylor, S. S. (1985) *J. Biol. Chem.* 260: 6640–6448.
42. Robinson-Steiner, A. M., Beebe, S. J., Rannels, S. R., and Corbin, J. D. (1984) *J. Biol. Chem.* 259:10596–10605.
43. Weldon, S. L., Mumby, M. C., Beavo, J. A., and Taylor, S. S. (1983) *J. Biol. Chem.* 258:1129–1135.
44. Weldon, S. L., and Taylor, S. S. (1985) *J. Biol. Chem.* 260:4203–4209.
45. De Crombrugghe, B., Busby, S., and Buc, H. (1983) In *Biological Regulation and Development*, Vol. III, edited by K. Yamamoto. Plenum Press, New York.

46. McKay, D. B., Weber, I. T., and Steitz, T. A. (1982) *J. Biol. Chem.* 257:9518–9524.
47. Weber, I. T., Takio, K., Titani, K., and Steitz, T. A. (1982) *Proc. Natl. Acad. Sci USA* 79:7679–7693.
48. Kerlavage, A. R., and Taylor, S. S. (1980) *J. Biol. Chem.* 255:8483–8488.
49. Bubis, J., and Taylor, S. S. (1987) *Biochemistry* 26:3478–3486.
50. Rannels, S. R., and Corbin, J. D. (1981) *J. Biol. Chem.* 256:7871–7877.
51. Doskeland, S. O., Ogreid, D., Ekanger, R., Sturm, P. A., Miller, J. P., and Suva, R. H. (1983) *Biochemistry* 22:1094–1101.
52. Saraswat, L. D., Filutowics, M., and Taylor, S. S. (1986) *J. Biol. Chem.* 261:11091–11096.
53. Weber, I. T., Steitz, T. A., Bubis, J., and Taylor, S. S. (1987) *Biochemistry* 26:343–351.

39
Conformation and Dynamics of Oligomeric Enzymes

C. L. Tsou

The effect of structural changes on the activity of enzymes have been extensively studied; however, those investigators have largely concentrated on the chemical modification of the primary structures of the enzymes. Although the importance of conformational integrity for the activity of an enzyme is generally recognized and there have been extensive studies on the unfolding of enzyme molecules by heat, acid, and denaturants, few attempts have been made to correlate the conformational changes of an enzyme to its enzymatic activity. It is the purpose of the present series of investigations to compare the conformational and activity changes during unfolding of several enzyme molecules by guanidine and urea. The results obtained suggest that the active site of an enzyme is usually situated in a limited region of the molecule more fragile to denaturants than is the molecule as a whole.

Rapid Inactivation and Slow Unfolding of Creatine Kinase During Guanidine and Urea Denaturation

The extent of inactivation of creatine kinase increases with increasing urea or guanidine concentrations. A comparison (1) of the activity changes in urea solutions of various concentrations with the extent of unfolding of the molecule, as indicated by the exposure of buried SH groups and the red shift of the emission maximum of the intrinsic protein fluorescence, shows that low concentrations of urea inactivate without causing significant conformational changes of the enzyme molecule (Fig. 39.1). Similar results have been obtained during guanidine denaturation.

Few studies designed to monitor the rate of enzymic inactivation during the unfolding of the molecules have been reported. This small number is due partly to the lack of a suitable method to measure quantitatively rapid inactivation rates of enzymes during conformational changes brought about by denaturating agents. A kinetic method has been proposed (2, 3) and verified experimentally (4) that follows the rapid activity changes during irreversible modification of enzyme activity by measuring the substrate reac-

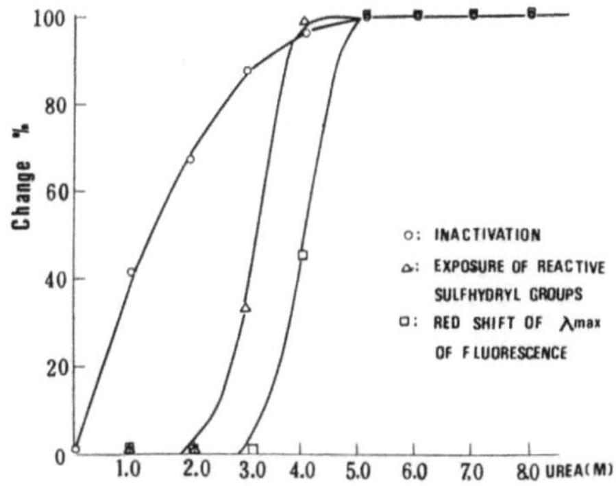

FIG. 39.1. Inactivation and unfolding of creatine kinase in urea solutions. The enzyme solutions were incubated with urea for 24 h in 0.1 *M* glycine-NaOH buffer, pH 9.0, before determination of the remaining activity, the exposure of SH groups, and the red shift of the fluorescence emission maxima. The enzyme was stable in the absence of urea. Final concentrations of the enzyme were 0.2 μM for activity assay, 1.1 μM for fluorescence measurements, and 5.8 μM for SH determinations.

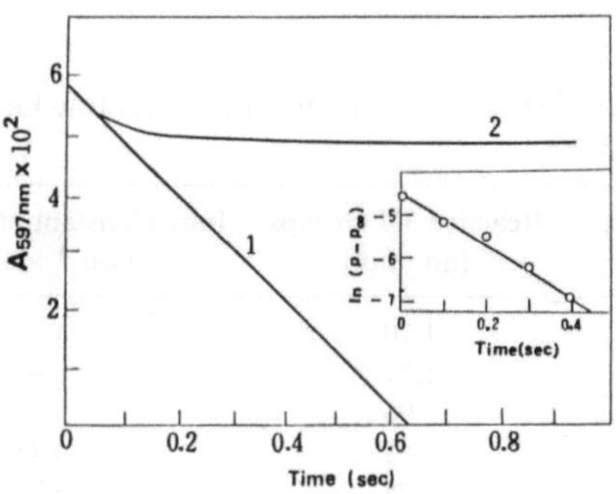

FIG. 39.2. Inactivation kinetics of creatine kinase in 3 *M* guanidine. The reactions were followed by a pH indicator method monitored at 597 nm in a stopped-flow apparatus with an enzyme concentration of 3.6 μM in 0.1 *M* glycine-NaOH buffer, pH 9.0, at 25°C. Curves 1 and 2 show, respectively, the substrate reaction in the absence and presence of guanidine. *Inset*: Semilogarithmic plot for the determination of the rate constant.

tion in the presence of the modifier. Using this method we have obtained the inactivation rate constants of creatine kinase during denaturation induced by guanidine or urea. The result obtained for the inactivation kinetics in 3 *M* guanidine is given in Fig. 39.2 (5).

The conformational changes of this enzyme in guanidine or urea of various concentrations have been followed by ultraviolet absorbance, fluorescence and circular dichroism (CD) changes, and exposure of buried SH groups. The effects of increasing the guanidine concentration on the CD spectra of creatine kinase in the ultraviolet range are given in Fig. 39.3 (6). The data indicate a

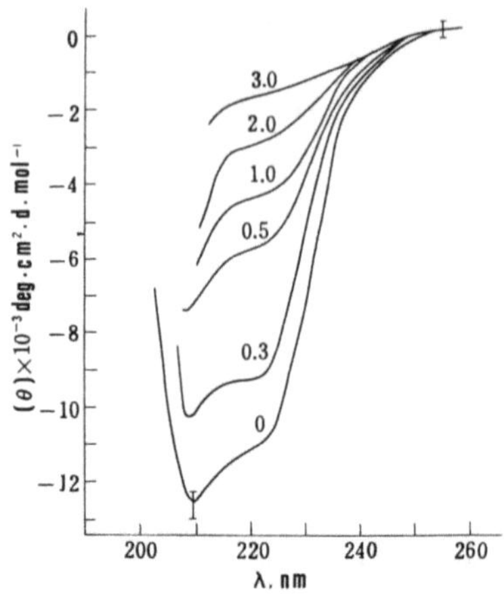

FIG. 39.3. CD spectra of creatine kinase denatured in guanidine solutions of various concentrations. The enzyme, 0.34 μM in tris buffer at pH 8.3, was incubated with guanidine for 18 h at 25°C before CD measurements in 0.1 mm cuvettes. Numbers indicate the molar concentrations of guanidine solutions.

TABLE 39.1. Exposure of SH groups of creatine kinase in urea solutions.

Urea Conc. (M)	Reactive SH Groups (no./mol)	Rate Constant of Exposure ($sec^{-1} \times 10^3$)
0	1.70	—
1	1.77	—
2	1.96	—
3	5.52	0.45
5	6.78	5.75

Rates of exposure of SH groups were measured by taking aliquots from an incubation mixture of 5.5 μM creatine kinase in 0.1 M glycine-NaOH buffer, pH 9.0, at 25°C and assaying for reactive SH groups with 5,5′-dithiobis-2-nitrobenzoic acid (DTNB). For other details, see ref. 6.

gradual decrease in ordered secondary structure with increasing guanidine concentration. Table 39.1 summarizes the number of reactive SH groups and the rates of exposure of the buried sulfhydryl groups in the presence of various urea concentrations. At low guanidine or urea concentrations, the activity of creatine kinase is seriously affected without significant conformational changes of the enzyme molecule. The rates obtained for the conformational changes as followed by the above-mentioned methods have been compared with its inactivation rates under similar conditions. The results thus obtained, as summarized in Table 39.2 for guanidine denaturation, clearly show that the inactivation rate constants of the enzyme in guanidine are several orders of

TABLE 39.2. Comparison of the denaturation and inactivation rate constants of creatine kinase in guanidine solutions.

	Denaturation Rate (sec^{-1})[a]				Inactivation	
Guanidine (M)	Absorbance	Fluorescence	CD	SH Exposure	Rate[a] (sec^{-1})	Residual Activity (%)
0.5	0.0043	0.0036	0.0008	0.0014	3.6	15
1.0	0.053	0.040	0.017	—	4.3	0

[a] As first-order rate constant.
Absorbance chance was measured at 287 nm (5), fluorescence change at 345 nm (5), ellipticity change at 220 nm (6), and the exposure of SH as indicated in Table 39.1 (6). The inctivation rates were measured by following the substrate reaction in the presence of guanidine (5). See the relevant references for details of experimental conditions.

magnitude faster than the rate constants of conformational changes measured, particularly at the lower concentrations of guanidine (6).

As creatine kinase is a dimeric enzyme and it has been suggested (7) that the enzyme must be in the dimeric state to be enzymatically active, the above results could be explained by a fast dissociation of the dimer followed by a slow conformational change of the monomer. However, after the enzyme has been largely inactivated in 0.5 M guanidine, gel filtration with Sephadex G-150, analytical ultracentrifugation, and viscosity determinations show clearly that the inactivated enzyme is still in the dimeric state. Consequently, the fast inactivation reaction cannot be ascribed to dissociation of the dimeric enzyme to the monomeric state.

We suggest instead that the active site is located in a limited region of the molecule that is more sensitive to guanidine or urea than the enzyme as a whole. A slight disturbance to the correct spatial geometry of the functional groups responsible for the catalytic mechanism of the enzyme destroys its activity before any gross conformational changes can be detected by the methods employed.

Conformational Change at the Active Site of D-Glyceraldehyde-3-Phosphate Dehydrogenase

A probe introduced at the active site would be useful for revealing conformational change in this region compared to that in the molecule as a whole. We have previously reported that by ultraviolet irradiation of active site carboxymethylated glyceraldehyde-3-phosphate dehydrogenase a fluorescent derivative of NAD is formed covalently linked to the enzyme presumably at its active site (8), as shown in Fig. 39.4. This fluorescent NAD derivative has now been employed as a probe to follow the conformational change at the active site during guanidine denaturation of this enzyme.

Denaturation of lobster muscle enzyme in guanidine solutions leads to a

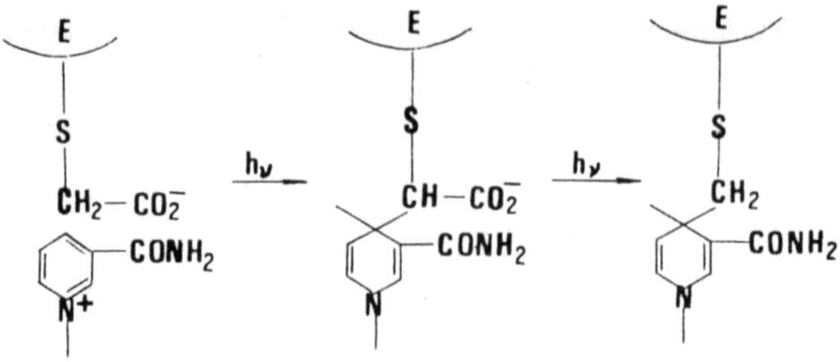

FIG. 39.4. Reactions leading to the formation of the fluorescent NAD derivative of glyceraldehyde-3-phosphate dehydrogenase.

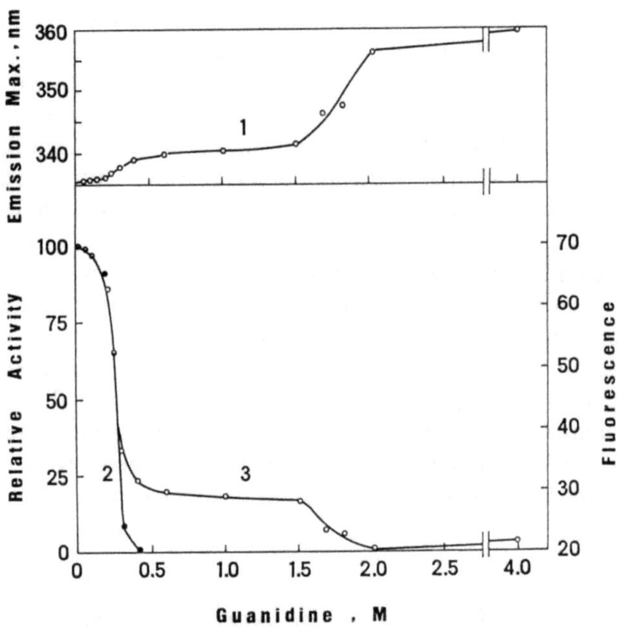

FIG. 39.5. Fluorescence and activity changes of glyceraldehyde-3-phosphate dehydrogenase in guanidine solutions of various concentrations. The lobster enzyme, 0.3 μM, was incubated in 0.1 M phosphate buffer, pH 7.0, containing 1 mM EDTA and guanidine at the desired concentration at about 20°C for 24 h. At that time, when the denaturation reached completion, aliquots were taken for activity and fluorescence measurements with an excitation wavelength of 295 nm. The enzyme was stable under the above conditions in the absence of guanidine. Curves 1 to 3 represent, respectively, the effects of guanidine on emission maximum, relative activity (left ordinate), and emission intensity (right ordinate).

decrease in intensity and a red shift of its intrinsic fluorescence. Both of these changes occur in two stages, as shown in Fig. 39.5. A marked decrease in intensity takes place at about 0.2 to 0.4 *M* guanidine together with a slight red shift from 335 to 340 nm. Further changes in the fluorescence properties occur at guanidine concentrations higher than 1.5 *M* with a slight decrease in emission intensity but a marked red shift from 340 to 360 nm. The extent of inactivation in guanidine solutions is also recorded in Fig. 39.5; and as can be clearly seen, it coincides almost exactly with the extent of fluorescence changes in the initial stage. The enzyme is completely inactivated in 0.4 *M* guanidine.

Changes in fluorescence properties of the enzyme carrying the fluorescent NAD derivative in guanidine solutions of various concentrations are shown in Fig. 39.6. Denaturation by guanidine produces a decrease in emission intensity of both the intrinsic protein fluorescence as well as the fluorescence of the covalently attached NAD derivative (410 nm) together with red shifts of the

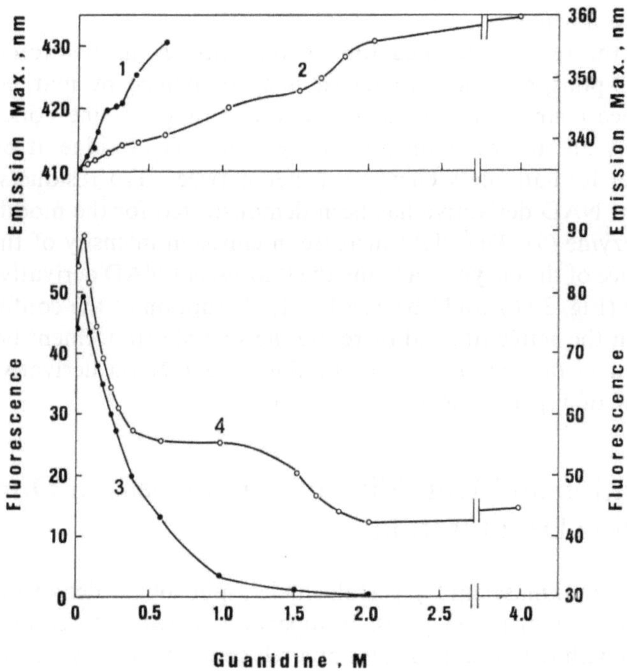

FIG. 39.6. Changes in fluorescence properties of glyceraldehyde-3-phosphate dehydrogenase carrying the fluorescent NAD derivative in guanidine. Conditions were as for Fig. 39.5 except that the enzyme carrying the fluorescent NAD derivative was employed. For the measurement of fluorescence at 410 nm, an excitation wavelength of 325 nm was used. Curves 1 to 4 represent, respectively, the effects of guanidine on emission maxima of the fluorescent derivative (left ordinate) and the intrinsic fluorescence (right ordinate), emission intensities of the fluorescent derivative (left ordinate), and intrinsic fluorescence (right ordinate).

respective emission maxima. These changes in the fluorescence properties of the covalently linked NAD derivative strongly suggest its exposure from a buried environment. It is also to be noted that the greater part of the decrease in its fluorescence takes place at guanidine concentrations of $<0.5\ M$ largely in coincidence with the initial stage of the intrinsic fluorescence changes of the native enzyme.

As this NAD derivative is formed in situ at the active site of this enzyme, it appears that the fluorescence change at low guanidine concentrations results from the unfolding of the active site region of the enzyme molecule. In other words, the same low concentrations of guanidine produce not only complete inactivation of the enzyme but also a conformational change at the active site as revealed by the changes in the fluorescence of the covalently bound NAD. Much higher guanidine concentrations are required for further conformational changes. It is clear, therefore, that the active site of the lobster muscle glyceraldehyde-3-phosphate dehydrogenase as represented by the NAD binding site is also much more sensitive to guanidine denaturation than the molecule as a whole.

Unlike the results obtained for creatine kinase, inactivation of glyceraldehyde-3-phosphate dehydrogenase is accompanied by marked intrinsic fluorescence changes at the same low range of guanidine concentrations presumably due to environmental changes of a Trp residue at or near the active site. Radiationless energy transfer between Trp residue(s) and the fluorescent NAD derivative has been demonstrated for the modified rabbit muscle enzyme (9). The slight increase in emission intensity of the intrinsic fluorescence of the enzyme carrying the fluorescent NAD derivative at $0.1\ M$ guanidine (Fig. 39.6) could be ascribed to disruption of the conformational integrity at the active site and therefore the spatial arrangement between the substituted nicotinamide ring of the fluorescent NAD derivative and the indole ring of Trp at or near the active site.

Inactivation and Unfolding of Ribonuclease a During Guanidine Denaturation

Both creatine kinase and glyceraldehyde-3-phosphate dehydrogenase are fairly large oligomeric enzymes. It appears worthwhile to compare the inactivation with the unfolding rates of a relatively small enzyme, active in the monomeric state and with its native conformation stabilized by multiple disulfide linkages. In this respect, bovine pancreatic ribonuclease A appears to be ideally suitable, as the course of its unfolding during denaturation has been extensively studied in the literature (10).

The unfolding of ribonuclease A during guanidine denaturation has been followed by fluorescence emission changes at 305 nm and absorbance changes at 287 nm due to the exposure of buried Tyr residues. As far as conformational

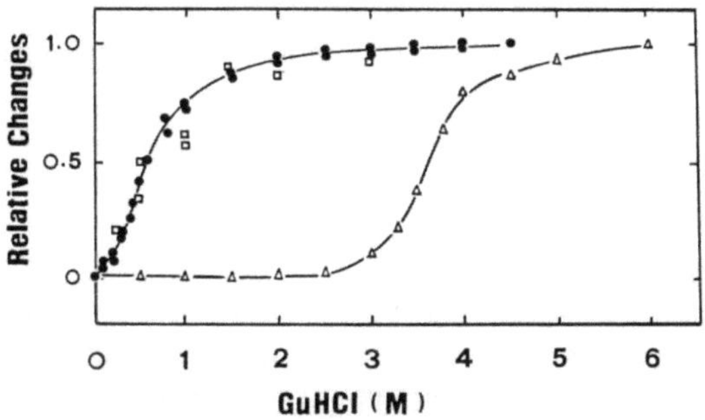

FIG. 39.7. Inactivation and unfolding of ribonuclease A in guanidine solutions. Enzyme activities were measured by the hydrolysis of cCMP (•), 2′-CMP binding by absorbances at 260 nm (□), and unfolding of the enzyme followed by fluorescence emission at 305 nm (△). All measurements were made in 0.1 *M* cacodylate buffer pH 6.0 at 10°C containing various concentrations of guanidine.

changes are concerned, this enzyme is fairly resistant to guanidine, and unfolding becomes significant at 3 *M*, approaching completion at 6 *M* guanidine. However, its activity is nearly as sensitive to guanidine as the oligomeric enzymes studied above. As depicted in Fig. 39.7, the enzyme is 50% inactivated in 0.5 *M* guanidine; in 3 *M* guanidine, when the molecule starts to unfold as revealed by the fluorescence changes, the enzyme has already lost essentially all of its activity.

It is known that binding of a competitive inhibitor, 2′-CMP, to ribonuclease A is accompanied by a change in A_{260}. This property has been used to study recovery of the ability of the unfolded enzyme to bind this competitive inhibitor—hence the reformation of the substrate binding site during refolding of the denatured enzyme (10). The effects of guanidine on 2′-CMP binding are also shown in Fig. 39.7. The data reveal that the lost of 2′-CMP binding capacity is essentially parallel to the decrease in activity. It is to be noted that at higher guanidine concentrations the completely inactivated enzyme can still bind this competitive inhibitor to a slight extent. It should be pointed out, however, that the absorbance change in this region could be due to either a decrease in 2′-CMP binding or a decrease in the difference absorbance coefficient upon formation of the enzyme–2′-CMP complex in guanidine solution, or both. Taking the latter possibility into consideration, the enzyme probably binds significantly more 2′-CMP than indicated in Fig. 39.7, and complete inactivation of the enzyme occurs at lower guanidine concentrations than does destruction of the substrate binding site. This is not surprising, as catalysis of the enzyme would require more than the conformational intactness of the substrate binding site alone.

It appears that, as in the case of the oligomeric enzymes, the active site of a small enzyme, stabilized by multiple disulfide linkages like ribonuclease A, is also situated in a region much more fragile to perturbation by denaturants than the molecule as a whole. Similar results have been obtained for chymotrypsin.

Discussion

X-ray crystallographic studies have shown that the active sites of enzymes are usually situated at the hinge regions covering two structural domains (11). It seems likely, therefore, that the active sites of most enzymes are situated in a region that is relatively more flexible or mobile than the molecule as a whole and hence more sensitive to denaturants. It is known that in some proteins there are exposed peptide segments usually without ordered secondary structure, free to move about in solution, and probably nonessential for the biological activities of the proteins concerned. This flexibility or, rather, mobility of enzyme active sites could only be relative and within certain limits. Nevertheless, its functional implications should be seriously considered as this mobility may well be dictated by the requirements for the expression of enzymic activity. The functional importance of protein flexibility at the active site has been suggested based on results obtained from an entirely different angle (12). It is also known that substrate binding would lead to the relative movements of the active site region that could be important for the catalysis of some enzymes (13).

References

1. Yao, Q., Tian, M., and Tsou, C-L. (1984) *Biochemistry* 23:2740–2744.
2. Tsou, C. L. (1965) *Acta Biochim. Biophys. Sin.* 5:398–408.
3. Tsou, C. L. (1965) *Acta Biochim. Biophys. Sin.* 5:409–417.
4. Tian, W. X., and Tsou, C. L. (1982) *Biochemistry* 21:1028–1032.
5. Yao, Q., Zhou, H., Hou, L., and Tsou, C-L. (1982) *Sci. Sin.* 25B:1296–1302.
6. Yao, Q., Liang, S., Tian, M., and Tsou, C-L. (1985) *Sci. Sin.* 28B:484–493.
7. Degani, C., and Degani, Y. (1980) *J. Biol. Chem.* 255:8221–8228.
8. Tsou, C. L., Xu, G. Q., Zhou, J. M., and Zhao, K. Y. (1983) *Biochem. Soc. Trans.* 11:425–429.
9. Ho, Y. S., and Tsou, C. L. (1979) *Sci. Sin.* 22:478–489.
10. Kim, P. S., and Baldwin, R. L. (1982) *Annu. Rev. Biochem.* 51:459–489.
11. Schulz, G. E., and Schirmer, R. H. (1979) *Principles of Protein Structure*, pp. 95–97. Springer-Verlag, New York.
12. Artymiuk, P. J., Blake, C. C. F., Grace, D. E. P., Oatley, S. J., Phillips, D. C., and Sternberg, M. J. E. (1979) *Nature*, 280:563–568.
13. Chothia, C., and Lesk, A. M. (1985) *Trends Biochem. Sci.* 10:116–118.

40
Coenzyme Binding Site of NADP-Dependent Isocitrate Dehydrogenase

ROBERTA F. COLMAN

When elucidating the relation between structure and function for a two-substrate enzyme, e.g., a dehydrogenase, a major objective is to map the binding site of the coenzyme and to understand how each amino acid residue contributes to specificity and catalysis. A variety of techniques are available to explore the coenzyme site of an enzyme while it is in solution, including fluorescence and ultraviolet absorption spectroscopy, ^{31}P and ^{1}H-NMR (nuclear magnetic resonance) spectroscopy, as well as affinity labeling with reactive coenzyme analogues. The NADP-dependent isocitrate dehydrogenase provides an example in which these approaches have been used in a complementary manner to yield information about the nature of the pyridine nucleotide binding site.

Pig heart NADP-dependent isocitrate dehydrogenase has a single type of polypeptide chain of molecular weight 58,000 (1) and under many conditions exists as a dimer of identical subunits (2, 3). The enzyme has a high affinity for the coenzymes NADP and NADPH, and the coenzyme fragments 2′-phosphoadenosine 5′-diphosphoribose (PADPR) and 2′,5′-adenosine diphosphate (2′,5′-ADP) have been shown to function as effective competitive inhibitors with respect to NADP (4, 5). Because these coenzyme fragments lack the nicotinamide and the nicotinamide ribose phosphate moieties, respectively, it is apparent that these portions of the coenzyme are not essential for interaction with the enzyme, although they do contribute to the binding energy (5).

In order to compare their environments on the enzyme, the absorption spectra of the binary complexes of nucleotides with isocitrate dehydrogenase were measured at pH 7.6. Figure 40.1 indicates that the spectral perturbations observed for all three complexes are remarkably similar and do not depend on the buffer. A minimum at about 250 nm and positive bands occurring in the 270 to 280 nm region may be attributed to perturbation of the electronic transitions of the purine ring resulting from interaction with a cationic or a combination of cationic and nonpolar residues on the enzyme (6). Altered protein chromophores could also contribute to spectral changes in this region, whereas the characteristic minimum at about 290 nm might be ascribed to the

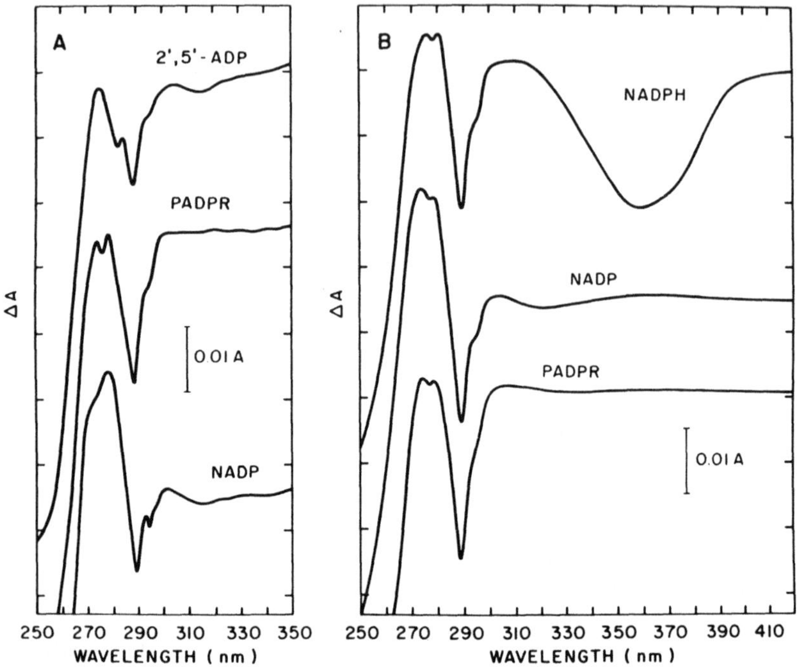

FIG. 40.1. Difference absorption spectra of the binary complexes of isocitrate dehydrogenase with coenzymes and coenzyme fragments. *A*: Spectra in 50 m*M* MOPS buffer, pH 7.6, containing 10% glycerol and 0.5 m*M* EDTA, at 25°C. *B*: Spectra in 50 m*M* triethanolamine chloride buffer, pH 7.6, containing 10% glycerol and 0.5 m*M* EDTA at 25°C.

perturbation of an indole chromophore entering a more polar environment (7). In addition, for NADPH the negative band with a minimum at 360 nm provides evidence for a direct interaction of the reduced nicotinamide ring with entities in the enzyme able to produce electron shifts in the nicotinamide chromophore (8). Whatever the origin of these spectral bands, the nearly identical ultraviolet difference spectra from 250 to 290 nm of the binary complexes of enzyme with NADPH, NADP, PADPR, and 2′,5′-ADP strongly suggest that there are similar interactions of the adenosine ribose moiety with the same binding site on the enzyme.

^{31}P-NMR spectroscopy of the binary enzyme–nucleotide complexes also indicates that the coenzymes and coenzyme fragments bind in the same way to a common site on the enzyme. The spectrum of free NADPH at pH 7.5 exhibits two ^{31}P resonances: The downfield resonance at 4.22 ppm corresponds to the 2′-phosphate, whereas that at −10.60 is due to the two phosphates of the pyrophosphate moiety (9). Figure 40.2 shows ^{31}P-NMR spectra of the binary complex of enzyme with NADPH at pH 7.5 at various coenzyme/enzyme ratios. The sharp peaks that are seen at an NADPH/

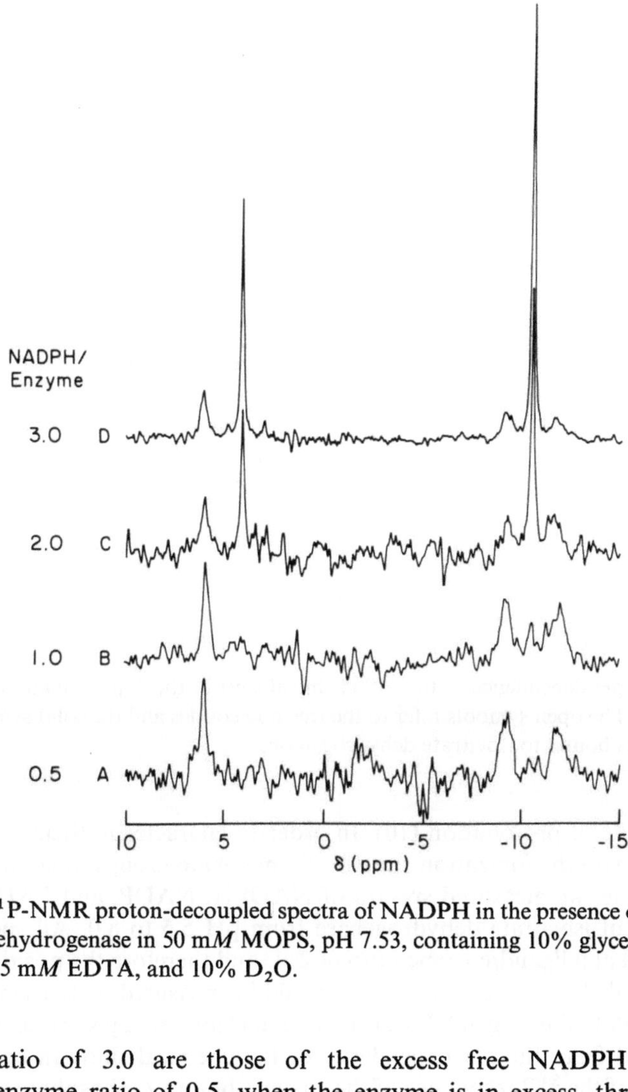

FIG. 40.2. ^{31}P-NMR proton-decoupled spectra of NADPH in the presence of 0.3 m*M* isocitrate dehydrogenase in 50 m*M* MOPS, pH 7.53, containing 10% glycerol, 0.1 *M* Na_2SO_4, 0.5 m*M* EDTA, and 10% D_2O.

enzyme ratio of 3.0 are those of the excess free NADPH. At an NADPH/enzyme ratio of 0.5, when the enzyme is in excess, three broad resonances are observed: a resonance at 5.97 ppm corresponding to the 2′-phosphate of bound nucleotide and two signals in the pyrophosphate region of the spectrum, centered at −9.2 and −11.7 ppm. Apparently, upon binding to the enzyme the two phosphoryl groups of the pyrophosphate experience different environments. The occurrence of two sets of signals, corresponding to free and bound NADPH, is characteristic of slow exchange of nuclei, on the NMR time scale, between free and bound states. Remarkably similar ^{31}P spectra were observed for the complexes of isocitrate dehydrogenase with the oxidized coenzyme NADP and with PADPR (9).

Chemical shifts of ^{31}P resonances of phosphate groups are very sensitive

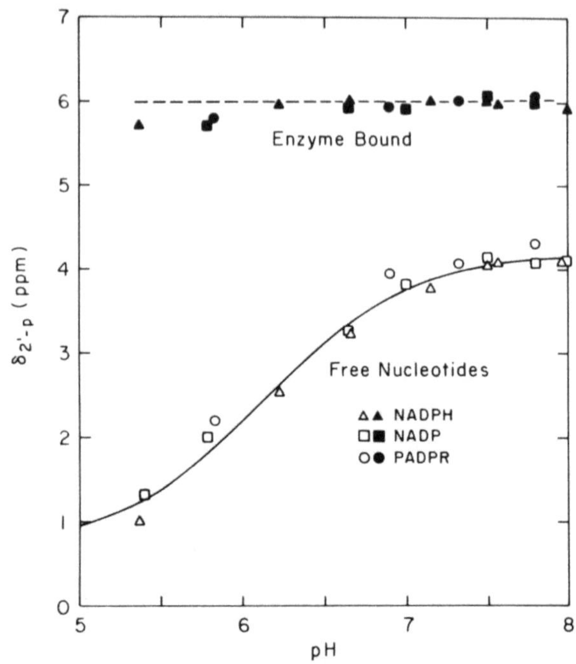

FIG. 40.3. pH dependence of the ^{31}P chemical shift of the 2′-phosphate resonance of NADPH. The open symbols refer to the free nucleotides and the solid symbols to the nucleotides bound to isocitrate dehydrogenase.

to the state of protonation (10). In order to characterize their environment and compare the ionization state of 2′-phosphate groups in complexes with the enzyme, we measured spectra of NADPH, NADP, and PADPR in the presence of isocitrate dehydrogenase from pH 5.5 to 8.0. All spectra were measured at a ligand/enzyme ratio of 2:1, and therefore the positions of free and bound 2′-phosphate resonances could be measured on the same samples and are plotted in Fig. 40.3. For the free nucleotides, a pK value of 6.13 was calculated from the pH dependence of the chemical shift. In contrast, the position of the ^{31}P resonance of the enzyme-bound 2′-phosphate is invariant at 6 ppm over this pH range, indicating that the pK of the 2′-phosphate is much lower in the nucleotide complex with enzyme than in the free nucleotides. The decrease in pK of the 2′-phosphate could be the result of an interaction with a positively charged amino acid within the nucleotide binding site. The resonance of the bound 2′-phosphate occurs at 6 ppm and is shifted by 1.8 ppm with respect to the 4.22 ppm characteristic of the unprotonated form of free nucleotide. This comparison suggests that other factors must contribute to the observed change in chemical shift, perhaps distortion of the O-P-O angles as a result of binding to the enzyme.

Proton NMR studies have revealed additional information about the con-

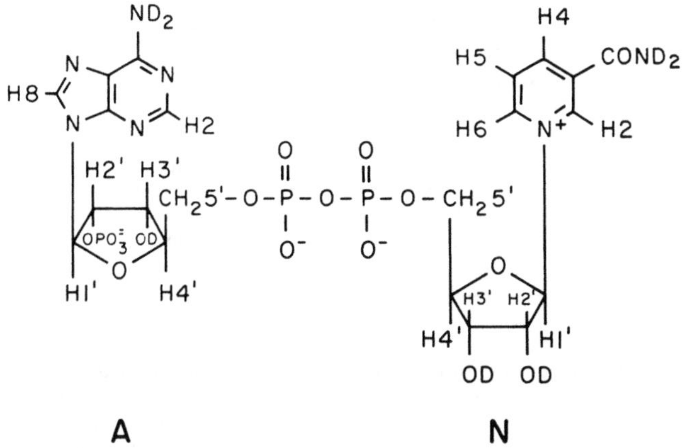

FIG. 40.4. Structure of NADP drawn with adenine in the syn conformation and nicotinamide in the anti conformation.

formation of nucleotides as bound to isocitrate dehydrogenase and the types of amino acids that interact with the bound coenzymes. The conformation of nucleotides was investigated by means of the transferred nuclear Overhauser effect (11, 12) in which a given proton of the free nucleotide is irradiated. A nuclear Overhauser effect (NOE), or influence on the amplitude of the signal of one proton by the irradiation of a different proton, occurs only when the two protons are quite close (<3.2 Å). This technique can be used to ascertain which protons of the bound nucleotide are close together and, by deduction to yield an indication of the conformation of the bound nucleotide. Figure 40.4 is a drawing of NADP with adenine in the syn conformation and with nicotinamide in the anti conformation. We have ascertained that this model best describes NADP bound to isocitrate dehydrogenase, as we have observed NOEs (and hence proximity) between the H8 of the adenine and the H1′ of the adenine ribose, and between the H2 of the nicotinamide and the H1′ of the nicotinamide ribose (13). The reduced coenzyme NADPH also exhibits a syn conformation about the adenine–ribose bond and an anti conformation about the nicotinamide–ribose bond, as determined from the NOEs. Furthermore, both PADPR and 2′,5′-ADP bind with a predominantly syn conformation about the adenine ribose bond. Thus the coenzyme and coenzyme fragments (for those parts that they have in common) bind with the same conformation to the enzyme.

With the exception of the aromatic region, the ^{1}H-NMR spectrum of isocitrate dehydrogenase is relatively featureless. The failure to observe many distinct peaks probably arises from the relatively slow rotation of the isocitrate dehydrogenase dimer (MW 116,000), which causes broadening. At pH 7.5 three narrow peaks are observed in the region 7.5 to 8.0 ppm. The position of these peaks changes with pH, suggesting that they arise from the C4

TABLE 40.1. pK values and chemical shifts of histidines of isocitrate dehydrogenase.

Peak	pK	$\delta_{\text{protonated}}$ (ppm)	$\delta_{\text{unprotonated}}$ (ppm)
1	5.68	8.74	7.84
2	5.35	8.79	7.79
3	5.75	8.76	7.68

TABLE 40.2. Coenzyme binding sites on isocitrate dehydrogenase.

Ligand	Technique	Maximum No. of Sites per Enzyme Subunit
NADPH	Enhancement of coenzyme fluorescence	1.03
NADPH	UV difference spectra	1.00
NADP	UV difference spectra	1.00
NADP	Ultrafiltration	0.97
PADPR	Ultrafiltration	1.03

histidine protons (13). The pH dependence of these chemical shifts can be fit to a standard titration curve, with the pK values and chemical shifts for fully protonated and unprotonated forms summarized in Table 40.1. As measured at pH 7.50, peak 2 (with a pK of 5.35) moves upfield by 0.07 ppm in the presence of NADP or NADPH, whereas peaks 1 and 3 do not change position (data not shown). In contrast, saturating concentrations of PADPR and 2′, 5′-ADP fail to produce any changes in the observed proton NMR spectrum of isocitrate dehydrogenase. Thus the histidine giving rise to peak 2 is likely to be in the vicinity of the nicotinamide ring, rather than near the adenine ring or the phosphates (13). In addition, NOEs between enzyme protons and nucleotide protons give evidence of proximity between the adenine ring and a nontitrating histidine or an aromatic residue (13).

The number of coenzyme binding sites has been evaluated by three techniques, as summarized in Table 40.2. The fluorescence intensity of NADPH is increased sevenfold when bound to isocitrate dehydrogenase as compared to free NADPH (5). Fluorescence titration experiments in which increasing concentrations of NADPH are added to a constant amount of enzyme allow determination of dissociation constants as well as the maximum number of NADPH sites (1.03) per peptide chain of enzyme. Furthermore, ultraviolet difference titrations, based on the difference spectra typified by Fig. 40.1, yield values of one site per enzyme subunit for NADPH and NADP. Ultrafiltration experiments (Table 40.2), performed as an independent method for determining the number of nucleotide binding sites for NADP and PADPR, also give a stoichiometry of one ligand site per enzyme subunit.

Nucleotides have been shown to quench the protein fluorescence of many dehydrogenases. In contrast, an increase of about 50% is observed in the 350 nm fluorescence of the pig heart isocitrate dehydrogenase induced by the interaction with NADP, PADPR, and 2′,5′-ADP. We have attributed this large enhancement to a local conformational change of the enzyme caused by binding of nucleotides. Experiments were conducted in which increasing concentrations of NADP and PADRP were added to a constant amount of enzyme, and the protein fluorescence was monitored. These studies revealed a maximum increase in protein fluorescence upon saturation of 0.5 sites per average enzyme subunit (5). Scatchard plots derived from this type of data for NADP and PADPR also indicate apparent values of 0.5 mol of nucleotide per peptide chain. These results suggest that binding of ligand to one subunit of an enzyme dimer is responsible for the total observed change, whereas binding to the second subunit is not detectable by protein fluorescence.

Interaction between enzyme subunits is also apparent in the affinity labeling of the coenzyme site of pig heart isocitrate dehydrogenase by reactive nucleotide analogues. Because pig heart isocitrate dehydrogenase does not accept NAD as a coenzyme but binds tightly NADP, PADPR, and 2′,5′-ADP, it seemed that when designing an affinity label that would react at the coenzyme binding site of isocitrate dehydrogenase it was critical to include a 2′-phosphate moiety.

We have synthesized 2-[(4-bromo-2,3-dioxobutyl)thio]-1,N^6-ethenoadenosine 2′,5′-diphosphate (BDB-TεADP), which is closely related to the adenine nucleotide structure and retains the required 2′-phosphate (3). This reagent, which is shown in Fig. 40.5, is water-soluble, negatively charged at neutral pH,

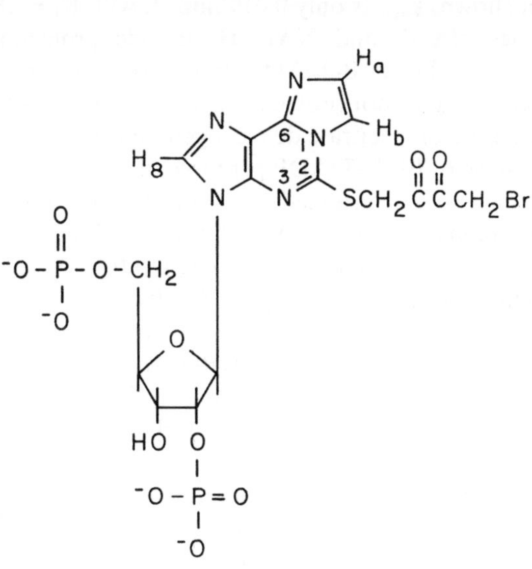

FIG. 40.5. Structure of 2-[(4-bromo-2,3-dioxobutyl)thio]-1,N^6-ethenoadenosine 2′,5′-diphosphatē (BDB-TεADP).

and potentially reactive with several nucleophiles found in proteins, including cysteine, lysine, histidine, and glutamic and aspartic acids (14). In addition, the dioxo group lends the possibility of reaction with arginine residues (15, 16). Furthermore, the reagent is fluorescent, with an emission peak at 428 nm and an excitation maximum at 302 nm. This property allows easy measurement of the extent of reagent incorporation into protein, as well as providing a convenient means of introducing a fluorescent probe into a putative NADP-binding site of a protein.

The overall synthetic scheme involves both chemical and enzymatic steps. Starting with NADP, PADPR was generated enzymatically. The PADPR was converted to the fluorescent εPADPR by reaction with chloroacetaldehyde. Treatment with NaOH followed by reaction with carbon disulfide yielded 2-thio-1,N^6-ethenoadenosine 2′,5′-diphosphate (TεADP). Condensation of TεADP with 1,4-dibromobutanedione gave the final product BDB-TεADP (3).

Incubation of NADP-specific isocitrate dehydrogenase with 100 μM BDB-TεADP at pH 7.0 and 25°C resulted in time-dependent inactivation of the enzyme, as indicated in Fig. 40.6. The biphasic inactivation kinetics observed could be described in terms of a fast initial phase of inactivation resulting in partially active enzyme of 8 to 10% residual activity, followed by a slower phase leading to total inactivation. For both phases we observed a nonlinear dependence of the rate constant of inactivation on the reagent concentration. The data shown in Fig. 40.7 for the fast phase indicate the initial formation of a reversible enzyme reagent complex, with $K_I = 56$ μM, prior to irreversible modification. For the fast phase $k_{max} = 0.55$ min^{-1}; whereas for the slow phase (data not shown) k_{max} is only 0.019 min^{-1}, with $K_I = 58$ μM.

The coenzymes NADP and NADPH provide protection against the inactivation produced by BDB-TεADP. In contrast, NAD, which lacks the critical 2′-phosphate, does not protect, suggesting that the site of attack is a natural coenzyme site of isocitrate dehydrogenase (3).

The incorporation of BDB-TεADP per mole of subunit in the absence and presence of ligands was determined by comparing the fluorescence of the modified enzyme (denatured in 5*M* guanidine hydrochloride) to that of BDB-TεADP measured under the same conditions. Up to 0.99 mol of reagent per mole subunit is incorporated, with only 2.5% activity remaining when enzyme is incubated with BDB-TεADP in the absence of protecting ligands. There is a limited extent of incorporation of BDB-TεADP into isocitrate dehydrogenase, which is an essential criterion of an affinity label. Incorporation of the first 0.5 mol/mol enzyme subunit correlates with the rapid phase of inactivation.

Pig heart isocitrate dehydrogenase exists as a dimer under the conditions used for these experiments (3). Therefore the incorporation of 0.5 mol BDB-TεADP per mole of peptide chain in each of the two distinguishable phases of the reaction can be interpreted as modification of one subunit per enzyme dimer. Because modification of one of the two subunits results in enzyme of 8

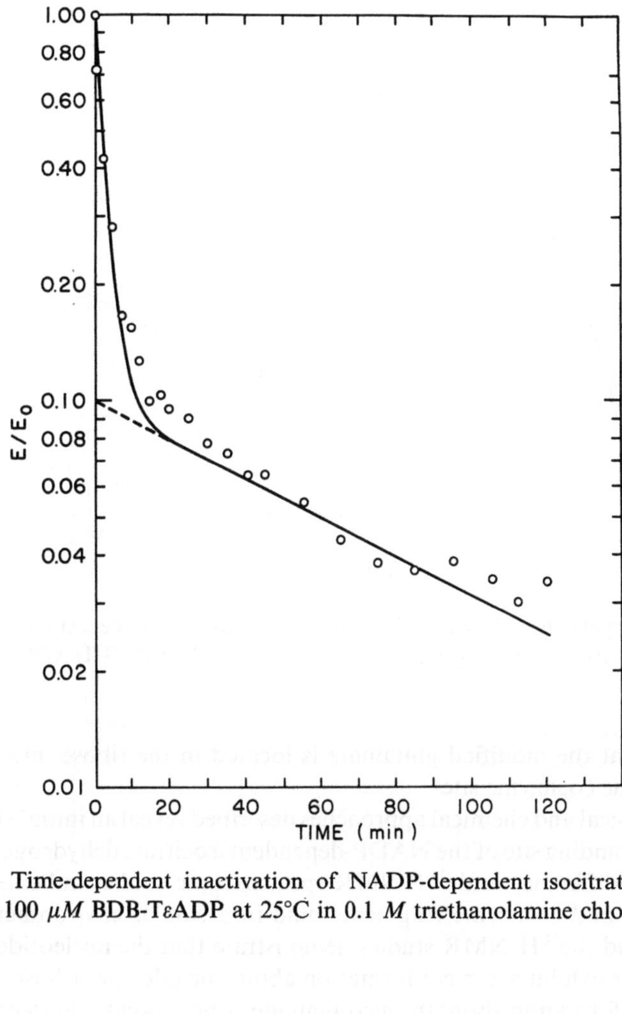

FIG. 40.6. Time-dependent inactivation of NADP-dependent isocitrate dehydrogenase by 100 μM BDB-TεADP at 25°C in 0.1 M triethanolamine chloride buffer, pH 7.0.

to 10% activity, rather than the expected 50% residual activity, there must be interactions between the two subunits. Modification and inactivation of the active site of one subunit must indirectly decrease substantially the catalytic function of the active site on the second subunit. These results can best be explained by a model in which BDB-TεADP, in the absence of ligands, reacts rapidly at the coenzyme binding site on one of two subunits of the dimeric enzyme causing a conformational change, which decreases the rate of modification by BDB-TεADP on the second subunit.

We have isolated a tryptic peptide from enzyme inactivated by BDB-TεADP. The evidence indicates that the reaction site is a glutamate residue (17). Because of the syn conformation about the purine–ribose bond of the natural nucleotides bound to isocitrate dehydrogenase, and the location of the reactive bromodioxobutyl group adjacent to the 2 position of the purine, it is

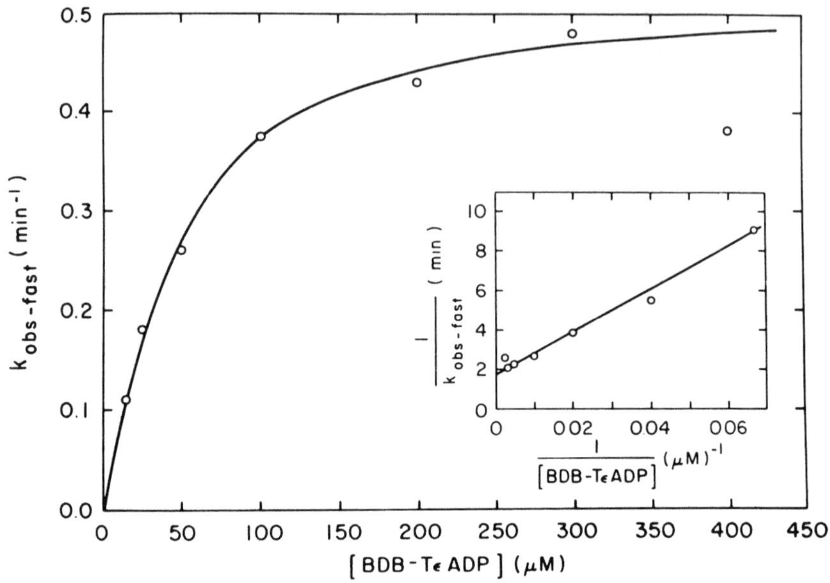

FIG. 40.7. Dependence of pseudo-first-order rate constant for the fast phase of inactivation of isocitrate dehydrogenase on the concentration of BDB-TεADP.

possible that the modified glutamate is located in the ribose–nicotinamide region of the coenzyme site.

The physical and chemical approaches described reveal an initial view of the coenzyme binding site of the NADP-dependent isocitrate dehydrogenase. The ultraviolet difference and ^{31}P-NMR spectroscopic studies indicate that the coenzymes and coenzyme fragments bind in a similar environment on the enzyme, and the ^{1}H-NMR studies demonstrate that the nucleotides in their bound state exhibit a syn conformation about the adenine–ribose bond and an anti conformation about the nicotinamide–ribose bond. The decreased pK of the 2′-phosphate suggests proximity of a positively charged amino acid in this part of the coenzyme site. The effect of NADP and NADPH on the chemical shift of a histidine imply the presence of such a residue near the nicotinamide moiety. Finally, affinity labeling of the purine nucleotide site with BDB-TεADP has yielded evidence for a glutamate in the vicinity of the nicotinamide or adjacent ribose. The continued application of these varied approaches should lead to a more sharply defined picture of the coenzyme interacting with the amino acid participants of the binding site of isocitrate dehydrogenase.

Acknowledgment. This research was supported by USPHS Grant DK 39075.

References

1. Colman, R. F., Szeto, R. C., and Cohen, P. (1970) *Biochemistry* 9:4945–4949.
2. Kelly, J. H., and Plaut, G. W. E. (1981) *J. Biol. Chem.* 256:330–334.
3. Bailey, J. M., and Colman, R. F. (1985) *Biochemistry* 24:5367–5377.
4. Ehrlich, R. S., and Colman, R. F. (1978) *Eur. J. Biochem.* 89:575–587.
5. Mas, M. T., and Colman, R. F. (1985) *Biochemistry* 24:1634–1646.
6. Subramanian, S., Ross, J. B. A., Ross, P. O., and Brand, L. (1981) *Biochemistry* 20:4086–4093.
7. Ananthanarayanan, V. S., and Bigelow, C. C. (1969) *Biochemistry* 8:3717–3723.
8. Fisher, H. F., Adija, D. L., and Cross, D. G. (1969) *Biochemistry* 8:4424–4431.
9. Mas, M. T., and Colman, R. F. (1984) *Biochemistry* 23:1675–1683.
10. Cohn, M., and Rao, B. D. N. (1979) *Bull. Magn. Reson.* 1:38–60.
11. Clore, G. M., and Gronenborn, A. M. (1982) *J. Magn. Reson.* 45:402–417.
12. Clore, G. M., and Gronenborn, A. M. (1983) *J. Magn. Reson.* 53:423–442.
13. Ehrlich, R. S., and Colman, R. F. (1985) *Biochemistry* 24:5378–5387.
14. Hartman, F. C. (1977) *Methods Enzymol.* 46:130–153.
15. Yankeelov, J. A., Jr. (1970) *Biochemistry* 9:2433–2439.
16. Riordan, J. F. (1973) *Biochemistry* 12:3915–3923.
17. Bailey, J. M. and Colman, R. F. (1987) *J. Biol. Chem.*, in press.

41
Significance of Domain Structure of Calmodulin on the Activation of Ca^{2+}–Calmodulin-Requiring Enzymes

K. Yagi, M. Yazawza, O. Minowa, M. Ikura, T. Hiraoki, K. Hikichi, and H. Toda

Calmodulin is known as a multifunctional protein factor that can activate more than 10 enzymes. In order to activate these enzymes, calmodulin ought to be in the Ca^{2+}-bound state. Ca^{2+} is known to work as a common trigger for many cellular processes, e.g., muscle contraction and cell division. Calmodulin is an intracellular receptor of the Ca^{2+} signal, and each of the calmodulin-required enzymes participates to develop a particular physiological function. Therefore, much attention is devoted to the activation mechanism of calmodulin for various enzymes. The question then arises as to how calmodulin distinguishes these different enzymes.

Calmodulin binds 4 mol Ca^{2+} per mole. Ca^{2+} binding data of calmodulin were well explained by Haiech et al. (1) as the sequential-ordered binding mechanism. If the binding order of Ca^{2+} to the four binding sites of calmodulin is inherently fixed, calmodulin can exist in several conformations depending on the number of bound Ca^{2+}. According to the sequential model, it is possible to assume that calmodulin in different conformations matches different enzymes, respectively.

Our results, however, are not in accord with the explanation of Haiech et al., and the Ca^{2+} binding occurs depending on the binding constant of each site and the cooperativity between the sites (2). It leads to the view that various enzymes recognize a certain conformation of calmodulin (probably the Ca^{2+}-saturated state), but the way to distinguish the enzymes remains to be proved.

Kretsinger and Nockolds (3) determined the crystal structure of parvalbumin, which is a Ca^{2+}-binding protein found in cytosol of vertebrate white muscle. They designated the characteristic structure of the Ca^{2+}-binding site as "EF-hand structure." The primary structure of calmodulin suggests the presence of two pairs of EF-hand structures in it (4), which may participate in Ca^{2+} binding in the amount of 4 mol/mol (5, 6). The crystal structures of two homologous Ca^{2+}-binding proteins, troponin C (7) and calmodulin (8), have been determined. Both crystals reveal two globular domains connected by a long exposed α helix, and the existence of EF-hand structure was ascertained in these proteins. This dumbbell-shaped structure of troponin C and calmodulin is rather unusual, but our observations on calmodulin can be explained by this apparently curious shape of calmodulin.

Two Structural Domains

We have investigated the structure of calmodulin and its fragments by measuring ^{1}H-NMR (nuclear magnetic resonance). Assignments of resonances appearing in the aromatic and high-field methyl regions were made, and the Ca^{2+} dependence of the spectra was investigated (9). The resonance were assignable only to residues in the *C*-terminal-half region (*C*-domain) and little information about the *N*-terminal-half region (*N*-domain) was obtained. Therefore ^{1}H-NMR studies of the tryptic fragments of calmodulin were investigated. Tryptic digestion of scallop testis calmodulin in the presence of Ca^{2+} produced two main fragments, Ala-1–Lys-75 (F12) and Asp-78–Lys-148 (F34).

Spectral comparison of ^{1}H-NMR of F12, F34, and intact calmodulin was made in the absence and presence of Ca^{2+} (9). Chemical shifts of His 107 and Tyr 138 of F34 in the absence and presence of Ca^{2+} are identical with those reported with intact calmodulin, respectively, (10, 11). Resonances of two among nine phenylalanines and one valine appear in the spectrum of F34; and three phenylalanines, one valine, and one isoleucine appear in F12. Simple superposition of the spectra of F12 and F34 was nearly identical with that of intact calmodulin.

Our investigation on the tryptic fragments indicates that F12 and F34 hold tertiary structures nearly identical to those of the corresponding domains—*N*-domain and *C*-domain, respectively—of the intact calmodulin in the absence and presence of Ca^{2+}.

The same conclusion was deduced from the measurements of circular dichroism with two tryptic fragments and calmodulin (12). Because it is demonstrated from NMR experiments that *N*-domain and *C*-domain in calmodulin molecule are structurally independent from each other, Ca^{2+} binding to one domain may not be influenced by the binding to the other; that is, the affinity of Ca^{2+} to each domain can also be independent. This view is in contrast to the sequential model proposed for calmodulin by Haiech et al (1).

Ca^{2+} binding to F12, F34, and calmodulin was measured using flow dialysis and equilibrium dialysis methods (13). The existence of two Ca^{2+} binding sites in each of F12 and F34 was demonstrated. Half-saturating concentrations of the Ca^{2+} titration curves were 11 μM for F12 and 3.2 μM for F34, indicating that F34 confines the two high-affinity sites. Strong positive cooperative Ca^{2+} binding was observed with F34, and Hill coefficients were 1.14 and 1.84 for F12 and F34, respectively. The sum of two titration curves of F12 and F34 fit well to the Ca^{2+}-binding data to whole calmodulin. It indicates that F12 and F34, which are the proteolytic products, maintained the Ca^{2+}-binding properties of calmodulin. We then conclude that the *N*-domain and *C*-domain behave independently in intact calmodulin. We further suggest from our results that, although *C*-domain was first saturated by Ca^{2+}, a small amount of Ca^{2+} binding may also occur independently at *N*-domain at the same time.

Possible Functions of the Two Domains

We isolated calmodulin from various animals and plants, and the primary structures of eight calmodulins have been determined so far. The amino acid sequences of vertebrate and invertebrate calmodulins were identical from *N*-terminal acetylalanine to leucine at position 69, which nearly corresponds to *N*-domain. This situation has occurred probably because *N*-domain, which contains lower affinity sites for Ca^{2+}, should have a certain crucial function of calmodulin.

In the course of Ca^{2+} titration of F12, a single peak of the ^{1}H-NMR spectrum was observed at an average position of the two chemical shifts of Ca^{2+} -free and Ca^{2+}-bound F12, and the peak position changes with varying content of Ca^{2+}. On the other hand, F34 showed two separate peaks at positions of resonances in Ca^{2+}-free and Ca^{2+}-bound states. These results indicate that the conformational change between two states of F12 with low-affinity sites for Ca^{2+} occurs much more rapidly ($>300\ s^{-1}$) than that of F34 with high-affinity sites ($<50\ s^{-1}$). The results suggest that the exchange rate of Ca^{2+} in *N*-domain of intact calmodulin occurs probably about 100 times more rapidly than that in *C*-domain. Ca^{2+} stays in the Ca^{2+}-binding sites of *C*-domain under a low exchange rate, but Ca^{2+} at the binding sites of *N*-domain exchanges rapidly. *N*-domain may thus be able to show a rapid response to the change in Ca^{2+} concentration. We then assume that the *C*-domain participates in the association with target proteins (enzymes) and the *N*-domain in the regulation in response to the concentration change in Ca^{2+} signal.

Activation of Enzymes by One Domain?

It has been reported that the tryptic fragments of calmodulin could activate myosin light-chain kinase (14), phosphorylase kinase, and cAMP-phosphodiesterase (15) when 100 to 3000 times more than the intact calmodulin in weight were added. However, the question has been raised whether the preparations of tryptic fragments were free from even a small amount of the mother protein, calmodulin. In our experiments the extent of calmodulin activation of myosin light-chain kinase decreased by incubation with trypsin, and the activity was nearly abolished after 120 min of incubation under the same conditions. Analysis of the tryptic products by urea polyacrylamide gel electrophoresis (PAGE) showed that the residual enzyme activity appears not to be correlated with the amounts of F12 and F34 but, rather, was parallel with the remaining amount of intact calmodulin. In order to remove a small amount of contaminant calmodulin (if present), preparations of F12 and F34 were purified using reverse-phase high performance liquid chromatography (HPLC). The HPLC-purified fragments (F12 and F34) were used to test whether one domain is able to activate enzymes. Myosin light-chain kinase and cAMP phosphodiesterase were used as the target enzymes. Under the same conditions in which a few nanomolar of calmodulin are able to produce full activation of these enzymes, no enzyme activity was observed after adding the purified F12 or F34 in as much as 10 μM in place of the calmodulin.

On the other hand, a large amount of F34 showed inhibition on the calmodulin activation of phosphodiesterase (16) and myosin light-chain kinase despite the fact that F12 could not show any inhibition. It suggests that the *C*-domain of calmodulin first binds Ca^{2+} and then participates in the association with target proteins.

Conclusion

Cellular responses to chemical signals are generally rapid and transient. In the contraction of striated muscle, the Ca^{2+} signal stimulates the muscle fibers within milliseconds, and the fibers return to the original level rapidly. In order to produce rapid stimulation for the cell function, it is desirable that calmodulin can be saturated by Ca^{2+} in a narrow range of Ca^{2+} concentration around 1 μM. As described above, F34 can easily be saturated by Ca^{2+} by its strong cooperativity, and it is compatible with the proposed function of the *C*-domain. Cooperativity of the Ca^{2+} binding to F12 was also observed, but it was not remarkable. However, the cooperative Ca^{2+} binding to the lower binding sites (in the *N*-domain) of calmodulin became remarkable by the addition of myosin light-chain kinase. It renders *N*-domain possible to bind Ca^{2+} in response to a small increase in the Ca^{2+} concentration. The increased Ca^{2+} stimulates the built-in Ca^{2+} pump of the membrane to exclude the accumulated Ca^{2+} to the outside.

The molecular model of calmodulin to activate enzymes in response to the transmembrane Ca^{2+} signal is described as follows.

1. The C-domain is saturated by Ca^{2+} at first, and it gains high affinity to enzymes.
2. The *C*-domain binds to a certain enzyme.
3. The *N*-domain connected to *C*-domain becomes close to the enzyme (probably because of its hydrophobic environment), which provides a strong cooperativity to the Ca^{2+} binding of the *N*-domain.
4. The *N*-domain is saturated by Ca^{2+}, and it binds to the activation site of the enzyme.
5. The enzyme becomes active.
6. Exclusion of Ca^{2+} by the Ca^{2+} pump decreases the Ca^{2+} concentration, and the *N*-domain rapidly releases the Ca^{2+}.

References

1. Haiech, J., Klee, C. B., and Demaille, J. G. (1981) *Biochemistry* 20:3890–3897.
2. Yagi, K., Yazawa, M., Yoshida, M., and Minowa, O. (1984) *Calcium Regulation in Biological System*, edited by S. Ebashi et al., pp. 17–27. Academic Press, New York.

3. Kretsinger, R. H., and Nockolds, C. E. (1973) *J. Biol. Chem.* 248:3313–3326.
4. Watterson, D. M., Sharief, F., and Vanaman, T. C. (1980) *J. Biol. Chem.* 255:962–975.
5. Teo, T. S., and Wang, J. H. (1973) *J. Biol. Chem.* 248:5950–5955.
6. Crouch, T. H., and Klee, C. B. (1980) *Biochemistry* 19:3692–3698.
7. Herzberg, C., and James, M. N. G. (1985) *Nature* 313:653–659.
8. Babu, Y. S., Sack, J. S., Greenhough, T. J., Bugg, C. E., Means, A. R., and Cook, W. J. (1985) *Nature* 315:37–40.
9. Ikura, M., Hiraoki, T., Hikichi, K., Minowa, O., Yamaguchi, H., Yazawa, M., and Yagi, K. (1984) *Biochemistry* 23:3124–3128.
10. Ikura, M., Hiraoki, T., Hikichi, K., Mikuni, T., Yazawa, M., and Yagi, K. (1983) *Biochemistry* 22:2568–2572, 2573–2579.
11. Seamon, K. B. (1980) *Biochemistry* 19:207–215.
12. Abe, Y., Yazawa, M., and Yagi, K. (1985) *Seikagaku* 8:928.
13. Minowa, O., and Yagi, K. (1984) *J. Biochem.* 96:1175–1182.
14. Perry, S. V. (1980) *Muscle Contraction. Its Regulatory Mechanisms*, edited by S. Ebashi et al., pp. 207–220. Springer-Verlag, New York.
15. Kuznicke, J., Grabarek, Z., Brzeska, H., Drabikowski, W., and Cohen, P. (1981) *FEBS Lett.* 130:141–145.
16. Newton, D. L., Oldewurtel, M. D., Krink, M. H., Shiloach, J., and Klee, C. B. (1984) *J. Biol. Chem.* 259:4419–4426.

42
Modification of Leucyl-tRNA Synthetase by Affinity Labeling and Limited Proteolysis

J. P. SHI, S. X. LIN, S. T. HUANG,
F. MIAO, and Y. L. WANG

Aminoacyl-tRNA synthetases catalyze the transfer of amino acid to its cognate tRNA. Generally this reaction consists of two successive steps:

1. Activation of amino acid

$$\text{E (enzyme)} + \text{AA (amino acid)} + \text{ATP} \rightleftharpoons \text{E}\cdot\text{AA} \sim \text{AMP}$$

2. Aminoacylation of tRNA

$$\text{E}\cdot\text{AA} \sim \text{AMP} + \text{tRNA} \rightleftharpoons \text{AA} - \text{tRNA} + \text{AMP} + \text{E}$$

We have begun a study of the structure and function of leucyl-tRNA synthetase (LeuRS) of *Escherichia coli* K12. The enzyme has been purified and characterized. It is composed of a single polypeptide chain having a molecular weight of 102,000. In this chapter we describe the modification of LeuRS by affinity labeling and limited proteolysis. Preliminary results on the structural and kinetic analysis of the modified LeuRS are presented.

Materials and Methods

Purification and characterization of *E. coli* LeuRS are reported elsewhere (1). *E. coli* tRNA nucleotidyltransferase was partially purified as described elsewhere (2). Pancreatic ribonuclease was from Dong-feng Biochemicals Factory of Shanghai Institute of Biochemistry. *E. coli* tRNA^{leu} was a kind gift of Dr. P. Guillemaut and was further purified by polyacrylamide gel electrophoresis. $\text{tRNA}^{\text{leu}}_{\text{ox}}$ was prepared as described elsewhere (3). [8-^{3}H]ATP was from Shanghai Institute of Nuclear Research. TPCK-treated trypsin was from Sigma. Dimethylaminoazobenzene isothiocyanate was from Fluka. Polyamide sheet was from Huangyan Chemical Experiment Factory. All chemicals were of analytical or chromatography grade. Routine methods (1) were used for amino acid activation and aminoacylation assays.

1. *[^{3}H]ATP and tRNA^{leu} isotopic exchange reaction* (4). [^{3}H]ATP 160 nmol (460 cpm/pmol) and 40 nmol of tRNA^{leu}, both dried with P_2O_5, were dis-

solved in 0.3 ml buffer containing 25 m*M* Tris-HCl (pH 8.3), 5 m*M* $MgCl_2$, 3 m*M* pyrophosphate (PPi), and 20 m*M* 2-mercaptoethanol. The mixture was incubated with tRNA nucleotidyltransferase at 37°C. The isotopic exchange reaction was stopped by addition of an equal volume of phenol. The aqueous phase was extracted twice with ether.

2. *Schiff's base formation and reduction.* The reaction between $tRNA^{leu}_{ox}$ (4 μM) and LeuRS (2 μM) was carried out at 37°C in 25 m*M* barbitone buffer pH 8.8 containing 10 m*M* $MgCl_2$ and 5 m*M* sodium cyanoborohydride. At various time intervals aliquots were pipetted out to determine the residual activity of LeuRS.

3. *Pancreatic ribonuclease digestion.* The above mixture was extensively digested by pancreatic RNase (0.2 mg/ml) at 37°C for 24 h. The digested sample was loaded on a Sephadex G-75 column (0.8 × 16.0 cm), which was equilibrated with 10 m*M* Tris-HCl buffer pH 7.4 containing 10 m*M* 2-mercaptoethanol, 0.1 m*M* phenylmethylsulfonyl fluoride (PMSF), and 10% (v/v) glycerol and eluted with the same buffer into 0.5 ml fractions. A 0.1 ml aliquot of elution from each fraction was placed on filter paper, dried, and its radioactivity determined.

4. *Limited trypsin digestion.* LeuRS (0.2–0.5 mg/ml in 0.1 *M* phosphate buffer (pH 8.0)) was incubated with trypsin 3 to 5 μg/ml at 30°C in the presence or absence of Leu 0.5 m*M*, ATP 1 m*M*, and $MgCl_2$ 2 m*M*. At timed intervals 20 μl aliquots were taken, and PMSF was added to a final concentration of 1 m*M* to stop tryptic hydrolysis. The quenched enzyme solution (5 μl) was diluted and assayed. The remaining solution was precipitated by 10% TCA. After centrifugation the precipitate was collected, washed with acetone, and dried.

5. *Sodium dodecyl sulfate polyacrylamide gel electrophoresis (SDS-PAGE).* The procedure was performed according to Douglas et al. (4). Protein 5 to 10 μg was loaded in each slot. The DABITC method was used for *N*-terminus determination (6). Tryptic products were isolated by Sephacryl S-200 column and lyophilized. LeuRS (3–5 nmol) was used for each analysis.

Results

Affinity Labeling of LeuRS with $tRNA^{leu}_{ox}$

The procedures for affinity labeling include preparation of $[^3H]tRNA^{leu}$, periodate oxidation of the tRNA, and Schiff's base formation and reduction.

1. *Preparation of* $[^3H]tRNA^{leu}$. There are two alternatives for preparing $[^3H]tRNA^{leu}$. The enzyme used can be either snake venom phosphodiesterase (6) or tRNA nucleotidyltransferase (TNT) (3). We used the latter. At pH 8.0 TNT catalyzes the exchange reaction between [³H]ATP and 3′-terminal adenosine of tRNA:

$$tRNA_{pA} + PP_i + [^3H]ATP \underset{}{\overset{pH\ 8.0}{\rightleftharpoons}} tRNA_{p[^3H]A} + PP_i + ATP$$

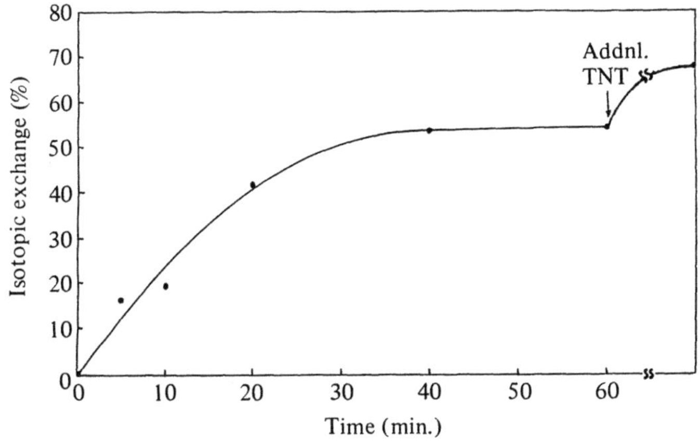

FIG. 42.1. Reaction of isotope exchange between 3′-terminal adenosine of $tRNA^{leu}$ and $[^3H]ATP$ catalyzed by tRNA nucleotidyltransferase (TNT).

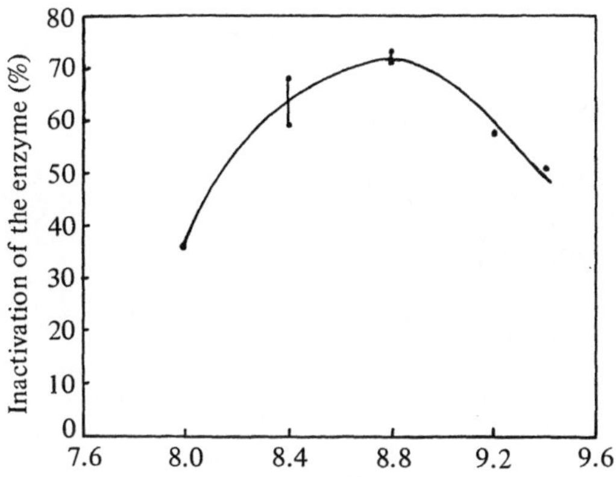

FIG. 42.2. Effect of pH on the reaction of the enzyme (2 μM) with $tRNA^{leu}$ dialdehyde (4 μM).

Figure 42.1 shows the time course of this exchange reaction. After 40 min the exchange reached 55%. If additional TNT was added, the exchange could be increased to 70%, with the specific radioactivity at 200 cpm/pmol and an amino acid accepting capacity of 1400 pmol/A_{260} unit.

2. *Schiff's base formation and reduction*. Schiff's base formation is a nucleophilic reaction. High pH is advantageous to nucleophilic attack. We studied the effect of pH on the reaction. As shown in Fig. 42.2, the optimal pH for Schiff's base formation was pH 8.8. Other buffer systems were also compared. Results indicated that barbitone buffer was better than phosphate and Tris for Schiff's base formation (data not shown).

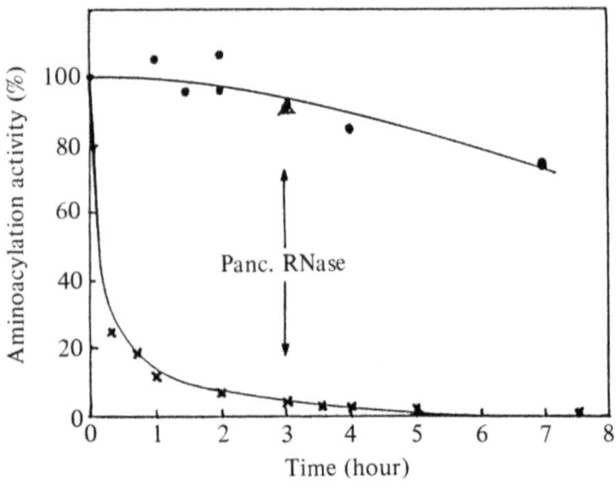

FIG. 42.3. Kinetics of inactivation of aminoacylation activity of Leu-tRNA synthetase by native $tRNA^{leu}$ (top curve) and $tRNA^{leu}$ dialdehyde (bottom curve).

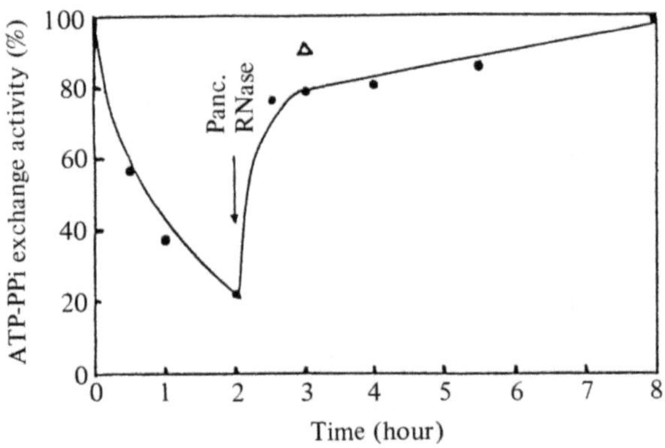

FIG. 42.4. Effects of $tRNA^{leu}$ dialdehyde and RNase on $ATP\text{-}PP_i$ exchange activity of Leu-tRNA sythetase.

3. *Effect of affinity labeling and RNase digestion on enzyme activities.* During the course of Schiff's base formation, both the activation and aminoacylation activities decreased gradually, but the inactivation rates were not parallel (Figs. 42.3 and 42.4). Aminoacylation activity was almost totally lost after 3 h and >90% of LeuRS had been converted to the covalent complex LeuRS-$tRNA^{leu}_{ox}$ (Figs. 42.3 and 42.4). In contrast, $ATP\text{-}PP_i$ exchange activity decreased to 20% of the original activity and then stopped its decline.

Extensive digestion of covalently bound tRNA with pancreatic RNase (0.2 mg/ml, 37°C, 24 h) resulted in regeneration of the activity of $ATP\text{-}PP_i$

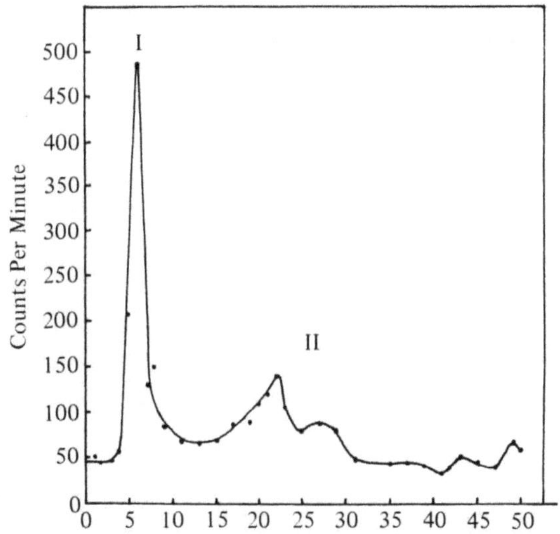

FIG. 42.5. Sephadex G-75 gel filtration chromatography of LeuRS-[^{3}H]tRNA$^{leu}_{ox}$ digested by RNase.

TABLE 42.1. Kinetic constants of various LeuRS forms evaluated by the pyrophosphate exchange reaction.

Enzyme Form	Substrate	Km (m*M*)	k_{cat} (s^{-1})
LeuRS	Leu	0.081	119
	ATP	1.0	122
LeuRS-tRNA$_{ox}$	Leu	0.22	47
	ATP	2.36	66
LeuRS-tRNA$_{ox}$ digested by RNase	Leu	0.082	109
	ATP	1.0	124

exchange (Fig. 42.4) but not aminoacylation. Isolation of RNase digestive products on a Sephadex G-75 column showed two radioactive peaks (Fig. 42.5). Peak I was identified as LeuRS with the ratio corresponding to nearly 1 mol of ^{3}H-labeled ribose per mole of enzyme.

4. *Kinetic parameters of tRNA$^{leu}_{ox}$-modified LeuRS.* Table 42.1 summarizes the kinetic constants for three types of LeuRS. So far as the ATP-PP$_i$ exchange reaction was concerned, introduction of tRNA$^{leu}_{ox}$ into LeuRS resulted in an increase in Km and a decrease in k_{cat} for both leucine and ATP. Hence the specificity constant k_{cat}/Km, was decreased by 4.6- to 6.5-fold. After extensive digestion by RNase, the kinetic parameters were restored to the same level as the native enzyme.

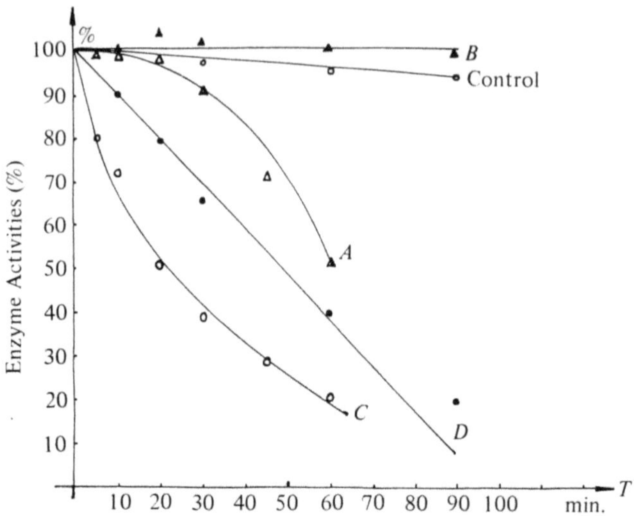

FIG. 42.6. Effects of proteolysis on activities of LeuRS in the absence and presence of substrates. ATP-PP_i exchange: (A) in the absence of substrates; (B) 0.5 m*M* Leu, 1 m*M* ATP, and 10 m*M* $MgCl_2$. Aminoacylation: (C) in the absence of substrates; (*D*) 0.5 m*M* Leu, 1 m*M* ATP, and 10 m*M* $MgCl_2$.

Limited Tryptic Digestion of LeuRS

TIME-DEPENDENT MODIFICATION OF ACTIVITIES AND MOLECULAR SIZE BY TRYPSIN

Figure 42.6 shows the time course of the reaction of LeuRS with trypsin. Samples taken from proteolysis reaction mixtures after various times of incubation were assayed for both activities of the enzyme. In the absence of substrates trypsin modification resulted in a rapid loss of the activity of aminoacylation whereas the exchange activity remained constant for 30 min. At 30 min the residual aminoacylation activity was $<40\%$. In the presence of 0.5 m*M* leucine, 1 m*M* ATP, and 10 m*M* Mg^{2+} the tRNA charging activity decreased at a relatively slow rate, but the exchange activity was fully protected from inactivation.

The sizes of the reaction products at corresponding times were directly monitored by SDS-PAGE. As can be seen in Fig. 42.7, in the presence of leucine and ATP proteolysis by 1/60 (w/w) trypsin appears to remove a small part of the enzyme and produces only a single stable peptide product of about 95,000 daltons. When the amount of trypsin was increased to 1/30 of LeuRS, half of the exchange activity remained after proteolysis for 90 min at 37°C. The result suggests that the structure of the truncated LeuRS of 95,000 daltons is in a tight core that is not easily hydrolyzed.

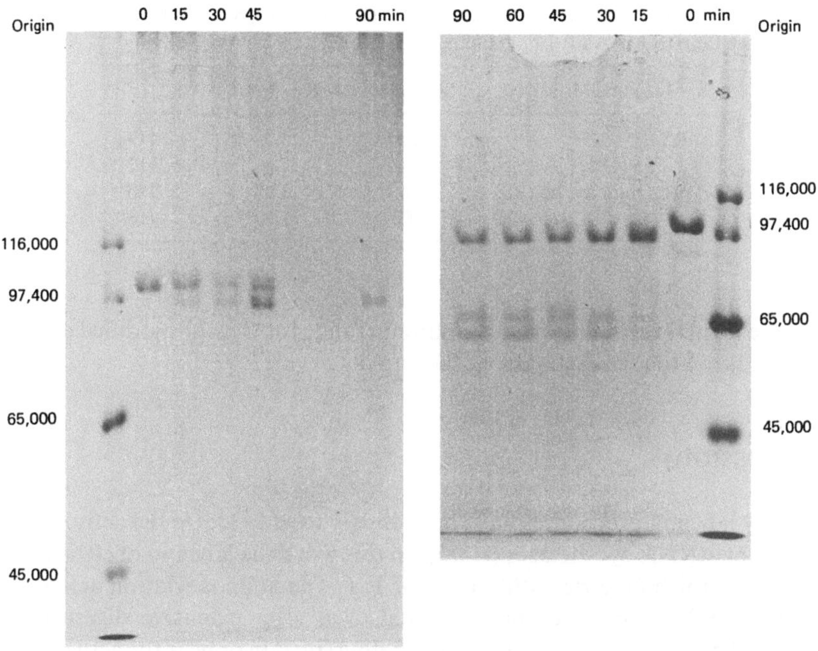

FIG. 42.7. SDS-PAGE of native and trypsin-modified LeuRS/trypsin: 60.

NH_2-TERMINAL DETERMINATION

We used dimethylaminoazobenzene isothiocyanate (DABITC) as the reagent to detect *N*-termini of the native and trypsin modified LeuRS. Results showed that for both enzymes no DABITC-amino acid could be detected on the thin-layer chromatography (TLC) sheet, even though the procedure was carried out successively for four cycles. The results indicate that *N*-termini of both native and trypsin-modified LeuRS were blocked, and the tryptic digestion removed a small part of the *C*-terminus of LeuRS.

COMPARISON OF THE KINETIC CONSTANTS OF NATIVE AND TRYPSIN-MODIFIED LEURS

Tryptic digestion was stopped by adding PMSF. The reaction mixtures were applied to a Mono Q column of the FPLC system (Pharmacia). Trypsin-modified LeuRS was isolated as a single peak, and some of its properties were investigated. As shown in Table 42.2, the kinetic constants of native and trypsin-modified LeuRS for the ATP-PP_i exchange reaction were apparently the same. We also studied the dependence of exchange activity on the magnesium ion concentration and pH. There was no difference between the two

TABLE 42.2. Kinetic constants of native and trypsin-modified LeuRS for ATP-PP_i exchange reaction.

Enzyme Form	Substrates	Km (mM)	k_{cat} (s^{-1})
LeuRS	Leu	0.08	119
	ATP	1.00	112
Trypsin-modified LeuRS	Leu	0.08	120
	ATP	1.20	116

enzyme forms except that the pH optimum range for trypsin-modified enzyme was slightly broader than that of the native enzyme.

Discussion

Affinity labeling procedures have been widely used to locate the active site of aminoacyl-tRNA synthetases (8–15). In this work the labeling of $tRNA^{leu}_{ox}$ to LeuRS is stoichiometric with a ratio of 1 : 1. The aminoacylation activity of labeled LeuRS could not be recovered even after extensive digestion by RNase. These results suggest that the labeling site is in or near the active site of LeuRS.

The independence of two activities, i.e., activation of amino acid and aminoacylation of tRNA, has been found in some of the aminoacyl-tRNA synthetases (16–21). *E. coli* LeuRS also has the double-domain structure. Modification of LeuRS by affinity labeling or by limited proteolysis resulted in the loss of aminoacylation while retaining the activity of amino acid activation. This finding suggests that the structural domains responsible for the two activities should be distinguished from each other.

As a useful approach to structural analysis, dissection of aminoacyl-tRNA synthetases with appropriate proteolytic enzymes has been reported (22–24). In the case of LeuRS, trypsin removes a small peptide of 7000 dalton from the *C*-terminus of LeuRS. The truncated enzyme is fully active for amino acid activation with similar kinetic constants as the intact enzyme. This result shows that the structural domain for amino acid activation is in the *N*-terminal part of the enzyme, and the small part of the *C*-terminus, corresponding to 5% of the molecular weight of the whole enzyme, plays a crucial role in the catalysis of charging of tRNA.

An earlier report (25) proposed a structural model for *E. coli* LeuRS. It consisted of two identical subunits linked by a small peptide that was susceptible to tryptic digestion. Limited proteolysis removed the small peptide and dissociated the enzyme into two subunits of 56,000 dalton, as determined by SDS-PAGE. Under similar conditions we were not able to reproduce these data. Our results, is agreement with the others (26), indicate that there is no subunit structure in LeuRS.

Summary

LeuRS was specifically labeled with a 3′-oxidized $tRNA^{leu}$ involving a Schiff's base formation and was subsequently reduced by sodium cyanoborohydride. The catalytic properties of the labeled enzyme were studied. Stoichiometric inactivation of aminoacylation was achieved with the incorporation of 1 mol of $tRNA^{leu}_{ox}$ per mole of enzyme, and the pyrophosphate (PP_i) exchange activity was inhibited reversibly. After RNase digestion, this activity almost fully recovered.

We also conducted limited proteolysis of LeuRS. In the presence of substrate leucine and ATP, trypsin (1/60 weight of LeuRS) cleaved the intact enzyme (102,000 daltons) into a 95,000-dalton fragment. This fragment lost aminoacylation activity but retained PP_i exchange activity, with the same kinetic parameters as the intact enzyme.

Acknowledgment. We thank You-xin Jin for his help in the purification of $tRNA^{leu}$ and Yue-han Tang and Ji-kang Fang for their help in the *N*-terminal determination. The technical assistance of Bing Li is greatly appreciated. The project was supported by the Science Fund of the Chinese Academy of Sciences

References

1. Shi, J. P., Lin, S. X., Huang, S. T., Miao, F., and Wang, Y. L. (1987) *Acta Biochim et Biophys. Sinica 19*, in press
2. Miller, J. P., and Phillipps, G. R. (1971) *J. Biol. Chem.* 246:1274–1279.
3. Waterson, R. M., Gufferman, A. W., Youngblood, P., Put, T. D., Beyersdorf, S. R., and Schusterman, M. (1979) *J. Biol. Chem.* 254:8982–8987.
4. Strenbach, H., and Cramer, F. (1977) *Hoppe Seylers Z. Physiol. Chem.* 358:312.
5. Douglas, M., Finkelstein, D., and Butow, R. A. (1979) *Methods Enzymol.* 56:61–66.
6. Chang, J. Y. (1983) *Methods Enzymol.* 91:455–466.
7. Preiss, J., Dieckmann, M., and Berg, P. (1961) *J. Biol. Chem.* 236:1748–1757.
8. Bruton, C. J., and Hartly, B. S. (1970) *J. Mol. Biol.* 52:165–178.
9. Santi, D. V., Marchant, W., and Yarus, M. (1973) *Biochem. Biophys. Res. Commun.* 51:370–375.
10. Schoemaker, H. J. P., and Schimmel, P. R. (1974) *J. Mol. Biol.* 84:503–513.
11. Budzik, G. P., Lam, S. S. M., Schoemaker, H. J. P., and Schimmel, P. R. (1975) *J. Biol. Chem.* 250:4433–4439.
12. Schoemaker, H. J. P., Budzik, G. P., Girge, R., and Schimmel, P. R. (1975) *J. Biol. Chem.* 250:4440–4444.
13. Piszkiewicz, D., Duval, J., and Rostas, S. (1977) *Biochemistry* 16:3538–3543.
14. Yue, V. T., and Schimmel, P. R. (1977) *Biochemistry* 16:4678–4684.
15. Akhverdyan, V. Z., Kisselev, L. L., Knorre, D. G., Lavrik, O. I., and Nevinsky, G. A. (1977) *J. Mol. Biol.* 113:475–501.

16. Waye, M. M. Y., Winter, G., Wilkinson, A. J., and Fersht, A. R. (1983) *EMBO J.* 2:1827–1829.
17. Barker, D. G., and Winter, G. (1982) *FEBS Lett.* 145:191–193.
18. Zelwer, C., Risler, J. L., and Brunie, S. (1982) *J. Mol. Biol.* 155:63–81.
19. Putney, S. D., Sauer, R. T., and Schimmel, P. R. (1982) *J. Biol. Chem.* 256:198–204.
20. Eidlic, L., and Neidhardt, F. C. (1965) *J. Bacteriol.* 89:706–711.
21. Nazario, M., Kinsey, J. A., and Ahmad, M. (1971) *J. Bacteriol.* 105:121–126.
22. Lee, M. L. (1974) *Biochemistry* 13:4747–4752.
23. Piszkiewicz, D., and Goitein, R. K. (1974) *Biochemistry* 13:2505–2511.
24. Epely, S., Gros, C., Labouesse, J., and Lemaire, G. (1976) *Eur. J. Biochem.* 61:139–146.
25. Rouget, P., and Chapeville, F. (1971) *Eur. J. Biochem.* 23:459–467.
26. Renaud, M., Fasiolo, F., Baltzinger, M., Boulanger, Y., and Remy, P. (1982) *Eur. J. Biochem.* 123:267–274.

43
NAD Metabolism in Eukaryotic Cells: Purification and Characterization of NMN Adenylyltransferase from Baker's Yeast

G. Magni, P. Natalini, I. Santarelli, A. Vita, N. Raffaelli, and S. Ruggieri

Since the discovery that nicotinamide mononucleotide adenylyltransferase (NMN:ATP adenylyltransferase; EC 2.7.7.1) is exclusively located in the cell nucleus (1), various authors have suggested that NAD fulfills a role in addition to that of respiratory coenzyme in eukaryotic cells. Morton (2) hypothesized that NAD might play a key role in the regulation of cell proliferation in that the nucleus may sense changes in intracellular NAD concentrations. Thereafter the discovery that another nuclear enzyme, poly(ADP)ribose synthetase, utilizes NAD as a substrate for the modification of nuclear proteins (3) suggested that such an enzyme may provide the mechanism by which NAD concentrations are sensed by the nucleus. The involvement of this modification in DNA metabolism in eukaryotes in fact is now generally accepted (4). In prokaryotes NAD is known to be a substrate in the DNA ligase-catalyzed joining of DNA fragments (5). Both ADP ribosylation and DNA ligase reactions result in the breakdown of the NAD molecule, yielding free nicotinamide and nicotinamide mononucleotide, respectively, which are then recycled back to dinucleotide through appropriate salvage pathways known as "pyridine nucleotide cycles" (6, 7). Figure 43.1 represents the overall pathway of NAD metabolism taking place in eukaryotic cells.

The occurrence of these reaction cycles results in surprisingly high turnover rates for the NAD molecule, with half-lives in the order of 2 h in *Escherichia coli* and rat liver, 1 h in HeLa D98/AH2 cells, and 25 min in *Physarum polycephalum* (6). Moreover, NAD turnover is highly suppressed in enucleated cells (8), where the NAD half-life increases to 10 h, so that nuclear events seem to be responsible for both breakdown and resynthesis of NAD. Although the biological significance of NAD turnover taking place at the nuclear level is as yet unclear, it seems to be of fundamental importance to study the individual enzymes involved in this phenomenon. A central role in NAD salvage synthesis is played by NMN adenylyltransferase.

In partially purified preparations, the same enzyme has been shown to catalyze the formation of the dinucleotide by using either NMN or deamido-NMN as the substrate (9, 10); such an enzyme, first described by Kornberg (11), reversibly catalyzes the synthesis of NAD from nicotinamide mono-

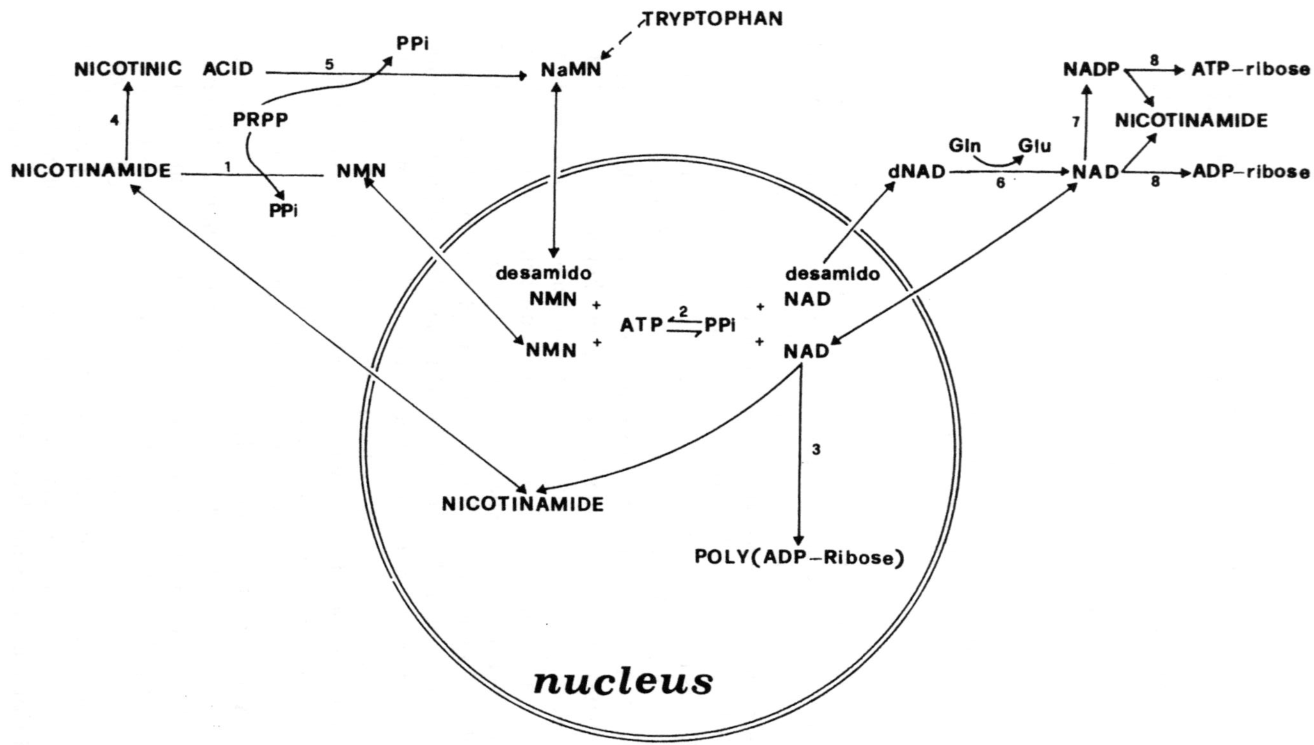

FIG. 43.1. Overall pathway of NAD metabolism in eukaryotic cells.

1 = NMN pyrophosphorylase (EC 2.4.2.12)
2 = NMN adenylyltransferase (EC 2.7.7.1)
3 = Poly (ADP-ribose) polymerase (EC 3.2.2.5)
4 = Nicotinamide deamidase (EC 2.4.2.11)
5 = NaMN pyrophosphorylase (EC 2.4.3.11)
6 = NAD synthetase (EC 6.3.3.1)
7 = NAD kinase (EC 2.7.1.23)
8 = Microsomal NAD glycohydrolase (EC 3.2.2.5)

nucleotide and ATP with the concomitant formation of pyrophosphate. The deamido forms of pyridine nucleotides also seem to serve as substrates for NMN adenylyltransferase. Unfortunately, no studies of the homogeneous enzyme have been reported so far owing principally to difficulties in the purification procedure. The present contribution deals with the purification procedure and the major molecular and enzymatic properties of the homogeneous enzyme.

Results

Purification of the Enzyme

NMN adenylyltransferase from bakers' yeast has been purified using conventional protein purification methods, as shown in Table 43.1. The details of the purification procedure have been published elsewhere (12). Aliquots of the active fractions from the last step of the purification procedure were individually subjected to polyacrylamide gel electrophoresis (PAGE) to check for their homogeneity after dialysis against 20 m*M* Hepes buffer pH 7.5. Homogeneous fractions were pooled and dialyzed against 20 m*M* Hepes pH 7.5. A sample of this pool was analyzed on PAGE. A single band was demonstrated by Coomassie staining, whose relative migration (0.44) corresponded to that of the activity assayed on 2-mm slices of the gel.

Stability

After 2 weeks at 4°C, the pure enzyme retained 70% of its original activity in 20 m*M* Hepes buffer pH 7.5. After 1 month at −20°C the activity was 80% in the same buffer.

Absorption Spectrum

The absorption spectrum of pure enzyme shows a maximum at 280 nm and a minimum at 253 nm. The A_{280}/A_{260} ratio is 1.86. The percent absorption coefficient (E_{280}) is 4.91.

TABLE 43.1. Purification of yeast NMN adenylyltransferase.[a]

Step	Total Protein (mg)	Total Activity (units)	Specific Activity (units/mg)	Purification (-fold)	Yield (%)
KCl extract	36,625.0	35.6	0.00098	—	100.0
Ammonium sulfate	14,650.0	28.5	0.00190	1.9	80.0
Polyethylenimine	2,100.0	20.4	0.00970	9.9	57.3
DE-52	246.0	18.1	0.07300	74.5	50.8
Green A	4.3	12.1	2.80000	2,857.0	34.0

[a] For details of the experimental procedure, see Natalini et al. (12).

Molecular Weight

The molecular weight of the native enzyme estimated by both gel filtration and sucrose gradient ultracentrifugation is about 200,000. The gel filtration experiment was conducted by including 0.2% bovine serum albumin (BSA) in the buffer. In the absence of BSA $<1\%$ activity could be recovered. The protective action exerted by BSA is under investigation. A single band of molecular weight 50,000 was observed upon sodium dodecyl sulfate (SDS) PAGE. Identical behavior was shown in the presence and the absence of 2-mercaptoethanol. Therefore the enzyme appears to be composed of four electrophoretically indistinguishable subunits not linked by disulfide bridges.

Isoelectric Point

The isoelectric point is at pH 6.2 in the presence of 4 *M* urea. In the absence of urea, multiple pIs are observed in the acidic range, i.e., at pH 4.2, 4.9, and 6.2, with a total peak activity ratio of approximately 1:1:3. A similar pattern is obtained after resubmitting individual peaks to isoelectrofocusing, suggesting the formation of aggregates in the absence of urea.

Amino Acid Composition

Amino acid analysis (Table 43.2) shows a relative excess of acidic amino acid residues, aspartic and glutamic. In addition, the enzyme contains seven half-cystine residues per subunit (M_r 50,000).

Sulfhydryl Group Content

Analysis of sulfhydryl group content with 5,5′-dithiobis-(2-nitrobenzoic acid) (DTNB), performed with and without previous reduction with sodium borohydride, revealed in both cases the presence of 28 mol of sulfhydryl groups per mole of enzyme, (M_r 200,000), indicating the absence of disulfide bridges.

Carbohydrate and Phosphate Content

The enzyme was shown to be a glycoprotein by periodic acid-Schiff staining following PAGE. In order to measure carbohydrate content, an aliquot of pure enzyme was analyzed by the phenol-sulfuric acid method after hydrolysis in 2 *N* H_2SO_4; the results indicated that the preparation contained about 2% carbohydrate by weight using glucose as a standard and assuming a 3.4% (w/v) carbohydrate content for ovalbumin taken as the standard. Total phosphate and alkali-labile phosphate analysis were also conducted on pure NMN adenylyltransferase by using the ammonium molybdate-malachite green procedure. A total of 2 mol of total phosphate was detected per mole

TABLE 43.2. Amino acid composition of yeast NMN adenylyltransferase.

Amino Acid	Residues per Subunit[a]
Aspartic acid, asparagine	35
Glutamic acid, glutamine	52
Serine	35[b]
Histidine	4
Glycine	39
Threonine	10[b]
Arginine	19
Alanine	29
Tyrosine	17
Methionine	7
Valine	35[c]
Phenylalanine	16
Isoleucine	23[c]
Leucine	26[c]
Lysine	25
Proline	12
Half-cystine	7[d]
Tryptophan	5
Total	396

[a] All calculations are based on a molecular weight of 50,000. Figures represent the nearest integer and are the average of five preparations.
[b] Values were extrapolated to zero time of hydrolysis.
[c] Values were extrapolated to infinite time of hydrolysis.
[d] Determined as cysteic acid after dimethylsulfoxide oxidation.

(M_r 200,000) of enzyme. The same result was obtained when alkali-labile phosphate was analyzed; therefore the enzyme appears to contain 2 mol of bound, alkali-labile phosphate per mole (M_r 200,000).

Adenine Derivatives Content

Pure enzyme 100 μg was analyzed for the presence of adenine derivatives using glyoxal hydrate trimer according to Yuki et al. (13). The results show the presence of 0.5 nmol of adenine derivatives, i.e., 1 mol/mol of enzyme (M_r 200,000).

pH Optimum and Effect of Temperature

The enzyme activity was found to be maximal between pH 7.2 and 8.4 as determined using an overlapping-buffers system containing 20 m*M* each of acetic acid, succinic acid, glycylglycine, and glycine adjusted to suitable pH with KOH. The activity was determined by radioassay.

The influence of temperature on enzyme activity was investigated in the range 20° to 55°C. The enzyme activity continuously increased in this range; however, when data were plotted as the logarithm of enzyme activity versus the reciprocal of absolute temperature (Arrhenius plot), a break point occurred at 42°C. The activation energies were 8.3 and 1.0 kcal/mol below and above this point, respectively.

Enzyme Kinetics

NMN adenylyltransferase exhibits linear kinetics with respect to both NMN and ATP, the K_m values being 0.19 and 0.11 m*M*, respectively. When deamido-NMN (sodium nicotinate mononucleotide, NaMN) was used as the substrate, a K_m value of 5 m*M* was obtained. Pyrophosphate (PP_i, a product in the forward direction of the reaction exhibits linear noncompetitive inhibition with respect to ATP or NMN. This finding is consistent with an ordered Bi-Bi kinetic mechanism, but the possibility of a random Bi-Bi mechanism with dual abortive complexes cannot be ruled out. K_i for PP_i is 3.5 m*M* as calculated from intercept replots. Table 43.3 summarizes the properties of yeast NMN adenylyltransferase.

TABLE 43.3. Properties of yeast NMN adenylyltransferase.

Property	Data
Enzyme location	Yeast chromatin
Reaction catalyzed	NMN + ATP = NAD + PP_i
Molecular properties	
MW, native	200,000
MW, subunit	50,000
Isoelectric point	pH 6.2
Spectral properties	
Absorption maximum	280 nm
Absorption minimum	253 nm
A_{280}/A_{260}	1.86
% Absorption coefficient	4.91
Chemical composition	
–SH groups per subunit	7
Adenine moiety per mole	1
Moles of alkali labile phosphate per mole	2
Sugar content (%)	2
Catalytic properties	
pH optimum	Broad: from 7.8 to 8.4
Km for ATP	0.15 m*M*
Km for NMN	0.33 m*M*
Km for NaMN	5 m*M*
Metal requirement	Mg^{2+}
Product inhibition	Pyrophosphate; noncompetitive to both NMN and ATP
Reaction mechanism	Ordered bi-bi

Discussion

NMN adenylyltransferase activity has been observed in a variety of sources and found to be localized in the nucleus (3, 14–16). In our laboratory we have found that the yeast enzyme is almost completely associated to chromatin fractions (17). We described here a reproducible and relatively simple purification procedure that leads for the first time to a homogeneous enzyme preparation.

The protein has a molecular weight, as estimated by gel filtration and sucrose gradient centrifugation, of about 200,000. It is composed of four apparently identical subunits of 50,000 daltons each. SDS-PAGE conducted in the absence of reducing agents reveals the absence of interchain disulfide groups. Molecular weights for the enzyme from other sources range from 200,000 to 300,000 (10, 18, 19). NMN adenylyltransferase exhibits an absorption spectrum typical of that expected for a protein with no visible-absorbing cofactor. Amino acid analysis indicates the presence of at least 396 residues per subunit with a slight excess of aspartic and glutamic acids, in agreement with the isoelectric pH values of 6.2, observed by submitting the enzyme to isoelectrofocusing in the presence of 4 *M* urea. In the absence of urea, two additional peaks at pH 4.2 and 5.0 are observed, similar to that described for the chicken enzyme (10). This phenomenon can be interpreted as the result of aggregation of the enzyme molecules as reported for other nuclear proteins (20). Amino acid analysis reveals the presence of seven half-cystine residues per subunit, which can be completely titrated by DTNB in denaturating conditions.

NMN adenylyltransferase from yeast displays linear kinetics. The enzyme is inhibited by the reaction product pyrophosphate. The data show noncompetitive inhibition by this compound, suggesting a kinetic mechanism of the ordered Bi Bi type, according to the nomenclature adopted by Cleland, even though more detailed product inhibition experiments are needed before a final conclusion can be drawn. The observation that the Km value for NaMN is one order of magnitude higher than that for NMN suggests that the amido pathway is predominant in yeast with respect to the deamido pathway as postulated for prokaryotes (6). The effect of temperature on the enzyme's activity, analyzed by plotting the data according to Arrhenius, shows biphasic behavior similar to that described for the chicken erythrocyte enzyme (10). Chemical characterization of NMN adenylyltransferase reveals that it contains 2 mol of alkali-labile phosphate per mole, 2% sugar, and 1 mol of adenine moiety per mole. This observation deserves some attention, and further studies should be undertaken in order to investigate the possibility that this nuclear enzyme is ADP-ribosylated. It is tempting to speculate that ADP ribosylation of this enzyme may be a sort of regulatory mechanism by which the level of NAD in the nucleus is mediated.

Acknowledgment. This work was partially supported by grants from Ministero della Pubblica Istruzione (Italy) and Regione Marche (Italy).

References

1. Hogeboom, G., and Schneider, W. (1952) *J. Biol. Chem.* 197:611–620.
2. Morton, R. N. (1958) *Nature* 181:540–543.
3. Chambon, P., Weill, J. D., Doly, J., Strosser, M. T., and Mandel, P. (1966) *Biochem. Biophys. Res. Commun.* 25:638–643.
4. Purnell, M. R., Stone, P. R., and Whish, W. J. D. (1980) *Biochem. Soc. Trans.* 8:215–225.
5. Olivera, B. M., Hall, Z. W., and Lehman, I. R. (1968) *Proc. Natl. Acad. Sci USA* 61:237–244.
6. Hillyard, D., Rechsteiner, N., Manlapaz Ramos, P., Imperial, J. S., Cruz, L. J., and Olivera, B. M. (1981) *J. Biol. Chem.* 256:8491–8497.
7. Gholson, R. K. (1966) *Nature* 212:933–935.
8. Rechsteiner, M., and Catanzarite, V. (1974) *J. Cell. Physiol.* 84:409–421.
9. Dahmen, W., Webb, B., and Preiss, J. (1967) *Arch. Biochem. Biophys.* 120:440–450.
10. Cantarow, W., and Stollar, B. D. (1977) *Arch. Biochem. Biophys.* 180:26–34.
11. Kornberg, A. (1948) *J. Biol. Chem.* 176:1475–1476.
12. Natalini, P., Ruggieri, S., Raffaelli, N., and Magni, G. (1986) *Biochemistry* 25:3725–3729.
13. Yuki, H., Sempuku, C., Park, M., and Takiura, K. (1972) *Anal. Biochem.* 46:123–128.
14. Atkinson, M. R., Jackson, J. F., and Morton, R. K. (1961) *Nature* 192:946–948.
15. Ferro, A., and Kuehl, L. (1973) *Fed. Proc.* 32:554 (abstract).
16. Siebert, G., and Humphrey, G. B. (1965) *Adv. Enzymol.* 27:239–288.
17. Ruggieri, S., and Magni, G. (1982) *Physiol. Chem. Phys.* 14:315–322.
18. Kono, K., Shimizu, C., Matsui, T., and Matsuura, F. (1978) *Bull. Jpn. Soc. Sci. Fisheries* 44:379–384.
19. Ferro, A., and Kuehl, L. (1975) *Biochim. Biophys. Acta* 410:285–298.
20. Adamietz, P., Klapproth, K., and Hilz, H. (1979) *Biochem. Biophys. Res. Commun.* 91:1232–1238.

44
Specific β-D-Fucosidase from *Aspergillus phoenicis*

ZENG YU-CHENG, GU YA-JUN, and ZHANG SHU-ZHENG

β-D-Fucosidase (β-D-fucoside fucohydrolase, EC 3.2.1.38; β-fuc) has been shown to be always associated with high β-galactosidase (β-gal) and β-glucosidase (β-glc) activities (1–6). The relatively specific β-fuc came from a marine snail, which has a β-fuc/β-gal/β-glc activity ratio of 100:20:65(5).

We isolated and purified a highly specific β-fuc from the solid culture extract of *Aspergillus phoenicis* whose β-fuc/β-gal/β-glc ratio was about 100:0.9:4.9. The purification and characterization of this enzyme are reported herein.

Results

Purification of β-D-Fucosidase

The crude extract of wheat bran culture of *A. phoenicis* contains many glycosidase activities, as shown in Table 44.1. Purification proceeded through

TABLE 44.1. Glycosidase activities of crude extracts.

Enzyme	Activity (units/ml)
α-L-Fuc	0
β-D-Fuc	0.49
α-Glc	0.03
β-Glc	0.53
α-Gal	0.61
β-Gal	0.22
α-Man	0.01
β-Man	0.07
β-Xyl	1.06
α-GalNAc	0.02
β-GalNAc	0.35
β-GlcNAc	0.56

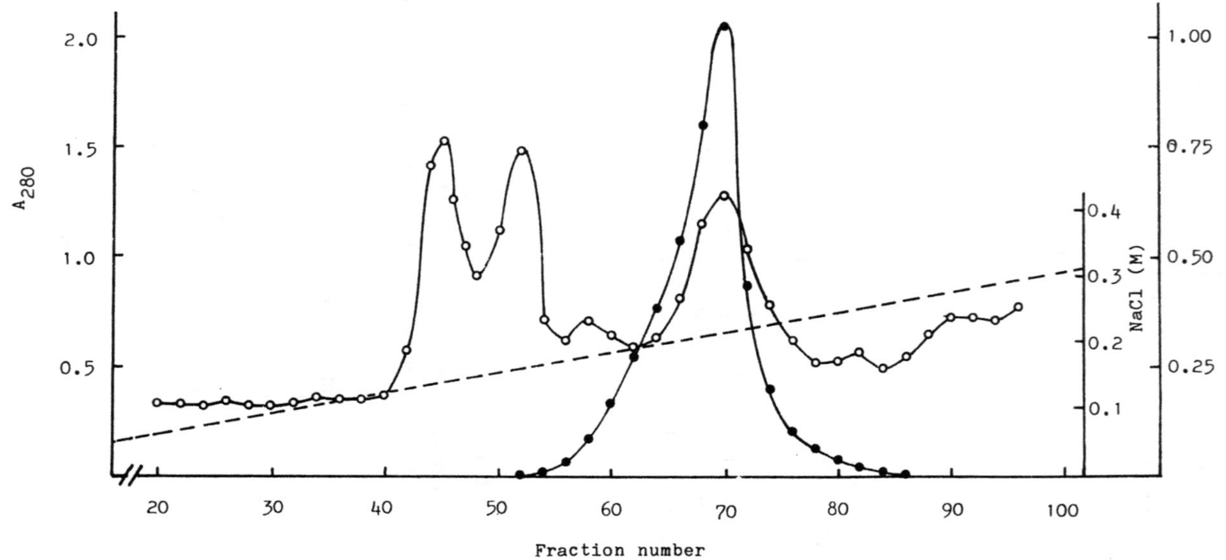

FIG. 44.1. Chromatography of β-D-fucosidase on DEAE-Sephadex A-50. The desalted ammonium sulfate precipitate was applied to a DEAE-Sephadex A-50 column (2.8 × 24 cm) equilibrated with 20 m*M* acetate buffer pH 6.0, and eluted with 0 to 0.3 *M* NaCl at a flow rate of 15 ml/h; 10-ml fractions were collected. (o——o) A_{280}; (●——●) β-D-fucosidase activity; (– – – –)NaCl gradient.

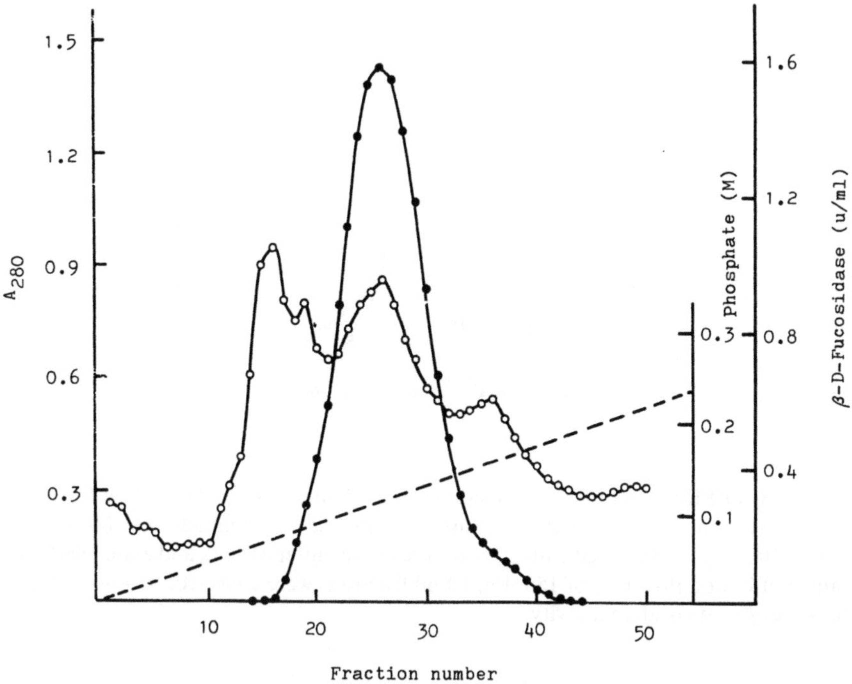

FIG. 44.2. Chromatography of β-D-fucosidase on hydroxylapatite. The active fractions from the DEAE-Sephadex A-50 column were applied to a hydroxylapatite column (1.7 × 22 cm) equilibrated with 1m*M* phosphate buffer pH 6.0, and eluted with 0 to 0.3 *M* phosphate buffer pH 6.0 at a flow rate of 15 ml/h; 10-ml fractions were collected. (○——○) A_{280}; (●——●) β-D-fucosidase activity; (----) phosphate buffer gradient.

five steps:

1. Biphasic separation in potassium phosphate-PEG 6000
2. Ammonium sulfate precipitation (60% saturation)
3. DEAE-Sephadex A-50 ion-exchange chromatography (Fig. 44.1)
4. Chromatography on hydroxylapatite (Fig. 44.2)
5. Gel filtration on Sephadex G-100 (Fig. 44.3)

The β-D-fucosidase active peak from gel filtration on Sephadex G-100 was shown to be homogeneous on polyacrylamide gel electrophoresis (PAGE) (Fig. 44.4). The purification procedure is summarized in Table 44.2. The overall purification factor is 507 with a recovery rate of 1.8%.

Properties

Some properties of the purified β-D-fucosidase are summarized in Table 44.3 and compared with those of β-galactosidase and β-glucosidase purified sepa-

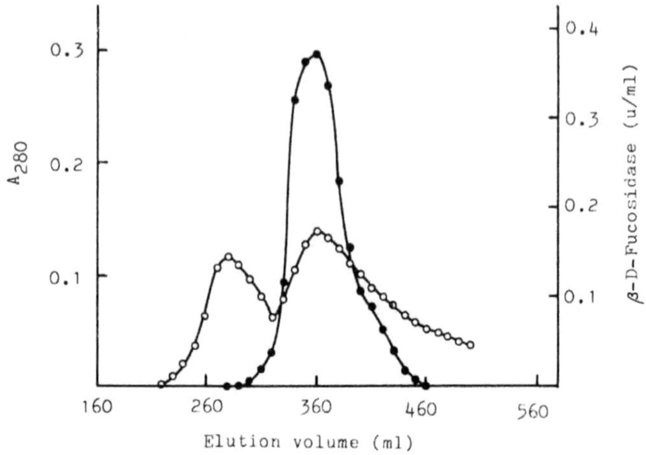

FIG. 44.3. Gel filtration of β-D-fucosidase on Sephadex G-100. The active fractions (30 ml) from the hydroxylapatite column were applied to a Sephadex G-100 column (3.0 × 100 cm), equilibrated with 20 m*M* phosphate buffer pH 6.0, and eluted with the same buffer at a flow rate of 15 ml/h; 10-ml fractions were collected. (○——○) A_{280}; (●——●) β-D-fucosidase activity.

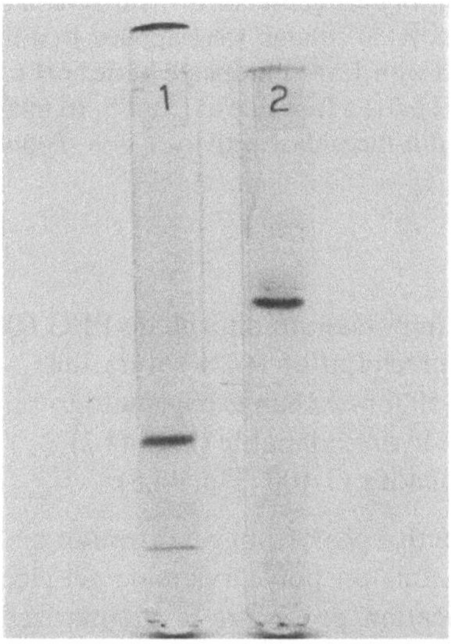

FIG. 44.4. PAGE of purified β-D-fucosidase on 7% (1) and 10% (2) polyacrylamide gel. Enzyme protein was 15 μg.

TABLE 44.2. Purification of β-D-fucosidase.

Step	Protein (mg)	Total Activity (units)	Specific Activity (units/mg)	Purification Factor	Recovery (%)
Crude extract	69,000.0	1,040.0	0.015	1.0	100.0
Biphase separation	24,300.0	959.0	0.039	2.6	92.2
$(NH_4)_2SO_4$ precipitation	8,803.0	407.0	0.046	3.1	39.1
DEAE-Sephadex A-50	175.0	204.0	1.170	78.0	19.6
Hydroxylapatite	7.2	32.4	4.500	300.0	3.1
Sephadex G-100	2.5	18.8	7.600	507.0	1.8

TABLE 44.3. Properties of purified β-glycosidases.

Properties	β-Fuc	β-Gal	β-Glc
Optimal pH	6.0	4.0	5.0
Optimal temp. (°C)	35–40	60	60
pH stability	5.5–6.5	4.5–9.0	4.0–7.5
Thermostability[a]	35°	60°	60°
Molecular weight (kD)			
SDS-PAGE	57	115	118
Gradient-PAGE	50–60[b]	118	195

[a] Loss of 50% of activity in 30 min.
[b] This value was estimated by gel filtration on Sephadex G-100.

rately from the same crude extract. It is clear that they are quite different enzymes.

Substrate Specificity

Purified β-D-fucosidase showed high activity toward PNP-β-fucoside and no activity toward PNP-β-galactoside and PNP-β-glucoside under the usual assay conditions. However, at higher enzyme concentration and longer incubation time, it showed low activity toward PNP-β-galactoside and PNP-β-glucoside, and no activity toward any other glycoside. β-Galactosidase showed high activity toward PNP-β-galactoside and low activity toward PNP-β-fucoside. β-Glucosidase hydrolyzed only various kinds of β-glucoside (Table 44.4).

Kinetic Parameters of β-D-Fucosidase

The kinetic parameters were derived from the Lineweaver-Burk plot for the hydrolysis of PNP-β-glycosides by β-D-fucosidase. The enzyme shows much higher V_{max} toward PNP-fucoside, although the K_m values for various substrates were of the same order (Table 44.5).

TABLE 44.4. Substrate specificity of glycosidases.

Substrate	β-Fuc	β-Gal	β-Glc
PNP-β-fuc	100.0	10.3	0
PNP-β-gal	0.9	100.0	0
PNP-β-glc	4.9	0	100.0
PNP-α-L-fuc	0	0	0
PNP-α-gal	0	0	0
PNP-α-glc	0	0	0
PNP-β-man	0	0	0
PNP-β-xyl	0	0	0
Lactose	0	63.1	0
Cellubiose	5.7	0	180.2
Salicin	0	0	67.3
Sucrose	0	0	0

TABLE 44.5. Kinetic parameters for the hydrolysis of PNP-β-glycosides by β-D-fucosidase.

Substrates	K_m (mM)	V_{max} (μmol/min/mg)	k_{cat} (s^{-1})
PNP-β-fuc	2.4	12.800	12.200
PNP-β-gal	2.6	0.077	0.073
PNP-β-glc	1.8	0.480	0.460

TABLE 44.6. Effects of various reagents on β-D-fucosidase activity.

Reagents	Activity (%)
$AgNO_3$ (1 mM)	1.8
$HgCl_2$ (1 mM)	4.1
$AlCl_3$ (1 mM)	91.7
$BaCl_2$ (1 mM)	86.9
$CaCl_2$ (1 mM)	99.7
$MnCl_2$ (1 mM)	95.3
$CuSO_4$ (1 mM)	80.5
$ZnSO_4$ (1 mM)	91.5
$PbAc_2$ (1 mM)	86.9
EDTA (1 mM)	90.7
SDS (10 mM)	0
SDS (1 mM)	80.1
Mercaptoethanol (10 mM)	147.7
Serum albumin (2 mg/ml)	155.8

TABLE 44.7. Effects of carbohydrates on glycosidase activity.

Carbohydrate[a]	β-Fuc	β-Gal	β-Glc
D-Arabinose	106.9	96.6	102.9
L-Arabinose	93.5	94.6	99.6
D-Ribose	96.7	92.9	97.1
D-Xylose	216.2	95.5	101.5
L-Fucose	97.4	95.9	97.2
D-Galactose	130.5	18.2	100.0
D-Glucose	343.5	98.9	24.1
D-Mannose	103.2	96.1	98.5
D-Lactose	87.8	83.4	89.4
Cellubiose	42.9	96.7	15.0
Melibiose	187.5	89.5	94.9
Salicin	197.8	112.5	65.5
Xylitol	138.8	82.8	99.7
Dulcitol	88.4	83.4	100.6
Inositol	179.5	89.0	118.4
Glycerol	209.8	65.8	93.2
D-Galacturonic acid	90.8	109.8	101.9
D-Galactono-γ-lactone	93.4	7.8	95.6
D-Glucono-γ-lactone	105.2	90.5	77.2

[a] All carbohydrate concentrations were 10 m*M*, except for D-galactono-γ-lactone, the concentration of which was 1 m*M*.

Effectors

Metal ions such as Ag^+, Hg^{2+}, and SDS had a potent inhibitory effect on β-D-fucosidase, whereas mercaptoethanol and serum albumin enhanced the activity of the enzyme (Table 44.6). Carbohydrates such as melibiose, D-glucose, D-xylose, xylitol, inositol, glycerol, and salicin enhanced β-D-fucosidase activity significantly. D-Galactono-γ-lactone and D-glucono-γ-lactone had an inhibitory effect on β-galactosidase and β-glucosidase, respectively, but had no effect on β-D-fucosidase (Table 44.7).

Chemical Modification

When incubated with protein side chain modification reagents (EDC, DEP, IAA, PCMB, and NBS), β-D-fucosidase was inactivated considerably. PCMB and DEP, even in very low concentration, inactivated β-D-fucosidase almost completely, whereas PCMB and DEP had little effect on β-galactosidase and β-glucosidase (Table 44.8; Fig. 44.5). The results of chemical modification showed that the carboxyl, sulfhydryl, imidazole, and indole groups are essential for its activity.

TABLE 44.8. Activity of glycosidases after chemical modification.

Reagent[a]	Conc. (M)	β-Fuc	β-Gal	β-Glc
EDC[b]	4×10^{-2}	21.7	45.1	56.6
DEP	4×10^{-2}	4.3	70.1	86.3
IAA	10^{-2}	4.1	83.6	89.7
PCMB	10^{-4}	0	90.2	96.9
	2.5×10^{-7}	0	—	—
	2.5×10^{-8}	17.2	—	—
	2.5×10^{-9}	92.8	—	—
NEM	10^{-2}	6.0	106.7	99.2
	10^{-3}	30.4	—	—
	10^{-4}	79.9	—	—
NBS	10^{-5}	0	31.8	3.8

(—) Not determined.
[a] EDC = 1-ethyl-3-dimethylaminopropyl carbodiimide. DEP = diethyl pyrocarbonate. IAA = iodoacetate. PCMB = *p*-chloromercuribenzoate. NEM = *N*-ethylmaleimide. NBS = *N*-bromosuccinimide.
[b] EDC was incubated in 50 mM TEMED, pH 4.75, at 25°C for 30 min. The others were incubated in 20 mM phosphate or citrate-phosphate at 25°C for 30 min.

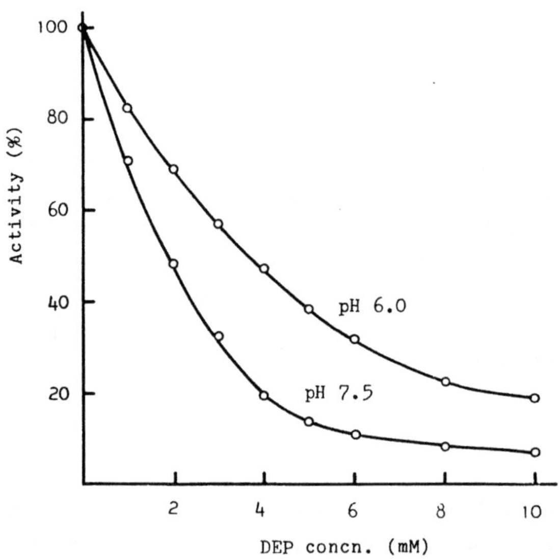

FIG. 44.5. DEP modification of β-D-fucosidase. β-D-Fucosidase (5 μg) was incubated in 20 mM phosphate buffer, pH 6.0 or 7.5, with the indicated concentration of DEP in a total volume of 400 μl at 15°C for 1 h.

Discussion

From the crude extract of wheat bran culture of *A. phoenicis* we purified a specific β-D-fucosidase. It showed high activity toward PNP-β-fucoside but negligible activity toward PNP-β-galactoside and PNP-β-glucoside. Its substrate specificity and other properties are quite different from those of β-galactosidase and β-glucosidase purified from the same crude extract. The molecular weight of *A. phoenicis* β-D-fucosidase was only 57,000, much less than that of the snail *Littorina littorea* (300,000) (5) and *Achatina belteata* (110,000 and 300,000) (6). The enzyme is susceptible to heat, pH change, dilution, and freezing, so it is not as stable as β-D-fucosidase from other sources.

So far as we know, this β-D-fucosidase is the first one isolated from a microorganism and is a new enzyme with a specificity much higher than those hitherto reported.

Acknowledgment. We are grateful to Prof. Yu-teh Li of Tulane University, New Orleans, LA, USA, for his suggestion for this work and his generous gift of various substrates.

References

1. Levvy, G. A. (1960) *Nature* 187:1027.
2. Wiederschain, G. Y., and Prokopenkov, A. A. (1973) *Arch. Biochem. Biophys.* 158:539–543.
3. Rodringuez, J. A., Cabezas, J. A., and Calvo, P. (1982) *Int. J. Biochem.* 14:539–543.
4. Calvo, P., Santamaria, M. G., Melgar, M. J., and Carbezas, J. A. (1983) *Int. J. Biochem.* 14:685–693.
5. Calvo, P., Melgar, M. J., and Cabezas, J. A. (1983) In *Proceedings of the 7th International Symposium on Glycoconjugates, Lund-Rouneby.*
6. Colas, B. (1980) *Biochim. Biophys. Acta* 613:448–458.
7. Colas, B. (1978) *Biochim. Biophys. Acta* 527:150–158.

45
Formation of Active Aspartate Transcarbamoylase from Defective Polypeptide Chains Produced by Site-Directed Mutagenesis

YING R. YANG, SUSAN R. WENTE, and H. K. SCHACHMAN

Although the genetics literature is replete with exquisite examples of interallelic complementation (1–3) in bacterial strains containing two different mutant copies of the same gene, there is little precise structural information illustrating how functional enzyme complexes are formed by the noncovalent interaction between defective polypeptide chains. For a protein chemist or an enzymologist, interpretation of the results of complementation experiments in terms of molecular mechanisms requires both the availability of suitable mutants containing known defects as well as knowledge of the tertiary and quaternary structures of the wild-type enzyme. Indeed, as Fincham (1) stated in 1966 in the preface to his book: "The present consensus may be summarized by saying that allelic complementation is basically irrelevant to primary gene action, except insofar as it confuses the investigator, but that it is of considerable importance in providing an insight into the structure and functions of multimeric protein molecules." Because so much is known about the regulatory enzyme aspartate transcarbamoylase (ATCase, aspartate carbamoyltransferase, carbamoylphosphate: L-aspartate carbamoyltransferase, EC 2.1.3.2) from *Escherichia coli*, it can serve as a useful model system for mechanistic studies of interallelic complementation. A large variety of complementing mutants are available (4), and the three-dimensional structure of the wild-type enzyme is known from x-ray crystallography (5, 6). It seemed likely therefore that in vivo and in vitro complementation experiments with various mutations in *pyrB*, which encodes the catalytic chains of ATCase, would provide valuable information about the folding of the chains into domains, help to identify amino acid residues critical for catalysis, and contribute to our understanding of the nature of the active sites in the enzyme. In this chapter we describe the in vitro formation of active enzyme from defective mutants, each of which contains a single amino acid replacement produced by site-directed mutagenesis. With knowledge of the precise three-dimensional location of the amino acid substitutions in the catalytic chains of ATCase, it is possible to account for the formation of partially active enzyme in terms of the tertiary and quaternary structures of the catalytic subunits in the holoenzyme (7, 8).

Conceptually, the simplest way to form an active enzyme from two inactive mutant polypeptide chains involves the formation of a hybrid molecule in which the functional portion of one polypeptide chain either displaces a defective segment or substitutes for a missing region of the other chain. Such a *chain displacement* mechanism is particularly relevant for an enzyme such as ATCase in which the folding of the catalytic chains leads to two clearly recognizable structural domains (9, 10). A second mechanism can be visualized for an oligomeric enzyme in which the active site is at an interface between two polypeptide chains, and residues from the adjacent chains are required for the formation of a *shared active site*. For such enzymes, active sites can be generated by forming interchain hybrids from the two inactive parental mutants. It is also possible for activity to be generated in hybrid oligomers if quaternary interactions between the dissimilar chains cause alterations in the folding of at least one of the chains. This type of interallelic complementation by *conformational correction* can be demonstrated with mutants in which the defects in the individual parental proteins result from the introduction of bulky or charged residues, which cause incorrect folding rather than the replacement of critical active site residues.

Experimental Approach

By exploiting the development of a positive selection technique (11) for inactive mutants of ATCase, Jenness and Schachman (4) were able to isolate and characterize 29 mutant strains harboring defects in the structural gene *pyrB* of *E. coli.* Interallelic complementation experiments with organisms containing one allele on the chromosome and the other on an episome suppressed the pyrimidine auxotrophy exhibited by the individual strains containing only one allele and showed that the mutants can be grouped into four complementation units, which divided the fine-structure map into distinct segments (4). Those defining the α and β complementation groups are missense mutants with amino acid substitutions (determined from the nucleotide sequences) located, respectively, in the polar and equatorial domains of the folded catalytic chains in ATCase (5, 6). These strains containing the missense mutations produced fully assembled ATCase-like molecules composed of six catalytic and six regulatory chains; and the isolated catalytic trimers, like the holoenzymes, were devoid of enzyme activity (12). The other mutants, comprising the γ and δ complementation units, are chain-termination mutants; they are not be considered here because so far the defective chains have not been isolated and therefore in vitro experiments are not yet feasible. Although the merodiploid strains containing two *pyrB* alleles from the α and β complementation groups exhibited the phenotype of pyrimidine-independent growth (4), quantitative estimates of the enzyme activity in the strains is lacking. Because the complementation assay is so nonquantitative and so sensitive, it is not known whether the active complexes formed in these strains have as much as 10% or as little as 0.01% of the specific activity of the wild-type enzyme. It is necessary, therefore, to isolate the defective enzymes (and their catalytic subunits) and perform in vitro experiments to determine quantitatively the enzyme activity generated in hybrid molecules formed from the defective polypeptide chains.

Thus far in vivo complementation experiments, which presumably are based on chain displacement, have been performed only with mutants, one or both of which involve a significant deletion of the polypeptide chain (7). Because these mutant chains are incapable of assembling into ATCase-like holoenzymes, which are readily isolated as defective proteins, and the incomplete polypeptide chains are so readily degraded in vivo through the action of proteolytic enzymes, in vitro experiments aimed at testing the mechanism have not been performed. In contrast, defective holoenzymes containing missense mutations have been produced by both random and site-directed mutagenesis, and it has been possible through in vitro hybridization experiments with the isolated catalytic subunits to generate partially active enzymes. Figure 45.1 shows the tertiary structure of a catalytic chain in wild-type ATCase that is segmented into two recognizable folded regions (9, 10), designated the polar domain comprising approximately the *N*-terminal 140 amino acids and the equatorial domain consisting of residues 150 to 284. Also

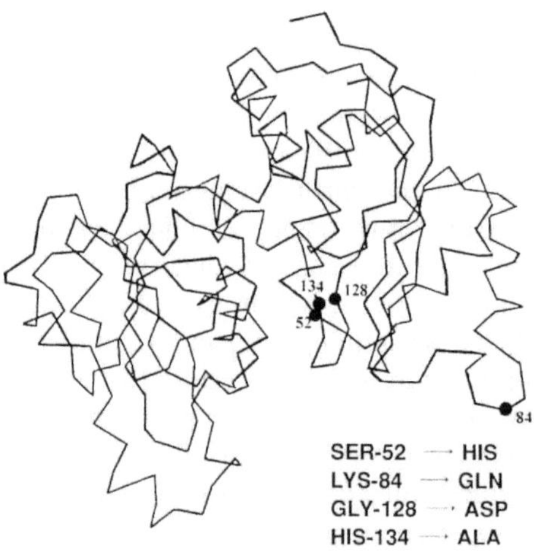

FIG. 45.1. Locations of mutational alterations in the catalytic chain of ATCase. The drawing represents a trace of the polypeptide backbone of a single catalytic chain within ATCase. This view, taken from an angle off the two-fold symmetry axis, emphasizes the separation of the polar and equatorial domains. The former, encompassing residues 1 to 140, is on the right; and the latter, including residues 150 to 284, is on the left. The locations of the amino replacements at positions 52, 84, 128, and 134 are illustrated by •. Each of the mutants was generated by oligonucleotide-directed mutagenesis.

shown in Fig. 45.1 are the positions 52, 84, 128, and 134, which in the wild-type enzyme are Ser, Lys, Gly, and His, respectively. Converting Ser-52 to His, Lys-84 to Gln, and Gly-128 to Asp by site-directed mutagenesis leads to three virtually inactive holoenzymes (10^{-4} to 10^{-5} the activity of wild-type enzyme), and the inactive catalytic subunits isolated from them are also almost devoid of enzyme activity.[1] Replacing His-134 by Ala causes a loss of about 95% of the catalytic activity in both the holoenzyme and the isolated catalytic subunit (12).

The availability of these four mutants with known defects afforded the opportunity to test complementation mechanisms based on conformational correction and on shared active sites. In all the experiments described below, the holoenzymes were dissociated into subunits by treating the proteins with the mercurial neohydrin, and the catalytic subunits were purified by procedures established earlier in this laboratory for wild-type ATCase and other mutants (13, 14).

Complementation by Conformational Correction

Treatment of cells with diethyl sulfate led to a variety of pyrimidine auxotrophs (4) from which one inactive mutant ATCase was isolated containing Phe substituted for Ser-52 in the catalytic chains of the enzyme. The purified enzyme produced by this random mutagenesis procedure had $< 10^{-5}$ the activity of the wild-type enzyme, and no binding of the bisubstrate analogue, *N*-(phosphonacetyl)-L-aspartate (PALA), to the mutant enzyme could be detected by difference spectroscopy. Similarly, His-52 ATCase produced by site-directed mutagenesis was essentially devoid of catalytic activity and exhibited no detectable binding of PALA. In both of these mutants a relatively small amino acid side chain was replaced by a much larger one that may affect adversely the proper folding of the polypeptide chain, thereby leading to a loss in activity. It is, of course, possible that the hydroxyl group of Ser-52 is needed for substrate binding or for catalysis, and the replacement of Ser-52 by either Phe or His leads directly to inactivation. According to crystallographic studies (6), Ser-52 may be implicated in the binding of PALA, but as yet its direct involvement in catalysis has not been demonstrated.

An equally inactive mutant produced by random mutagenesis involved the

[1] The inactive mutant in which Ser-52 was replaced by Phe was produced by random mutagenesis (4, 7). Because the catalytic subunit isolated from the holoenzyme tended to aggregate in solution, site-directed mutagenesis was used to substitute His for Ser-52, thereby yielding an equally inactive holoenzyme from which a soluble catalytic subunit was isolated (Y. R. Yang and H. K. Schachman, unpublished data). The Gly-128 → Asp mutant was described earlier (13, 14), but large quantities of this inactive protein were not available. Therefore, site-directed mutagenesis of wild-type *pyrB* was used to produce large quantities of this mutant.

replacement of Gly-128 by Asp. Again, no binding of PALA to this purified enzyme could be detected. For this mutant, in which a large, charged side chain replaces a hydrogen atom, it seems likely that inactivation is the indirect consequence of altered folding of the polypeptide chain in the region of the active site.

As seen in Fig. 45.1, positions 52 and 128 are in the same region of the polar domain, and the α carbons for those residues are separated by only about 10 Å. Nonetheless, in vivo experiments with a strain containing the two alleles responsible for the Ser-52 → Phe and the Gly-128 → Asp mutations exhibited pyrimidine-independent growth. Each strain containing only one allele was auxotrophic for pyrimidines. How is this complementation achieved? What is the level of enzyme activity in the strain containing the two alleles? To answer these questions we turned to in vitro hybridization experiments in an attempt to form active molecules with the Ser-52 → His mutant because the isolated catalytic subunit does not aggregate like its Phe counterpart.

Mixtures of the His-52 and Asp-128 catalytic subunits were incubated in 5 *M* urea; at various intervals aliquots were withdrawn, the urea was removed, and the samples were assayed for enzyme activity. As seen in Table 45.1, there was a progressive increase in activity with time, reaching a plateau at about 2 h. Although the enzyme activity was low (about 0.5% that of a comparable amount of wild-type catalytic subunit), the increase (compared to that for the individual mutants) was > 100-fold. The patterns in Fig. 45.2, obtained from polyacrylamide gel electrophoresis experiments, showed that reconstitution

TABLE 45.1. Complementation by conformational correction.

	Enzyme Activity (μmol/h/μg)			
Time (min)	His-52	Asp-128	Mixture of His-52 and Asp-128	Wild-type
15	~[a]	~[a]	6×10^{-3}	1.8×10^{-2}
45	~[a]	~[a]	3×10^{-3}	1.1×10^{-2}
75	~[a]	~[a]	1×10^{-2}	0.57
105	~[a]	~[a]	1.2×10^{-2}	1.8
135	10^{-4}	10^{-5}	3.1×10^{-2}	3.2
165	10^{-5}	10^{-4}	2.2×10^{-2}	4.3
225	~[a]	~[a]	2.5×10^{-2}	5.1

Catalytic subunits were incubated at 0°C for 15 min in 5 *M* urea solutions in a pH 8.0 Tris buffer containing 0.1 *M* DTT, 0.2 m*M* EDTA, and 2 m*M* 2-mercaptoethanol. All solutions were then diluted 1:50 in the Tris buffer containing 0.2 m*M* EDTA and 2 m*M* 2-mercaptoethanol. The proteins were concentrated by filtration and then assayed for enzyme activity at the times indicated. The original concentrations for the mutants were about 0.50 and 0.02 mg/ml for the wild-type catalytic subunit. Enzyme activity is given in micromoles of carbamoyl asparate formed per hour per microgram of protein.

[a] No enzyme activity could be detected.

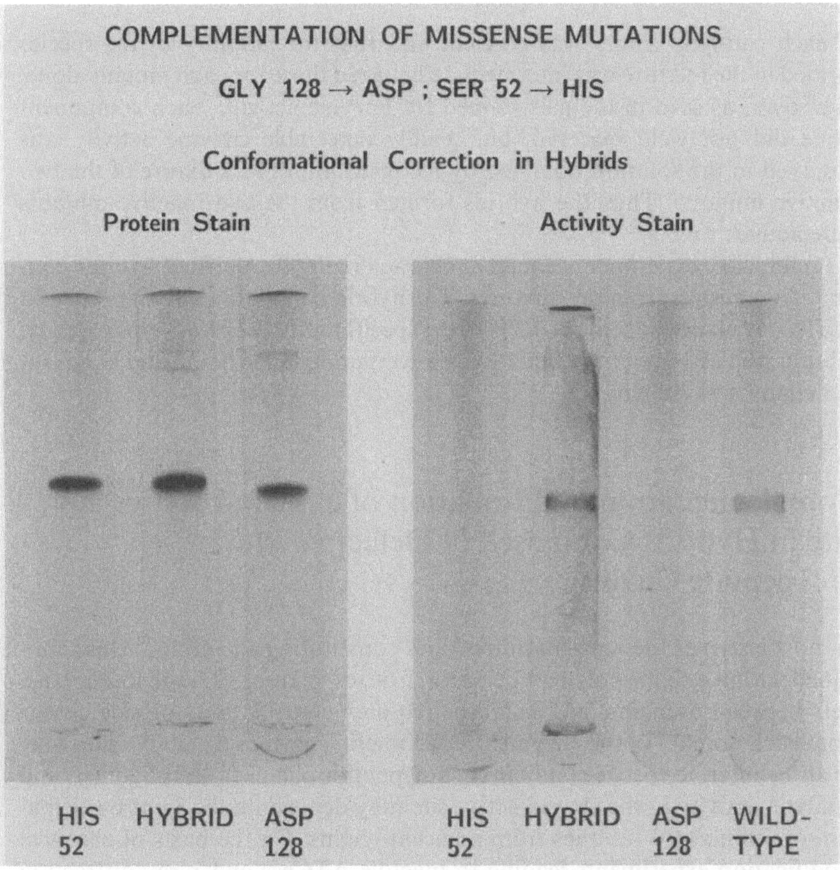

FIG. 45.2. Formation of partially active catalytic trimers in hybrids formed from inactive mutants. The mutants in which Ser-52 was replaced by His and Gly-128 was replaced by Asp were incubated separately and as a mixture in 5 *M* urea for 15 min. At that time the solutions were diluted 1:50; the proteins were concentrated by filtration and then subjected to electrophoresis in polyacrylamide gels. The patterns on the left were stained for protein, showing that reconstitution of trimers was efficient. On the right are the patterns in which the gels were stained for enzyme activity. Each of the mutants alone exhibited no detectable enzyme activity, whereas the gel for the reconstituted mixture exhibited significant activity. On the extreme right is the pattern for the wild-type catalytic trimer at a protein concentration that was 4% of that used for the mutant proteins. The mobility of the active species formed by hybridization of the two inactive mutant trimers is similar to that observed for the reconstituted wild-type catalytic subunit.

of each catalytic trimer was efficient and that the mobility of the species formed in the mixture was intermediate between those for each mutant alone. Moreover, as seen in the gels stained for enzyme activity, each component alone did not yield material, but readily detectable enzyme activity was observed in the solution obtained by reconstitution of a mixture of the two inactive mutants. Thus the hybrids formed from the two inactive mutants were active.

Preliminary experiments aimed at characterizing the purified hybrids confirmed the results from the mixtures. The hybrid containing one Asp-128 and two His-52 chains had about 0.5% of the specific activity of wild-type catalytic subunit; the activity of the other hybrid containing one His-52 and two Asp-128 chains was similar.

Complementation by Formation of a Shared Active Site in Hybrids Composed of Defective Mutant Polypeptide Chains

In most enzymes the amino acid residues constituting an active site are contained within a single polypeptide chain. For some, these sites are located in a cleft between domains, and catalysis requires contributions of side chains from each domain in the enzyme. In oligomeric enzymes the active sites may be at the interface between adjoining polypeptide chains in the oligomer, and catalytic activity in the shared active site may depend on the joint participation of amino acid residues from adjacent chains. On the basis of chemical modification experiments leading to inactive ATCase and x-ray diffraction studies on the wild-type enzyme, it has been suggested that the active sites are at the interfaces between adjacent catalytic polypeptide chains (6, 9, 10). Robey and Schachman (8) described a direct approach aimed at demonstrating the existence of shared active sites in oligomeric enzymes, and they obtained strong evidence for their presence in the catalytic trimers of ATCase. This approach, which is extended here, is based on the premise that hybrid molecules formed from inactive derivatives would have high specific activities if the defects in the two parental oligomers were on opposite sides of the interfaces between the chains.

The rationale for demonstrating the presence of a shared active site is illustrated for a trimer in Fig. 45.3. A functional active site requires both a △ and a □ at the interface between the adjacent chains, and the unmodified trimer would have three such sites. In one derivative (or mutant) the defective region is illustrated by ▲ and in the other by ■. When these two mutant trimers are incubated under conditions that promote interchain exchange (by dissociation and random reassociation of the trimers), four species would be formed. Two of the resulting species, containing no active sites, are identical to the original parental trimers; the other two, representing the hybrids, would

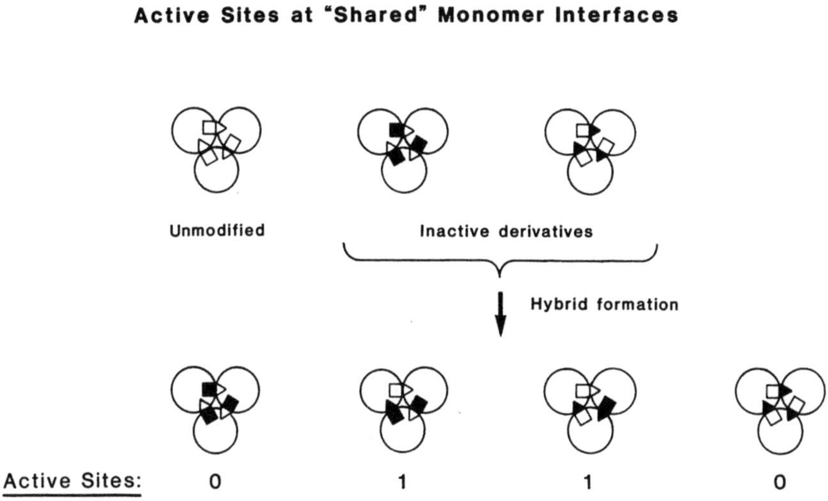

FIG. 45.3. Schematic view of the use of hybrids to demonstrate shared active sites in the catalytic trimer of ATCase. Large circles correspond to individual catalytic chains, and □ and △ represent unmodified (wild-type) active-site residues. The corresponding modified residues caused by mutations that cause inactivation are designated by ■ and ▲. An active site requires adjoining □ and △. For the shared site model, the active sites are at the interfaces between chains. In one mutant, inactivation is caused by the conversion of □ to ■, and in the other, △ is converted to ▲. Hybrid formation from the two inactive mutants leads to the production of four species, two of which correspond to the inactive parental species and two of which contain one active site per trimer.

each contain one functional site. Thus the demonstration that each of the hybrids (constructed from inactive parental mutants) had 33% the specific activity of the wild-type catalytic trimer constitutes strong evidence for shared active sites.

In the course of a study of the effects of mutational alterations on the catalytic and regulatory properties of ATCase (12), we found that the replacement of Lys-84 by Gln led to almost a complete loss of enzyme activity ($<10^{-4}$ that of the wild-type enzyme). Similarly, the substitution of Ala for His-134 yielded a mutant with only 5% the activity of wild-type ATCase. Moreover, the Km for the isolated catalytic subunit of the mutant was increased about six-fold over its wild-type counterpart. The location of these two amino acid residues in the catalytic chains is shown in Fig. 45.4, which presents the α carbon backbone structure of the three chains in a catalytic trimer within ATCase (5). Lys-84 and His-134 are near the interfaces between the chains, and the crystallographic study of the ATCase–PALA complex indicates that they are both implicated in binding of the bisubstrate analogue (6).

When a mixture of the two mutants was incubated under conditions promoting hybrid formation, the enzyme activity increased about 100-fold and

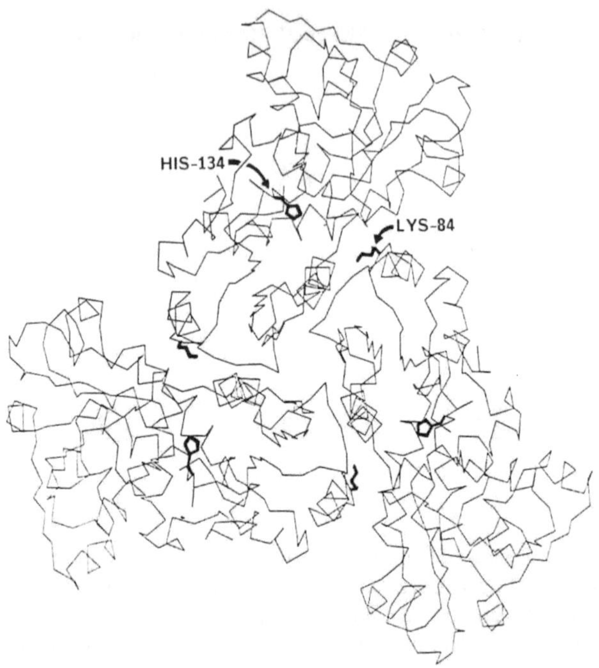

FIG. 45.4. Structure of a catalytic trimer within intact unliganded ATCase. The α carbon trace of each chain is shown as light lines, with the side chains of Lys-84 and His-134 represented by dark lines. The view is from the center of the ATCase molecule [based on the structure determination of Ke et al. (5)].

TABLE 45.2. Complementation by formation of shared active sites.

Type of Trimer	K_m of Asparate (m*M*)	V_{max} (μmol/h/μg)	Relative Activity (%)
Wild-type	7.2	34.000	100
C_{LLL}	7.5	0.003	8×10^{-3}
C_{LLH}	5.6	11.100	33
C_{LHH}	7.4	13.500	39
C_{HHH}	45.0	2.000	6

Each of the mutants was prepared by oligonucleotide-directed mutagenesis, and the catalytic trimers were isolated from the holoenzymes. The mutant trimer, C_{LLL}, in which Lys-84 was replaced by Gln, was acylated by treatment with 3,4,5,6-tetrahydrophthalic anhydride to increase its negative charge (15); it was incubated with the mutant C_{HHH} in which His-134 was replaced by Ala. Hybridization was achieved in a Tris-acetate buffer at pH 8.3 containing 0.2 m*M* EDTA, 2 m*M* 2-mercaptoethanol, and 20 m*M* sodium pyrophosphate. The four species were separated by ion-exchange chromatography on DEAE-Sephadex, and the purified catalytic trimers were deacylated by dialysis at pH 6.1 in 40 m*M* potassium phosphate containing 0.2 m*M* EDTA and 2 m*M* 2-mercaptoethanol. The purification procedure was monitored by electrophoresis on cellulose polyacetate strips, and assays for enzyme activity were then performed on each of the four species. Results for the wild-type catalytic subunit are included as a reference.

attained a limiting value corresponding to about 25% that of pure wild-type trimers. This value would be expected if there was a completely random reassortment of the chains in the two mutants and each hybrid, as shown in Fig. 45.3, had one active site. On the basis of this observation, we isolated each of the hybrids by procedures developed in this laboratory and measured their specific activities. The results summarized in Table 45.2. (15) show clearly that the hybrid containing two Gln-84 chains and one Ala-134 chain had 33% the activity of a wild-type catalytic trimer as predicted by the scheme in Fig. 45.3. Similarly, the hybrid containing two Ala-134 chains and one Gln-84 chain had 39% the specific activity of the wild-type trimer. Moreover, the Km values for each of the hybrids were close to that determined for the wild-type subunit, indicating that the active sites generated from the virtually inactive mutants corresponded closely to those in the wild-type catalytic trimer.

It could be argued that the regeneration of active species from the mutants resulted from conformational corrections in the hybrids rather than from the formation of a shared site, and that it was fortuitous that each hybrid had about 33% the activity of the wild-type protein. This interpretation cannot account for an additional experiment which showed that a hybrid formed from one wild-type chain and two chains containing both the Lys-84 → Gln and the His-134 → Ala replacements had virtually no activity. This negative complementation experiment demonstrates that the shared active site requires functional residues on both sides of the interface between the chains. We conclude, therefore, that the active sites in the catalytic trimers of ATCase are shared and require the joint participation of amino acid residues from each of the adjacent chains at the interface.

Summary

Interallelic complementation experiments with missense, chain-termination, and deletion mutations in *pyrB*, the structural gene encoding the catalytic chain of *E. coli* aspartate transcarbamoylase (ATCase), have demonstrated that enzymically active complexes are formed in vivo by the noncovalent association of defective polypeptide chains. However, the observations could not be interpreted in terms of the structure of the enzyme until in vitro experiments were conducted with defective mutants containing precisely located amino acid substitutions. Accordingly, site-directed mutagenesis was employed to generate appropriate inactive mutants, which were used in hybridization experiments to form partially active oligomers. Hybrid molecules with about 0.5% the specific activity of the wild-type catalytic subunit of ATCase were obtained from two inactive mutants in which the bulky, charged amino acids His and Asp replaced, respectively, Ser-52 and Gly-128.

Because the inactivity of the mutants is probably due to incorrect folding of the polypeptide chains, rather than the loss of critical residues, we attribute the 100-fold increase in activity of the hybrids (relative to the parental mutants) to a *conformational correction* in the oligomers comprising different polypeptide chains. In contrast to the low activity of these hybrids, extremely active heterooligomers were formed from an inactive mutant in which Lys-84 is replaced by Gln and another mutant of low activity in which Ala was substituted for His-134. The catalytic activity of these hybrids, amounting to 33% that of the wild-type trimer, is interpreted in terms of a *shared active site* at the interface between two polypeptide chains and the joint participation of functional amino acid residues from the adjacent chains.

Acknowledgment. This investigation was supported by U.S. Public Health Service Research Grant GM 12159 from the National Institute of General Medical Sciences and by National Science Foundation Research Grant PCM 85-02131

References

1. Fincham, J. R. S. (1966) *Genetic Complementation.* Benjamin, New York.
2. Fincham, J. R. S. (1977) *Carlsberg Res. Commun.* 42:421–430.
3. Zabin, I., and Villarejo, M. R. (1975) *Annu. Rev. Biochem.* 44:295–313.
4. Jenness, D. D., and Schachman, H. K. (1983) *J. Biol. Chem.* 258:3266–3279.
5. Ke, H-M., Honzatko, R. B., and Lipscomb, W. N. (1984) *Proc. Natl. Acad. Sci. USA* 81:4037–4040.
6. Krause, K. L., Volz, K. W., and Lipscomb, W. N. (1985) *Proc. Natl. Acad. Sci. USA* 82:1643–1647.
7. Schachman, H. K., Pauza, C. D., Navre, M., Karels, M. J., Wu, L., and Yang, Y. R. (1984) *Proc. Natl. Acad. Sci. USA* 81:115–119.
8. Robey, E. A., and Schachman, H. K. (1985) *Proc. Natl. Acad. Sci. USA* 82:361–365.
9. Monaco, H. L., Crawford, J. L., and Lipscomb, W. N. (1978) *Proc. Natl. Acad. Sci. USA* 75:5276–5280.
10. Honzatko, R. B., Crawford, J. L., Monaco, H. L., Ladner, J. E., Edwards, B. F. P., Evans, D. R., Warren, S. G., Wiley, D. C., Ladner, R. C., and Lipscomb, W. N. (1982) *J. Mol. Biol.* 160:219–263.
11. Jenness, D. D., and Schachman, H. K. (1980) *J. Bacteriol.* 141:33–40.
12. Robey, E. A., Wente, S. R., Markby, D., Flint, A., Yang, Y. R., and Schachman, H. K. (1986) *Proc. Natl. Acad. Sci. USA* 83:5934–5938.
13. Wall, K. A., Flatgaard, J. E., Gibbons, I., and Schachman, H. K. (1979) *J. Biol. Chem.* 254:11910–11916.
14. Wall, K. A., and Schachman, H. K. (1979) *J. Biol. Chem.* 254:11917–11926.
15. Gibbons, I., and Schachman, H. K. (1976) *Biochemistry* 15:52–60.

46
Catalytic Mechanisms Revealed by Protein Engineering of Tyrosyl-tRNA Synthetase

ALAN R. FERSHT

Enzymology has been revolutionized by genetic engineering: Any amino acid residue in a protein may be replaced by mutation of its gene using variations of oligodeoxynucleotide-directed mutagenesis (1) or cassette mutagenesis (2). Thus given a gene for an enzyme, a means for its expression, and detailed knowledge of its three-dimensional structure, the relation between its structure and activity may be studied in a rational and systematic manner by preparing mutants and examining them by kinetics and spectroscopy. The first enzyme to be analyzed in such a manner is the tyrosyl-tRNA synthetase from *Bacillus stearothermophilus* (3). After 4 years of intensive efforts by my research group and that of G. Winter in preparing and analyzing mutants, what have we learnt about catalysis in general and the tyrosyl-tRNA synthetase in particular?

Simple Strategy

In order for the results of protein engineering to be mechanistically interpretable at our present level of knowledge, the mutations that are induced must cause minimal perturbation of protein structure. One of the safer ways of doing it is to replace amino acid side chains with ones that are slightly smaller, as proteins can often tolerate the existence of small holes in their structures. We have therefore used a simple approach of replacing amino acid side chains that interact with the substrate with smaller ones that lack the specific interactions. This step enables the importance of the binding energies of the side chains in catalysis to be studied. The results of such experiments can be analyzed using a theory developed in 1974 (4) for the interaction of modified substrates with a single enzyme. These equations allow calculation of apparent group binding energies and analyze how these energies can be used to provide an increased catalytic rate rather than increased affinity of the enzyme for its substrates (4, 5). Most of our results thus relate to the use of binding energy in catalysis and specificity—in particular, the role of hydrogen bonding. Because biological specificity is, in general, determined by complementary binding interactions (6) and enzymic catalysis results from the use of binding energy, it is a profitable approach.

Basic Properties of Tyrosyl-tRNA Synthetase

The enzyme is a symmetrical dimer of subunit molecular mass 47 kD that catalyzes the aminoacylation of tRNA in two steps. Tyrosine is first activated to form a tightly bound complex of tyrosyl adenylate with the enzyme (Eq. 1). There is then the transfer to tRNA (Eq. 2).

$$\mathrm{E + Tyr + ATP \leftrightharpoons E \cdot Tyr\text{-}AMP + PP_i} \qquad [1]$$

$$\mathrm{E \cdot Tyr\text{-}AMP + tRNA \rightarrow Tyr\text{-}tRNA + E + AMP} \qquad [2]$$

The stability of the enzyme-bound tyrosyl adenylate complex has enabled its crystal structure to be solved (7) in addition to the structure of the native enzyme (8). The enzyme consists of two domains. The one that activates tyrosine (residues 1–319) is well resolved in the x-ray map. Unfortunately, the tRNA-binding domain (residues 320–419, the *C*-terminus) is too mobile to be resolved. The *N*-terminal domain has been constructed by deletion mutagenesis, and its kinetic properties have been shown to be identical to full-length enzyme in reaction 1 (9). The crystal structure of the truncated enzyme has also been shown to be identical to that of the wild-type enzyme (10). Interestingly, the enzyme shows half-of-the-sites reactivity (11). Note that the active site of the enzyme has a large number and variety of hydrogen bonds that interact with the substrate (Fig. 46.1).

Biological Specificity and Intermolecular Interactions

It has been assumed since the days of Emil Fischer and Paul Ehrlich that biological specificity is achieved by complementary interactions between the molecules concerned. In current terms, the hydrophobic portions of each molecule are matched, as are the polar or charged groups (6). There is a major fundamental difficulty in calculating the contributions of hydrogen bonds and paired charges (salt bridges) to specificity in that calculations of energetics in vacuo are not relevant. The problem is that the binding energy between two molecules (A and B) in solution is the *difference* in energy between the two molecules being bound to each other and the two molecules being surrounded by solvent (Eq. 3). The concentration of water in water is 55 *M*. Therefore

$$\mathrm{A}\cdot\cdot(n\mathrm{H_2O}) + \mathrm{B}\cdot\cdot(m\mathrm{H_2O}) \leftrightharpoons \mathrm{A}\cdot\cdot\mathrm{B}(x\mathrm{H_2O}) + (n + m - x)\mathrm{H_2O} \qquad [3]$$

water, being also polar and forming strong hydrogen bonds, competes for any hydrogen bonding or charged sites on A and B. Just as the hydrophobic bond is described by the purist as an "effect" because it is not a true bond, so hydrogen bonding (and charge pairing) in biological systems should be simi-

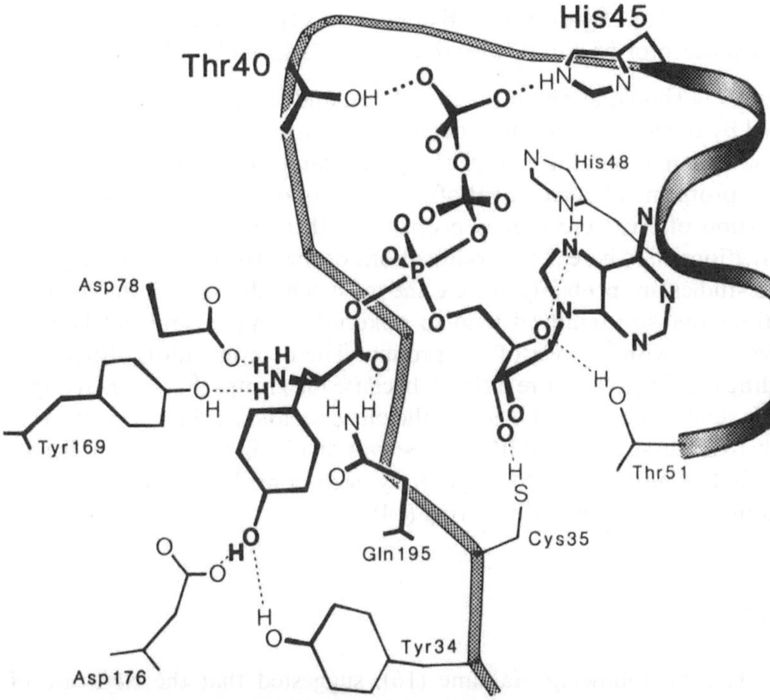

FIG. 46.1. Transition state for the formation of tyrosyl adenylate built into the structure of the tyrosyl-tRNA synthetase by adding pyrophosphate to the known structure of the E · Tyr-AMP complex. (Reprinted from Leatherbarrow, R. J., Fersht, A. R., and Winter, G. [1985] *Proc. Natl. Acad. Sci. USA* 82:7840–7844; with permission.)

larly described as it represents the difference in energy between hydrogen bonds between the two molecules with each other and hydrogen bonds of each with water. In addition, there are important entropy terms arising from the rearrangement of the solvent.

The energetics of the hydrogen bonds at the active site of the tyrosyl-tRNA synthetase were measured by constructing mutants with changes such as Tyr → Phe-34, Tyr → Phe-169, Cys → Gly-35, etc. These changes remove groups that primarily just form hydrogen bonds with the ATP and tyrosine substrates. Kinetic measurements on the mutant enzymes then give experimentally the net contribution of the bond to binding energy—the *apparent* group binding energy (12, 13). The results of mutating many such bonds may be summarized thus (12):

1. Removal of hydrogen bond donor or acceptor to leave an unpaired but *uncharged* acceptor or donor in an enzyme–substrate complex weakens binding by only 2 to 7 kJ mol^{-1}.

2. Removal of a hydrogen bond donor or acceptor to leave an unpaired and *charged* acceptor or donor loses some 12 to 25 kJ mol^{-1} of binding energy.

These data should apply to other systems, and similar results are now being found by other workers. More detailed discussion is in ref. 12.

The apparent binding energy of a group can be obscured by artifacts. The major problem is that removal of a side chain of an enzyme could lead to a distortion of structure that is propagated through the protein and affects interactions elsewhere. The modifications of the hydrogen bonding groups in these studies are probably among the most reliable because the hydrophilic portions that are removed tend to poke into solution and not have many interactions with the rest of the protein. The determination of hydophobic binding is going to be more difficult because the groups that form hydophobic bonds tend to be partly buried in the enzyme, and so their removal is more likely to affect structure. With this reservation in mind, we find that the β- and γ-methyl groups of Thr-51 apparently each contribute 3 to 4 kJ mol^{-1} of binding energy with the ribose ring (14).

Catalysis

Pauling (15), following Haldane (16), suggested that the structure of an enzyme is complementary to the structure of the transition state of the substrate rather than the substrate itself, so that as the reaction proceeds the increase in binding energy lowers the energy difference between the transition state and the substrate and hence lowers the apparent activation energy. Evidence for this idea has arisen classically from experiments with transition state analogues, which mimic the changes in bonds immediately around the reacting part of the molecule. Also, experiments on proteases show that the occupation of subsites on the enzyme can lead to increases in rate rather than improvements in the affinity for the substrate (see ref. 5 for a review). Direct evidence for the use of the binding energy to increase the rate has come from detailed rapid reaction studies on mutants that lack hydrogen bonding side chains (17). For example, the site chain of Cys-35 contributes no apparent binding energy when binding ATP in the enzyme–substrate complex but stabilizes the transition state by 5 kJ mol^{-1}. Several sites have been found with similar phenomenology (13, 17–19).

Mechanism of Activation of Tyrosine

Much effort has been put in over the years on the 20 aminoacyl-tRNA synthetases in the search for nucleophilic, general acidic, and general basic groups that are involved in catalysis—with little success. The reason now is clear. The enzyme appears to use binding energy only for catalysis. In particular, we have located a site for the γ-phosphate of ATP that contributes no binding energy whatsoever until the transition state is reached (19). This action involves hydrogen bonds from residues Thr-40 and His-45.

Enzyme–Intermediate Complementarity

The complete free energy profile for the activation of tyrosine has been determined for wild-type enzyme and some dozen mutants (13, 17, 20). It has enabled the contribution of each of the side chains to binding to be calculated thoughout the activation reaction. Surprisingly, it was found that *some* regions of the enzyme (Tyr-34, Gly-35, and His-48) are complementary in structure to the intermediate Tyr-AMP and contribute more binding energy with this state than with the transition state. There are, in fact, good biological reasons why the intermediate should be bound extremely tightly (18). First, Tyr-AMP is a potent acylating agent and must be sequestered by the enzyme and not allowed to dissociate from the enzyme. Second, we have measured the free energy of hydrolysis of Tyr-AMP and found that the equilibrium constant for its formation in solution is extremely low (21) ([Tyr-AMP][PP_i]/[Tyr][ATP] $= 4 \times 10^{-7}$). On the enzyme, however, the analogous constant ([E · Tyr-AMP · PP_i]/[E · Tyr · ATP]) is 2.3. It is necessary to have a favorable equilibrium constant to allow the intermediate to accumulate. The enzyme changes the equilibrium constant for the enzyme-bound reagents by binding the products more tightly than the starting materials. [Note that the free energy of hydrolysis of Tyr-AMP was measured enzymatically using mutant enzymes of suitable characteristics (21).]

Evolution of Optimal Rate

Fine tuning of enzyme activity is achieved by using binding energy (4, 5, 22). The ways that an enzyme can use binding energy to alter the energy levels of enzyme-bound substrates, transition states, intermediates, and products have been classified as: (a) uniform binding, where a change in structure affects the level of each state equally; (b) differential binding, where a change alters substrates, intermediates, and products in different ways; and (c) catalysis of the elementary step, whereby the transition state is preferentially stabilized (22). (The last is really a subclass of differential binding.) All three postulated classes of mutations have been found (20). It was also proposed that K_m values should evolve to match substrate concentrations found in vivo (4). Optimization of rate requires high K_m values. We have generated a family of mutant enzymes that bind ATP better than does the wild-type enzyme (14). However, it is at the expense of k_{cat}, and so the mutant enzymes are less active than the wild-type enzyme at the high concentrations of ATP found in vivo.

References

1. Smith, M. (1982) *Trends Biochem. Sci.* 7:440–442.
2. Wells, J. A., Vasser, M., and Powers, D. B. (1985) *Gene* 34:315–323.
3. Winter, G., Fersht, A. R., Wilkinson, A. J., Zoller, M., and Smith, M. (1982) *Nature* 299:756–758.
4. Fersht, A. R. (1974) *Proc. R. Soc. Lond.* [*Biol.*] 187:397–407.
5. Fersht, A. R. (1985) *Enzyme Structure and Mechanism*, 2nd ed. W. H. Freeman, New York.
6. Fersht, A. R. (1985) *Trends Biochem. Sci.* 9:145–147.
7. Rubin, J., and Blow, D. M. (1981) *J. Mol. Biol.* 145:489–500.
8. Blow, D. M., and Brick, P. (1985) In *Biological Macromolecules and Assemblies: Nucleic Acids and Interactive Proteins*, Vol. 2, edited by F. Jurnak, and A. McPherson, pp. 442–469. Wiley, New York.
9. Waye, M. M. Y., Winter, G., Wilkinson, A. J., and Fersht, A. R. (1983) *EMBO J.* 2:1827–1829.
10. Brick, P. (1986) *Philos. Trans. R. Soc. Lond.* [*A*] 317:439–441.
11. Jakes, R., and Fersht, A. R. (1975) *Biochemistry* 14:3344–3350.
12. Fersht, A. R., Shi, J-P., Knill-Jones, J., Lowe, D. M., Wilkinson, A. J., Blow, D. M., Brick, P., Carter, P., Waye, M. M. Y., and Winter, G. (1985) *Nature* 314:235–238.
13. Fersht, A. R., Leatherbarrow, R. J., and Wells, T. N. C. (1986) *Philos. Trans. R. Soc. Lond.* [*A*] 317:305–320.
14. Fersht, A. R., Wilkinson, A. J., Carter, P., and Winter, G. (1985) *Biochemistry* 24:5858–5861.
15. Pauling, L. (1946) *Chem. Engin. News* 24:1375–1377.
16. Haldane, J. B. S. (1930) *Enzymes*, p. 182. Longmans, New York.
17. Wells, T. N. C., and Fersht, A. R. (1985) *Nature* 316:656–657.
18. Wells, T. N. C., and Fersht, A. R. (1986) *Biochemistry* 25:1881–1886.
19. Leatherbarrow, R. J., Fersht, A. R., and Winter, G. (1985) *Proc. Natl. Acad. Sci. USA* 82:7840–7844.
20. Ho, C. K., and Fersht, A. R. (1986) *Biochemistry* 25:1891–1897.
21. Wells, T. N. C., Ho, C. K., and Fersht, A. R. (1986) *Biochemistry* 25:6603–6608.
22. Albery, W. J., and Knowles, J. R. (1977) *Angew. Chem.* 16:285–293.

47
Site-Directed Mutagenesis of Alkaline Phosphatase

DEBRA A. KENDALL and E. T. KAISER

The redesign of active site and structural regions of enzymes by means of site-directed mutagenesis provides a powerful tool with which the determinants of biological function can be elucidated. Systematic replacement of amino acid residues by other residues, together with the examination of the resulting structure-function relations, can clarify the importance of functional groups and specific residues in maintaining the catalytic activity and structural integrity of enzymes. By means of site-directed mutagenesis we have undertaken the redesign of *Escherichia coli* alkaline phosphatase to: (a) change the active site serine in order to initiate investigations of how specific functional groups affect catalysis; and (b) in separate studies, make multiple residue changes in the signal sequence designed to test the role of conformation in precursor processing and transport of the enzyme.

Active Site Mutations

In our work on mutations at the active site, our initial effort has focused on establishing whether the nucleophilic group Ser-102 can be replaced by other nucleophilic residues. In a number of studies prior to our work, through either chemical modification (1, 2) or site-directed mutagenesis (3) the nucleophilic Ser residues of several serine proteases had been replaced by Cys residues. Generally, what was observed with these acyl transfer catalysts was that the catalytic activity of the mutated enzymes was lowered considerably relative to the native enzymes except in the cases of the most highly activated substrates. The reasons for this lowering of catalytic activity have not been completely elucidated, but one possibility is that alteration of the active site nucleophile may cause substantial retardation in the rate of one or more important proton transfer steps. In the catalytic action of alkaline phosphatase it is not known if pentacoordinate intermediates intervene along the reaction pathway to and from the phosphoryl enzymes species, which are known to be formed transiently. Even if such pentacoordinate species were formed, however, they should be unstable and their breakdown is not likely to require rapid proton transfer. By examining the results of changing the active-site nucleophile in alkaline phosphatase, we could probe if a group transfer reaction not requiring rapid proton transfer in a step affecting the rate would reflect in a major way the effects of changing the nucleophile.

The site-directed mutagenesis approach we employed for the conversion of the Ser-102 residue to a Cys group has already been described (4). We found that the thiol alkaline phosphatase acted as an effective catalyst in the hydrolysis of phosphate monoesters containing good leaving groups. Typically, in 1 *M* Tris employing a substrate concentration of 10 m*M*, the wild-type alkaline phosphatase hydrolyzes a variety of phosphate monoesters including 4-nitrophenylphosphate, 2-naphthylphosphate, and 2,4-dinitrophenylphosphate at equal rates despite the considerably different leaving group tendencies of the product alcohols. On the other hand, the action of thiol alkaline phosphatase is strongly affected by the leaving group tendency of the product alcohol. Thus the thiol enzyme shows a substantially higher rate of catalysis in its action on 4-nitrophenylphosphate than on 4-methylumbelliferylphosphate and 2-naphthylphosphate. At pH 7.5 in 50 m*M* Tris the thiol enzyme exhibits a k_{cat} value for the hydrolysis of 2,4-dinitrophenylphosphate comparable to that seen in the catalytic action of the wild-type enzyme. However, using 4-nitrophenylphosphate, the thiol enzyme shows a k_{cat} value that is at least three times less than the k_{cat} value in the case of the wild-type enzyme. We have not as yet been able to obtain direct evidence that a phosphoryl enzyme intervenes in the action of thiol alkaline phosphatase on phosphate monoester substrates. However, if the pathway for the action of the thiol enzyme is analogous to that which has been established for the wild-type species (5), a reasonable explanation for the kinetic results we obtained is that, in contrast to the wild-type species where the rate-limiting step in catalysis of phosphate monoester hydrolysis occurs subsequent to the formation of the phosphoryl enzyme, for the thiol enzyme the rate-determining step may well be formation of the phosphoryl enzyme. This proposal would be in accord with expectations based on the chemical reactivity of the thiol group, which is present in the Cys-102 alkaline phosphatase. It would be anticipated that the attack of the thiol function on a phosphoryl group would be appreciably slower than that by the corresponding hydroxyl of the naturally occurring Ser residue. Part of the retardation in the effectiveness of the thiol group in attacking the phosphoryl center might be due to the possible coordination of the sulfhydryl to a neighboring zinc ion that is present in the active site. Additionally, once the phosphoryl group of the substrate had been transferred to the sulfur of the Cys-102 residue, it would be expected that the resultant thiol phosphoryl species would further hydrolyze quite rapidly, possibly even more rapidly than the serine phosphate formed as an intermediate in the wild-type reaction. Thus the differences seen in the catalytic action of the thiol enzyme and the wild-type species are reasonably explained in terms of a change in the rate-controlling step in the action of the enzyme. The reason the thiol enzyme works effectively with phosphate monoesters where the product alcohol is a good leaving group is that, although the attack step of the thiol enzyme is not as effective as that of the wild type, once the phosphoryl enzyme species is formed it breaks down rapidly in the case of the thiol enzyme.

In the next stages of our investigations on the active site of alkaline phosphatase we are probing the reactivity of other mutant forms. A particularly interesting case is the enzyme species containing a Glu residue instead of the Ser residue at position 102. After we have probed a number of such active site mutants, it is our intention to examine the effects of changes in the groups liganding the active site zinc ions. A fundamental question we hope to probe is if it is possible by appropriate changes in the active site nucleophiles and in the metal ion binding ligands to convert a zinc-containing enzyme such as alkaline phosphatase into a species capable of catalyzing other types of reactions such as those catalyzed by the zinc-containing enzymes carbonic anhydrase and carboxypeptidase A.

Signal Sequence Mutations

Like other *E. coli* proteins destined for extracytoplasmic locations, alkaline phosphatase is synthesized as a precursor containing an amino-terminal signal sequence (6). This signal peptide mediates translocation of the enzyme across the *E. coli* inner membrane. It is then cleaved, and the mature protein resides in the periplasmic space. The alkaline phosphatase signal sequence is representative of signal peptides in general: signal peptides share little primary sequence homology but are typically characterized by a positively charged amino-terminus, a central hydrophobic core, and a region recognized by the signal peptidase (7). Additionally, the CD spectra of several chemically synthesized signal peptides indicate that they exhibit α-helical structure in nonpolar environments (8–10).

To evaluate the role of a hydrophobic α-helix in alkaline phosphatase signal peptide function, we have constructed a mutant signal sequence that contains a core region optimized for formation of this structural unit. It involved making amino acid substitutions by site-directed mutagenesis to produce a sequence containing nine consecutive leucines (mutant 13A). During mutagenesis, an additional mutant (63A) was isolated that contains an insertion of six amino acids. This insertion resulted in a sequence that includes 14 leucines with a serine at the center of the hydrophobic core region. The sequences of the wild-type (WT) and mutant signal peptides are as follows:

WT: MKQSTIALALLPLLFTPVTKA
13A: MKQSTILLLLLLLLLTPVTKA
63A: MKQSTILLLLLLLSLLLLLLLLTPVTKA

Cell fractionation studies with cells that harbored the wild-type or mutant plasmids indicate that mature alkaline phosphatase is correctly localized within the periplasm in each case. A comparison of whole cell samples, however, shows some accumulation of wild-type and mutant 63A precursor.

In contrast, no precursor accumulation is observed in the whole cell sample from mutant 13A. This finding demonstrates that mature alkaline phosphatase is exported efficiently when synthesized as a precursor containing the mutant 13A signal sequence, and formation of a hydrophobic α-helix may be a key event in signal peptide function.

Pulse-chase experiments have been carried out to assess the rate at which the signal peptide is cleaved to yield the mature form of the enzyme. These studies further indicate that maturation of the mutant 13A precursor is accelerated relative to the wild-type precursor. In parallel experiments some wild-type precursor is evident up to 1 min after chase initiation, but no precursor is observed from mutant 13A. These results indicate that a signal peptide core segment, optimized for hydrophobic α-helix formation, complements the translocation machinery and that specific residues are not required in the hydrophobic domain of this signal sequence. In contrast, the kinetics of maturation for the 63A gene product are slower than those for the wild-type gene, and some mutant precursor is evident over the duration of the experiment (45 min). Although this mutant sequence is rich in leucine, it contains a serine at the center of the otherwise hydrophobic domain, and it is substantially longer than the wild-type sequence. In future experiments replacement of this serine with leucine should permit clear correlations between the optimal length and α-helix forming potential of this segment.

To distinguish clearly between the role of the primary sequence and the secondary structure, models must be tested that possess little sequence homology with the native peptide and that maximize the potential for formation of the secondary structural feature in question. We have described a mutant signal sequence designed to test the involvement of a hydrophobic α-helix in alkaline phosphate transport. Although this mutant differs from the wild-type sequence by four amino acids, it retains full biological activity. This approach can now be extended to other models to examine the specific requirements for hydrophobicity, length, and conformation.

References

1. Polgar, L., and Bender, M. L. (1966) *J. Am. Chem. Soc.* 88:3153–3154.
2. Neet, K. E., and Koshland, D. E., Jr. (1966) *Proc. Natl. Acad. Sci. USA* 56:1606–1611.
3. Sigal, I. S., DeGrado, W. F., Thomas, B. J., and Petteway, S. R., Jr. (1984) *J. Biol. Chem.* 259:5327–5332.
4. Ghosh, S. S., Bock, S. C., Rokita, S. E., and Kaiser, E. T. (1986) *Science* 231:145–148.
5. Coleman, J. E., and Gettins, P. (1983) *Adv. Enzymol.* 55:381–451.
6. Michaelis, S., and Beckwith, J. (1982) *Annu. Rev. Microbiol.* 36:435–465.
7. Watson, M. E. E. (1984) *Nucleic Acids Res.* 12:5145–5164.
8. Briggs, M. S., and Gierasch, L. M. (1984) *Biochemistry* 23:3111–3114.
9. Reddy, G. L., and Nagaraj, R. (1985) *Biochim. Biophys. Acta* 831:340–346.
10. Shinnar, A. E., and Kaiser, E. T. (1984) *J. Am. Chem. Soc.* 106:5006–5007.

Index

MIX
Paper | Supporting responsible forestry

FSC
www.fsc.org
MIX
Papier aus verantwortungsvollen Quellen
Paper from responsible sources
FSC® C105338

If you have any concerns about our products,
you can contact us on
ProductSafety@springernature.com

In case Publisher is established outside the EU,
the EU authorized representative is:
Springer Nature Customer Service Center GmbH
Europaplatz 3, 69115 Heidelberg, Germany

Printed by Libri Plureos GmbH
in Hamburg, Germany